Quantum Mechanics

Designed for a two-semester advanced undergraduate or graduate-level course, this distinctive and modern textbook provides students with the physical intuition and mathematical skills to tackle even complex problems in quantum mechanics with ease and fluency. Beginning with a detailed introduction to quantum states and Dirac notation, the book then develops the overarching theoretical framework of quantum mechanics, before explaining physical quantum-mechanical properties such as angular momentum and spin. Symmetries and groups in quantum mechanics, important components of current research, are covered at length. The second part of the text focuses on applications, and includes a detailed chapter on quantum entanglement, one of the most exciting modern applications of quantum mechanics, and of key importance in quantum information and computation. Numerous exercises are interspersed throughout the text, expanding upon key concepts and further developing students' understanding. A fully worked solutions manual and lecture slides are available for instructors.

Arjun Berera is Professor in Physics at the University of Edinburgh. He has held research positions in the USA before joining the faculty at Edinburgh. His research is focused on quantum field theory and statistical physics and he has published extensively in these areas. He has over two decades of teaching experience at both undergraduate and graduate level across a range of courses, including quantum mechanics and statistical mechanics.

Luigi Del Debbio is Professor in Physics at the University of Edinburgh. He has held research positions in the UK, France, Italy and at CERN, before joining the faculty at Edinburgh. His research is centred on non-perturbative aspects of field theories and their application to particle physics. He has extensive teaching experience at both undergraduate and graduate level and has been the Programme Coordinator for the Mathematical Physics degree at Edinburgh for a number of years.

'Berera and Del Debbio do a wonderful job of walking the reader through the mathematics of quantum theory, never shying away from the necessary complexities while keeping things as simple as they can be.'

David Tong, University of Cambridge

'Today's quantum mechanics students should learn not only the harmonic oscillator and the hydrogen atom, but also entanglement and quantum information. This book treats the old and the new with great clarity, including a first look at quantum computation through the Deutsch and Grover algorithms. Another highlight is the collection of well-crafted problems.'

Matthew Reece, Harvard University

'This text promises to be useful to a wide audience, from intermediate-level undergraduates to beginning graduate students. It is pedagogical and rather complete, and attempts to guide readers with different backgrounds via gentle yet precise mathematical asides. The chapter on quantum entanglement is the most comprehensive and complete discussion of the topic in a broad quantum mechanics textbook and will play an important role in introducing twenty-first-century undergraduates to the contemporary and rapidly growing field of quantum computing.'

André de Gouvêa, Northwestern University

'Quantum mechanics is difficult to teach as it defies intuitions of everyday experience. The textbook by Berera and Del Debbio grounds the subject by laying its mathematical foundations first, giving students a coherent framework distilled from a century of teaching experience. Filled with helpful exercises and modern topics including quantum information theory, this book would be an excellent text for both undergraduate and graduate courses.'

Maxim Lavrentovich, University of Tennessee, Knoxville

'One of the best features of this book is the substantial chapter on quantum entanglement, quantum computing and information theory (Bell's inequality, no-cloning theorem, quantum teleportation). It is based on early introduction of the mathematical foundations and Dirac notation. Beyond the standard topics, students will appreciate the inclusion of symmetry groups, applications involving multi-electron systems, the WKB(J) method, the discussion of the Dirac equation, and an in-depth treatment of quantum scattering.'

Russell Herman, University of North Carolina

'A concise yet complete introduction to quantum mechanics at the undergraduate level. The authors do a great job of exploring the formal and qualitative aspects of the theory, as well as more modern topics of interest such as quantum computation.'

Christopher Aubin, Fordham University

'Arjun Berera and Luigi Del Debbio's *Quantum Mechanics* is an exceptional textbook. It, of course, offers superb coverage of the requisite material for a standard two-semester upper-level undergraduate quantum mechanics (QM) course. Nonetheless, the textbook is much richer than that. Two features make it unique: Chapter 1, "Stories and Thoughts about Quantum Mechanics", presents a detailed and lively history of "the tortuous path that led to the formation of the theory as we know it". This reading both is entertaining and sets the groundwork for the chapters that follow. Also of special note is Chapter 15, "Quantum Entanglement". While quantum entanglement has long raised theoretical questions regarding freedom and the nature of reality, within the last two decades its applications have burst forth not just in physics, but also in engineering, computing, encryption and communications. Through very informative chapter readings and well-chosen problems, students will master associated key concepts, such as calculating and interpreting Bell's inequality; information theory and Shannon and von Neumann entropies; the no-cloning theorem, quantum teleportation and superdense coding; and quantum register and logic gates, Deutsch's algorithm and Grover's algorithm. This textbook sets the standard for quantum entanglement coverage. I ardently promote Arjun Berera and Luigi Del Debbio's *Quantum Mechanics*.'

Gerald B. Cleaver, Baylor University

'Lucid introduction to a complex subject accessible to advanced undergraduate and beginning graduate students in science and engineering. The authors combine physical insight and mathematical rigour in explaining abstract concepts with a variety of examples. Covers all the quintessential topics in the field, as well as a valuable modern introduction to quantum entanglement.'

Kaladi Babu, Oklahoma State University

Quantum Mechanics

ARJUN BERERA

University of Edinburgh

LUIGI DEL DEBBIO

University of Edinburgh

CAMBRIDGE
UNIVERSITY PRESS

University Printing House, Cambridge CB2 8BS, United Kingdom

One Liberty Plaza, 20th Floor, New York, NY 10006, USA

477 Williamstown Road, Port Melbourne, VIC 3207, Australia

314–321, 3rd Floor, Plot 3, Splendor Forum, Jasola District Centre, New Delhi – 110025, India

103 Penang Road, #05–06/07, Visioncrest Commercial, Singapore 238467

Cambridge University Press is part of the University of Cambridge.

It furthers the University's mission by disseminating knowledge in the pursuit of education, learning, and research at the highest international levels of excellence.

www.cambridge.org
Information on this title: www.cambridge.org/9781108423335
DOI: 10.1017/9781108525848

First published 2022

Printed in the United Kingdom by TJ Books Limited, Padstow, Cornwall

A catalogue record for this publication is available from the British Library.

Library of Congress Cataloging-in-Publication Data
Names: Berera, Arjun, 1964- author.
Title: Quantum mechanics / Arjun Berera, Luigi Del Debbio.
Description: New York : Cambridge University Press, 2021. | Includes bibliographical references and index.
Identifiers: LCCN 2021024878 (print) | LCCN 2021024879 (ebook) | ISBN 9781108423335 (hardback) | ISBN 9781108525848 (epub)
Subjects: LCSH: Quantum theory. | BISAC: SCIENCE / Physics / Quantum Theory
Classification: LCC QC174.12 .B4555 2021 (print) | LCC QC174.12 (ebook) | DDC 530.12–dc23
LC record available at https://lccn.loc.gov/2021024878
LC ebook record available at https://lccn.loc.gov/2021024879

ISBN 978-1-108-42333-5 Hardback

Additional resources for this publication at www.cambridge.org/bereradeldebbio

For our parents

Contents

Part II Applications

Preface

Quantum mechanics is one of the most successful scientific theories ever developed and is responsible for the vast majority of modern technology including computers and smartphones, lasers and telecommunications and magnetic resonance imaging. In its most basic form, quantum mechanics tells us that at the microscopic level, matter can behave both as a particle and as a wave. Such a simple concept leads to a fundamental shift in our understanding of how the world works.

The identification of mathematical physics as a distinct discipline at the University of Edinburgh can be traced back to 1922. Since then, notable physicists such as Charles Galton Darwin, Max Born, Nick Kemmer and Peter Higgs (to mention a few) have taught this programme, and contributed not only to its high-quality teaching but also to developing and updating course lecture notes. Today, as the current custodians of the quantum mechanics course, we have taken this accumulated knowledge of teaching this course and turned it into this textbook, whose key features are:

1. The inclusion of key, modern chapters on symmetries in quantum mechanics and quantum entanglement, as well as an extensive chapter on scattering.
2. The introduction of Dirac notations from the beginning to explain complex equations.
3. The adoption of a modern approach to introducing quantum mechanics via the mathematical underpinnings (Hilbert spaces, linear operators, etc.).
4. Detailed in-text examples and numerous end-of-chapter problems, which will enable students to better understand and practise the concepts discussed.

The market for this textbook is intermediate and senior undergraduate physics students. We appreciate that students' mathematical background can vary and so we have deliberately chosen to discuss these topics without assuming any previous subject knowledge. The goal is simply to enable the reader to acquire the basic principles that are covered in each chapter and be able to do basic calculations. Admittedly, some sections are treated in more detail than others, making them appeal also to graduate students and researchers. For instance, the time-independent perturbation theory chapter covers both higher-order perturbations and second-order degenerate perturbation theory. For time-dependent perturbation theory the first-order results are analysed in detail and the formal expressions at higher orders are derived. In the WKBJ section, a full treatment utilising this approximation is given for the double well; and in the quantum scattering chapter, a derivation of scattering starting with a wave packet is carried out, showing the reader how that can then be replaced by a plane wave state. The units throughout the book are Gaussian (cgs).

Overall though, this textbook encompasses all the standard topics covered in a one-year course. This now also includes an extensive chapter on quantum entanglement, which is becoming a standard topic needed within a quantum mechanics course. Instructors will also be able to access online a set of lecture slides and a solution manual to all the end-of-chapter problems.

Book Organisation

- **Chapter 1: Stories and Thoughts about Quantum Mechanics.** This is an introductory chapter, which summarises the historical development of quantum mechanics. It can be read in isolation as a source of trivia on the early days of the discipline. It also gives the reader an idea of the tortuous path that led to the formulation of the theory as we know it. In the rest of the book we have decided to opt for a systematic treatment of quantum mechanics, where the principles are stated from the very beginning and their consequences are then analysed, without spending too much time discussing how these principles emerged. We believe that this approach makes it easier for the reader to understand the logical framework of quantum mechanics. This first chapter is a necessary reminder of the fact that revolutionary ideas in physics are the result of a long process of trial and error, during which theoretical ideas are compared and challenged by experimental results and vice versa.

- **Chapter 2: Quantum States.** This chapter lays the foundation of our description of the states of a quantum system as vectors in complex vector spaces. We decided to introduce complex vectors, and Dirac's notation, from the very beginning since they provide the correct mathematical framework to discuss quantum mechanics. We aim to provide a self-consistent presentation, with mathematical details presented in separate *Mathematical Asides*. We study in detail the case of *two-state systems* and *one-dimensional quantum systems*. The first is a simple example of a vector space of quantum states, which is actually relevant in the description of numerous physical systems. The latter explains the connection between the description of quantum states using vectors and the one using wave functions.

- **Chapter 3: Observables.** The second ingredient that is needed in order to set up a theory of quantum phenomena is the concept of observables. Observables in quantum mechanics are treated in a peculiar and very precisely defined manner. Here we use the word 'peculiar' to emphasise the difference from the intuitive idea of an observable that we have from our everyday experience and from classical mechanics. In particular, this chapter introduces the probabilistic nature of measurements in quantum mechanics and discusses the far-reaching consequences of it. When discussing a quantum-mechanical experiment we need to adhere strictly to the postulates of the theory that specify what can be predicted about the outcome of these experiments. A central role is played by Hermitian operators acting in the vector space of physical states. These new concepts are illustrated in this chapter, together with the mathematical tools that are necessary for their implementation. As in the previous chapter, we aim to give sufficient mathematical details so that the reader can follow all the manipulations that are discussed.

- **Chapter 4: Dynamics.** This chapter presents the postulates that define the time evolution of a quantum state, i.e. the changes of the state vectors as a function of time. The time evolution is dictated by Schrödinger's equation, where the Hamiltonian (i.e. the operator associated with the energy of the system) determines the change of the state vector during an infinitesimal amount of time. We present a generic method to solve for the time evolution of a system once the Hamiltonian is known, and then focus on the time evolution of the wave function for a one-dimensional system. We present some general properties of the solution of Schrödinger's equation in a one-dimensional potential.

- **Chapter 5: Potentials.** Building on the material presented in Chapter 4, this chapter is devoted to the solution of the Schrödinger equation for a number of simple potentials. These are good examples, which can be solved explicitly, and provide some insight on the dynamics of quantum systems. The emphasis in this chapter is on two aspects. First, the technical solution of the equations that appear in the examples. These are typical manipulations that the reader needs to become familiar with and therefore we provide a fair amount of detailed discussion. Second, once we have the solution of the equations, we need to learn how to extract their physical interpretation, according to the principles that were laid out in the early chapters. Both aspects are important and need to be 'digested' by the reader. Exercises and problems are designed to build and consolidate the mathematical tools needed.

- **Chapter 6: Harmonic Oscillator.** The harmonic oscillator plays a central role in physics, since every potential close to its minimum can be approximated by a quadratic potential. This chapter presents the quantum-mechanical treatment of a system that evolves in a harmonic potential. We set up the Schrödinger equation for this system and present its solution using both algebraic methods and a *brute force* solution of the second-order differential equation. The algebraic solution allows us to introduce creation and annihilation operators, which are important concepts that are used in many applications.

- **Chapter 7: Systems in Three Spatial Dimensions.** This chapter presents the generalisation of concepts that were already discussed in previous chapters to the case of systems in three-dimensional space. While this is a crucial step for the description of most quantum systems in nature, going from one-dimensional to three-dimensional space is a straightforward procedure, which allows us to revise some of the material that has already been encountered. In particular, we discuss again the relation between complex vectors and wave functions. We discuss a new technique, known as separation of variables, and apply it to study the harmonic oscillator in three dimensions. We also see for the first time an explicit example of degenerate eigenvalues.

- **Chapter 8: Angular Momentum.** Angular momentum plays a central role in describing the properties of a quantum state under rotations. In this chapter we introduce a Hermitian operator that we associate with the orbital momentum of a quantum system and study its properties. We compute the commutation relations that characterise the Cartesian components of angular momentum, and express the corresponding operators as differential operators acting on the wave function. We identify a set of compatible observables and search for their common eigenstates. We find that angular momentum is quantised, and compute the corresponding eigenfunctions. All the basic features of angular momentum are introduced in this chapter, which is a prerequisite for the following four chapters.

- **Chapter 9: Spin.** This chapter introduces the concept of intrinsic angular momentum, i.e. a type of angular momentum that is not related to the spatial dependence of the wave function. This intrinsic angular momentum is known as *spin* and is a peculiar property of quantum mechanics. We summarise the properties that spin shares with the orbital angular momentum introduced in the previous chapter, and then focus on spin's peculiarities. We discuss in detail the properties of spin-$\frac{1}{2}$ systems that are relevant to discuss the elementary constituents of matter like the electrons.

- **Chapter 10: Addition of Angular Momenta.** In this brief chapter we explain the rules for adding angular momenta in quantum mechanics. These are useful when considering the total angular momentum of a system that is made up of multiple subsystems, or when adding the orbital angular momentum and the spin of a particle. We keep this chapter short, focusing on giving a set of practical rules, rather than aiming at deriving them. This chapter is the final part of the discussion of angular momentum that we started in Chapters 8 and 9.

- **Chapter 11: Central Potentials.** Central potentials are characterised by the fact that they only depend on the distance from the origin. As such, they describe the dynamics of systems that are symmetric under rotations in three spatial dimensions. Working in spherical coordinates it is possible to reduce the Schrödinger equation for these systems to a one-dimensional problem for the radial dependence of the wave function. Building on our previous discussion of orbital angular momentum, this chapter presents in full detail the mathematical steps that lead to the one-dimensional formulation of the problem. These methods are general and applicable to any radial potential. The chapter includes the study of the quantum rotator and the central square well as examples of central potentials. The content of this chapter is an essential prerequisite for the discussion of the hydrogen atom.

- **Chapter 12: Hydrogen Atom.** Pulling together all the tools that have been introduced so far in the book, we are in a position to discuss the energy levels of the hydrogen atom. This is one of the great successes of quantum mechanics, solving one of the problems that had exposed the limits of validity of classical physics. Once again, the emphasis is on presenting the detailed calculation so that the reader can follow the derivations step-by-step and check their understanding of the basic principles. A physical interpretation of the mathematical results is presented at the end of the chapter.

- **Chapter 13: Identical Particles.** Because quantum mechanics does not allow us to define the 'trajectory' of a particle – we cannot have states that are localised in one point in space with a given momentum because of the Heisenberg uncertainty principle – identical particles need to be treated as indistinguishable entities. The mathematical translation of this statement is that the state vector is unchanged (more precisely, only changes by a phase factor) under a permutation of the two particles. This seemingly logical statement has deep consequences, which are summarised in the spin-statistics theorem and the Pauli exclusion principle.

- **Chapter 14: Symmetries in Quantum Mechanics.** Symmetries play a central role in our description of natural phenomena. In this chapter we define the concept of symmetry transformations in both classical and quantum mechanics, and discuss how the set of symmetry transformations of a physical system naturally forms a mathematical structure called a *group*. We discuss in detail two important examples, namely the group of translations and the group of rotations. We introduce the *generators* of these groups and identify them with the operators respectively associated with momentum and angular momentum. We conclude the chapter with a discussion of the general properties of the set of generators, showing that it has the structure of an algebra. Even though we do not have the opportunity to fully develop the mathematical formalism in this book, the reader should get an idea of some of the fundamental underlying ideas, which have shaped the contemporary understanding of physical phenomena.

- **Chapter 15: Quantum Entanglement.** One of the challenges in writing this book has been including a textbook-level chapter on quantum entanglement. This is a growing subject area not only in physics but also in engineering, computing and communication, and one that physics students taking quantum mechanics should have a sound understanding of.

  This chapter presents a modern account of quantum entanglement and is aimed at all physics undergraduates in their intermediate years and requires no previous knowledge of the subject. The chapter has select topics, which all combined provide the reader with a solid foundation yet can be covered within a few weeks. The chapter fits seamlessly with the other more traditional topics in quantum mechanics presented in the book. Of course, we do appreciate that some of the more mathematical aspects that the chapter includes might be better suited for advanced undergraduate physics students or graduate students. However, the goal is making sure that – once the reader has completed this chapter – they will know the key concepts and be able to do basic calculations relevant to the subject.

  The *main topics* covered are the Einstein–Podolsky–Rosen (EPR) paradox, Bell's inequality, quantum teleportation, quantum computing, including the Deutsch and Grover algorithms. We also introduce quantum information theory, including a brief background to Shannon entropy and classical information theory, followed by a discussion of von Neumann entropy and entanglement entropy. For those interested in knowing more about this subject, the literature included in the References and Further Reading sections at the end of the chapter offers some good tips. Note also that since the topics discussed in this chapter are still evolving and emerged relatively recently compared to other topics in quantum mechanics, we have cited as footnotes some of the key seminal papers. This is the only chapter where some original journal articles are cited.

- **Chapter 16: Time-Independent Perturbation Theory.** This is one of the most well-used approximation methods in quantum mechanics. There is a derivation of nondegenerate perturbation theory, done up to third order in the energy correction. In addition, the normalisation and orthogonality of the eigenstates is done up to second order. Degenerate perturbation theory is then derived and developed up to second order in the energy correction. Application of perturbation theory to hydrogen fine structure is treated along with an Appendix that derives these fine-structure terms from the relativistic Dirac theory. The first-order perturbation theory correction to the ground-state energy of helium is also treated.

- **Chapter 17: Calculation Methods Beyond Perturbation Theory.** The Rayleigh–Ritz variational method is presented for estimating the ground-state energy of a system. The first excited state of the helium atom is treated by perturbation theory followed by a broader discussion for treating multi-electron atoms. The Born–Oppenheimer approximation is discussed. This method is useful for problems where there are heavy and light masses in a system. An application of this method is given for the hydrogen molecular ion. The Hellmann–Feynman method is given, which takes derivatives of the Hamiltonian operator with respect to some judiciously chosen parameters to extract results on expectation values of the system. Finally, the Wenzel–Kramers–Brillouin–Jeffreys (WKBJ) approximation is addressed for treating problems where the potential is slowly varying relative to the variation of the wave function. The WKBJ treatment of the potential well and barrier, as well as the symmetric double well, are presented in detail.

- **Chapter 18: Time-Dependent Perturbation Theory.** In this chapter, we examine problems where the Hamiltonian depends on time. The formalism of time-dependent perturbation theory is developed as an approximation method to compute transitions between a specified set of basis states. The general first-order expression is given and specific examples are given of a constant perturbation switched on at some given time and a harmonic perturbation. Interaction of radiation with a quantum system is treated at the lowest non-trivial dipole approximation. Absorption, stimulated emission and spontaneous emission of photons by a quantum system are considered, and the Einstein coefficients are obtained. Selection rules based on symmetry considerations are derived. The time-dependent perturbation theory formalism is derived to all orders.

- **Chapter 19: Quantum Scattering Theory.** The scattering of a particle with a potential is examined. Expressions for the scattering amplitude and *scattering cross-section* are derived for the lowest-order Born approximation and the Born series. The scattering of a wave-packet is treated to understand how it can then be replaced by a plane wave. Partial wave analysis, useful for low-energy scattering, is treated as a complementary method to the Born series, which is most useful at high energy.

Acknowledgements

This book has arisen from our lectures and the accompanying notes we have used. These notes include contributions from past lecturers within this programme, most notably from the most recent past lecturers of this subject: Ken Bowler and Brian Pendleton. Some of their lecture notes no doubt have been absorbed into the text of this book and for that we give our acknowledgement and gratitude. We also wish to pay our tributes to Ken Bowler, who passed away in 2019. He was a dedicated member of our programme, a great source of knowledge and an inspiring teacher. His lecture notes are coveted by all lecturers in our department. This book is all too lucky to contain his contributions dispersed throughout. Other lecturers within our programme have also contributed in direct or indirect ways to the development of this book, including Roger Horsley, Graeme Ackland, Tony Kennedy and also Richard Ball, Martin Evans, Einan Gardi, Richard Kenway, David McComb, Donal O'Connell, John Peacock, Jennifer Smillie and Roman Zwicky. The biggest thanks go to all past students on our quantum mechanics courses, whose questions and comments have continued to shape our lectures.

Alongside some of our own approaches, the lectures over the years have absorbed and modified what our programme has considered the best explanations from other classic textbooks on quantum mechanics, including Bransden and Joachain, Cohen-Tannoudji, Gasiorowicz, Griffiths, Landau and Lifshitz, Liboff, Merzbacher, Messiah, Sakurai, Schiff, Shankar, and many others.

We thank Jaime Calderón Figueroa and Michael Wilson for carefully proofreading the draft book, assisting with making the figures and helping us make the lecture slides for the supplementary material. We thank all the staff at Cambridge University Press for their support and guidance, in particular Nicholas Gibbons, Ilaria Tassistro and Victoria Parrin. We also thank Suresh Kumar Kuttappan and the other staff at CUPTeXSupport.

The entanglement chapter arose from lectures given by AB for the quantum mechanics course. He thanks Maurizio Ferconi, Tony Kennedy and Brian Pendleton for helpful discussions in the initial phases of development of this chapter. He especially thanks David Jennings for carefully reading this chapter and providing input in the intermediate and final stages of development and for several insightful suggestions. From his university education, he thanks Eugene Commins and Eyvind Wichmann for teaching him quantum mechanics. For support and encouragement, he thanks his parents, Jagmohan and Kailash, and his other family and friends, Geeta, Sonny, Neeta, Siddharth, Anika and Maurizio. Above all, he thanks Sarah for her care and, during the writing of this book, her good-natured patience.

LDD would like to thank Adriano Di Giacomo and Pietro Menotti. Their undergraduate courses on quantum mechanics at the University of Pisa have provided a unique environment that has fascinated several generations of students, and shaped their understanding of this subtle subject. Their teaching is fondly remembered, but any mistakes and omissions in this book are solely LDD's responsibility.

PART I

BASICS

Stories and Thoughts about Quantum Mechanics

Quantum mechanics emerged as a natural extension of classical mechanics. As physics probed into the microscopic realm, it could be argued it would be almost impossible not to discover quantum mechanics. The spectra of atoms, the blackbody spectra, the photoelectric effect and the behaviour of particles through an array of slits had characteristically non-classical features. These phenomena were waiting their time for a theory to explain them. That does not diminish from the huge scientific insights of the founders of the subject. In physics, the great accomplishments come more often than not from insight rather than foresight. Knowing what will be the right physics 50 years into the future is a game of speculation. Recognising what is the important physics in the present and being able to explain it is the work of scientific insight. Thus, whereas we might say Democritus had great foresight millennia ago to envision the discrete nature of particles, it was Albert Einstein, Max Planck, Niels Bohr, Erwin Schrödinger, Werner Heisenberg, Paul Dirac, Max Born and Wolfgang Pauli who had the insights to develop quantum mechanics. And since their foundational work, our understanding of the physical world grew dramatically like never before.

The ideas leading to quantum mechanics started with the realisation that phenomena which up to then were considered wave-like in nature, could also have particle properties and vice versa. In the classical picture, particle properties mean the phenomena have energy and momentum that are localised in space, whereas wave properties mean the energy and momentum are coherently distributed over space. Radiation, which classically was considered to be wave-like, was the first phenomenon to be found to also have a particle nature. This was first noticed in blackbody radiation. In the classical theory, based on a wave description of radiation, the theoretical expectation near the end of the nineteenth century was that at low wavelength, the power per unit area radiated by a blackbody became infinite, the Rayleigh–Jeans law. This was in contradiction to experiment, which found the power spectrum to go to zero. In what might be the most significant curve fit in the history of physics, in 1900 Planck observed that he could arrive at a function that explained the observed blackbody spectrum, but it required the energy of the radiation to take discrete values. In particular, at a given wavelength λ, the energy needed to take values $E_\lambda = 0, \epsilon, 2\epsilon, \ldots, n\epsilon, \ldots$, where $\epsilon = hc/\lambda$. Planck introduced the proportionality constant h between energy ϵ and wavelength λ, thus necessarily the dimensions of angular momentum, which is now known as Planck's constant. The value of **Planck's constant**,

$$h = 6.626196 \times 10^{-27} \text{ erg s},$$

sets the typical scale of quantum phenomena. The **rationalised Planck's constant** is defined as

$$\hbar = h/(2\pi) \, .$$

Planck attempted to reconcile this strange functional form that he had obtained with classical physics but achieved limited success.

The full realisation that light possesses a dual particle-like property came in 1905 by Einstein. It was known from experiments by then that when light strikes a clean metal surface, electrons get emitted. Increasing the intensity of the light increased in proportion the number of electrons emitted, but the kinetic energy of the electrons would remain the same, contrary to classical expectation. Einstein applied the Planck blackbody results to assert that quantisation of light was an inherent property. His picture was that light was made up of discrete quanta of energy, using Planck's expression, $E = hc/\lambda$. The idea that radiation was composed of particles was further confirmed in 1923 by Arthur Compton in studies of electrons scattered by radiation, where he concluded that the scattering process could easily be explained as a scattering between the electron with another particle of energy $h\nu$, where ν is the frequency of the radiation. In other words, as if the radiation has particles of this quanta of energy.

The classical picture of radiation also posed a problem for atoms. By 1913, experiments by Ernest Rutherford had established the structure of atoms as having a very tiny nucleus of positively charged particles. Niels Bohr, who had been working in the Rutherford laboratory at the time of his exciting discoveries, suggested the picture of the electrons as orbiting around this nucleus, bound to it by electrostatic attraction. In this classical picture, as the electrons are orbiting, as such accelerating, they are radiating energy. In such a picture the electron orbit should diminish as its energy goes into radiation, and within a fraction of a second the electron should spiral into the nucleus. Bohr showed in 1913 how this dilemma could be resolved by postulating that electrons only move in certain stable orbits of certain energy where they don't radiate, and in jumping between orbits they radiate energy equal to the difference between the energies of the two levels. Within this classical picture for an electron moving in a circular orbit at distance r from the nucleus acted on by a Coulomb field, since in that case the kinetic energy of the electron would equal half the magnitude of the potential energy, the total energy would be

$$E = -\frac{Ze^2}{2r} \, . \tag{1.1}$$

Thus the energy radiated by the atom as the electron moves from an orbit r_i to an orbit r_f under Bohr's assumption is

$$E_{\text{radiated}} = E_i - E_f = -\frac{Ze^2}{2}\left(\frac{1}{r_i} - \frac{1}{r_f}\right) \, . \tag{1.2}$$

Bohr argued his quantisation ideas in a few different ways, but perhaps the most far-reaching was his postulate that the angular momentum of the electron in the atom is quantised:

$$mvr = \frac{nh}{2\pi} \, , \tag{1.3}$$

where n is an integer and h is Planck's constant. From this expression was implied

$$E_{\text{radiated}} = \frac{2\pi^2 Z^2 m e^4}{ch^3} \left(\frac{1}{n_f^2} - \frac{1}{n_i^2} \right). \tag{1.4}$$

A few years later, in 1916, Sommerfeld and Wilson developed a generalised quantisation rule that the integral of a dynamical variable over its conjugate variable around a cycle had to be equal to an integer of the Planck constant. This could be applied for example to angular momentum as the dynamical variable, leading to angular momentum quantisation or momentum and position, leading to energy quantisation. The work on radiation and the Bohr atom constituted what has come to be called the old quantum mechanics. This was an attempt at patching up experimental results within what was still believed to be an underlying classical picture of nature.

With the idea of radiation, which classically had been pictured as wave-like, successfully demonstrating particle-like properties, in 1924 Louis de Broglie suggested that perhaps matter, in particular electrons, which classically were pictured as particles, possess dual wave-like properties. He postulated that an electron of energy E and momentum p has a wavelength associated with it of

$$\lambda = \frac{h}{p} \tag{1.5}$$

and frequency

$$f = \frac{E}{h}. \tag{1.6}$$

A cursory examination of de Broglie's idea might suggest that it was a simple-minded guess and he just happened to be lucky enough to be right. This view might be further supported by the fact that de Broglie had completed his university studies in history with plans for a career in the civil service, to then go back and study science. This interpretation of de Broglie's work is in fact misleading. What is less well remembered is that Louis de Broglie had a physicist older brother, Maurice de Broglie, who was an experimental physicist studying X-rays, a research area at the cutting edge of the then developing quantum mechanics. Maurice helped initiate his brother's interest in physics, including sharing certain prevailing views advocated by W. H. Bragg, that X-rays display characteristics that are both wave- and particle-like. Louis had considerable knowledge of Maurice's views and had an in-depth understanding of the developments underway in physics at the time. Louis was by no means just an outsider who made a good guess. He had studied deeply the issues and views of his time, and it was more than a decade after he entered his studies in science that he came forth with his ideas on the wave nature of particles.

Verification of these properties was confirmed in 1927 in experiments on diffraction of electrons in crystals by Davisson and Germer and thin films by G. P. Thomson (Davisson and Thomson shared the 1937 Nobel Prize for discovering these wave properties of electrons). This was 31 years after J. J. Thomson, G. P.'s father, was awarded the Nobel Prize for discovering the electron as a particle. (An example from physics why children should not always listen to their parents!) This timeline may suggest a tidy picture of theoretical insight followed by experimental verification, but reality in physics is seldom so clean. In fact, experiments on electron beams through material had been done for many years previously, displaying evidence of their wave-like property. In particular, Davisson's experiments on electron scattering dated back before 1920, with results indicating patterns akin to diffraction. But these results needed proper interpretation, which seemed only to come once a compelling theoretical perspective was put forth.

De Broglie's ideas that a wave is associated with a particle led to the expectation that there must be an equation that governs the wave. It was Schrödinger who arrived at the equation for such waves, which now is called the Schrödinger equation. There were other attempts at obtaining such a wave equation, notably by Peter Debye and separately Erwin Madelung, but they all failed in one way or another and are forgotten to the mainstream history on physics. Schrödinger was motivated by de Broglie's interpretation of the Sommerfeld quantisation condition that this can be viewed as the number of wavelengths which cover the exact orbit of the electron around the nucleus. To Schrödinger, this looked like an eigenvalue problem, which he was familiar with from his early work in continuous media. Schrödinger's key step was to replace in the Hamilton equation for classical mechanics

$$H\left(q, \frac{\partial S}{\partial q}\right) = E \tag{1.7}$$

S by $k \ln \psi$, with ψ the wave function. He imposed the conditions on this wave function that it must be real, single-valued and twice continuously differentiable. The reality condition he subsequently relinquished in deriving what now is called the time-dependent Schrödinger equation.

The question that perplexed physicists initially was what exactly does this wave function ψ represent. In particular, did this wave function represent a single particle or a beam of particles? If it were representing a beam of particles, then the resulting diffraction patterns that were observed could be argued as some kind of collective effect.

When just a single particle goes through a diffraction apparatus, such as an array of slits, and deposits its signature on the observation screen, in any single case the individual particle appears to leave the diffractive apparatus in a random fashion. However, when carried out many times, each time with a single particle through the diffraction apparatus, the result from all those cases is the screen showing an interference pattern. Such an experiment implies that the wave property attached to the particle applies to a single particle and that it has a probabilistic meaning. It was Max Born who, in 1926 (he was at Göttingen at the time but later became part of the faculty at Edinburgh), first put forth the probabilistic interpretation of the wave function. At that time, experiments to this level of sophistication were not available. In fact, even the wave nature of matter was not entirely on an experimental footing. Born was guided by his work on applying Schrödinger's wave function to the scattering of particles. He was trying to find an interpretation for his quantum scattering ideas. He liked the particle interpretation of scattering. The only way he could see to adhere to this picture and reconcile it with the Schrödinger wave function was to assert a probability interpretation of the wave function. Thus, he arrived at the conclusion that when particles scatter at the quantum level, one cannot ascribe a definite prediction where the particle will go but simply predict probabilities for the different occurrences.

The Schrödinger approach to quantum mechanics is now the one most commonly used. But in fact a year before Schrödinger discovered his equation, it was Heisenberg who had formulated his matrix approach to quantum mechanics. It was from this line of development that the formal underlying mathematical structure of quantum mechanics started to be understood. Heisenberg's work led in 1926 to the paper by Born and Pascual Jordan and then the famous paper by Born, Heisenberg and Jordan, which were the first to set out a logical mathematical foundation for quantum mechanics based on the matrix approach. Subsequently Schrödinger, and separately Pauli, showed that the wave and matrix approaches were in fact equivalent. The final phase in the development of quantum mechanics was the formalisation of ideas. Up to even the end of the 1920s quantum mechanics was a collection of ideas with different equations seeming to work for different physical situations. What was missing was a formal understanding of the mathematical structure of the theory. It is in fact this work at the final phases that has since been the main point explained and studied in the many textbooks of quantum mechanics written from then to now. Several people were involved in this step of formalising the theory, including of course Schrödinger and Heisenberg but also Born, Pauli and Paul Dirac and most notably Jordan, David Hilbert and John von Neumann. The culmination of this work by the end of the 1920s was the formulation of quantum mechanics in Hilbert space with the concepts of representation and transformation.

Further Reading

Jammer, M. (1966). *The Conceptual Development of Quantum Mechanics*. McGraw-Hill, New York.

Van Der Waerden, B. L. (Ed.) (1968). *Sources of Quantum Mechanics*. Dover Publications, New York.

2 Quantum States

Quantum mechanics describes the behaviour of matter and light at the atomic scale, where physical systems behave very differently from what we experience in everyday life – the laws of physics of the quantum world are different from the ones we have learned in classical mechanics. Despite this 'unusual' behaviour, the principles of scientific inquiry remain unchanged: the only way we can access natural phenomena is through experiment; therefore our task in these first chapters is to develop the tools that allow us to *compute predictions for the outcome of experiments starting from the postulates of the theory*. The new theory can then be *tested* by comparing theoretical predictions to experimental results. Even in the quantum world, *computing* and *testing* remain the workhorses of physics.

Because quantum mechanics is so different from our intuitive, 'classical' description, the procedure we adopt for describing the quantum world needs to be defined carefully; the physical description of quantum phenomena must adhere very strictly to this procedure. In the first part of this book we will focus on presenting the basic postulates, and developing the mathematical tools needed for studying elementary systems. In time we want to be able to relate the mathematical constructs to actual physical processes. As we venture on this journey, we can only develop some intuition by practice; solving problems is an essential component for understanding quantum mechanics.

Experimental results at the beginning of the twentieth century first highlighted that the behaviour of physical systems at atomic scales is inconsistent with classical mechanics. As discussed in the previous chapter, the combined efforts of many eminent physicists led to a consistent picture of the new dynamics. Here we will not follow the historical route; unsurprisingly, the latter was rather tortuous. Instead we will start by stating the new principles, and then work towards deriving their consequences.

Despite its unintuitive aspects, quantum mechanics describes very concrete features of the world as we know it (e.g. the spectrum of the hydrogen atom). By now, many of its predictions have been tested to great accuracy, and have always been found in agreement with experiments. This book's ambition is to set the foundations for developing a description of natural phenomena as observed in nature, i.e. to connect the abstract to the experiment. At every stage we will try to provide examples of physical systems described by the techniques that are being discussed.

It is useful to keep in mind the following disclaimer in Feynman's lectures:[1]

> In this subject we have, of course, the difficulty that the quantum mechanical behavior of things is quite strange. Nobody has an everyday experience to lean on to get a rough,

---

[1] R. P. Feynman, R. B. Leighton, M. Sands, *The Feynman Lectures on Physics – Quantum Mechanics*. Addison-Wesley, New York, 1965.

intuitive idea of what will happen. So there are two ways of presenting the subject: We could either describe what can happen in a rather rough physical way, telling you more or less what happens without giving the precise laws of everything; or we could, on the other hand, give the precise laws in their abstract form. But, then because of the abstractions, you wouldn't know what they were all about, physically. The latter method is unsatisfactory because it is completely abstract, and the first way leaves an uncomfortable feeling because one doesn't know exactly what is true and what is false. [...] Here, we will try to find a happy medium between the two extremes.

Any first course in quantum mechanics suffers from the same problem, and this textbook makes no exception. We chose to start with an abstract definition of the laws, making use when possible of abstract mathematical notation. The explicit examples should allow the reader to appreciate the power of the mathematical tools, while making the mathematical notation more meaningful.

In order to start our journey into quantum mechanics, we need to introduce two main concepts:

1. the *state* of the system, and later in this book the dynamical laws that determine its time evolution;
2. the *observables* (i.e. the possible outcomes of experiments), which allow us to probe the state of the system.

This chapter is devoted to the introduction of the mathematical structure needed to characterise the state of a quantum system, while the study of observables is deferred to the following chapter in the book. Here we will see that physical states are associated with vectors in complex vector spaces, and we will define the main concepts that are going to be used to deal with physical states. These are the foundations of the theory and will be used throughout the book. Different physical systems will have states belonging to different vector spaces, but the general mathematical framework will remain the same for *all* quantum systems. The tools that we introduce in this chapter will then be used for all subsequent discussions. Examples and exercises should help you in building the skills needed to manipulate vectors in cases that are relevant for quantum mechanics.

2.1 States of a Quantum System

In classical mechanics, the state of a point-like system is described by the position and the momentum of its components. For an elementary particle in our usual three-dimensional space, the state of the system is defined by a six-dimensional **real** vector:

$$\left(\underline{x}, \underline{p}\right). \tag{2.1}$$

Underlined letters indicate three-dimensional vectors, $\underline{x}$ and $\underline{p}$ being respectively the position and momentum of the particle. The knowledge of these two vectors at any moment in time allows us to compute all other properties of the system, for example its energy or its angular momentum. Moreover, once the dynamical laws are given, the state of the system at any other time can also be computed by solving the equations of motion.

We shall now introduce the postulates that allow us to describe the state of a quantum system. It is useful here to focus on the peculiarities of the quantum description. We will try to highlight these peculiarities as we encounter them and to reinforce our understanding through examples and exercises. We should always be able to connect the abstract concepts that are going to be introduced to physics, i.e. use the mathematical tools to compute theoretical predictions for the outcome of experiments. This process needs to be carefully defined: we will need to specify exactly what quantities can be computed and how these calculations are performed.

The first postulate of quantum mechanics defines the **states** of a quantum system.

Postulate 1   Quantum states are identified by vectors in **complex** vector spaces.

It is important to emphasise that we deal with complex vector spaces, instead of the real vector spaces that are more familiar from classical mechanics. This entails some differences that will be discussed in this chapter. A first comment arises by comparing the state vector in classical and quantum mechanics. As we discussed in the example above, the state of a classical point-like system is described by its position and its momentum, a six-dimensional real vector. In this case the components of this six-dimensional vector are also observables: the position and the momentum of the particle can be measured in an experiment. In quantum mechanics the picture is more subtle. Experiments do not measure complex vectors and we will have to put in some more work to define what can be predicted about the outcome of experiments.

For now, let us begin by introducing the concepts from linear algebra that are needed to work with the elements of complex vector spaces. This is standard material to be found in linear algebra courses, which we summarise in order to have a consistent set of definitions and notations. The chapter is dotted with examples and simple exercises that will hopefully help the reader to check that they understand and are able to apply the concepts that are being introduced. Their connection to physics will be the subject of following chapters. As we introduce physical concepts later in this book, we will continuously refer to the material presented in this chapter.

2.1.1 Kets

Following Dirac's notation, vectors will be denoted by **kets**, so for example the elements of a vector space $\mathcal{H}$ are written as

$$|v\rangle \in \mathcal{H}. \tag{2.2}$$

More generally we will use the notation $|\ldots\rangle$, where $\ldots$ can be any combination of symbols used to identify the vector. Sometimes we will insert just one letter, e.g. $|\psi\rangle$, other times we will use one or more indices, e.g. $|n\rangle$, or $|n, m\rangle$, with n, m integers. Finally we will encounter instances where the kets are identified by a continuous index, e.g. $|\xi\rangle$ with $\xi \in \mathbb{R}$. We will use the words state, vector and ket interchangeably.

The defining property of a vector space is that linear combinations of vectors also belong to the vector space:

$$\forall |u\rangle, |v\rangle \in \mathcal{H}, \forall \alpha, \beta \in \mathbb{C}, \quad \alpha|u\rangle + \beta|v\rangle \in \mathcal{H}. \tag{2.3}$$

The fact that $\mathcal{H}$ is a complex vector space is reflected in the fact that the coefficients α and β are complex numbers. It is straightforward to generalise this result to the linear combination of a set of vectors indexed by some integer n:

$$\forall \{|u_n\rangle\} \subset \mathcal{H}, \forall \{\alpha_n\} \subset \mathbb{C}, \quad \sum_n \alpha_n |u_n\rangle \in \mathcal{H}. \tag{2.4}$$

In Eq. (2.4) we have deliberately not specified the range of values taken by the index n. Depending on the system under study we may deal with either finite or infinite sums. We will come back to this issue later in the chapter.

Mathematical Aside: Properties of Vector Spaces
Here are some useful rules for manipulating vectors. In the following formulae, kets indicate generic elements of $\mathcal{H}$ and coefficients are generic complex numbers. Some of these properties may look familiar to readers who have some experience with linear algebra.

- There exists a '+' operation between vectors:

$$\forall |\psi\rangle, |\phi\rangle \in \mathcal{H}, \quad |\psi\rangle + |\phi\rangle \in \mathcal{H}. \tag{2.5}$$

We have already implicitly used this property in Eq. (2.3).

- The sum of vectors is associative and commutative:

$$|\psi\rangle + (|\phi\rangle + |\xi\rangle) = (|\psi\rangle + |\phi\rangle) + |\xi\rangle, \tag{2.6}$$
$$|\psi\rangle + |\phi\rangle = |\phi\rangle + |\psi\rangle. \tag{2.7}$$

Therefore the brackets in the first line are superfluous, and one can write simply

$$|\psi\rangle + |\phi\rangle + |\xi\rangle. \tag{2.8}$$

- Multiplication of a vector by a complex number yields another vector:

$$\forall |\psi\rangle \in \mathcal{H}, \forall \alpha \in \mathbb{C}, \quad \alpha|\psi\rangle \in \mathcal{H}. \tag{2.9}$$

We will typically write the complex coefficient to the left of the vector, but this is just a convention:

$$\alpha|\psi\rangle = |\psi\rangle\alpha. \tag{2.10}$$

- The multiplication by complex numbers is associative and distributive:

$$\alpha(\beta|\psi\rangle) = (\alpha\beta)|\psi\rangle, \tag{2.11}$$
$$\alpha(|\psi\rangle + |\phi\rangle) = \alpha|\psi\rangle + \alpha|\phi\rangle, \tag{2.12}$$
$$(\alpha + \beta)|\psi\rangle = \alpha|\psi\rangle + \beta|\psi\rangle. \tag{2.13}$$

Let us examine in more detail the operations that are described in the first of these equations. Note that, on the left-hand side of Eq. (2.11), we first perform the multiplication of a vector, $|\psi\rangle$, by a complex number β. The result of this operation is a vector, which is then multiplied by the complex number α. The final result on the left-hand side is therefore a vector. On the right-hand side, we first multiply the two complex numbers α and β. The result of this operation is another complex number, $(\alpha\beta)$, which multiplies the vector $|\psi\rangle$. The result on the right-hand side is also a vector, as one would expect if the equality has to hold. As a consequence of Eq. (2.11), the expression $\alpha\beta|\psi\rangle$ can be used without ambiguities. It is useful to perform this kind of 'grammatical' analysis of the equations, until the reader feels completely familiar with the operations performed with vectors in complex vector spaces.

- Existence of a null vector $|0\rangle$, such that

$$|\psi\rangle + |0\rangle = |\psi\rangle, \tag{2.14}$$

$$0|\psi\rangle = |0\rangle, \tag{2.15}$$

$$\alpha|0\rangle = |0\rangle. \tag{2.16}$$

- For each vector $|\psi\rangle$, there exists an *inverse* vector with respect to the '+' operation, denoted $-|\psi\rangle$:

$$|\psi\rangle + (- |\psi\rangle) = |0\rangle. \tag{2.17}$$

Vector spaces are ubiquitous in physics. The most familiar examples being the real vector spaces used in classical mechanics. Despite the abstract definitions above, vector spaces encode familiar properties. It is worthwhile to discuss a simple example of a complex vector space, which allows us to illustrate the main concepts.

Example 2.1 Consider the set of complex column vectors of size 2, i.e. the set of column vectors that can be written as

$$|v\rangle \equiv \begin{pmatrix} v_1 \\ v_2 \end{pmatrix}, \tag{2.18}$$

where v_1 and v_2 are complex numbers. This space is called $\mathbb{C}^2$. If we define the sum of vectors and the multiplication by a complex scalar λ respectively as

$$\begin{pmatrix} v_1 \\ v_2 \end{pmatrix} + \begin{pmatrix} w_1 \\ w_2 \end{pmatrix} = \begin{pmatrix} v_1 + w_1 \\ v_2 + w_2 \end{pmatrix}, \tag{2.19}$$

$$\lambda \begin{pmatrix} v_1 \\ v_2 \end{pmatrix} = \begin{pmatrix} \lambda v_1 \\ \lambda v_2 \end{pmatrix}, \tag{2.20}$$

then $\mathbb{C}^2$ is a vector space. It is important to notice that the '+' sign on the left-hand side of Eq. (2.19) defines the sum of two column vectors. The '+' sign on the right-hand side indicates that the components of each column vector are summed, and therefore is simply the sum of complex numbers that should be familiar to the reader. The reader can check that the elements of $\mathbb{C}^2$ satisfy all the properties listed above.

Superposition Principle

The vector space structure automatically implements the **superposition principle**, which states that any linear combination of quantum states with complex coefficients defines a new, valid quantum state of the system. This property will recur multiple times in the rest of the book.

Alternatively, one could state the superposition principle as the first fundamental postulate and deduce that the states of a quantum system need to be elements of a vector space.

2.1.2 Scalar Product

We will always be concerned with vector spaces on which a **scalar product** is defined. The scalar product, which is sometimes also called internal product, is a function that takes two vectors, $|u\rangle$ and $|v\rangle$, and returns a complex number, which we will denote by $\langle u|v\rangle$. We can rephrase this sentence using a mathematical notation and define the scalar product as:

$$\mathcal{H} \times \mathcal{H} \to \mathbb{C}$$

$$|u\rangle, |v\rangle \mapsto \langle u|v\rangle.$$

The notation above simply indicates that the scalar product is a function that takes two vectors and returns a complex number. It satisfies the following properties:

1. Complex conjugation

$$\forall |u\rangle, |v\rangle \in \mathcal{H}, \quad \langle u|v\rangle = \langle v|u\rangle^*. \tag{2.21}$$

2. Linearity

$$\forall |u\rangle, |v\rangle, |w\rangle \in \mathcal{H}, \text{ and } \forall \alpha, \beta \in \mathbb{C}, \quad \langle u|\big[\alpha|v\rangle + \beta|w\rangle\big] = \alpha\langle u|v\rangle + \beta\langle u|w\rangle. \tag{2.22}$$

Two comments are in order here. First, the scalar product is not symmetric with respect to the permutation of its arguments, $\langle u|v\rangle \neq \langle v|u\rangle$. Second, note that because of Eq. (2.21) above, the scalar product is antilinear in its first argument, viz.

$$\big[\alpha\langle u| + \beta\langle v|\big]|w\rangle = \alpha^*\langle u|w\rangle + \beta^*\langle v|w\rangle. \tag{2.23}$$

Vector spaces equipped with a scalar product are called **Hilbert spaces**.

Example 2.2 Given two vectors

$$\begin{pmatrix} v_1 \\ v_2 \end{pmatrix}, \begin{pmatrix} w_1 \\ w_2 \end{pmatrix} \in \mathbb{C}^2, \tag{2.24}$$

the scalar product can be defined as

$$\langle v|w\rangle = v_1^* w_1 + v_2^* w_2. \tag{2.25}$$

Check that this definition satisfies the properties above.

2.1.3 Bras

A linear **functional** Φ is a function defined on $\mathcal{H}$ that associates with each vector a complex number. Using a mathematical notation, we can write

$$\Phi : \mathcal{H} \to \mathbb{C}$$
$$|v\rangle \mapsto \Phi\left(|v\rangle\right) = z.$$

Linearity means that Φ satisfies

$$\Phi(\alpha|v\rangle + \beta|w\rangle) = \alpha\Phi\left(|v\rangle\right) + \beta\Phi\left(|w\rangle\right), \tag{2.26}$$

for arbitrary vectors $|u\rangle, |v\rangle$ and complex coefficients α, β.

For a given vector $|u\rangle$, the scalar product associates a complex number with any vector $|v\rangle$ through the correspondence

$$|v\rangle \mapsto z = \Phi_{|u\rangle}\left(|v\rangle\right) = \langle u|v\rangle, \tag{2.27}$$

and therefore defines a *bona fide* functional $\Phi_{|u\rangle}$ acting on $\mathcal{H}$. The linearity of the functional in this case is guaranteed by Eq. (2.22). It can be shown that all bounded functionals acting on $\mathcal{H}$ can be defined as in Eq. (2.27) – the latter result is known as the **Riesz theorem**, whose proof is beyond the scope of these lectures. The one-to-one correspondence between functionals and vectors led Dirac to introduce the **bra** notation, $\langle u|$, to denote the functional defined in Eq. (2.27). The notation $\langle u|v\rangle$ can also be seen as the application of the functional $\langle u|$ to the vector $|v\rangle$. The space of functionals is called the **dual** of the vector space $\mathcal{H}$, and is denoted by $\mathcal{H}^*$.

Example 2.3 We can look once again at the $\mathbb{C}^2$ vector space, and understand in detail the action of a functional. In order to have an explicit example, let us choose the vector

$$|u\rangle = \begin{pmatrix} i/\sqrt{2} \\ -1/\sqrt{2} \end{pmatrix} \in \mathbb{C}^2, \tag{2.28}$$

and consider the functional $\langle u|$ defined in Eq. (2.27). Acting with $\langle u|$ on

$$|v_1\rangle = \begin{pmatrix} 1 \\ 0 \end{pmatrix} \tag{2.29}$$

yields

$$\langle u|v_1\rangle = -i/\sqrt{2} \in \mathbb{C}. \tag{2.30}$$

Take a moment to make sure you understand that the functional acts on a vector and returns a scalar. Check that you can reproduce the result in the above equation. What is the origin of the minus sign on the right-hand side?

Acting with $\langle u|$ on a generic vector

$$|v\rangle = \begin{pmatrix} v_1 \\ v_2 \end{pmatrix}, \quad v_1, v_2 \in \mathbb{C} \tag{2.31}$$

we obtain

$$\langle u|v\rangle = -\frac{i}{\sqrt{2}}v_1 - \frac{1}{\sqrt{2}}v_2. \tag{2.32}$$

Note that the result of the action $\langle u|$ on a vector returns a complex number, as expected for a functional.

2.1.4 Norm

As usual for spaces with a scalar product, the norm of a vector $|v\rangle$ will be given by the scalar product of the given vector with itself, $||v||^2 = \langle v|v\rangle$. For reasons that we will discuss later, quantum states are associated with vectors with unit norm, $\langle v|v\rangle = 1$. Vectors with non-unit norm can be normalised:

$$|v\rangle \mapsto \frac{1}{\sqrt{\langle v|v\rangle}}|v\rangle. \tag{2.33}$$

You should practise the use of Eq. (2.33), as it will be used numerous times in the rest of the book, and in real-life applications.

Example 2.4 For all vectors

$$\begin{pmatrix} v_1 \\ v_2 \end{pmatrix} \in \mathbb{C}^2, \tag{2.34}$$

the norm obtained using the scalar product defined above is

$$||v||^2 = v_1^* v_1 + v_2^* v_2. \tag{2.35}$$

You should convince yourself that the norm is a positive real number. It is instructive to compute explicitly the norm of a few vectors in $\mathbb{C}^2$.

Exercise 2.1.1 The vector

$$|v\rangle = \begin{pmatrix} 1 \\ 2i \end{pmatrix} \in \mathbb{C}^2 \tag{2.36}$$

has norm squared $\langle v|v\rangle = 5$. Normalise the vector so that it has unit norm.

We will encounter vectors of infinite norm, which require us to extend the properties of the scalar product in order to be able to deal with them. In particular, if $|\xi\rangle$ is a vector of infinite norm, we shall assume that its scalar product with a vector $|v\rangle$ of finite norm is finite.

The norm satisfies the following **triangular inequalities**:

$$||u + v|| \leq ||u|| + ||v||, \tag{2.37}$$

$$\left|||u|| - ||v||\right| \leq ||u + v||. \tag{2.38}$$

Exercise 2.1.2 Consider the vectors

$$|u\rangle = \begin{pmatrix} 1 \\ 0 \end{pmatrix}, \quad |v\rangle = \begin{pmatrix} 1 \\ 2i \end{pmatrix} \in \mathbb{C}^2 \tag{2.39}$$

and check that the triangular inequalities are satisfied. Make sure you can add and subtract vectors, and understand how the norm is computed in this simple case.

Rays and Phase Ambiguity

A more sophisticated wording to describe the quantum-mechanical states is to say that they are associated with **rays** in the vector space. Note that the condition $\langle v|v\rangle = 1$ leaves the vector undefined up to a phase factor $e^{i\varphi}$. We have the following important result: vectors that differ by a global phase factor describe exactly the **same** quantum state. As we will see in the following chapters, *physical* predictions are always independent of the phase of the ket.

2.1.5 Orthogonality

Two vectors $|u\rangle$ and $|v\rangle$ are **orthogonal** if $\langle u|v\rangle = 0$. Two subspaces $\mathcal{H}_1, \mathcal{H}_2$ in $\mathcal{H}$ are **orthogonal subspaces** if

$$\forall |u\rangle \in \mathcal{H}_1, |v\rangle \in \mathcal{H}_2, \langle u|v\rangle = 0. \tag{2.40}$$

Given a subspace $\mathcal{H}_1$ in $\mathcal{H}$, the set of vectors orthogonal to $\mathcal{H}_1$ is also a subspace, called the **complementary subspace** to $\mathcal{H}_1$. We shall denote the complementary subspace by $\mathcal{H}_{1,\perp}$. For each vector $|v\rangle \in \mathcal{H}$ there is a unique decomposition:

$$|v\rangle = |v_1\rangle + |v_\perp\rangle, \quad |v_1\rangle \in \mathcal{H}_1, |v_\perp\rangle \in \mathcal{H}_{1,\perp}. \tag{2.41}$$

The vector $|v_1\rangle$ is called the **projection** of $|v\rangle$ in $\mathcal{H}_1$. A pictorial representation of the projection of a vector in different subspaces is shown in Fig. 2.1.

2.1.6 Operators

Following the formalism that we have set up so far, we expect that a change in the state of a system is represented by a change in the state vector. Therefore we need to introduce

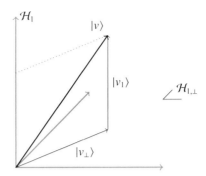

A vector space $\mathcal{H}_1$, and its complementary subspace $\mathcal{H}_{1,\perp}$. The vector $|v\rangle$ is decomposed into the sum of $|v_1\rangle$, its projection onto $\mathcal{H}_1$, and $|v_\perp\rangle$, its projection onto the complementary subspace $\mathcal{H}_{1,\perp}$.

a mathematical tool that transforms the vectors in the Hilbert space. Such a tool is called an operator.

More precisely, an **operator** is a function that acts on a vector and returns a vector

$$\hat{O} : \mathcal{H} \to \mathcal{H}$$
$$|v\rangle \mapsto |w\rangle = \hat{O}|v\rangle. \tag{2.42}$$

Both $|v\rangle$ and $|w\rangle$ are vectors in $\mathcal{H}$. Note that the result of acting on a state vector with an operator produces a *new* state vector. Also note the difference between an operator and a functional. Both act on vectors, i.e. on elements of $\mathcal{H}$. The action of an operator on a vector returns a vector, while the action of a functional on a vector returns a complex number.

A quick comment about conventions is in order. We will denote operators with 'hats', like $\hat{O}$ in the example above. Unless otherwise specified, we will *always* use the convention that operators act from the left on to kets.

In quantum mechanics we will deal with **linear operators**. A linear operator satisfies

$$\hat{O}\big(c_1|\psi_1\rangle + c_2|\psi_2\rangle\big) = c_1\hat{O}|\psi_1\rangle + c_2\hat{O}|\psi_2\rangle, \tag{2.43}$$

where c_1 and c_2 are complex numbers.

Commutators

The product of two operators is defined as

$$\hat{O}_1\hat{O}_2|\psi\rangle = \hat{O}_1\left(\hat{O}_2|\psi\rangle\right). \tag{2.44}$$

Note that the order in which the operators are applied to the state is important! The **commutator** of two operators is

$$\left[\hat{O}_1, \hat{O}_2\right] = \hat{O}_1\hat{O}_2 - \hat{O}_2\hat{O}_1. \tag{2.45}$$

In general the commutator does NOT vanish; it defines an operator acting on quantum states:

$$\left[\hat{O}_1, \hat{O}_2\right]|\psi\rangle = \hat{O}_1\hat{O}_2|\psi\rangle - \hat{O}_2\hat{O}_1|\psi\rangle. \tag{2.46}$$

Functions of Operators

The product of operators can be extended to the case of n factors in a straightforward manner:

$$\hat{O}^n|\psi\rangle = \underbrace{\hat{O}\left(\ldots\left(\hat{O}\ |\psi\rangle\right)\ldots\right)}_{n\ \text{factors}}. \tag{2.47}$$

Having defined monomials of arbitrary order, we can define a function of a generic operator using the function's Taylor expansion. If $f(x)$ is a function such that

$$f(x) = \sum_n c_n x^n, \tag{2.48}$$

then we define the action of the function on the operator

$$f(\hat{O}) = \sum_n c_n \hat{O}^n, \tag{2.49}$$

where the coefficients c_n are the same coefficients that appear in Eq. (2.48). As expected, Eq. (2.49) defines an operator acting on vectors in $\mathcal{H}$.

Example 2.5 The exponential has a well-known Taylor expansion:

$$e^x = \sum_n \frac{1}{n!} x^n = 1 + x + \frac{x^2}{2!} + \ldots. \tag{2.50}$$

Following the above discussion, we define the exponential of an operator as

$$\exp\left(\hat{O}\right) = \sum_n \frac{1}{n!} \hat{O}^n = 1 + \hat{O} + \frac{\hat{O}^2}{2!} + \ldots. \tag{2.51}$$

Note that each term in the sum is an operator, and therefore the exponential of an operator is also an operator.

Exercise 2.1.3 Compute the first two terms in the definition of

$$\log\left(1 + \hat{O}\right). \tag{2.52}$$

You may pause for a moment and think about the conditions under which the Taylor expansion is convergent, and therefore defines a *bona fide* operator.

2.1.7 Bases

The vectors in a given set $\{|e_k\rangle : k = 1, \ldots, n\}$ are called **linearly independent** if

$$\sum_k c_k |e_k\rangle = |0\rangle \quad \Longleftrightarrow \quad c_k = 0, \forall k, \tag{2.53}$$

i.e. none of the vectors in the set can be written as a linear combination of the others. The **dimensionality** of a vector space is the largest number of linearly independent vectors.

The dimensionality of the vector spaces that we will encounter can be either finite or infinite, depending on the physical system under study.

A **basis** is a set of linearly independent vectors in $\mathcal{H}$ such that any vector $|v\rangle$ can be expanded as a linear combination of the basis vectors. If we denote $\mathcal{B} = \{|e_k\rangle : k = 1, \ldots\}$ the set of vectors in the basis, then

$$\forall |v\rangle \in \mathcal{H}, \; \exists\{v_k\} \subset \mathbb{C}, \; |v\rangle = \sum_k v_k |e_k\rangle, \tag{2.54}$$

i.e. every vector $|v\rangle$ can be written as a linear combination of the basis vectors with complex coefficients. Clearly the coefficients depend on the vector $|v\rangle$ that is expanded. Note that the sum in Eq. (2.54) runs over a finite set of vectors for a finite-dimensional space, while the sum becomes an infinite sum for an infinite-dimensional space. The complex coefficients v_k are the **coordinates** of the vector $|v\rangle$ in the basis $\mathcal{B}$. Note at this stage the complete analogy between the complex case, and the more familiar case of real vector spaces. Compare for instance Eq. (2.54) with the familiar expansion of a vector in a real three-dimensional vector space $\mathcal{V}$:

$$\forall \underline{v} \in \mathcal{V}, \; \underline{v} = v_1 \underline{e}_1 + v_2 \underline{e}_2 + v_3 \underline{e}_3, \tag{2.55}$$

where v_i are real numbers, the *usual* coordinates of $\underline{v}$ in the basis $\{\underline{e}_1, \underline{e}_2, \underline{e}_3\}$. We will use Eq. (2.54) everywhere in this book.

We will often work with **orthonormal** bases, so that for any pair of vectors in the basis

$$\langle e_k | e_l \rangle = \delta_{kl}. \tag{2.56}$$

Under this hypothesis the coordinates v_k can be readily computed:

$$v_k = \langle e_k | v \rangle, \tag{2.57}$$

and therefore we can write Eq. (2.54) as

$$|v\rangle = \sum_k \langle e_k | v \rangle |e_k\rangle = \sum_k |e_k\rangle\langle e_k | v \rangle. \tag{2.58}$$

Remember that $\langle e_k | v \rangle$ is a complex number, and therefore can be written equivalently to the left or to the right of the vector $|e_k\rangle$, as discussed above in Eq. (2.10).

Example 2.6 The vectors

$$\begin{pmatrix} 1 \\ 0 \end{pmatrix}, \begin{pmatrix} 0 \\ 1 \end{pmatrix} \tag{2.59}$$

form an orthonormal basis in $\mathbb{C}^2$. Every vector in $\mathbb{C}^2$ can be expanded as

$$\begin{pmatrix} v_1 \\ v_2 \end{pmatrix} = v_1 \begin{pmatrix} 1 \\ 0 \end{pmatrix} + v_2 \begin{pmatrix} 0 \\ 1 \end{pmatrix}. \tag{2.60}$$

The norm of the vector $|v\rangle$ can be computed starting from Eq. (2.58):

$$||v||^2 = \langle v|v\rangle \tag{2.61}$$

$$= \left(\sum_{k'}\langle e_{k'}|\langle v|e_{k'}\rangle\right)\left(\sum_{k}\langle e_k|v\rangle|e_k\rangle\right) \tag{2.62}$$

$$= \sum_{k,k'}\langle e_k|v\rangle\langle v|e_{k'}\rangle\langle e_{k'}|e_k\rangle \tag{2.63}$$

$$= \sum_{k}v_k^* v_k = \sum_{k}|v_k|^2 . \tag{2.64}$$

Equation (2.64) confirms that the norm is a positive number, which vanishes only when $|v\rangle$ is the null vector. Note that the complex conjugate $v_k^* = \langle v|e_k\rangle$ is necessary to guarantee that the norm is real and positive, as discussed above.

The second equality in Eq. (2.58) illustrates nicely how Dirac's notation captures features of linear vector spaces in a suggestive graphical manner. The object $|e_k\rangle\langle e_k|$ can be recognised as the projector over the basis vector $|e_k\rangle$. We can act with $|e_k\rangle\langle e_k|$ to the left of a vector $|v\rangle$ in order to get its projection in the direction of $|e_k\rangle$. It is instructive to translate this sentence into an equation:

$$\left(|e_k\rangle\langle e_k|\right)|v\rangle = |e_k\rangle\left(\langle e_k|v\rangle\right) = \left(\langle e_k|v\rangle\right)|e_k\rangle. \tag{2.65}$$

Equation (2.65) shows that the parentheses are unnecessary in this expression, and indeed we will not use them unless we want to emphasise a specific interpretation of the formula. We shall denote by

$$\hat{\mathcal{P}}_k = |e_k\rangle\langle e_k| \tag{2.66}$$

the **projector** on a state $|e_k\rangle$. Projectors verify the usual properties:

$$\hat{\mathcal{P}}_k^2 = \hat{\mathcal{P}}_k \qquad \hat{\mathcal{P}}_k\hat{\mathcal{P}}_\ell = 0, \quad \text{if } k \neq \ell. \tag{2.67}$$

Exercise 2.1.4   Verify that you recognise the properties in Eq. (2.67) by considering the more familiar case of projectors in three-dimensional Euclidean space.

Exercise 2.1.5   Write the relations in Eq. (2.67) using Dirac's notation for the projectors.

Because it is valid for any vector $|v\rangle \in \mathcal{H}$, Eq. (2.58) can be rewritten as

$$\sum_{k}|e_k\rangle\langle e_k| = 1, \tag{2.68}$$

where 1 here denotes the identity operator acting on $\mathcal{H}$. Equation (2.68) is called a **completeness relation**. The vectors of a basis are said to be a **complete set** in $\mathcal{H}$. In a vector space of finite dimension n, any set of n linearly independent vectors is complete. If $\mathcal{B}_1$ is a subset of $\mathcal{B}$ spanning a subspace $\mathcal{H}_1$, then the projector on this subspace is

$$\mathcal{P}_1 = \sum_{|e_k\rangle\in\mathcal{B}_1}|e_k\rangle\langle e_k|. \tag{2.69}$$

Exercise 2.1.6   Show that $\mathcal{P}_1^2 = \mathcal{P}_1$, as expected for a projection operator.

Solution:

$$
\begin{aligned}
\mathcal{P}_1^2 &= \left\{ \sum_{|e_k\rangle \in \mathcal{B}_1} (|e_k\rangle\langle e_k|) \right\} \left\{ \sum_{|e_k'\rangle \in \mathcal{B}_1} \left(|e_k'\rangle\langle e_k'| \right) \right\} \\
&= \sum_{|e_k\rangle \in \mathcal{B}_1} \sum_{|e_{k'}\rangle \in \mathcal{B}_1} (|e_k\rangle\langle e_k|)\,(|e_{k'}\rangle\langle e_{k'}|) \\
&= \sum_{|e_k\rangle \in \mathcal{B}_1} \sum_{|e_{k'}\rangle \in \mathcal{B}_1} |e_k\rangle\,(\langle e_k|e_{k'}\rangle)\,\langle e_{k'}| \\
&= \sum_{|e_k\rangle \in \mathcal{B}_1} \sum_{|e_{k'}\rangle \in \mathcal{B}_1} |e_k\rangle\,\delta_{kk'}\,\langle e_{k'}| \\
&= \sum_{|e_k\rangle \in \mathcal{B}_1} |e_k\rangle\langle e_k| \\
&= \mathcal{P}_1.
\end{aligned}
$$

Having chosen a basis $\mathcal{B}$, any vector is uniquely identified by its coordinates. A vector $|v\rangle$ can be represented as a column vector made up of the coordinates $\langle e_k|v\rangle$:

$$
|v\rangle \rightarrow \begin{pmatrix} \langle e_1|v\rangle \\ \langle e_2|v\rangle \\ \vdots \end{pmatrix}. \tag{2.70}
$$

We have used an arrow instead of an equals sign to underline the fact that the vector is represented by its coordinates only in a given basis. When there is no ambiguity, we may use the column vector notation to represent a state, but it is necessary to keep in mind the underlying choice of basis. It is useful to repeat the statement once again: a vector is identified by a column vector *in a given basis*. Clearly the same vector $|v\rangle$ can be expanded in different bases; let us assume that we have a second basis $\mathcal{B}' = \{|e_k'\rangle\}$, the coordinates of the vector will be different in each basis:

$$
|v\rangle = \sum_k v_k |e_k\rangle = \sum_k v_k' |e_k'\rangle. \tag{2.71}
$$

As it is customary when performing a change of basis in a vector space, the different coordinates are related to each other. The relation between different sets of coordinates can be readily worked out, the derivation being a standard result in linear algebra. Starting from the expression for the coordinates v_k':

$$
v_k' = \langle e_k'|v\rangle, \tag{2.72}
$$

and using the completeness of $\mathcal{B}$:

$$
v_k' = \sum_l \langle e_k'|e_l\rangle\langle e_l|v\rangle \tag{2.73}
$$

$$
= \sum_l U_{kl} v_l, \tag{2.74}
$$

where $U_{kl} = \langle e'_k | e_l \rangle$ defines the matrix that relates the coordinates in the two bases. It is important to realise that Eq. (2.73) is nothing but the usual expression for the change of coordinates of a vector written using bras and kets. The same relation in terms of column vectors can be written as

$$\begin{pmatrix} v'_1 \\ v'_2 \\ \vdots \end{pmatrix} = \begin{pmatrix} U_{11} & U_{12} & \cdots \\ U_{21} & U_{22} & \cdots \\ \vdots & \vdots & \ddots \end{pmatrix} \begin{pmatrix} v_1 \\ v_2 \\ \vdots \end{pmatrix}, \tag{2.75}$$

which should be familiar from linear algebra.

Exercise 2.1.7 If $\mathcal{H}$ is a finite vector space of dimension n, and the basis vectors are all orthonormal, show that U_{kl} in Eq. (2.74) is a unitary $n \times n$ matrix.

Finally let us consider two vectors $|v\rangle$ and $|w\rangle$, which can be respectively expanded in a given orthonormal basis as

$$|v\rangle = \sum_k v_k |e_k\rangle, \tag{2.76}$$

$$|w\rangle = \sum_k w_k |e_k\rangle, \tag{2.77}$$

then their scalar product is given by

$$\langle w | v \rangle = \sum_k w_k^* v_k. \tag{2.78}$$

The proof of this statement is left as an exercise for the reader. It is important to note that the basis vectors need to be orthogonal for the formula to hold. Also, because we are dealing with complex vector spaces, the scalar product is not invariant under the exchange of the bra and the ket, which is reflected in the fact that the coordinates of the bra (w_k) in Eq. (2.78) are complex conjugated.

Operators as Matrices

We can conclude this section by discussing how the action of an operator can be implemented at the level of the coordinates of the vector. We consider a vector $|v\rangle \in \mathcal{H}$ and a linear operator $\hat{O}$. Having chosen an orthonormal basis $\{|e_k\rangle : k = 1, \ldots\}$, we can expand $|v\rangle$ in this basis:

$$|v\rangle = \sum_l v_l |e_l\rangle, \tag{2.79}$$

where v_k are the complex coordinates of $|v\rangle$. When acting with $\hat{O}$ on $|v\rangle$, we obtain a new vector

$$|v'\rangle = \hat{O}|v\rangle, \tag{2.80}$$

which can also be expanded in the *same* basis:

$$|v'\rangle = \sum_k v'_k |e_k\rangle. \tag{2.81}$$

The coordinates of $|v'\rangle$ are computed as discussed above:

$$v'_k = \langle e_k|v'\rangle = \sum_l v_l\langle e_k|\hat{O}|e_l\rangle. \tag{2.82}$$

Note that in order to get the second equality, we used the fact that $\hat{O}$ is a linear operator. Equation (2.82) is the usual matrix vector multiplication; if we define the **matrix elements** of the operator $\hat{O}$ between two basis vectors as

$$O_{kl} = \langle e_k|\hat{O}|e_l\rangle, \tag{2.83}$$

then the relation between the coordinates of $|v\rangle$ and $|v'\rangle$ is given by the familiar matrix–vector multiplication:

$$v'_k = \sum_l O_{kl}v_l. \tag{2.84}$$

After having chosen a basis in $\mathcal{H}$, a linear operator can be represented as a matrix. We will use an arrow to associate a matrix with an operator, in order to emphasise again that the matrix representation depends on the basis being used:

$$\hat{O} \rightarrow \begin{pmatrix} O_{11} & O_{12} & \cdots \\ O_{21} & \ddots & \\ \vdots & & \end{pmatrix}. \tag{2.85}$$

Note that we will use matrix elements of operators not only between basis vectors but also between generic states. The matrix element of an operator $\hat{O}$ between the states $|\psi\rangle$ and $|\phi\rangle$ is the scalar product

$$\langle\psi|\hat{O}|\phi\rangle. \tag{2.86}$$

2.1.8 Tensor Product

Let us consider two vector spaces $\mathcal{H}_1$ and $\mathcal{H}_2$. For any pair of vectors, $|v^{(1)}\rangle \in \mathcal{H}_1$ and $|v^{(2)}\rangle \in \mathcal{H}_2$, we define the product

$$|v^{(1)}v^{(2)}\rangle = |v^{(1)}\rangle|v^{(2)}\rangle = |v^{(2)}\rangle|v^{(1)}\rangle \tag{2.87}$$

satisfying the following properties:

$$|v^{(1)}\rangle = \alpha_1|w_1^{(1)}\rangle + \alpha_2|w_2^{(1)}\rangle \implies |v^{(1)}v^{(2)}\rangle = \alpha_1|w_1^{(1)}v^{(2)}\rangle + \alpha_2|w_2^{(1)}v^{(2)}\rangle, \tag{2.88}$$

$$|v^{(2)}\rangle = \alpha_1|w_1^{(2)}\rangle + \alpha_2|w_2^{(2)}\rangle \implies |v^{(1)}v^{(2)}\rangle = \alpha_1|v^{(1)}w_1^{(1)}\rangle + \alpha_2|v^{(1)}w_2^{(2)}\rangle. \tag{2.89}$$

Under these conditions, the set of linear combinations of the vectors

$$\left\{|v^{(1)}v^{(2)}\rangle; |v^{(1)}\rangle \in \mathcal{H}_1, |v^{(2)}\rangle \in \mathcal{H}_2\right\}$$

spans a vector space, called the **tensor product** of $\mathcal{H}_1$ and $\mathcal{H}_2$, which we denote as

$$\mathcal{H} = \mathcal{H}_1 \otimes \mathcal{H}_2. \tag{2.90}$$

Given a basis $\mathcal{B}_1 = \left\{|e_k^{(1)}\rangle\right\}$ in $\mathcal{H}_1$, and a basis $\mathcal{B}_2 = \left\{|e_l^{(2)}\rangle\right\}$ in $\mathcal{H}_2$, then the set $\mathcal{B} = \left\{|e_k^{(1)}e_l^{(2)}\rangle\right\}$ is a basis in $\mathcal{H}$, i.e.

$$\forall |\psi\rangle \in \mathcal{H}, \exists c_{kl} \in \mathbb{C} : |\psi\rangle = \sum_{kl} c_{kl} |e_k^{(1)}e_l^{(2)}\rangle. \tag{2.91}$$

Clearly, if $\mathcal{H}_1$ and $\mathcal{H}_2$ have finite dimensions d_1 and d_2, respectively, then $\mathcal{H}$ is also finite-dimensional, and

$$\dim \mathcal{H} = d_1 \times d_2. \tag{2.92}$$

Knowing that $\mathcal{H}$ is a vector space, we can introduce all the notions discussed above (scalar product, norm, bras, operators…).

Tensor products are useful to describe the space of states of a composite system made up of two subsystems denoted 1 and 2. If the spaces of states of the subsystems are denoted by $\mathcal{H}_1$ and $\mathcal{H}_2$, respectively, then the space of states of the whole system is

$$\mathcal{H} = \mathcal{H}_1 \otimes \mathcal{H}_2. \tag{2.93}$$

Operators Acting on the Tensor Product Space

Operators can be introduced in the tensor product $\mathcal{H}$ as functions that transform vectors into vectors, along the lines discussed above. It is interesting to note that with each operator $\hat{O}^{(1)}$ acting in $\mathcal{H}_1$ we can associate an operator acting in $\mathcal{H}$. If the action of $\hat{O}^{(1)}$ is such that

$$\hat{O}^{(1)}|v^{(1)}\rangle = |w^{(1)}\rangle, \tag{2.94}$$

where $|v^{(1)}\rangle, |w^{(1)}\rangle \in \mathcal{H}_1$, then we can define the action of the operator on tensor-product kets of the form $|v^{(1)}v^{(2)}\rangle$ as follows:

$$\hat{O}^{(1)}|v^{(1)}v^{(2)}\rangle = |w^{(1)}v^{(2)}\rangle. \tag{2.95}$$

Using the fact that these are linear operators, we can extend the definition of $\hat{O}^{(1)}$ to the entire space $\mathcal{H}$. Strictly speaking, the extension of the operator $\hat{O}^{(1)}$ to the tensor product space $\mathcal{H}$ defines a *new* operator. Nonetheless we will use the same symbol to denote both the original operator, acting in $\mathcal{H}_1$, and its extension, acting in $\mathcal{H}$. Similarly, for an operator acting in $\mathcal{H}_2$:

$$\hat{O}^{(2)}|v^{(2)}\rangle = |w^{(2)}\rangle, \tag{2.96}$$

where $|v^{(2)}\rangle, |w^{(2)}\rangle \in \mathcal{H}_2$, the extension of the operator to product states is given by

$$\hat{O}^{(2)}|v^{(1)}v^{(2)}\rangle = |v^{(1)}w^{(2)}\rangle. \tag{2.97}$$

Example 2.7 Let us consider an operator $\hat{O}^{(1)}$ defined by specifying its action on the basis vectors $\mathcal{B}_1$:

$$\hat{O}^{(1)}|e_k^{(1)}\rangle = \sum_m |e_m^{(1)}\rangle\langle e_m^{(1)}|\hat{O}^{(1)}|e_k^{(1)}\rangle, \tag{2.98}$$

$$= \sum_m |e_m^{(1)}\rangle O_{mk}^{(1)}. \tag{2.99}$$

Note that in the first line of the equation above we have used the completeness relation of the basis in $\mathcal{H}_1$, while in the second line we have used the definition of the matrix element given above in Eq. (2.83). Working in the basis defined in Eq. (2.91), we can write a generic vector in $\mathcal{H}$ as

$$|v\rangle = \sum_{kl} v_{kl}|e_k^{(1)}e_l^{(2)}\rangle, \tag{2.100}$$

where v_{kl} are the coordinates of the vector in the basis $\mathcal{B}$. Using the linearity properties of the operator yields

$$|v'\rangle = \hat{O}^{(1)}|v\rangle = \hat{O}^{(1)} \sum_{kl} v_{kl}|e_k^{(1)}e_l^{(2)}\rangle, \tag{2.101}$$

$$= \sum_{kl} v_{kl} \sum_m |e_m^{(1)}e_l^{(2)}\rangle O_{mk}^{(1)}, \tag{2.102}$$

$$= \sum_{ml} \left(\sum_k O_{mk}^{(1)}v_{kl}\right)|e_m^{(1)}e_l^{(2)}\rangle. \tag{2.103}$$

If we denote by v'_{ml} the coordinates of the vector $|v'\rangle$ in the basis $\mathcal{B}$:

$$|v'\rangle = \sum_{ml} v'_{ml}|e_m^{(1)}e_l^{(2)}\rangle, \tag{2.104}$$

we see explicitly that

$$v'_{ml} = \sum_k O_{mk}^{(1)}v_{kl}. \tag{2.105}$$

In the algebraic manipulations above, pay attention to the use of dummy indices in the sums! The reader should repeat this exercise for the case of the operator $\hat{O}^{(2)}$ discussed above.

2.2 Two-State Systems

Before going further in setting up the mathematical tools needed for the description of states in quantum mechanics, it is time to consider a few physical examples that allow us to apply the concepts introduced so far.

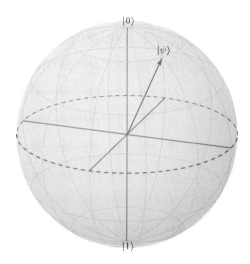

Fig. 2.2 Representation of states of a two-state system as points on the surface of the Bloch sphere.

A simple (yet non-trivial) example of a complex vector space is a two-dimensional vector space $\mathcal{H}$. Given a basis $\mathcal{B} = \{|1\rangle, |2\rangle\}$, any state in $\mathcal{H}$ can be written as

$$|v\rangle = v_1|1\rangle + v_2|2\rangle, \tag{2.106}$$

where $v_1, v_2 \in \mathbb{C}$. We are now going to illustrate some of the properties discussed in the previous section in this simple example.

State vectors need to be normalised, hence

$$|v_1|^2 + |v_2|^2 = 1. \tag{2.107}$$

Because quantum states are always defined up to a complex phase we require without loss of generality that v_1 is real. Denoting by φ the relative phase between v_1 and v_2, we can parametrise the state of the system in terms of two angles:

$$v_1 = \cos \vartheta/2, \quad v_2 = e^{i\varphi} \sin \vartheta/2, \tag{2.108}$$

where we have introduced the angle $\vartheta/2$ to implement the normalisation condition. The parametrisation above allows us to identify the quantum states with the points on the surface of a two-dimensional unit sphere, called the *Bloch sphere* (see Fig. 2.2).

Example 2.8 It is instructive to look at a change of basis in this simple setting. Having chosen a basis $\mathcal{B} = \{|1\rangle, |2\rangle\}$, the vector $|v\rangle$ in Eq. (2.106) is identified by the two complex numbers v_1 and v_2. Let us now consider a new orthonormal basis, $\mathcal{B}'$, whose elements are

$$|1'\rangle = V_{11}|1\rangle + V_{12}|2\rangle, \tag{2.109}$$
$$|2'\rangle = V_{21}|1\rangle + V_{22}|2\rangle, \tag{2.110}$$

where the coefficients V_{ij} are complex numbers. Orthonormality of the new basis vectors requires

$$\langle 1'|1'\rangle = \left(V_{11}^*\langle 1| + V_{12}^*\langle 2|\right)\left(V_{11}|1\rangle + V_{12}|2\rangle\right), \tag{2.111}$$

$$= V_{11}V_{11}^\dagger\langle 1|1\rangle + V_{12}V_{21}^\dagger\langle 2|2\rangle + \cdots + \langle 1|2\rangle + \cdots + \langle 2|1\rangle, \tag{2.112}$$

$$= 1. \tag{2.113}$$

In the equations above we have used $V_{ij}^\dagger = V_{ji}^*$, and we have not written explicitly the coefficients in front of the terms $\langle 1|2\rangle$ and $\langle 2|1\rangle$ because both these terms vanish due to the orthonormality of the vectors in $\mathcal{B}$.

Going back to Eq. (2.112), and using $\langle 1|1\rangle = \langle 2|2\rangle = 1$, we obtain

$$\sum_{j=1,2} V_{1j}V_{j1}^\dagger = 1. \tag{2.114}$$

Similar relations can be obtained by considering the scalar products $\langle 1'|2'\rangle$, $\langle 2'|1'\rangle$ and $\langle 2'|2'\rangle$. The four relations can be summarised as

$$\sum_{j=1,2} V_{ij}V_{jk}^\dagger = \delta_{ik}, \quad \text{for } i,k = 1,2. \tag{2.115}$$

This is nothing but the statement that V_{ij} is a unitary 2×2 matrix. Equations (2.109) and (2.110) can readily be inverted:

$$|1\rangle = V_{11}^\dagger|1'\rangle + V_{12}^\dagger|2'\rangle, \tag{2.116}$$

$$|2\rangle = V_{21}^\dagger|1'\rangle + V_{22}^\dagger|2'\rangle, \tag{2.117}$$

and therefore

$$|v\rangle = \left(v_1 V_{11}^\dagger + v_2 V_{21}^\dagger\right)|1'\rangle + \left(v_1 V_{12}^\dagger + v_2 V_{22}^\dagger\right)|2'\rangle, \tag{2.118}$$

$$= v_1'|1'\rangle + v_2'|2'\rangle, \tag{2.119}$$

which yields the coordinates of the *same* vector $|v\rangle$ in the new basis $\mathcal{B}'$.

This is a simple 2×2 realisation of the more general discussion that follows Eq. (2.71). Check that you can match the two arguments. (Note that with the notation used here we have $V_{kl} = U_{kl}^*$.)

Example 2.9 Two-state systems also provide a simple framework to construct explicitly the tensor product of two vector spaces. Let us consider two spaces $\mathcal{H}_1$ and $\mathcal{H}_2$, whose elements are the states of two independent two-state systems. Let us denote by $\mathcal{B}_1 = \left\{|1^{(1)}\rangle, |2^{(1)}\rangle\right\}$ and $\mathcal{B}_2 = \left\{|1^{(2)}\rangle, |2^{(2)}\rangle\right\}$ their respective bases. The set

$$\mathcal{B} = \left\{|1^{(1)}1^{(2)}\rangle, |1^{(1)}2^{(2)}\rangle, |2^{(1)}1^{(2)}\rangle, |2^{(1)}2^{(2)}\rangle\right\} \tag{2.120}$$

is a basis of the four-dimensional vector space $\mathcal{H} = \mathcal{H}_1 \otimes \mathcal{H}_2$. A generic vector in $\mathcal{H}$ can be written as

$$|\psi\rangle = c_{11}|1^{(1)}1^{(2)}\rangle + c_{12}|1^{(1)}2^{(2)}\rangle + c_{21}|2^{(1)}1^{(2)}\rangle + c_{22}|2^{(1)}2^{(2)}\rangle, \tag{2.121}$$

where $c_{11}, c_{12}, c_{21}, c_{22}$ are complex coefficients. Setting

$$|V_1\rangle = |1^{(1)}1^{(2)}\rangle, \quad |V_2\rangle = |1^{(1)}2^{(2)}\rangle, \tag{2.122}$$

$$|V_3\rangle = |2^{(1)}1^{(2)}\rangle, \quad |V_4\rangle = |2^{(1)}2^{(2)}\rangle, \tag{2.123}$$

$$C_1 = c_{11}, C_2 = c_{12}, \quad C_3 = c_{21}, C_4 = c_{22}, \tag{2.124}$$

allows us to rewrite the relation above as

$$|\psi\rangle = \sum_{k=1}^{4} C_k |V_k\rangle, \tag{2.125}$$

which shows clearly that $\mathcal{H}$ is a four-dimensional vector space.

Let us now take one state $|\phi^{(1)}\rangle \in \mathcal{H}_1$ and one state $|\phi^{(2)}\rangle \in \mathcal{H}_2$. These states can be written as linear combinations of the basis vectors in their respective vector spaces:

$$|\phi^{(1)}\rangle = c_1^{(1)}|1^{(1)}\rangle + c_2^{(1)}|2^{(1)}\rangle, \quad \left|c_1^{(1)}\right|^2 + \left|c_2^{(1)}\right|^2 = 1, \tag{2.126}$$

$$|\phi^{(2)}\rangle = c_1^{(2)}|1^{(2)}\rangle + c_2^{(2)}|2^{(2)}\rangle, \quad \left|c_1^{(2)}\right|^2 + \left|c_2^{(2)}\right|^2 = 1. \tag{2.127}$$

We can build the tensor product of these two states, which is an element of $\mathcal{H}$:

$$|\phi\rangle = |\phi^{(1)}\phi^{(2)}\rangle \tag{2.128}$$

$$= c_1^{(1)}c_1^{(2)}|1^{(1)}1^{(2)}\rangle + c_1^{(1)}c_2^{(2)}|1^{(1)}2^{(2)}\rangle + c_2^{(1)}c_1^{(2)}|2^{(1)}1^{(2)}\rangle + c_2^{(1)}c_2^{(2)}|2^{(1)}2^{(2)}\rangle. \tag{2.129}$$

The expansion of $|\phi\rangle$ is of the general form of Eq. (2.121). However, it is clear that not all vectors in $\mathcal{H}$ can be written as a product like in Eq. (2.128).

Despite their apparent simplicity, two-state systems describe numerous physical situations. They enter in the description of systems where only some degrees of freedom are considered, e.g. when discussing the spin of the electron, or the polarisation of the photon, independently of other properties of these particles. Another case where two-state systems are relevant is when describing the transitions between two atomic levels in a setting where other transitions are suppressed. Last but not least, two-state systems are the building blocks of quantum computing, thereby providing the foundations for one of the most exciting new developments of quantum mechanics.

2.3 The Wave Function of a One-Dimensional System

Let us now consider the description of a quantum system in a one-dimensional space – here we refer to actual space, so for example a point particle moving on a line, not to the dimension of the Hilbert space of states. In this case, the state of the system is described by the **wave function**, i.e. a complex function $\psi(x)$ of the coordinate x of the system. All the information about the state of the system is encoded in the wave function. We need to reconcile the concept of a wave function with our claim that quantum states are vectors in a Hilbert space.

Fig. 2.3 Discretised one-dimensional system with six sites. The particle can only be in one of the six points denoted by 0, ..., 5. The distance between successive points is ϵ.

2.3.1 Discretised System

In order to understand better how the information about the state of the system is encoded in the wave function, we shall start with a simpler version of the one-dimensional system, namely with a point particle in discretised space. Such a particle can only be in a finite number of positions along the real axis, as shown in Fig. 2.3. In this particular example, the particle can be in one of six[2] points along the real axis labelled $0, \ldots, 5$. The lattice spacing (i.e. the distance between two points) is denoted ϵ.

Let us assume that the state of the system is described by a wave function ψ; then, at each point x_i, the wave function of the system takes some complex value $\psi(x_i)$. If we define

$$\psi_i = \sqrt{\epsilon}\,\psi(x_i), \quad i = 0, \ldots, 5, \tag{2.130}$$

then the state of the system is encoded in the values ψ_i of the wave function at the discrete points x_i, i.e. in a complex vector:

$$|\psi\rangle = (\psi_0, \ldots, \psi_5). \tag{2.131}$$

The reason for the factor of $\sqrt{\epsilon}$ in the definition of ψ_i will become clear below. The values of $\psi(x)$ at different spatial points are the coordinates of the state vector. Two remarks are in order:

1. The vector $|\psi\rangle$ in Eq. (2.131) is defined by giving its coordinates, and therefore this definition assumes that a basis has been specified.
2. The x dependence is not explicitly written in the ket. Indeed the value of x serves as a label for each coordinate of the vector, and in the discretised case it is traded for the index i labelling the discrete set of points on the line. When we refer to the vector we do not write the x dependence in the ket, just like we do not add a coordinate index when we denote a vector by underlining its corresponding symbol. We write $\underline{v}$ and not $\underline{v}_i$; similarly we write $|\psi\rangle$ and not $|\psi(x)\rangle$.

In the discretised system we see explicitly that the state of the system is represented by a vector. Because we considered a discretised space with six sites, we obtained a six-dimensional complex vector. In the general case with N sites, we would obtain an N-dimensional complex vector. This is an important concept to remember, as we will generalise it to the case of infinite-dimensional vector spaces.

---

[2] There is no particular significance in the fact that we have chosen six points here. The same example could have been worked out with two points, or any finite number of points.

Basis Vectors and Physical Interpretation

It is useful to introduce explicitly a basis such that the state vector $|\psi\rangle$ can be expanded as discussed above. The orthonormal base vectors are denoted

$$|\xi_0\rangle = (1, 0, 0, 0, 0, 0), \tag{2.132}$$

$$|\xi_1\rangle = (0, 1, 0, 0, 0, 0), \tag{2.133}$$

$$\vdots \tag{2.134}$$

$$|\xi_5\rangle = (0, 0, 0, 0, 0, 1). \tag{2.135}$$

The vector $|\psi\rangle$ can be written as

$$|\psi\rangle = \sum_{k=0}^{5} \psi_k |\xi_k\rangle, \tag{2.136}$$

where we see that the complex numbers ψ_k are precisely the coordinates of $|\psi\rangle$ in the basis $\{|\xi_0\rangle, \ldots, |\xi_5\rangle\}$. They are computed as usual by taking the scalar product of the vector $|\psi\rangle$ with the basis vectors

$$\psi_k = \langle \xi_k | \psi \rangle. \tag{2.137}$$

The physical interpretation of the basis vectors above is the following. Let us assume that the modulus squared of the value of the wave function at a given position x is the probability density of finding the system at x. We will discuss this assumption in detail later when we discuss observables. In our discretised example, $\left|\psi_k\right|^2$ yields the probability of finding the system at x_k. Note that the factor $\sqrt{\epsilon}$ introduced in Eq. (2.130) is precisely the factor that is necessary to convert the continuous probability density $\left|\psi(x)\right|^2$ into a finite probability. The state described by the basis vector $|\xi_k\rangle$ is a state in which the system has probability one of being at position x_k. For a generic state $|\psi\rangle$, the value of the wave function at x_k, ψ_k, is simply the projection of the ket describing the quantum state of the system onto a basis vector that describes a state being localised at x_k with probability one.

The scalar product of two generic vectors defined on the discretised line is

$$\langle \phi | \psi \rangle = \sum_k \phi_k^* \psi_k \, ; \tag{2.138}$$

once again notice that the coordinates of $|\phi\rangle$ are complex-conjugated when taking the scalar product. Following our convention, we write the norm as the scalar product of $|\psi\rangle$ with itself:

$$\|\psi\|^2 = \langle \psi | \psi \rangle = \sum_{k=0}^{5} \left|\psi_k\right|^2, \tag{2.139}$$

and the normalisation condition becomes

$$\sum_{k=0}^{5} \left|\psi_k\right|^2 = 1. \tag{2.140}$$

2.3.2 Continuum System

The wave function for a continuous system can be seen as the limit of the discretised case where the number of points goes to infinity, while the distance ϵ becomes infinitesimally small. In this limit, instead of a finite-dimensional vector, we obtain an infinite number of coordinates, encoded in a continuous function $\psi(x)$.

We shall still refer to the wave function as the state vector, bearing in mind that in this case the vector space is infinite-dimensional. Remember that the values $\psi(x)$ are interpreted as the coordinates of the vector $|\psi\rangle$: there is one coordinate $\psi(x)$ for each point x on the real axis, and there is an infinity of points along the real axis. So the vector space is infinite-dimensional. We continue to denote the state vector using a ket $|\psi\rangle$. It should be clear from the discussion above that the spatial coordinate x labels the components of the state vector, and that the basis vectors define a set of **continuum states** labelled by the variable x.

Interestingly, the concept of linear combination can also be extended to the case of kets that are labelled by a continuous index x. First let us define the kets

$$|x_k\rangle = \frac{1}{\sqrt{\epsilon}}|\xi_k\rangle. \tag{2.141}$$

A linear combination of the $|\xi_k\rangle$ vectors can be written as

$$\sum_k \psi_k|\xi_k\rangle = \sum_k \sqrt{\epsilon}\,\psi(x_k)\sqrt{\epsilon}\,|x_k\rangle, \tag{2.142}$$

$$\stackrel{\epsilon\to 0}{=} \int dx\,\psi(x)\,|x\rangle \in \mathcal{H}, \tag{2.143}$$

where in the second line we have taken the limit $\epsilon \to 0$, and we should remember that $\psi(x)$ is a complex function of x. The ket $|x\rangle$ is defined by taking the limit in Eq. (2.141).

The completeness relation can be written as

$$\int dx\,|x\rangle\langle x| = 1\,; \tag{2.144}$$

note that here we have a sum over an infinite number of projectors over the continuum basis vectors, we will come back to this point in the next chapter.

Similarly the norm of the state vector can be written as the limit of the norm of the finite-dimensional vector for $\epsilon \to 0$. Starting from Eq. (2.139), and taking the limit

$$\langle\psi|\psi\rangle = \lim_{\epsilon\to 0}\sum_k \epsilon\left|\psi(x_k)\right|^2 = \int dx\,\left|\psi(x_k)\right|^2, \tag{2.145}$$

where the sum over the discrete sites is replaced by the integral over the continuous values of x. A state described by the wave function ψ is called **normalisable** if the integral in Eq. (2.145) is finite. More generally, the scalar product of two wave functions can be defined as the limit of the discrete case:

$$\langle\phi|\psi\rangle = \int dx\,\phi(x)^*\,\psi(x). \tag{2.146}$$

It is instructive to summarise these relations and compare them with the ones we defined for discrete systems:

	Discrete system	Continuous system
$\lvert\psi\rangle =$	$\sum_k \psi_k \lvert\xi_k\rangle$	$\int dx\, \psi(x)\lvert x\rangle$
$1 =$	$\sum_k \lvert\xi_k\rangle\langle\xi_k\rvert$	$\int dx\, \lvert x\rangle\langle x\rvert$
$\lVert\psi\rVert^2 =$	$\sum_k \lvert\psi_k\rvert^2$	$\int dx\, \lvert\psi(x)\rvert^2$
$\langle\phi\lvert\psi\rangle =$	$\sum_k \phi_k^* \psi_k$	$\int dx\, \phi(x)^* \psi(x)$

Once we understand that the wave function encodes the set of coordinates of a vector in an infinite-dimensional space, the connection with the previous discussion about Hilbert spaces and the related tools becomes fairly natural.

Mathematical Aside: Square-Integrable Functions on the Real Axis
The function $\psi(x) = e^{-x^2/2}$ is an example of a normalisable function. Its norm is

$$\lVert\psi\rVert^2 = \int dx\, e^{-x^2} = \sqrt{\pi}, \tag{2.147}$$

and therefore the *normalised wave function* is

$$\psi(x) = \frac{1}{\pi^{1/4}} \exp[-x^2/2]. \tag{2.148}$$

However, the function $e^{x^2/2}$ is non-normalisable, and therefore does not represent a physical state. The set of normalisable wave functions is the set of square-integrable functions on the real line, denoted in the mathematical literature as $\mathcal{L}_2(\mathbb{R})$.
In general, if $\int dx\, \lvert\psi(x)\rvert^2 = c$, then the normalised wave function is $\frac{1}{\sqrt{c}}\psi(x)$.

It is interesting to consider what happens to the scalar product of two of the basis states introduced in Eq. (2.132) ff. when we take the continuum limit, i.e. the limit where $\epsilon \to 0$:

$$\langle x\lvert x'\rangle = \lim_{\epsilon\to 0} \frac{1}{\epsilon}\langle\xi\lvert\xi'\rangle. \tag{2.149}$$

Following the convention introduced earlier in this chapter, Greek letter ξ is used to denote the states of the discretised system, while Roman letter x stands for the states in the continuum limit. We chose the states $\lvert\xi\rangle$ to be normalised to unity, as discussed just above Eq. (2.132). Therefore $\langle x\lvert x'\rangle$ vanishes when $x \neq x'$, while it diverges for $x = x'$. Using our definitions, we have

$$\langle x\lvert\psi\rangle = \int dx'\, \psi(x')\langle x\lvert x'\rangle, \tag{2.150}$$

$$= \lim_{\epsilon\to 0} \sum_k \epsilon\, \psi(\xi_k)\frac{1}{\epsilon}\langle\xi\lvert\xi_k\rangle, \tag{2.151}$$

$$= \sum_k \psi(\xi_k)\delta_{\xi,\xi_k}, \tag{2.152}$$

$$= \psi(x). \tag{2.153}$$

The scalar product $\langle x|x'\rangle$ is the generalisation of the Kronecker delta to the case of continuum states, and is known as the **Dirac delta**. Some properties of the Dirac delta are summarised below.

Mathematical Aside: Dirac Delta

The delta function is the natural extension of the familiar Kronecker delta δ_{ij} to the case of continuous variables; it is defined as

$$\delta(x) = 0, \quad \text{if } x \neq 0, \tag{2.154}$$

and

$$\int_a^b dx\, \delta(x)g(x) = \begin{cases} g(0), & \text{if } a < 0 < b, \\ 0, & \text{otherwise}, \end{cases} \tag{2.155}$$

for any continuous function $g(x)$.

The following properties are often useful when dealing with delta functions:

$$\int_a^b dx\, \delta(x-c)g(x) = \begin{cases} g(c), & \text{if } a < c < b, \\ 0, & \text{otherwise}; \end{cases} \tag{2.156}$$

$$\delta(-x) = \delta(x); \tag{2.157}$$

$$\delta(ax) = \frac{1}{|a|}\delta(x); \tag{2.158}$$

$$g(x)\delta(x-y) = g(y)\delta(x-y); \tag{2.159}$$

$$x\delta(x) = 0; \tag{2.160}$$

$$\frac{d}{dx}\theta(x) = \delta(x), \quad \text{where } \theta(x) = \begin{cases} 1, & \text{if } x \geq 0, \\ 0, & \text{if } x < 0; \end{cases} \tag{2.161}$$

$$\delta(g(x)) = \sum_{i=1}^r \frac{1}{|g'(x_i)|}\delta(x-x_i), \quad \text{where } g(x_i) = 0; \tag{2.162}$$

$$\int_0^\infty dx\, g(x)\delta(x) = \tfrac{1}{2}g(0). \tag{2.163}$$

The delta function can be defined as the limit of ordinary functions. Here are some examples; you are encouraged to sketch these functions in order to visualise what happens when taking the limit:

$$\delta(x) = \lim_{\epsilon \to 0} \frac{1}{\sqrt{\pi}\epsilon} e^{-x^2/\epsilon^2}, \tag{2.164}$$

$$= \lim_{L \to \infty} \frac{\sin(Lx)}{\pi x}, \tag{2.165}$$

$$= \lim_{L \to \infty} \frac{\sin^2(Lx)}{\pi L x^2}, \tag{2.166}$$

$$= \lim_{\epsilon \to 0} \frac{1}{\pi} \frac{\epsilon}{\epsilon^2 + x^2}. \tag{2.167}$$

The following integral representation for the delta function is very useful:

$$\int_{-\infty}^{\infty} dx \, e^{ix(k-k')} = 2\pi\delta(k - k').$$

(2.168)

This completes the overview of the mathematical tools that will be used to characterise and manipulate the states of a quantum system. They are the backbone of our description of nature in the quantum realms. It is important to appreciate the role of linear algebra and complex vector spaces in the description of physical states. The physical states of every system described later in this book are described by vectors in Hilbert spaces – make sure that you identify these vector spaces as you read through the following chapters.

The next step in order to connect this abstract mathematical structure to the physical world is the definition of a set of rules that allow us to deal with observables. The next chapter is therefore devoted to the definition of observables in quantum mechanics and to the laws that ultimately will connect in a precise mathematical way the results of experiments to the predictions of the theory.

Summary

- Quantum states are represented by vectors in Hilbert spaces. (2.1.1)
- The superposition principle emerges as a consequence of the mathematical structure of vector spaces: a linear combination of state vectors with arbitrary complex coefficients yields a possible quantum state. (2.1.1)
- Functionals associate complex numbers with the vectors, and can be represented by bras according to Dirac's notation. (2.1.3)
- The scalar product in the Hilbert space allows us to define the norm of the vectors. We will require that vectors representing physical states are normalised to one. (2.1.4)
- Operators transform vectors into vectors. Acting with an operator on a quantum state changes the state. (2.1.6)
- Vectors can be described by coordinates, after having chosen a particular basis. The same vector is described by different coordinates in different bases. (2.1.7)
- Given two vector spaces, it is possible to construct a new vector space, called the tensor product of the two. Tensor products play a key role in the description of composite systems. (2.1.8)
- The two-state system is a simple example of a complex vector space, which describes the physical states of numerous real-life systems. The two-state system provides simple explicit examples where we can practise with the concepts introduced so far. (2.2)
- The wave function of a one-dimensional system can be viewed as a vector in an infinite-dimensional vector space, thereby unifying the description of quantum states as vectors. (2.3)

Problems

2.1   The energy levels for the *infinite one-dimensional square well* are given by

$$E_n = \frac{\hbar^2 \pi^2 n^2}{8ma^2} \quad \text{for } n = 1, 2, 3, \ldots, \infty. \tag{2.169}$$

Calculate the energies of the first ($n = 1$) and second ($n = 2$) levels for the case of an electron of mass 9.1×10^{-31} kg, confined to a box of atomic dimensions ($a = 10^{-10}$ m).

Calculate the wavelength of a photon emitted in a transition between these levels.

2.2   Consider a three-dimensional vector space, and an orthonormal basis:

$$\mathcal{B} = \{|1\rangle, |2\rangle, |3\rangle\}.$$

Normalise the vectors

$$|v\rangle = 2|2\rangle,$$
$$|\phi\rangle = i|1\rangle + 2|3\rangle,$$
$$|\psi\rangle = 2|1\rangle + i|2\rangle - 2(\cos\alpha + i\sin\alpha)|3\rangle,$$

where α is a real number.

2.3   Given two orthonormal basis vectors $|1\rangle, |2\rangle$, consider a new set of basis vectors

$$|1'\rangle = \frac{1}{\sqrt{2}}|1\rangle + \frac{i}{\sqrt{2}}|2\rangle,$$
$$|2'\rangle = -\frac{1}{\sqrt{2}}|1\rangle + \frac{i}{\sqrt{2}}|2\rangle.$$

Check that the new basis is orthonormal. Write down explicitly the 2×2 matrix that relates the coordinates of a vector in the basis $\{|1\rangle, |2\rangle\}$ to the coordinates of the same vector in the basis $\{|1'\rangle, |2'\rangle\}$, and check that it is a unitary matrix.

The vector $|v\rangle$ has coordinates $v_1 = i$, $v_2 = 3$ in the basis $\{|1\rangle, |2\rangle\}$. Find its coordinates in the basis $\{|1'\rangle, |2'\rangle\}$. Normalise the vector $|v\rangle$.

2.4   The following inequalities recur frequently when working with vector spaces.

(a) Using the positivity of the norm of the vector $|\psi\rangle - \lambda|\phi\rangle$, for all values of the complex number λ, deduce the **Schwarz inequality**

$$\left|\langle\psi|\phi\rangle\right| \leq ||\psi|| \, ||\phi||.$$

(b) Deduce the triangular inequalities, Eqs (2.37) and (2.38).

(c) Show that

$$||\psi + \phi||^2 + ||\psi - \phi||^2 = 2\left(||\psi||^2 + ||\phi||^2\right).$$

2.5 The action of a linear operator U on the elements of a vector space $\mathcal{H}$ is completely specified by the action of the operator on the vectors of one basis. Let us consider a generic vector of a two-state system:

$$|\psi\rangle = \psi_1|1\rangle + \psi_2|2\rangle,$$

then the action of U is given by

$$|\psi\rangle' = \hat{U}|\psi\rangle = \psi_1\hat{U}|1\rangle + \psi_2\hat{U}|2\rangle.$$

Both $\hat{U}|1\rangle$ and $\hat{U}|2\rangle$ are elements of $\mathcal{H}$, and therefore can be written as linear combinations of the basis vectors:

$$|1'\rangle = \hat{U}|1\rangle = U_{11}|1\rangle + U_{21}|2\rangle,$$
$$|2'\rangle = \hat{U}|2\rangle = U_{12}|1\rangle + U_{22}|2\rangle.$$

Find the coordinates ψ_1' and ψ_2' of the vector $|\psi'\rangle$.
Show that, if the basis is orthonormal,

$$U_{ij} = \langle i|\hat{U}|j\rangle, \quad i,j = 1,2.$$

2.6 Consider two spaces $\mathcal{H}_1$ and $\mathcal{H}_2$, whose elements are the states of two independent two-state subsystems denoted (1) and (2). Let us denote by $\mathcal{B}_1 = \left\{|1^{(1)}\rangle, |2^{(1)}\rangle\right\}$ and $\mathcal{B}_2 = \left\{|1^{(2)}\rangle, |2^{(2)}\rangle\right\}$ their respective orthonormal bases. The set

$$\mathcal{B} = \left\{|1^{(1)}1^{(2)}\rangle, |1^{(1)}2^{(2)}\rangle, |2^{(1)}1^{(2)}\rangle, |2^{(1)}1^{(2)}\rangle\right\} \tag{2.170}$$

is a basis of the four-dimensional vector space $\mathcal{H} = \mathcal{H}_1 \otimes \mathcal{H}_2$ that contains the states of the composite system made up of the two subsystems. The subsystem (1) is in the state

$$|\psi^{(1)}\rangle = \frac{1}{\sqrt{2}}|1^{(1)}\rangle + \frac{i}{\sqrt{2}}|2^{(1)}\rangle,$$

and the subsystem (2) is in the state

$$|\psi^{(2)}\rangle = \frac{1}{\sqrt{3}}|1^{(2)}\rangle + \frac{\sqrt{2}}{\sqrt{3}}|2^{(2)}\rangle,$$

find the coordinates of the state vector associated with the full system in the basis $\mathcal{B}$.

2.7 The action of a linear operator U on $\mathcal{H}_1$ and $\mathcal{H}_2$ is specified by the transformation properties of the basis vectors:

$$|1^{(1)'}\rangle = \hat{U}|1^{(1)}\rangle = U_{11}^{(1)}|1^{(1)}\rangle + U_{21}^{(1)}|2^{(1)}\rangle,$$
$$|2^{(1)'}\rangle = \hat{U}|2^{(1)}\rangle = U_{12}^{(1)}|1^{(1)}\rangle + U_{22}^{(1)}|2^{(1)}\rangle,$$

and

$$|1^{(2)'}\rangle = \hat{U}|1^{(2)}\rangle = U_{11}^{(2)}|1^{(2)}\rangle + U_{21}^{(2)}|2^{(2)}\rangle,$$
$$|2^{(2)'}\rangle = \hat{U}|2^{(2)}\rangle = U_{12}^{(2)}|1^{(2)}\rangle + U_{22}^{(2)}|2^{(2)}\rangle.$$

Find the transformation properties of the vector

$$|\psi\rangle = \sum_{i,j=1,2} \psi_{ij}|i^{(1)}\rangle|j^{(2)}\rangle.$$

2.8 (a) Compute

$$\int_{-\infty}^{+\infty} dx\ \sin(4x)\delta(x - x_0).$$

(b) Compute

$$\int_{-\infty}^{+\infty} dx\ \sin(x)\delta\left(\frac{x}{4} - x_0\right).$$

(c) Compute

$$f(k) = \int_{-\infty}^{+\infty} dx\ e^{ikx}\delta(x).$$

2.9 The first and third energy eigenstates of the one-dimensional harmonic oscillator are, respectively,

$$\psi_0(x) = C_0 \exp\left[-\alpha^2 x^2/2\right],\tag{2.171}$$

$$\psi_2(x) = C_2\left(4\alpha^2 x^2 - 2\right)\exp\left[-\alpha^2 x^2/2\right],\tag{2.172}$$

where $\alpha^2 = m\omega/\hbar$.

Calculate explicitly the normalisation constants for these two energy eigenstates, and verify that the eigenfunctions are orthogonal. Note that

$$\int_{-\infty}^{\infty} \exp\{-\alpha^2 x^2\}dx = (\pi/\alpha^2)^{\frac{1}{2}} \quad \text{and} \quad \int_{-\infty}^{\infty} x^2 \exp\{-\alpha^2 x^2\}dx = \tfrac{1}{2}\sqrt{\pi}/\alpha^3.$$

$$\tag{2.173}$$

2.10 Show that

$$\frac{1}{\hat{A}} - \frac{1}{\hat{B}} = \frac{1}{\hat{A}}\left(\hat{B} - \hat{A}\right)\frac{1}{\hat{B}}.\tag{2.174}$$

Remember that $\hat{A}$ and $\hat{B}$ are operators, and $1/\hat{A}$ denotes the inverse of $\hat{A}$.

2.11 Given a sequence of vectors $|v_k\rangle$, we define a new set of vectors $|u_k\rangle$ by the following recursion relation:

1. $|u_1\rangle = |v_1\rangle$ and $|e_1\rangle = \dfrac{1}{\sqrt{\langle u_1|u_1\rangle}}|u_1\rangle$.

2. $|u_k\rangle = |v_k\rangle - \displaystyle\sum_{j=1}^{k-1}|e_j\rangle\langle e_j|v_k\rangle$ and $|e_k\rangle = \dfrac{1}{\sqrt{\langle u_k|u_k\rangle}}|u_k\rangle$.

Show by an explicit computation that $|e_1\rangle, |e_2\rangle$ and $|e_3\rangle$ are orthonormal.

Show by induction that $\langle e_1|e_k\rangle = 0$ for all $k > 1$.

This procedure is called **Gram–Schmidt orthogonalisation** and is widely used in constructing orthonormal bases.

3 Observables

In this chapter we are going to set up the formalism to describe observables in quantum mechanics. This is an essential part of the formulation of the theory, as it deals with the description of the outcome of experiments. Beyond any theoretical sophistication, a physical theory is first and foremost a description of natural phenomena; therefore it requires a very precise framework that allows the observer to relate the outcome of experiments to theoretical predictions. As we will see, this is particularly true for quantum phenomena. The necessary formalism is very different from the intuitive one used in classical mechanics. A subtle point is that the state of the system is not an observable by itself. As seen in the previous chapter, the state of the system is specified by a complex vector. Given a quantum system in a state that is described by a ket, we can perform measurements in that particular state; the results of these measurements are what we call observables. Therefore we need a formalism that allows us to determine:

- the possible outcomes of a measurement;
- what can be predicted about these outcomes, assuming that we know the complex vector that describes the state of the system.

In order to achieve these goals we will need to set up a new mathematical structure, based on operators that act on the state vectors. In this chapter you will learn the exact relation between states, operators and the outcome of experiments. This is a crucial step in building a framework that can yield predictions for the outcome of experiments, i.e. to connect the theory to the experiment. This chapter is focused on the general formalism and a few initial examples. More observables, like the angular momentum, will be introduced in detail in following chapters.

3.1 Introducing Observables

We define an observable O to be the outcome of an experiment that is measured by a given apparatus. The description of observables needs to be specified very precisely. This precise definition is encoded in a series of postulates that we are going to introduce in this chapter and discuss in detail. The overall picture can be summarised as follows.

For a quantum system in a state described by some complex vector $|\psi\rangle$, the outcome of a measurement of the observable O is a stochastic variable. The theory can predict:

1. the probability distribution function (pdf), p_O, that describes the possible results of the measurement and their respective frequencies;
2. the state of the system after the measurement.

We should note in passing that the knowledge of the pdf is equivalent to the knowledge of all its moments. Remember that the result of a measurement needs to be treated as a random variable and depending on the context we may focus either on its pdf or on its moments, i.e. we will want to compute its mean value, its variance, etc.

Mathematical Aside: Moments of Probability Distributions

Let us consider a random real variable r, characterised by a probability distribution function $p(r)$. The probability of finding r in the interval $\mathcal{I} = [a, b]$ is

$$P(r \in \mathcal{I}) = \int_a^b dx\, p(x). \tag{3.1}$$

(Note that x is a dummy variable of integration, we could use any name to identify the integration variable. If this does not sound familiar, think about it. ....)

The nth moment of the distribution p is defined as

$$\mu_n = \int dr\, p(r)\, r^n. \tag{3.2}$$

For a properly normalised probability distribution function:

$$\mu_0 = 1. \tag{3.3}$$

The mean value of the random variable r is

$$\mu_1 = \langle r \rangle = \int dr\, p(r)\, r. \tag{3.4}$$

The variance of the variable r is

$$\mathrm{Var}[r] = \mu_2 - (\mu_1)^2. \tag{3.5}$$

Exercise 3.1.1 Compute the mean value and the variance of a Gaussian variable r, i.e. a random variable with probability distribution function

$$p(r) = \frac{1}{\sqrt{2\pi\sigma^2}}\, \exp\left[-\frac{(x - x_0)^2}{2\sigma^2}\right]. \tag{3.6}$$

Note that the Gaussian distribution is completely determined by the knowledge of its first two moments.

3.1.1 A First Example: Position Operator

As discussed in the previous chapter, the modulus squared of the wave function $|\psi(x)|^2$ describes the probability density of finding the system in the state $|\psi\rangle$ at x. So the outcome of a measurement of the position of the system, x, is a real stochastic variable and its pdf is $|\psi(x)|^2$. The mean value of the position of the system is therefore

$$\langle x \rangle = \int dx\, |\psi(x)|^2 x. \tag{3.7}$$

The crucial point here is the realisation that the mean value of the position of the system in the state $|\psi\rangle$, $\langle x \rangle$, can be written as the **expectation value** of a suitably defined operator $\hat{X}$ in the state $|\psi\rangle$, i.e. the matrix element of the operator $\hat{X}$ between $|\psi\rangle$ and $|\psi\rangle$.

The Hermitian operator $\hat{X}$ associated with the position of the system is such that the states $|x\rangle$ introduced in the previous chapter are eigenstates of $\hat{X}$ to the eigenvalue x – pay attention to the notation here! Using the fact that $|x\rangle$ are eigenstates of $\hat{X}$, we have

$$\langle x|\hat{X}|\psi\rangle = \left(\langle \psi|\hat{X}|x\rangle\right)^*, \tag{3.8}$$

$$= x\left(\langle \psi|x\rangle\right)^*, \tag{3.9}$$

$$= x\langle x|\psi\rangle; \tag{3.10}$$

with a slight abuse of notation, we can think of the operator $\hat{X}$ acting on the wave function as

$$\hat{X}\psi(x) = x\psi(x). \tag{3.11}$$

It is important to realise that Eq. (3.11) does not mean that $\psi(x)$ is an eigenfunction of $\hat{X}$ with eigenvalue x. The eigenvalue needs to be a constant that multiplies the state vector. The mean value in Eq. (3.7) can then be written as

$$\langle \hat{X} \rangle = \langle \psi|\hat{X}|\psi\rangle = \int dx\, \psi(x)^* \, x\psi(x) = \langle x \rangle. \tag{3.12}$$

As shown explicitly in the equation above, the expectation value in Eq. (3.12) depends on the state ψ. When there is no ambiguity, we shall omit the explicit ψ dependence, and write simply $\langle \hat{X} \rangle$, but remember that this is the mean value in a given state.

The mean value of the outcome of a measurement, $\langle x \rangle$, is therefore expressed as the expectation value of an operator acting in the space of physical states. We will now generalise this idea to a generic observable and discuss in detail the physical meaning of the mean value in the equation above.

3.1.2 Generic Observables

We define an **observable** to be a quantity that can be measured in an experiment. Experiments here are considered as black boxes: the system is prepared in some quantum state $|\psi\rangle$, then the experiment is performed and yields a value for the observable O. We will not dwell upon the details of the experimental settings, but we will assume that the experiment is reproducible, i.e. that it is possible to repeatedly prepare the system in the same state and perform the same measurement. Following the example discussed in Section 3.1.1 for the position of a quantum system, we are going to associate an operator with each observable. This is the **second postulate** of quantum mechanics.

Postulate 2   In quantum mechanics each observable O is associated with a *linear operator* $\hat{O}$ acting in the Hilbert space of physical states:

$$\hat{O}: \mathcal{H} \to \mathcal{H}$$
$$|\psi\rangle \mapsto |\psi'\rangle = \hat{O}|\psi\rangle. \tag{3.13}$$

This is a weird statement! The postulate associates an operator with each observable. However, be aware that the measurement is NOT associated with acting with the operator on the state. By analogy with the example of the position operator, we expect that the mean value of the observable O in a quantum state can be computed as the expectation value of the operator $\hat{O}$ in that state:

$$\langle O \rangle = \langle \psi | \hat{O} | \psi \rangle. \tag{3.14}$$

The link between operators and the outcome of experiments needs to be explained in detail and this will be the focus of this chapter.

Example 3.1 The operator associated with the energy of the system is the Hamiltonian. The Hamiltonian of the system may be written as $\hat{H} = \hat{T} + \hat{V}$, where the kinetic and potential energy operators are defined by

$$\hat{T} = \frac{\hat{P}^2}{2m}, \quad \hat{V} = V(\hat{X}). \tag{3.15}$$

The operator $V(\hat{X})$, which is a function of the position operator $\hat{X}$, acts as

$$V(\hat{X})\psi(x) = V(x)\psi(x). \tag{3.16}$$

Any other operator that is a function of $\hat{X}$ can be defined in the same way. We shall see later how to define the operator $\hat{P}$ associated with the momentum of the system, and more complicated operators. $\hat{H}$ acts on state vectors and returns state vectors.

3.2 Observing Observables

Let us now explain in more detail the process of measurement in quantum mechanics and in particular what we mean when we say that the result of an experiment is a stochastic variable. If the observable O is measured several times under identical conditions, the result is a set of values $\left\{ O^{(1)}, O^{(2)}, \ldots, O^{(n)} \right\}$, where the suffix labels each measurement. This is a characteristic feature of quantum mechanics: the results of several independent measurements performed on a system prepared always in the same state $|\psi\rangle$ are different; they are distributed according to some pdf that can be predicted by the theoretical framework. Hence the mean value of the observables in the given state can be computed by taking the average of the outcomes of the measurements, in the limit where the number of measurements becomes large. We see here the first explicit connection between the operator $\hat{O}$ and experimental results.

The expectation value of the operator $\hat{O}$ defined in Eq. (3.14) is equal to the average $\sum_{k=1}^{n} O^{(k)}/n$ in the limit where the number of measurements $n \to \infty$.

So far our examples have focused on the mean value of an observable, i.e. the first moment of its probability distribution. Moving beyond the first moment of the pdf, the theoretical framework of quantum mechanics allows us to predict:

1. the possible outcomes of the measurements above;
2. the probability of obtaining each of these possible outcomes.

It is important to realise that all questions in quantum mechanics need to be formulated in terms of possible outcomes of an experiment and their respective probabilities. These are the only quantities that are predicted by the theory. Trying to shortcut this procedure leads to well-known paradoxes.

Eigenvalues and Eigenstates

Our next step in building the theoretical framework is to specify the relation between the outcomes of measurements and some properties of the operator $\hat{O}$. This is encapsulated in the **third postulate** of quantum mechanics.

Postulate 3   The possible outcomes of experiments, O_k, are the **eigenvalues** of the operator $\hat{O}$, i.e. the solutions of the eigenvalue equation

$$\hat{O}|\psi_k\rangle = O_k|\psi_k\rangle, \tag{3.17}$$

where $|\psi_k\rangle$ is the **eigenstate** corresponding to the eigenvalue O_k. Note that an eigenstate is always associated with a specific eigenvalue.

Mathematical Aside: Eigenvalues and Degeneracies

An eigenvalue O_k of an operator $\hat{O}$ is called g-fold degenerate if there are exactly g linearly independent eigenvectors corresponding to the same eigenvalue:

$$\exists\, |u_k^{(n)}\rangle, \text{ such that } \hat{O}|u_k^{(n)}\rangle = O_k|u_k^{(n)}\rangle, \text{ for } n = 1, \ldots, g. \tag{3.18}$$

Note that any linear combination

$$\sum_{n=1}^{g} c_n |u_k^{(n)}\rangle, \text{ with } c_n \in \mathbb{C}, \tag{3.19}$$

is also an eigenstate of $\hat{O}$ with the same eigenvalue O_k. The space of eigenstates corresponding to the eigenvalue O_k is a g-dimensional vector space.

The eigenfunction represents the state in which the measurement of O yields the value O_k with probability 1. To check this statement, we can compute the variance of O in the state $|\psi_k\rangle$:

$$\text{Var}_k[O] = \langle O^2 \rangle - \langle O \rangle^2, \tag{3.20}$$
$$= \langle \psi_k | \hat{O}^2 | \psi_k \rangle - \langle \psi_k | \hat{O} | \psi_k \rangle^2,$$
$$= O_k^2 \langle \psi_k | \psi_k \rangle - \left(O_k \langle \psi_k | \psi_k \rangle \right)^2 = 0, \tag{3.21}$$

where we have used the fact that the eigenfunctions are normalised to one.

It is important in the equations above to distinguish the outcome of the ith measurement, denoted $O^{(i)}$, from the set of possible outcomes O_k. The latter are a property of the operator $\hat{O}$ and do not depend on the state of the system.

3.3 Hermitian Operators

We stated above that every observable is represented by an operator; ultimately we want to establish a correspondence

OBSERVABLE		OPERATOR
total energy	$\Longleftrightarrow$	$\hat{H}$
position	$\Longleftrightarrow$	$\hat{X}$
momentum	$\Longleftrightarrow$	$\hat{P}$
$\vdots$	$\Longleftrightarrow$	$\vdots$

Before we look into the details of associating observables with operators, we need to establish some properties of these operators. The main constraint comes from the fact that observables take **real** values only. Therefore we must require that the operators that represent observables have only **real eigenvalues**, since we want to identify the eigenvalues with the possible results of measurements. We can guarantee this if we only use **Hermitian** operators to represent observables.

3.3.1 Hermitian Conjugate

Let us define first the **Hermitian conjugate** $\hat{O}^\dagger$ of an operator $\hat{O}$.

Let $|\psi\rangle$ and $|\phi\rangle$ be arbitrary states in $\mathcal{H}$, then

$$\langle \phi | \hat{O}^\dagger | \psi \rangle = \left(\langle \psi | \hat{O} | \phi \rangle \right)^* . \tag{3.22}$$

In order to compute the matrix element of $\hat{O}^\dagger$ between $|\phi\rangle$ and $|\psi\rangle$, we need to compute the matrix element of $\hat{O}$ swapping the states and then taking the complex conjugate.

Mathematical Aside: A Comment on Hermitian Conjugation

Compare Eq. (3.22) with the more familiar expression from linear algebra:

$$O_{ij}^{\dagger} = O_{ji}^{*}. \tag{3.23}$$

The quantities in Eq. (3.22) are the matrix elements of the operator $\hat{O}$, just like O_{ij} are the matrix elements of a usual matrix O. The quantum states $|\phi\rangle$ and $|\psi\rangle$ are the 'indices' that label the matrix elements. Using this identification, many equations that we encounter in quantum mechanics become rather familiar.

Example 3.2 The operator $\hat{O} = \frac{d}{dx}$ is defined by specifying its action on the wave function of a quantum system:

$$\langle x|\hat{O}|\psi\rangle = \hat{O}\psi(x) = \frac{d}{dx}\psi(x). \tag{3.24}$$

Following our prescriptions, the matrix element of $\hat{O}$ between two states $|\psi\rangle$ and $|\phi\rangle$ is given by

$$\langle\psi|\hat{O}|\phi\rangle = \int dx\, \psi(x)^{*}\frac{d}{dx}\phi(x)\,; \tag{3.25}$$

then we can integrate by parts to obtain

$$\int_{-\infty}^{\infty} \psi(x)^{*}\frac{d}{dx}\phi(x)\,dx = \left[\phi(x)\psi(x)^{*}\right]_{-\infty}^{\infty} - \int_{-\infty}^{\infty}\phi(x)\frac{d}{dx}\psi(x)^{*}\,dx. \tag{3.26}$$

We can discard the constant term on the right-hand side, since physically acceptable wave functions vanish at $x = \pm\infty$, and if we then take the complex conjugate of the resulting equation we obtain

$$\left(\int_{-\infty}^{\infty} \psi(x)^{*}\frac{d}{dx}\phi(x)\,dx\right)^{*} = -\int_{-\infty}^{\infty}\phi(x)^{*}\frac{d}{dx}\psi(x)\,dx$$

$$\equiv \int_{-\infty}^{\infty}\phi(x)^{*}\left(\frac{d}{dx}\right)^{\dagger}\psi(x)\,dx$$

from the definition of Hermitian conjugate. Thus we can make the identification

$$\left(\frac{d}{dx}\right)^{\dagger} = -\frac{d}{dx}.$$

3.3.2 Hermitian Operators

We can now define a **Hermitian operator**. An operator $\hat{O}$ acting on a Hilbert space $\mathcal{H}$ is called **Hermitian** if

$$\hat{O}^{\dagger} \equiv \hat{O}. \tag{3.27}$$

Example 3.3 Equation (3.27) is clearly not true for all operators; $\frac{d}{dx}$ is NOT Hermitian since we have just shown that

$$\left(\frac{d}{dx}\right)^{\dagger} = -\frac{d}{dx},$$

whereas the operator $-i\hbar\,\frac{d}{dx}$ IS Hermitian; the proof is straightforward and is left as an exercise for the reader.

Strictly speaking, physical observables are associated with **self-adjoint** operators. In order to be able to define self-adjoint operators we first need to define the **domain**, $\mathcal{D}(\hat{O})$, of an operator $\hat{O}$. The domain $\mathcal{D}(\hat{O})$ is the subspace of $\mathcal{H}$ upon which the operator acts, i.e.

$$\hat{O}|\psi\rangle \in \mathcal{H} \quad \Longrightarrow \quad |\psi\rangle \in \mathcal{D}(\hat{O}). \tag{3.28}$$

A self-adjoint operator is a Hermitian operator such that

$$\mathcal{D}(\hat{O}^{\dagger}) = \mathcal{D}(\hat{O}). \tag{3.29}$$

While the Hermitian operators that we encounter in quantum mechanics are mostly automatically self-adjoint, this pedantic distinction becomes relevant in some physically meaningful problems. Typically, problems with non-trivial boundary conditions require some care to extend a Hermitian operator so that it is actually self-adjoint.

3.4 Properties of Hermitian Operators

Hermitian operators obey properties that are important for building the logical framework of quantum mechanics.

1. Hermitian operators have *real* eigenvalues. The eigenvalue equation is

$$\hat{O}|\psi_k\rangle = O_k|\psi_k(x)\rangle, k = 1, \ldots.$$

$|\psi_k\rangle$ are the eigenfunctions of $\hat{O}$, O_k are the eigenvalues. Then we have

$$\hat{O} = \hat{O}^{\dagger} \Longrightarrow O_k \in \mathbb{R}.$$

2. The eigenstates of a Hermitian operator that belong to different eigenvalues are orthogonal.
3. If $\hat{O}$ is a Hermitian operator acting on a vector space $\mathcal{H}$, there exists an orthogonal basis of $\mathcal{H}$ made up of eigenvectors of $\hat{O}$. In other words, every vector $|\psi\rangle$ can be expanded as

$$|\psi\rangle = \sum_k c_k|\psi_k\rangle, \tag{3.30}$$

where c_k are complex coefficients, computed by taking the projection of $|\psi\rangle$ onto the states $|\psi_k\rangle$:

$$c_k = \langle \psi_k | \psi \rangle.$$

As you can see from the equation above, for each state $|\psi\rangle$ there is a set of coefficients c_k; they are the *coordinates* of the function $|\psi\rangle$ in the basis $\{|\psi_k\rangle,\ k = 1,\dots\}$. Do not confuse the coefficients c_k with the eigenvalues O_k! The latter are a characteristic of the operator $\hat{O}$ and have nothing to do with the state ψ.

4. The commutator of two Hermitian operators is anti-Hermitian. While this property is trivial to prove, it is useful to keep in mind as a way to check your results. Every time you compute a commutator of Hermitian operators, you have a sanity check of your answer.

Mathematical Aside: Proof of the Properties

We shall now prove the first two properties above. The proofs are useful examples of manipulations involving operators acting on wave functions. Familiarity with these kinds of manipulations is essential for solving problems in quantum mechanics.

1. Hermitian operators have real eigenvalues.

Proof. Suppose $\hat{O}$ is a Hermitian operator so that $\hat{O}^\dagger = \hat{O}$, and let $\hat{O}$ have an eigenvalue O_k, with corresponding eigenfunction $|\psi_k\rangle$:

$$\hat{O} |\psi_k\rangle = O_k |\psi_k\rangle.$$

Then

$$\langle \psi_k | \hat{O} | \psi_k \rangle = O_k \langle \psi_k | \psi_k \rangle = O_k,$$

where in the last equality we have used the fact that the ket $|\psi_k\rangle$ is properly normalised to one.

If we take the complex conjugate of this equation, we obtain

$$\left(\langle \psi_k | \hat{O} | \psi_k \rangle \right)^* = O_k^*,$$

but if we make use of the definition of the Hermitian conjugate, we can rewrite the left-hand side of this equation in terms of $\hat{O}^\dagger$ and use the fact that $\hat{O}^\dagger = \hat{O}$ by hypothesis:

$$\left(\langle \psi_k | \hat{O} | \psi_k \rangle \right)^* = \langle \psi_k | \hat{O}^\dagger | \psi_k \rangle = \langle \psi_k | \hat{O} | \psi_k \rangle.$$

The right-hand side is now just the matrix element that appears in the first equation and is equal to O_k, so we have proved that

$$O_k^* = O_k,$$

thus showing that the eigenvalue O_k is real as stated.

2. The eigenfunctions of a Hermitian operator which belong to different eigenvalues are orthogonal.

Proof. Suppose that

$$\hat{O}|\psi_1\rangle = O_1|\psi_1\rangle \quad \text{and} \tag{3.31}$$

$$\hat{O}|\psi_2\rangle = O_2|\psi_2\rangle \quad \text{with } O_1 \neq O_2. \tag{3.32}$$

From Eq. (3.31) we have

$$\langle\psi_2|\hat{O}|\psi_1\rangle = O_1\langle\psi_2|\psi_1\rangle, \tag{3.33}$$

whereas from Eq. (3.32)

$$\langle\psi_1|\hat{O}|\psi_2\rangle = O_2\langle\psi_1|\psi_2\rangle. \tag{3.34}$$

Taking the complex conjugate of Eq. (3.34) yields on the left-hand side

$$\left(\langle\psi_1|\hat{O}|\psi_2\rangle\right)^* \equiv \langle\psi_2|\hat{O}^\dagger|\psi_1\rangle = \langle\psi_2|\hat{O}|\psi_1\rangle, \tag{3.35}$$

whereas the right-hand side gives

$$O_2^*\langle\psi_2|\psi_1\rangle = O_2\langle\psi_2|\psi_1\rangle, \tag{3.36}$$

using the fact that $O_2 = O_2^*$.

Comparing with Eq. (3.33) we see that

$$O_2\langle\psi_2|\psi_1\rangle = O_1\langle\psi_2|\psi_1\rangle, \tag{3.37}$$

which we can rearrange to yield the result

$$(O_2 - O_1)\langle\psi_2|\psi_1\rangle = 0. \tag{3.38}$$

Given that $O_2 \neq O_1$ by hypothesis, this implies that

$$\langle\psi_2|\psi_1\rangle = 0, \tag{3.39}$$

which is the desired result.

3.5 Spectral Decomposition

Using the fact that we can define a basis in the space of physical states made up of eigenstates of the operator $\hat{O}$ as shown in Eq. (3.30), we can formulate the rule to compute the probability of finding any given eigenvalue when performing a measurement. This is the **fourth postulate** of quantum mechanics.

Postulate 4   Given the decomposition in Eq. (3.30), the probability of finding the nondegenerate eigenvalue O_k when measuring O in the state described by the normalised vector $|\psi\rangle$ is given by

$$P_k = |c_k|^2. \tag{3.40}$$

Clearly the sum of probabilities should be properly normalised and therefore

$$\sum_k P_k = \sum_k |c_k|^2 = 1. \tag{3.41}$$

This is simply a confirmation of the fact that the state vector $|\psi\rangle$ is normalised to one.

Example 3.4 Note that the concept of superposition of states is very different from anything we have encountered in classical mechanics. Consider two quantum states $|A\rangle$ and $|B\rangle$, such that the measurement of an observable O yields the result a with probability 1 when the system is in the state $|A\rangle$, and the result b with probability 1 when the system is in the state $|B\rangle$. The superposition principle states that the state vector

$$|C\rangle = c_A|A\rangle + c_B|B\rangle, \tag{3.42}$$

where c_A and c_B are complex numbers such that $|c_A|^2 + |c_B|^2 \neq 0$, describes a possible physical state of the system. According to the postulates that we formulated so far, the measurement of the observable O in the state $|C\rangle$ can only yield the value a or b, with respective probabilities

$$p_a = \frac{|c_A|^2}{|c_A|^2 + |c_B|^2}, \quad p_b = \frac{|c_B|^2}{|c_A|^2 + |c_B|^2}. \tag{3.43}$$

No other results are possible for the measured value of O in the state $|C\rangle$.

Example 3.5 Let us consider again the discretised one-dimensional system in Fig. 2.3; we can have a state $|1\rangle$ where the particle is localised for example at site 1, and a state $|2\rangle$ where the particle is localised at site 2. Measuring the position of the particle in state $|1\rangle$ yields $x = 1$ with probability 1. Likewise we obtain $x = 2$ with probability 1 for a particle described by the state vector $|2\rangle$.

The state $\frac{1}{\sqrt{2}}|1\rangle + \frac{1}{\sqrt{2}}|2\rangle$ is an admissible quantum state. Measuring the position of the particle in this latter state, the outcome will be $x = 1$ or $x = 2$ with 50% probability. No other value is allowed in this state.

Degenerate Eigenvalues

In the case where the eigenvalue O_k is g-fold degenerate, the probability of O_k being the result of a measurement of the observable O in the state $|\psi\rangle$ is obtained by summing over the contributions from the whole subspace spun by the degenerate eigenvectors. Following the notation introduced in Eq. (3.18), the probability is

$$P_k = \sum_{n=1}^{g} \left| \langle u_k^{(n)} | \psi \rangle \right|^2. \tag{3.44}$$

3.6 Collapse of the State Vector

Given the state vector $|\psi\rangle$, the fourth postulate of quantum mechanics allows us to compute the probability of any given outcome for the measurement of an observable. However, after the measurement has been performed, the outcome of the experiment is known with probability 1, and therefore we must conclude that the state of the system has changed. This is an important aspect of measurement in quantum mechanics: the measurement must change the state of the system. This is summarised in the **fifth postulate** of quantum mechanics.

Postulate 5   Immediately after a measurement that gave the result O_k, where O_k is a nondegenerate eigenvalue, the system is in the state $|\psi_k\rangle$, the eigenvector of $\hat{O}$ associated with the eigenvalue O_k. The state vector has been *projected* onto the eigenstate by the process of performing the measurement.

We will discuss more precisely what we mean by *immediately after* when we discuss the time evolution of quantum systems in the next chapter.

We can express the same concept using the projection operator introduced in Eq. (2.66); immediately after a measurement yielding the value O_k, the state of the system is transformed according to

$$|\psi\rangle \mapsto \frac{1}{\sqrt{\langle \psi | \hat{\mathcal{P}}_k | \psi \rangle}} \hat{\mathcal{P}}_k |\psi\rangle, \qquad (3.45)$$

where we have explicitly normalised the projected vector to have unit norm. This is sometimes referred to as the *collapse of the state vector*. Clearly this is an idealised description of a much more complicated process where a 'classical' instrument and the quantum system interact. While a fully satisfactory description of quantum measurement is a subtle issue, the fifth postulate gives a practical recipe, which underlies the numerous successful predictions of quantum mechanics.

After a measurement that yielded the value O_k, the wave function of the system coincides with the eigenfunction $|\psi_k\rangle$. Then, as discussed below Eq. (3.17), if we perform immediately another measurement of O we will find the *same* value O_k with probability 1.

Conversely, if the wave function does not coincide with one of the eigenfunctions, then the observable O does not have a given value in the state $|\psi\rangle$. We can only compute the probability for each eigenvalue to be the outcome of the experiment.

Clearly these phenomena do not have a classical analogue. The description of a physical system in quantum mechanics is radically different from the classical one. You need to practice in order to get familiar with the quantum-mechanical framework.

Degenerate Eigenvalues

In the case of degenerate eigenvalues, the state vector before the measurement can be expanded as

$$|\psi\rangle = \sum_k \sum_{n=1}^{g_k} c_{kn} |u_k^{(n)}\rangle, \tag{3.46}$$

where the index k runs over all eigenvalues, and for each eigenvalue we have a sum over n running over the number of degenerate eigenvectors that correspond to the kth eigenvalue.

After the measurement the state of the system is still given by Eq. (3.45), but the projector $\hat{\mathcal{P}}_k$ needs to project on the subspace of degenerate eigenstates, i.e.

$$\hat{\mathcal{P}}_k = \sum_{n=1}^{g_k} |u_k^{(n)}\rangle\langle u_k^{(n)}|. \tag{3.47}$$

Exercise 3.6.1   Show that the projected (and normalised) state according to Eq. (3.47) is

$$\frac{1}{\sqrt{\sum_{n=1}^{g_k} |c_{kn}|^2}} \sum_{n=1}^{g_k} c_{kn} |u_k^{(n)}\rangle. \tag{3.48}$$

3.7 Compatible Observables

Suppose A and B are observables and we perform the following sequence of measurements in succession on a single system:

1. measure A 2. measure B 3. remeasure A

Then *if and only if* the result of 3 is *certain* to be the same as the result of 1, we say that A and B are **compatible observables**.

In general, this will not be the case: according to our previous discussion about the collapse of the state vector, the measurement of B will project the state of the system onto an eigenstate of B, and therefore 'spoil' the result of 1. Let us analyse this statement in a little more detail, following the ideas that we have introduced so far about observables and measurements. Suppose that A and B are represented by operators $\hat{A}$ and $\hat{B}$, respectively, with eigensystems

$$\hat{A}|u_i\rangle = A_i|u_i\rangle,$$
$$\hat{B}|v_i\rangle = B_i|v_i\rangle.$$

For simplicity we are going to consider the case of nondegenerate eigenvalues – taking into account degeneracies adds a little complexity for no intellectual gain. Measurement 1 must return one of the eigenvalues of the operator $\hat{A}$, A_j say, collapsing the system into the state $|u_j\rangle$. Measurement 2 yields an eigenvalue of $\hat{B}$, B_k, forcing the system into the state $|v_k\rangle$, so that measurement 3 is made with the system in the state $|v_k\rangle$. The only way that 3 is *certain* to yield the result A_j as obtained in 1 is if $|v_k\rangle \equiv |u_j\rangle$. For this to be true in all circumstances it must be the case that each eigenvector $|v_k\rangle$ of $\hat{B}$ is identical to some eigenvector $|u_j\rangle$ of $\hat{A}$. If there is no degeneracy this implies a one-to-one correspondence between the eigenvectors of $\hat{A}$ and the eigenvectors of $\hat{B}$. We say that $\hat{A}$ and $\hat{B}$ have *a common eigenbasis*. These properties are summarised in the so-called **compatibility theorem**.

Theorem 3.1   *Given two observables, A and B, represented by Hermitian operators $\hat{A}$ and $\hat{B}$, then any one of the following three statements implies the other two:*

1. *A and B are compatible observables;*
2. *$\hat{A}$ and $\hat{B}$ have a common eigenbasis;*
3. *the operators $\hat{A}$ and $\hat{B}$ commute: $[\hat{A}, \hat{B}] = 0$.*

Example Proof. Let us show, for instance, that $3 \Rightarrow 2$. We have

$$\hat{A}\,|u_i\rangle = A_i\,|u_i\rangle,$$
$$\hat{B}\,|v_i\rangle = B_i\,|v_i\rangle,$$

so that for *any* eigenvector of $\hat{A}$

$$\hat{A}\hat{B}\,|u_i\rangle = \hat{B}\hat{A}\,|u_i\rangle \quad \text{by virtue of 3}$$
$$= \hat{B}\,A_i\,|u_i\rangle$$
$$= A_i\,\hat{B}\,|u_i\rangle.$$

Thus $\hat{B}\,|u_i\rangle$ is an eigenvector of $\hat{A}$ belonging to the eigenvalue A_i. If we assume that the eigenvalues are nondegenerate, then $\hat{B}\,|u_i\rangle$ must be some multiple of $|u_i\rangle$:

$$\hat{B}\,|u_i\rangle = \rho\,|u_i\rangle, \quad \text{say.} \tag{3.49}$$

This just says that $|u_i\rangle$ is an eigenstate of $\hat{B}$ belonging to the eigenvalue ρ, and we must have that, for some j,

$$\rho = B_j \quad \text{and } |u_i\rangle = |v_j\rangle. \tag{3.50}$$

Thus any eigenstate of the set $\{|u_i\rangle\}$ coincides with some member of the set $\{|v_j\rangle\}$. The correspondence has to be one-to-one because both sets are orthonormal; if we assume that two states in one set coincide with a single state in the other set, we are led to a contradiction that two orthogonal vectors are identical to the same vector. By simply relabelling all the vectors in one set we can always ensure that

$$|u_1\rangle = |v_1\rangle, \ |u_2\rangle = |v_2\rangle, \ |u_3\rangle = |v_3\rangle, \dots \text{etc.,} \tag{3.51}$$

and this is the common eigenbasis. A more general proof, in the case where the eigenvalues are degenerate, is left as an exercise.

3.8 Complete Sets of Commuting Observables

Consider an observable A, and a basis made up of eigenstates of $\hat{A}$, $\{|u_1\rangle, |u_2\rangle, \dots\}$. If all the eigenvalues are nondegenerate, each eigenvalue identifies uniquely *one* eigenstate. Hence we can label the eigenstates by their eigenvalue; if

$$\hat{A}|u_n\rangle = a_n|u_n\rangle, \tag{3.52}$$

then we can rename

$$|u_n\rangle \equiv |a_n\rangle. \tag{3.53}$$

In this case, the observable A constitutes by itself a **complete set of commuting observables** (CSCO), i.e. the eigenvalues of $\hat{A}$ are sufficient to identify the eigenvectors that form a basis of the space of physical states.

However, this is no longer true if some of the eigenvalues are degenerate, since in this case there are several eigenstates corresponding to the same degenerate eigenvalue. In order to distinguish these eigenstates, we can use the eigenvalues of a second observable B, which commutes with A. According to the compatibility theorem, we can find a basis of common eigenstates of $\hat{A}$ and $\hat{B}$. If each pair of eigenvalues $\{a_n, b_p\}$ identifies uniquely *one* vector of the basis, then the set $\{A, B\}$ is a CSCO. If this is not the case, then there must be at least one pair $\{a_n, b_p\}$ for which there exists more than one eigenvector with these eigenvalues, i.e. there exist at least two vectors $|w_1\rangle$ and $|w_2\rangle$, such that

$$\hat{A}|w_1\rangle = a_n|w_1\rangle, \quad \hat{B}|w_1\rangle = b_p|w_1\rangle, \tag{3.54}$$

$$\hat{A}|w_2\rangle = a_n|w_2\rangle, \quad \hat{B}|w_2\rangle = b_p|w_2\rangle. \tag{3.55}$$

In this case specifying the values of a_n and b_p is not sufficient to identify uniquely one eigenvector, since any linear combination of $|w_1\rangle$ and $|w_2\rangle$ is also a simultaneous eigenvector of $\hat{A}$ and $\hat{B}$ with the same eigenvalues. Iterating the above procedure, we add to our set of observables one more quantity C, which commutes with both A and B, and we choose a basis made up of simultaneous eigenvalues of the three operators $\hat{A}, \hat{B}, \hat{C}$. If each eigenstate in the basis is uniquely identified by the set of eigenvalues $\{a_n, b_p, c_q\}$, then $\{A, B, C\}$ is a CSCO. If not, we need to add one more observable to our set, and so on.

A set of observables $A, B, C, \ldots$ is called a CSCO if:

i. all the observables commute by pairs;
ii. specifying the eigenvalues of all the operators in the CSCO identifies a unique common eigenvector.

Given a CSCO, we can choose a basis for the space of states made up of common eigenvectors of the operators associated with the observables. Each eigenvector is uniquely identified by the values of the eigenvalues to which it corresponds:

$$\hat{A}|a_n, b_p, c_q, \ldots\rangle = a_n|a_n, b_p, c_q, \ldots\rangle,$$
$$\hat{B}|a_n, b_p, c_q, \ldots\rangle = b_p|a_n, b_p, c_q, \ldots\rangle,$$
$$\hat{C}|a_n, b_p, c_q, \ldots\rangle = c_q|a_n, b_p, c_q, \ldots\rangle,$$
$$\ldots. \tag{3.56}$$

Given a CSCO, we can expand any generic wave function in the basis of common eigenstates labelled by the eigenvalues of the observables:

$$|\psi\rangle = \sum_{n,p,q} \psi_{n,p,q}|a_n, b_p, c_q\rangle.$$

The modulus square of the coefficients, $\left|\psi_{n,p,q}\right|^2$, yields the probability of finding simultaneously the values a_n, b_p, c_q if we measure A, B, C in the state $|\psi\rangle$.

3.9 Continuous Spectrum

Until now we have discussed a number of examples where operators have a discrete spectrum, i.e. where the eigenvalues are numbered by some integer index k.[1] However, there are operators that have a continuous spectrum, like the energy and the momentum of an unbound state, or the position operator $\hat{X}$. In order to deal with these cases, we need to generalise the formalism that we have introduced so far. As you will see below, the modifications are minimal, and rather straightforward.

3.9.1 Eigenvalue Equation

Let us denote by $\hat{f}$ the Hermitian operator associated with an observable with a continuous spectrum; the eigenvalue equation takes the form

$$\hat{f}|f\rangle = f|f\rangle. \tag{3.57}$$

Note that in Eq. (3.57) we have denoted the eigenstate simply by $|f\rangle$. The set of eigenvectors is a complete set, and we can write the completeness relation by substituting the sum over the eigenvalues with an integral, viz.

$$\int df \, |f\rangle\langle f| = 1. \tag{3.58}$$

Once again, using the completeness relation, a generic state can be expanded as a superposition of eigenstates:

$$|\psi\rangle = \int df \, \psi(f)|f\rangle \, ; \tag{3.59}$$

you should compare this expression with its analogue Eq. (3.30) in the case of discrete eigenvalues. Here we are implicitly assuming that the integral over vectors in $\mathcal{H}$ can be properly defined and yields another element of the same vector space $\mathcal{H}$. The coefficients in the expansion Eq. (3.59) are obtained by taking the scalar product

$$\psi(f) = \langle f|\psi\rangle. \tag{3.60}$$

Example 3.6 For a quantum system on a line, we can use the completeness relation

$$\int dx \, |x\rangle\langle x| = 1 \tag{3.61}$$

---

[1] In the case of a discrete spectrum, the total number of eigenvalues may well be infinite, however, the eigenvalues are labelled by integer numbers.

and rewrite Eq. (3.60) as

$$\langle f|\psi\rangle = \int dx\, \langle f|x\rangle\langle x|\psi\rangle = \int dx\, f(x)^*\psi(x). \tag{3.62}$$

The probabilistic interpretation of the state vector, i.e. the fourth postulate of quantum mechanics, can be generalised to the case of a continuous spectrum. The probability of finding a result between f and $f + df$ when measuring the observable f is given by

$$\left|\psi(f)\right|^2 df.$$

Thus we derive the normalisation condition

$$\int df\, \left|\psi(f)\right|^2 = 1. \tag{3.63}$$

The integral in Eq. (3.59) defines a normalisable state if and only if the function $\psi(f)$ is square-integrable.

3.9.2 Orthonormality

Following the steps performed to obtain Eq. (3.39), we have

$$\langle f'|\hat{f}|f\rangle = f\langle f'|f\rangle. \tag{3.64}$$

Taking the complex conjugate of Eq. (3.64), and using the fact that $\hat{f}$ is Hermitian, yields

$$\left(\langle f|\hat{f}|f'\rangle\right)^* = \langle f'|\hat{f}^\dagger|f\rangle, \tag{3.65}$$

$$= \langle f'|\hat{f}|f\rangle. \tag{3.66}$$

Combining the results above:

$$(f - f')\,\langle f'|f\rangle = 0, \tag{3.67}$$

and therefore, if $f \neq f'$,

$$\langle f'|f\rangle = 0. \tag{3.68}$$

Now comes a subtle point. The norm of the state $|\psi\rangle$ is given by

$$\langle\psi|\psi\rangle = \int df\, df'\, \psi(f)^*\psi(f')\langle f|f'\rangle, \tag{3.69}$$

$$= \int df\, \left|\psi(f)\right|^2 \int df'\langle f|f'\rangle, \tag{3.70}$$

where we used the orthogonality result Eq. (3.68). Now the integral over df' in Eq. (3.70) vanishes for any finite value of $\langle f|f\rangle$. Therefore we need to impose that $\langle f|f\rangle$ is infinite and normalised so that

$$\int df'\,\langle f'|f\rangle = 1. \tag{3.71}$$

The **delta function** introduced by Dirac satisfies precisely this condition, and we can require

$$\langle f'|f \rangle = \delta(f - f').$$ (3.72)

The eigenstates of the continuous spectrum have infinite norm, and therefore cannot be considered as genuine physical states. However, they provide a useful basis for expanding normalisable state vectors that correspond to physical states, as in Eq. (3.59).

3.9.3 Spectral Decomposition

For a generic operator $\hat{O}$, the eigenvalue spectrum can be made up of discrete eigenvalues O_n and continuous values f. In this case we need to consider both the eigenfunctions of the discrete spectrum and those of the continuous spectrum to have a basis to expand quantum states in:

$$|\psi\rangle = \sum_n \langle n|\psi\rangle \, |n\rangle + \int df \, \langle f|\psi\rangle \, |f\rangle.$$ (3.73)

Equation (3.73) summarises the completeness relation for both discrete and continuous spectra.

Taking the scalar product with eigenstates of the position eigenstates $|x\rangle$:

$$\psi(x) = \sum_n \psi_n u_n(x) + \int df \, \psi(f) u_f(x),$$ (3.74)

where

$$\psi_n = \langle n|\psi\rangle, \qquad u_n(x) = \langle x|n\rangle,$$ (3.75)
$$\psi(f) = \langle f|\psi\rangle, \qquad u_f(x) = \langle x|f\rangle.$$ (3.76)

Example 3.7 Gaussian wave functions. For a quantum system on a line, the position operator $\hat{X}$ has a continuous spectrum. The eigenvalue equation

$$\hat{X}|x\rangle = x|x\rangle$$ (3.77)

has a solution for each value of x. The eigenvectors are normalised to the Dirac delta:

$$\langle x|x'\rangle = \delta(x - x'),$$ (3.78)

and a generic vector can be expanded as

$$|\psi\rangle = \int dx \, |x\rangle\langle x|\psi\rangle = \int dx \, \psi(x)|x\rangle.$$ (3.79)

Consider now the state defined by the wave function

$$\psi(x) = \langle x|\psi\rangle = C \exp\left[\frac{i}{\hbar}p_0 x - \frac{(x - x_0)^2}{2\xi^2}\right].$$ (3.80)

The state has a finite norm:

$$\int dx \, |\psi(x)|^2 = |C|^2 \sqrt{\pi} \, \xi, \tag{3.81}$$

which yields the normalisation constant

$$C = \frac{1}{\pi^{1/4}\sqrt{\xi}}. \tag{3.82}$$

Having properly normalised the state vector, we can compute the mean value of x, i.e. the average of multiple measurements of the position of the system prepared in the state $|\psi\rangle$, specified by Eq. (3.80):

$$\langle x \rangle = \int dx \, |\psi(x)|^2 \, x = x_0. \tag{3.83}$$

The variance of the same ensemble of measurements is

$$\langle \Delta x^2 \rangle = \langle (x - x_0)^2 \rangle = \int dx \, |\psi(x)|^2 \, (x - x_0)^2 = \xi^2. \tag{3.84}$$

In the last two equations, the quantities on the left-hand side are obtained by preparing the system several times in the same state $|\psi\rangle$, performing measurements and taking averages. The quantities on the right-hand side, i.e. x_0 and ξ, are parameters that define the state of the system. By performing measurements, and looking at the moments of the observables, we can determine these parameters, i.e. we learn about the state of the system.

3.10 Momentum Operator

The *momentum operator* is defined as a differential operator acting on the wave function:

$$\hat{P}\psi(x) = \langle x|\hat{P}|\psi\rangle = -i\hbar\frac{d}{dx}\psi(x). \tag{3.85}$$

This can be seen as a realisation of de Broglie's duality hypothesis; according to wave–particle duality, with a particle with momentum p we can associate a wave with wavelength h/p. A wave with a fixed wavelength is a plane wave, described by the function

$$\psi_p(x) = C \exp[ipx/\hbar]. \tag{3.86}$$

When we act with the operator $\hat{P}$ defined in Eq. (3.85), we see that $\psi_p(x)$ is an eigenstate of $\hat{P}$ with eigenvalue p, and therefore we can associate the plane wave with a state with given momentum p.

Note that there are no restrictions on the possible values of p. The spectrum of $\hat{P}$ is therefore a continuous spectrum. The normalisation of the eigenstates is

$$
\begin{aligned}
\langle p'|p\rangle &= \int dx\,\langle p'|x\rangle\langle x|p\rangle \\
&= |C|^2 \int dx\,\exp\left[\frac{i}{\hbar}(p-p')x\right] \\
&= |C|^2 2\pi\hbar\delta(p-p').
\end{aligned}
\tag{3.87}
$$

Following the convention we set in Section 3.9.2, we can choose $C = 1/\sqrt{2\pi\hbar}$, and hence

$$
\psi_p(x) = \langle x|p\rangle = \frac{1}{\sqrt{2\pi\hbar}}\exp\left[ipx/\hbar\right].
\tag{3.88}
$$

Example 3.8 We have seen previously that the action of the position operator $\hat{X}$ is

$$
\hat{X}\psi(x) = x\psi(x),
\tag{3.89}
$$

i.e. the wave function is simply multiplied by the value of x. Consider the case $\hat{O}_1 = \hat{X}$, $\hat{O}_2 = \frac{d}{dx}$. Then

$$
\hat{O}_1\hat{O}_2\psi(x) = \hat{X}\left(\frac{d}{dx}\psi(x)\right),
\tag{3.90}
$$

$$
= x\frac{d}{dx}\psi(x),
\tag{3.91}
$$

while

$$
\hat{O}_2\hat{O}_1\psi(x) = \hat{O}_2\left(\hat{X}\psi(x)\right),
\tag{3.92}
$$

$$
= \hat{O}_2\left(x\psi(x)\right),
\tag{3.93}
$$

$$
= \frac{d}{dx}\left(x\psi(x)\right),
\tag{3.94}
$$

$$
= \psi(x) + x\frac{d}{dx}\psi(x).
\tag{3.95}
$$

Putting the two results together, we obtain for this particular choice of $\hat{O}_1$ and $\hat{O}_2$:

$$
\left[\hat{O}_1,\hat{O}_2\right]\psi(x) = -\psi(x),
\tag{3.96}
$$

i.e.

$$
\left[\hat{O}_1,\hat{O}_2\right] = -1.
\tag{3.97}
$$

From the example above we deduce the fundamental **canonical commutation relation**:

$$
\left[\hat{X},\hat{P}\right] = i\hbar.
\tag{3.98}
$$

3.10.1 Momentum as Generator of Translations

Consider the wave function of a one-dimensional system, and translate the coordinate x by some amount a. Expanding at first order in a:

$$\psi(x + a) = \psi(x) + a\frac{d}{dx}\psi(x) + O(a^2). \tag{3.99}$$

We want to describe the change in the wave function by the action of a Hermitian operator acting on it. As discussed above, the differential operator d/dx is anti-Hermitian; we can introduce a Hermitian operator by rewriting Eq. (3.99) as

$$\delta_a\psi(x) = \psi(x + a) - \psi(x)$$

$$= \frac{i}{\hbar}a\left(-i\hbar\frac{d}{dx}\right)\psi(x) + O(a^2)$$

$$= ia\left(\frac{\hat{P}}{\hbar}\right)\psi(x) + O(a^2). \tag{3.100}$$

The quantity in the bracket on the right-hand side of Eq. 3.100 is a Hermitian operator. It can easily be checked that it has dimensions of inverse length, and actually coincides with the momentum operator that we defined above in order to implement de Broglie's idea of duality.

If we only consider infinitesimal transformations and neglect the non-linear dependence on the transformation parameter, we can write

$$\delta_a\psi(x) = ia\hat{T}\psi(x). \tag{3.101}$$

The operator $\hat{T}$ is called the **generator** of translations. The momentum operator coincides with the generator of translations up to the factor $1/\hbar$, which takes care of dimensions.

3.10.2 Position and Momentum Representations

So far we have adopted a heuristic approach in order to define the position and momentum operators, based on physical considerations. It is worthwhile having a more formal introduction, which follows closely the ideas of vectors and bases that we have developed in this book.

The state of the quantum system is described by a vector $|\psi\rangle$. This vector can be expanded in a basis made up of eigenstates of the position operator $\{|x\rangle\}$:

$$|\psi\rangle = \int dx\,|x\rangle\langle x|\psi\rangle, \tag{3.102}$$

and the vector is fully specified by specifying its coordinates in that basis:

$$\psi(x) = \langle x|\psi\rangle. \tag{3.103}$$

This is called the **position representation** of the state.

The same vector can be expanded in eigenstates of momentum:

$$|\psi\rangle = \int dp\,|p\rangle\langle p|\psi\rangle. \tag{3.104}$$

In this case the vector is fully described by the wave function in the so-called **momentum representation**:

$$\tilde{\psi}(p) = \langle p | \psi \rangle. \tag{3.105}$$

Using the completeness relations allows us to write

$$\tilde{\psi}(p) = \int dx \langle p | x \rangle \langle x | \psi \rangle, \tag{3.106}$$

$$= \int \frac{dx}{\sqrt{2\pi\hbar}} \, e^{-i\frac{px}{\hbar}} \, \psi(x), \tag{3.107}$$

which shows that the two representations are related by Fourier transforms.

The vector $|\psi\rangle$ has finite norm, while the eigenstates of both position and momentum are states with an infinite norm and do not represent physical states. Nonetheless, they are useful as a basis to expand the state vector and define a pdf for observing given values of x and p in the state $|\psi\rangle$.

3.10.3 Wave Packets

Physical states, i.e. state vectors with a finite norm, can be constructed as linear superpositions of momentum eigenstates. These states are called **wave packets**:

$$|W\rangle = \int dp \, g(p) \, |p\rangle, \quad g(p) \in \mathbb{C}. \tag{3.108}$$

For the wave packet to be normalisable, we require

$$\langle W | W \rangle = \int dp' dp \, g(p')^* g(p) \, \langle p' | p \rangle$$

$$= \int dp \, |g(p)|^2 < \infty. \tag{3.109}$$

If we choose the function $g(p)$ such that $|g(p)|^2$ is peaked around some value p_0, then the probability of measuring a given value for the momentum of the system in the state $|W\rangle$ has a maximum at p_0 and rapidly goes to zero as we move away from p_0. While the wave packet is not an eigenstate of momentum, we can think of it as a physical realisation of a state with momentum approximately equal to p_0. Making the distribution more peaked around its maximum yields a more 'collimated' beam.

In this section we are going to focus on Gaussian wave packets, normalised to one, viz.

$$g(p) = \frac{1}{(2\pi\sigma^2)^{1/4}} \exp\left[-\frac{1}{4\sigma^2} (p - p_0)^2\right]. \tag{3.110}$$

Exercise 3.10.1   Check that the wave packet is properly normalised, $\langle W|W\rangle = 1$.
Compute the expectation values

$$\langle W|\hat{P}|W\rangle = \int dp'dp\, g(p')^*g(p)\,\langle p'|\hat{P}|p\rangle$$

$$= \int dp\,|g(p)|^2\,p = p_0,$$

$$\langle W|\hat{P}^2|W\rangle = p_0^2 + \sigma^2.$$

Using the results of the exercise above, we can determine the uncertainty on the momentum

$$\Delta p = \sqrt{\langle W|\hat{P}^2|W\rangle - \langle W|\hat{P}|W\rangle^2}$$

$$= \sigma. \qquad (3.111)$$

Starting from Eq. (3.108), we can readily compute

$$\langle p|W\rangle = \int dp'\, g(p')\langle p|p'\rangle = g(p), \qquad (3.112)$$

which is sometimes referred to as the wave function in momentum space.
   A slightly lengthier computation yields the wave function in position space:

$$\langle x|W\rangle = \int dp\,\frac{1}{(2\pi\sigma^2)^{1/4}}\exp\left[-\frac{1}{4\sigma^2}(p-p_0)^2\right]\frac{1}{(2\pi\hbar)^{1/2}}\exp\left[i\frac{px}{\hbar}\right], \qquad (3.113)$$

$$= \left(\frac{2\sigma^2}{\pi\hbar^2}\right)^{1/4}\exp\left[-\frac{\sigma^2 x^2}{\hbar 2}+i\frac{p_0 x}{\hbar}\right]. \qquad (3.114)$$

Exercise 3.10.2   Compute the integral yielding the wave function in Eq. (3.114). Check that
the modulus square of the wave function is properly normalised to one, as required
by the Parsefal identity.

Using the explicit expression for the wave function in position space yields

$$\langle W|\hat{X}|W\rangle = 0,$$

$$\langle W|\hat{X}^2|W\rangle = \frac{\hbar^2}{4\sigma^2}, \qquad (3.115)$$

and therefore the uncertainty on the position of the system in the state $|W\rangle$:

$$\Delta x = \frac{\hbar}{2\sigma}. \qquad (3.116)$$

3.11 The Uncertainty Principle

In classical mechanics the state of a particle in a one-dimensional world is completely
determined by the value of its position $x(t)$ and momentum $p(t)$, i.e. by its *trajectory*.

The situation is radically different in quantum mechanics. The probabilistic interpretation of the wave function implies that we can at best obtain the probability density for a particle to be at a given position x at time t. As a consequence, the concept of *classical trajectory* used in Newtonian mechanics does not make sense in quantum mechanics. The position and momentum of the particle can be defined, but their values cannot be measured simultaneously with arbitrary precision. As expected, when the scales in the problem are much larger than the Planck constant h, the classical results are recovered.

These two features are summarised in the so-called **uncertainty relations**, first derived by Heisenberg. The uncertainty relations state that the standard deviations Δx and Δp of the measured position and momentum in a given quantum state satisfy

$$\Delta x \cdot \Delta p \geq \frac{\hbar}{2}. \tag{3.117}$$

The derivation of this inequality is left as an exercise for the reader, see Problem 3.12 below. It is clear from Eq. (3.117) that if the position of the particle is known exactly, i.e. if the state is an eigenstate of position, then the knowledge of its momentum is completely lost, in the sense that the standard deviation of the measured momentum is divergent. In general, the product of the two uncertainties has to be greater than $\hbar/2$. Note that the Gaussian wave packet discussed in Section 3.10.3 saturates the bounds exactly.

It is important to appreciate that Heisenberg's inequalities reflect a physical limitation. The outcomes of measurements are stochastic variables, the state of the quantum system determines the expectation value and the standard deviation of these stochastic variables. A better experimental apparatus would **not** allow a higher precision to be obtained. The uncertainty is a quantum-mechanical property of the system.

The uncertainty principle also encodes the idea that in quantum mechanics the measurement of a quantity interferes with the state of the system. If we measure exactly the position of the particle, then we lose all knowledge of its momentum and vice versa. You should contrast this with the situation in classical mechanics, where we can assume that measurements do not perturb the state of the system.

Summary

- The basic concept introduced in this chapter is that observables are the result of experiments. While this may sound trivial, it is a very important statement. The only knowledge we have of physical systems is the outcomes of experiments. (3.1)
- In quantum mechanics, observables are associated with Hermitian operators acting on the space of physical states. Every time you want to consider an observable, make sure that you know which operator is associated with it. (3.1.2)
- The measurement of an observable can only yield one of the eigenvalues of the operator associated with the observable, independently of the state of the system. (3.2)

- Properties of the spectrum of Hermitian operators. These are extremely important in order to be able to manipulate operators. (3.3)
- The results of an experiment must be considered as a stochastic variable. The state of the system in which the experiment is performed determines the probability of the possible outcomes of measurements. (3.5)
- Measurements in a quantum-mechanical system change the state of the system. This phenomenon is sometimes called collapse of the wave function. (3.6)
- If the operators associated with two observables commute, then the observables are called compatible. Compatible observables have a common set of eigenvectors. (3.7)
- Some operators have a continuous spectrum, and we have introduced the tools to deal with those. Eigenstates of the continuous spectrum are not normalisable. (3.9)
- Definition of the momentum operator, interpretation of the momentum operator as the generator of translations. Construction of normalisable wave packets. (3.10)
- The uncertainty principle is a mathematical consequence of the canonical commutation relations between position and momentum. It is a property of quantum-mechanical systems that does not depend on the quality of the experimental apparatus. (3.11)

Problems

3.1 When acting on the wave function, the position operator, $\hat{X}$, corresponds simply to multiplication by x:

$$\hat{X}\psi(x) = x\psi(x).$$

Use the definition of Hermitian conjugation to show that $\hat{X}$ is Hermitian and hence that the potential energy operator $\hat{V} \equiv V(x)$ is also Hermitian.

3.2 Prove the following relations:

$$(\hat{f}^\dagger)^\dagger = \hat{f},$$

$$(\hat{f}\hat{g})^\dagger = \hat{g}^\dagger \hat{f}^\dagger,$$

$$\left[\hat{f}, \hat{g}\hat{h}\right] = \hat{g}\left[\hat{f}, \hat{h}\right] + \left[\hat{f}, \hat{g}\right]\hat{h},$$

$$\left[\hat{f}\hat{g}, \hat{h}\right] = \hat{f}\left[\hat{g}, \hat{h}\right] + \left[\hat{f}, \hat{h}\right]\hat{g}.$$

If $\hat{f}$, $\hat{g}$ are Hermitian, show that $\hat{f}\hat{g} + \hat{g}\hat{f}$ and $i\left[\hat{f}, \hat{g}\right]$ are also Hermitian. Show that for any operator $\hat{A}$:

$$\langle A^\dagger A \rangle \geq 0,$$

for any state.

3.3 The observables $\mathcal{A}$ and $\mathcal{B}$ are represented by operators $\hat{A}$ and $\hat{B}$ with eigenfunctions $\{u_i(x)\}$ and $\{v_i(x)\}$, respectively, such that

$$v_1(x) = \{\sqrt{3}\,u_1(x) + u_2(x)\}/2,$$
$$v_2(x) = \{u_1(x) - \sqrt{3}\,u_2(x)\}/2,$$
$$v_n(x) = u_n(x), \quad n \geq 3.$$

Verify that these relations are consistent with orthonormality of both bases. A certain system is subjected to three successive measurements:

  (i) a measurement of $\mathcal{A}$;
 (ii) a measurement of $\mathcal{B}$;
(iii) another measurement of $\mathcal{A}$.

    Show that if measurement (i) yields any of the values $A_3, A_4, \ldots$ then (iii) gives the same result, but that if (i) yields the value A_1 there is a probability of $\frac{5}{8}$ that (iii) will yield A_1 and a probability of $\frac{3}{8}$ that it will yield A_2. What may be said about the compatibility of $\mathcal{A}$ and $\mathcal{B}$?

3.4   Consider a *two-state system*. We denote the two orthonormal states by $|1\rangle$ and $|2\rangle$. In the general case, the Hamiltonian of the system can be written as a 2×2 matrix, where the elements of the matrix are given by

$$H_{ij} = \langle i|\hat{H}|j\rangle.$$

Let us consider the Hamiltonian

$$\hat{H} = \begin{pmatrix} E_0 & -\eta \\ -\eta & E_0 \end{pmatrix}, \quad E_0 \text{ real.}$$

Write the action of $\hat{H}$ on the states $|1\rangle$ and $|2\rangle$.
Show that η has to be real.
Compute the eigenvalues of $\hat{H}$, and the *normalised* eigenvectors.

3.5   A two-state quantum system describes a *qubit* in quantum computing. We shall assume that we have specified a basis in the space of physical states, and consider a qubit whose Hamiltonian is described by the matrix

$$\hat{H} = E_0 \begin{pmatrix} 1 & 0 \\ 0 & -1 \end{pmatrix}.$$

Two observables $\mathcal{A}$ and $\mathcal{B}$ are associated respectively with the Hermitian operators A and B that are represented by the matrices

$$\hat{A} = \begin{pmatrix} 0 & -i \\ i & 0 \end{pmatrix}, \quad \hat{B} = \begin{pmatrix} 2 & -\sqrt{2}i \\ \sqrt{2}i & 1 \end{pmatrix}.$$

Find the eigenvalues and eigenvectors for $\hat{A}$ and $\hat{B}$.
Are $\hat{A}$ and $\hat{B}$ compatible? Do they commute with the Hamiltonian?
Suppose that an observation of $\hat{A}$ has resulted in $A = 1$, what would be the results for $\hat{B}$, and what would be the respective probabilities?
What would be the probability of finding $A = 1$ if a second measurement is made immediately after the first one?

What is the probability of finding $A = 1$ if a measurement of A is made immediately *after* a measurement of B that yielded the larger eigenvalue of B?

3.6 Let us consider three operators $\hat{f}$, $\hat{g}$ and $\hat{h}$. Show that

$$\left[\hat{f},\left[\hat{g},\hat{h}\right]\right] + \left[\hat{g},\left[\hat{h},\hat{f}\right]\right] + \left[\hat{h},\left[\hat{f},\hat{g}\right]\right] = 0.$$

This result is known as the *Jacobi identity*.

3.7 Given the wave function

$$\psi(x) = \left(\frac{\pi}{\alpha^2}\right)^{-1/4} \exp\left(\frac{-\alpha^2 x^2}{2}\right),$$

calculate $\langle x^n \rangle$ and $\Delta x \equiv \sqrt{\langle x^2 \rangle - \langle x \rangle^2}$.

Now calculate the momentum space wave function associated with $\psi(x)$:

$$\tilde{\psi}(p) = \int \frac{dx}{\sqrt{2\pi\hbar}} e^{ipx/\hbar} \psi(x).$$

Using $\tilde{\psi}(p)$, calculate $\langle p^n \rangle$ and $\Delta p \equiv \sqrt{\langle p^2 \rangle - \langle p \rangle^2}$.

With the above results, what do you find for $\Delta x \Delta p$?

3.8 Let us consider two Hermitian operators $\hat{A}$, $\hat{B}$, such that

$$\left[\hat{A},\hat{B}\right] = 0,$$

and let us denote by $|\psi_k\rangle$ the eigenfunctions of $\hat{B}$:

$$\hat{B}|\psi_k\rangle = B_k|\psi_k\rangle.$$

Show that $\hat{A}|\psi_k\rangle$ is an eigenstate of $\hat{B}$ with eigenvalue B_k.

In general, there will be a finite number of eigenstates corresponding to the same eigenvalue B_k. We denote these orthonormal eigenstates by

$$\left\{|\psi_k^{(n)}\rangle, n = 1, 2, \ldots, g_k\right\}.$$

Using the result in (a), deduce that

$$\hat{A}|\psi_k^{(n)}\rangle = \sum_{m=1}^{g_k} A_{nm}|\psi_k^{(m)}\rangle,$$

where A_{nm} are complex numbers.

Show that $A_{nm} = A_{mn}^*$, i.e. that A is a $g_k \times g_k$ Hermitian matrix.

Any finite-dimensional Hermitian matrix can be diagonalised by a unitary transformation U:

$$U^\dagger A U = \begin{pmatrix} A_1 & 0 & \ldots & 0 \\ 0 & A_2 & \ldots & 0 \\ \vdots & \vdots & \ddots & \vdots \\ 0 & 0 & \ldots & A_{g_k} \end{pmatrix}.$$

We can choose the following set of orthonormal linear combinations of $|\psi_k^{(n)}\rangle$, for a given k: $|\phi_k^{(n)}\rangle = \sum_m U_{nm}^\dagger |\psi_k^{(m)}\rangle$.

Show that $\hat{A}|\phi_k^{(n)}\rangle = A_n|\phi_k^{(n)}\rangle$.

We have therefore found a basis of simultaneous eigenstates of $\hat{A}$ and $\hat{B}$:

$$\hat{A}|\phi_k^{(n)}\rangle = A_n|\phi_k^{(n)}\rangle,$$
$$\hat{B}|\phi_k^{(n)}\rangle = B_k|\phi_k^{(n)}\rangle.$$

By repeating the same argument for all eigenvalues B_k, we can explicitly construct a basis of simultaneous eigenvalues of $\hat{A}$ and $\hat{B}$. The two observables are called *compatible*.

3.9 The action of the momentum operator $\hat{P}$ on the wave function of a one-dimensional system is given by

$$\hat{P}\psi(x) = \langle x|\hat{P}|\psi\rangle = -i\hbar\frac{d}{dx}\psi(x).$$

Compute

$$\hat{P}\hat{X}^n\psi(x),$$
$$\hat{X}^n\hat{P}\psi(x).$$

Deduce that

$$\left[\hat{P},\hat{X}^n\right] = -i\hbar n\hat{X}^{n-1}.$$

The operator $\hat{V} = V(\hat{X})$ is defined via the Taylor expansion of the function V:

$$\hat{V} = V(\hat{X}) = \sum_k \frac{1}{k!}V^{(k)}(0)\hat{X}^k,$$

where $V^{(k)}(0)$ are the coefficients of the expansion, i.e. they are numbers computed by evaluating the kth derivative of the function V at $x = 0$. Show that

$$\left[\hat{P},\hat{V}\right] = -i\hbar\frac{d}{dx}V(\hat{X}).$$

3.10 Denoting by $|p\rangle$ the eigenstates of momentum with eigenvalue p, the momentum space wave function can be defined as

$$\tilde{\psi}(p) = \langle p|\psi\rangle$$
$$= \int dx\,\langle p|x\rangle\langle x|\psi\rangle$$
$$= \int dx\,e^{-ipx/\hbar}\psi(x).$$

A generic operator $\hat{O}$ acts on $\psi(p)$ according to

$$\hat{O}\tilde{\psi}(p) = \int dx\,e^{-ipx/\hbar}\hat{O}\psi(x).$$

Find the action of the momentum operator $\hat{P}$, and the position operator $\hat{X}$ on $\tilde{\psi}(p)$. Check that this new representation of $\hat{X}$ and $\hat{P}$ satisfies the canonical commutation relation.

3.11 *Hellmann–Feynman theorem.* Let us consider a system where the potential depends on an external parameter g, $\hat{V} = V(\hat{X}, g)$. Show that for the energy eigenvalues we have

$$\frac{\partial E_n}{\partial g} = \langle \psi_n | \frac{\partial V}{\partial g} | \psi_n \rangle.$$

Remember that the eigenfunctions also depend on g!

This result, known as the Hellmann–Feynman theorem, will be discussed in detail later in the book, in Chapter 17.

3.12 *Generalised uncertainty principle.* If ΔA and ΔB denote the uncertainties in the observables $\mathcal{A}$ and $\mathcal{B}$, respectively, in the state $\Psi(x, t)$, then the generalised uncertainty relation states that

$$\Delta A \, \Delta B \geq \tfrac{1}{2} \, ||\langle [\hat{A}, \hat{B}] \rangle||.$$

In order to prove this relation, consider the operators

$$\hat{X} = \hat{A} - \langle \hat{A} \rangle,$$
$$\hat{Y} = \hat{B} - \langle \hat{B} \rangle.$$

The uncertainties in the observables are given by

$$(\Delta A)^2 = \langle \hat{X}^2 \rangle,$$
$$(\Delta B)^2 = \langle \hat{Y}^2 \rangle.$$

For any real number λ, we can construct the state

$$|\phi\rangle = \hat{X}|\psi\rangle + i\lambda \hat{Y}|\psi\rangle.$$

The norm of the state $|\phi\rangle$ is positive by definition:

$$||\phi||^2 = \langle \phi | \phi \rangle.$$

Use this fact to prove the generalised uncertainty relation.

4 Dynamics

Having described the states and the observables of a quantum system, we shall now introduce the rules that determine their time evolution.

In classical mechanics, the evolution in time of the state of a single-particle system is described by the *classical trajectory*, i.e. by specifying the position of the particle as a function of time: $x(t)$.[1] The time dependence of the position is described by Newton's law:

$$\underline{f} = m\underline{a}, \tag{4.1}$$

where $\underline{a}(t) = d^2\underline{x}(t)/dt^2$ is the acceleration of the system, and $\underline{f}$ is the external force acting on it. Let us examine this equation and extract its physical content. On the left-hand side of Eq. (4.1), the acceleration measures the rate at which the velocity of the system is changing, i.e. it describes the change of state of the system. The right-hand side of the equation relates the change of state of the system to the forces acting on it. For a classical point-like particle, the state of the system is defined by a real vector $\underline{x}(t)$. Its change as a function of time is described by a second-order differential equation, which can be solved (at least in principle) once we know the initial conditions for both the position and the velocity of the particle at some given time t_0. Newton's law can be rewritten as a set of first-order differential equations involving the position and momentum of the particle:

$$\frac{d}{dt}x_k = \frac{\partial H}{\partial p_k}, \quad \frac{d}{dt}p_k = -\frac{\partial H}{\partial x_k}, \tag{4.2}$$

where H is the Hamiltonian of the system. These can be conveniently rewritten as

$$\frac{d}{dt}x_k = \{x_k, H\}, \tag{4.3}$$

$$\frac{d}{dt}p_k = \{p_k, H\}, \tag{4.4}$$

where we have introduced the Poisson bracket

$$\{f, g\} = \sum_i \left(\frac{\partial f}{\partial x_i} \frac{\partial g}{\partial p_i} - \frac{\partial f}{\partial p_i} \frac{\partial g}{\partial x_i} \right). \tag{4.5}$$

The six-dimensional vector $(\underline{x}, \underline{p})$ represents the state of the classical system, while Eqs. (4.3) and (4.4) state that the rate of variation of a classical state vector in an infinitesimal amount of time is obtained by taking the Poisson bracket of the state vector with the Hamiltonian of the system.

---

[1] If the particle evolves in more than one space dimension, then the position is identified by a vector $\underline{x}(t)$.

As discussed in the previous chapter, in quantum mechanics the concept of classical trajectory cannot be defined, and we cannot access simultaneously the position and the momentum of a quantum system. Instead, the state of the system is described by a vector in a complex Hilbert space. In this chapter we describe the quantum analogue of Newton's law, i.e. the equation that describes the time evolution of the state vector. Following the classical mechanics example, we will seek a dynamical equation that relates the variation of the state vector with respect to time on the left-hand side to some quantity that takes into account the dynamics of the quantum system under study on the right-hand side of the equation. As we will see, the Hamiltonian of the quantum system keeps playing an important role.

4.1 Schrödinger Equation

4.1.1 Equation of Motion

The sixth postulate dictates the dynamics of a quantum system. The time evolution of a quantum state is given by **Schrödinger's equation**.

Postulate 6

$$i\hbar \frac{d}{dt}|\Psi(t)\rangle = \hat{H}|\Psi(t)\rangle, \tag{4.6}$$

where

$$\hat{H} = \hat{T} + \hat{V} \tag{4.7}$$

is the Hamiltonian operator, i.e. the (Hermitian) operator that is associated with the energy of the system. The Hamiltonian dictates the change in time of the system, its specific form depends on the system under study. Note that in Eq. (4.6) we have explicitly indicated the time dependence of the state vector inside the ket.

In the expression above, $\hat{T}$ and $\hat{V}$ represent the operators associated with the kinetic and potential energy, respectively. Equation (4.6) replaces Newton's equation of classical mechanics.[2] The operator $\hat{H}$ is obtained from the classical Hamiltonian, by replacing the position and momentum with the corresponding operators $\hat{X}$ and $\hat{P}$ in Cartesian coordinates. In Eq. (4.6) we have indicated explicitly the time dependence of the state vector inside the ket. At times, we will refer to this equation as the **time-dependent Schrödinger equation** (TDSE).

Schrödinger's equation states that acting with the Hamiltonian operator on a state vector determines its variation in time. If we consider an infinitesimal time variation ϵ, we can cast Schrödinger's equation as

$$|\Psi(t+\epsilon)\rangle = |\Psi(t)\rangle - \frac{i\epsilon}{\hbar}\hat{H}|\Psi(t)\rangle + O(\epsilon^2), \tag{4.8}$$

---

[2] It was introduced by the Austrian physicist Erwin Schrödinger in 1926.

which shows explicitly the change of $|\Psi(t)\rangle$ as a result of the action of $\hat{H}$ on $|\Psi(t)\rangle$. We can rewrite the equation above as

$$\delta_\epsilon \Psi(t) = \Psi(t + \epsilon) - \Psi(t)$$

$$= -i\epsilon \left(\frac{\hat{H}}{\hbar} \right) \Psi(t) + O(\epsilon^2). \tag{4.9}$$

Comparing with Eq. (3.100), you see that (minus) the Hamiltonian is the generator of time translation, a concept that you may have encountered already in Hamiltonian dynamics courses.

Time Evolution of the Wave Function

For a one-dimensional system, the wave function is obtained by taking the scalar product of the state vector with eigenstates of position $|x\rangle$, as discussed in Section 2.3:

$$\Psi(x, t) = \langle x | \Psi(t) \rangle. \tag{4.10}$$

Note that the time dependence in the wave function comes from the time dependence of the state vector, while the dependence on the spatial coordinate x is the result of the scalar product of the state vector with eigenstates of the position.

Taking the scalar product of Eq. (4.6) with an eigenstate of position yields

$$\langle x | \left(i\hbar \frac{d}{dt} \right) | \Psi(t) \rangle = i\hbar \frac{\partial}{\partial t} \Psi(x, t), \tag{4.11}$$

$$= \langle x | \hat{H} | \Psi(t) \rangle, \tag{4.12}$$

$$= \hat{H} \Psi(x, t), \tag{4.13}$$

where in the last line we should consider $\hat{H}$ as a differential operator acting on the wave function $\Psi(x, t)$.

A Digression on Schrödinger's Equation

Let us discuss briefly a heuristic explanation of Schrödinger's equation by considering the propagation of a one-dimensional plane wave:

$$\Psi(x, t) = \text{const} \times e^{-i(\omega(k)t - kx)}, \tag{4.14}$$

with

$$\omega(k) = \frac{\hbar k^2}{2m}. \tag{4.15}$$

The variation in time of Ψ is described by the action of the time derivative on the function. Applying $i\hbar \frac{\partial}{\partial t}$ to Ψ we obtain:

$$i\hbar \frac{\partial}{\partial t} \Psi(x, t) = \hbar \omega \Psi(x, t) = h\nu \Psi(x, t) \quad (\omega = 2\pi\nu), \tag{4.16}$$

where $\hbar\omega$ is precisely the energy of a free particle with momentum $\hbar k$ and mass m, thereby implementing the **de Broglie hypothesis**, viz. that one can associate a wave with wavelength

$$\lambda = \frac{2\pi}{k} = \frac{h}{p} \tag{4.17}$$

with a particle with momentum p.

Moreover, we know that probabilities must be normalised to one. The mathematical translation of this property is that

$$\int dx\, \Psi(x,t)^* \Psi(x,t) = 1, \quad \text{for all } t. \tag{4.18}$$

Hence the derivative with respect to time of the integral above must vanish:

$$\int dx\left[\left(\frac{\partial}{\partial t}\Psi(x,t)^*\right)\Psi(x,t) + \Psi(x,t)^*\left(\frac{\partial}{\partial t}\Psi(x,t)\right)\right] = 0; \tag{4.19}$$

the latter equation shows that $\frac{\partial}{\partial t}$ must be represented by an anti-Hermitian operator. The Hamiltonian $\hat{H}$ is the operator associated with the energy of the system and therefore is Hermitian. This explains the factor of i that appears in Schrödinger's equation; it is necessary to connect $\frac{\partial}{\partial t}$ to a Hermitian operator. Finally, the factor $\hbar$ is needed on dimensional grounds. Other numerical factors (e.g. factors of 2, π, etc.) cannot be derived by these simple arguments, but are explained by the heuristic derivation above.

4.2 Eigenstates of the Hamiltonian

The eigenvalue equation for a given Hamiltonian $\hat{H}$ is

$$\hat{H}|\psi_n\rangle = E_n|\psi_n\rangle. \tag{4.20}$$

As usual, solving the eigenvalue equation means finding the eigenvalues E_n and the eigenvectors $|\psi_n\rangle$. If the Hamiltonian is time-independent, both the eigenvalues and the eigenstates are also independent of time. The eigenvalue equation for $\hat{H}$ is called the **time-independent Schrödinger equation** (TISE). Note that we are assuming here that the spectrum of $\hat{H}$ is discrete. If the spectrum is continuous, the eigenvalues E_n, indexed by the integer n, are replaced by a continuous variable E, and the sums over n are replaced by integrals over E as discussed in the previous chapter.

As for all other observables, the eigenvalues E_n of the operator $\hat{H}$ yield the possible values that can be obtained when measuring the energy of the system. The eigenstates $|\psi_n\rangle$ are states in which a measurement of the energy would return the value E_n with probability 1.

Time Evolution of Energy Eigenstates

The vector

$$|\Psi_n(t)\rangle = \exp\left(-\frac{i}{\hbar}E_n t\right)|\psi_n\rangle \tag{4.21}$$

is the unique solution of Schrödinger's equation with boundary condition $|\Psi_n(0)\rangle = |\psi_n\rangle$. This result can be shown by computing the left-hand side and right-hand side of Schrödinger's equation:

$$i\hbar\frac{d}{dt}|\Psi_n(t)\rangle = i\hbar\frac{d}{dt}\exp\left(-\frac{i}{\hbar}E_n t\right)|\psi_n\rangle = E_n|\Psi_n(t)\rangle, \tag{4.22}$$

$$\hat{H}|\Psi_n(t)\rangle = \exp\left(-\frac{i}{\hbar}E_n t\right)\hat{H}|\psi_n\rangle = E_n|\Psi_n(t)\rangle, \tag{4.23}$$

where in the second line we used the fact that $|\psi_n\rangle$ is an eigenstate of $\hat{H}$; this is clearly not true for a generic state. The boundary condition being trivially satisfied, we can conclude that $|\Psi_n(t)\rangle$ is indeed the unique solution mentioned above.

We can summarise the result above by saying that for a system that has been prepared in an eigenstate of the energy $|\psi_n\rangle$, the time evolution is simply given by the multiplication by a phase factor $e^{-iE_n t/\hbar}$. For this reason the energy eigenstates are also called **stationary states**.

Expectation Value of a Generic Observable

Let us consider a generic observable, with its associated operator $\hat{O}$. The expectation value of $\hat{O}$ in the state $|\Psi_n(t)\rangle$ is given by

$$O_n(t) = \langle\Psi_n(t)|\hat{O}|\Psi_n(t)\rangle, \tag{4.24}$$

$$= \langle\psi_n|e^{iE_n t/\hbar}\hat{O}e^{-iE_n t/\hbar}|\psi_n\rangle, \tag{4.25}$$

$$= \langle\psi_n|\hat{O}|\psi_n\rangle = O_n(0), \tag{4.26}$$

where in going from the first to the second line we used the explicit expression for the time evolution of $|\Psi_n(t)\rangle$, and then used the fact that $\hat{O}$ is a linear operator, and hence

$$\hat{O}\left(e^{-iE_n t/\hbar}\right)|\psi_n\rangle = \left(e^{-iE_n t/\hbar}\right)\hat{O}|\psi_n\rangle.$$

As expected, the expectation value of the observable in an eigenstate of the energy is independent of time.

Example 4.1 Consider a one-dimensional system, and take the expectation value of the projector onto an eigenstate of the position, i.e. consider the case where $\hat{O} = |x\rangle\langle x|$. As discussed above, the expectation value

$$\langle\Psi_n(t)|\hat{O}|\Psi_n(t)\rangle = \langle\Psi_n(t)|x\rangle\langle x|\Psi_n(t)\rangle = |\Psi_n(x,t)|^2$$

is independent of time. The physical interpretation of this statement is that the probability density of finding the system at position x does not change in time, if the system is in a stationary state.

4.3 Evolution of a Generic State

We shall start by writing a formal solution to Schrödinger's equation, and then discuss a practical way to find the time evolution of a generic state.

Time Evolution Operator

Let us define the following operator:

$$\hat{U}(t) = \exp\left[-\frac{i}{\hbar}\hat{H}t\right] = \sum_{k=0}^{\infty} \frac{1}{k!}\left(\frac{-it}{\hbar}\right)^k \hat{H}^k. \tag{4.27}$$

It is easy to verify that

$$i\hbar\frac{d}{dt}\hat{U}(t) = \hat{H}\,\hat{U}(t), \tag{4.28}$$

and therefore

$$|\Psi(t)\rangle = \hat{U}(t)|\phi\rangle \tag{4.29}$$

is the solution of Schrödinger's equation with boundary condition

$$|\Psi(0)\rangle = |\phi\rangle. \tag{4.30}$$

Even though Eq. (4.29) yields a formal solution of the Schrödinger equation, applying the time evolution operator and performing explicit calculations looks like a complicated task in general. However, a simple procedure can be established by working in a basis made of stationary states.

Expansion in Eigenstates of the Hamiltonian

Because the Hamiltonian is a Hermitian operator, we know that its eigenstates $\{|\psi_n\rangle\}$ form a basis of the vector space of physical states $\mathcal{H}$. Note that, for a time-independent Hamiltonian, the eigenvalues and eigenstates are static quantities, they characterise the Hamiltonian and are independent of time. A generic state, which evolves in time, can be expanded as a linear combination of stationary states:

$$|\Psi(t)\rangle = \sum_{n} c_n(t)|\psi_n\rangle, \tag{4.31}$$

where the coordinates $c_n(t)$ are computed according to the usual formula

$$c_n(t) = \langle\psi_n|\Psi(t)\rangle. \tag{4.32}$$

Because the state vector depends on time, the coordinates will also depend on time, and we have indicated this dependence explicitly. As time flows, the state of the system changes, and therefore the coordinates $c_n(t)$ change, while the basis vectors stay the same.

The state of a system at $t = 0$, $|\Psi(0)\rangle$, can be expanded as above:

$$|\Psi(0)\rangle = \sum_n c_n(0)|\psi_n\rangle, \tag{4.33}$$

and the action of the time evolution operator can readily be computed, which yields

$$\hat{U}(t)|\Psi(0)\rangle = \sum_n c_n(0)\hat{U}(t)|\psi_n\rangle, \tag{4.34}$$

$$= \sum_n c_n(0)e^{-i\frac{E_n t}{\hbar}}|\psi_n\rangle. \tag{4.35}$$

The evolution of the coordinates $c_n(t)$ is simple, and can be read from Eq. (4.35):

$$c_n(t) = c_n(0)e^{-i\frac{E_n t}{\hbar}}. \tag{4.36}$$

Each coefficient picks up a time-dependent phase. However, the phase depends on the energy value, and therefore the relative phase between the eigenstates changes as the system evolves.

Evolution of Expectation Values

Let us conclude this section by looking at the evolution of the expectation value of an observable O in a generic state $|\Psi(t)\rangle$. The evolution of the expectation value

$$O(t) = \langle\Psi(t)|\hat{O}|\Psi(t)\rangle \tag{4.37}$$

is obtained by considering the time derivative

$$\frac{d}{dt}O(t) = \left(\frac{d}{dt}\langle\Psi(t)|\right)\hat{O}|\Psi(t)\rangle + \langle\Psi(t)|\hat{O}\left(\frac{d}{dt}|\Psi(t)\rangle\right), \tag{4.38}$$

$$= -\frac{1}{i\hbar}\langle\Psi(t)|\hat{H}\hat{O}|\Psi(t)\rangle + \frac{1}{i\hbar}\langle\Psi(t)|\hat{O}\hat{H}|\Psi(t)\rangle, \tag{4.39}$$

$$= \frac{i}{\hbar}\langle\Psi(t)|\left[\hat{H},\hat{O}\right]|\Psi(t)\rangle. \tag{4.40}$$

As a consequence, the expectation value of O is conserved if and only if $\left[\hat{H},\hat{O}\right] = 0$.

Exercise 4.3.1 Check that you can reproduce all the steps leading to Eq. (4.40).

4.4 One-Dimensional System

Let us now consider again a one-dimensional system described by a wave function in position space $\psi(x) = \langle x|\psi\rangle$. The eigenvalue equation for the Hamiltonian can now be rewritten as a differential equation

$$\hat{H}\psi_n(x) = E_n\psi_n(x), \tag{4.41}$$

which yields the eigenvalues E_n, i.e. all possible values of the energy of the system, and the wave functions of the stationary states $\psi_n(x)$. Using the explicit definition of $\hat{H}$ in Eq. (3.15), and the definition of $\hat{P}$ in Eq. (3.85), we can write the time-independent Schrödinger equation as a second-order differential equation:

$$\left[-\frac{\hbar^2}{2m}\frac{d^2}{dx^2} + V(x)\right]\psi_n(x) = E_n\psi_n(x). \tag{4.42}$$

Eigenvalues and eigenstates are found by determining the normalisable solutions of the differential equation, taking into account the appropriate boundary conditions.[3]

Schrödinger's equation for the wave function of a one-dimensional system can be written as

$$i\hbar\frac{\partial}{\partial t}\Psi(x,t) = \left[-\frac{\hbar^2}{2m}\frac{\partial^2}{\partial x^2} + V(x)\right]\Psi(x,t). \tag{4.43}$$

The Free Particle

The simplest example of a one-dimensional system is the free particle, i.e. $V(x) = 0$. In this case the Hamiltonian is

$$\hat{H} = \frac{\hat{P}^2}{2m}, \tag{4.44}$$

and the TISE is the Helmholtz differential equation

$$\frac{-\hbar^2}{2m}\frac{d^2}{dx^2}\psi(x) = E\psi(x), \tag{4.45}$$

which we can rewrite as

$$\frac{d^2}{dx^2}\psi(x) = -k^2\psi(x), \tag{4.46}$$

where $k^2 = 2mE/\hbar^2$. A quick calculation shows that k has dimensions of inverse length. Equation (4.46) has *two* linearly independent solutions for each value of E. The free particle has a continuous spectrum, with eigenfunctions being labelled by the real positive number k:

$$\psi_{k,+}(x) = e^{ikx} \quad \text{and} \quad \psi_{k,-}(x) = e^{-ikx}. \tag{4.47}$$

Both solutions correspond to the same eigenvalue of the energy $E = \hbar^2 k^2/(2m)$, and therefore the energy eigenvalues are twofold degenerate. Going back to our previous discussion of complete sets of commuting observables, we see that in this case the energy alone does not provide a complete set, as its eigenvalues do not identify the eigenfunctions uniquely.

---

[3] The normalisability of the solutions will play an important role in several examples that we discuss in later chapters.

It is interesting to note that $\psi_{k,+}$ and $\psi_{k,-}$ are also eigenfunctions of $\hat{P}$ with respective eigenvalues $p = \hbar k$ and $-\hbar k$. The eigenvalues of $\hat{P}$ allow us to distinguish between the degenerate eigenstates, and hence the eigenvalues E and $\hbar k$ are sufficient to identify an eigenstate uniquely. We can write e.g.

$$\psi_{k,+}(x) = \langle x|E, \hbar k\rangle \quad \text{and} \quad \psi_{k,-}(x) = \langle x|E, -\hbar k\rangle. \tag{4.48}$$

Clearly $[\hat{H}, \hat{P}] = 0$, and hence the set $\{\hat{H}, \hat{P}\}$ constitutes a CSCO.

Parity Transformations

We define the **parity operator** $\hat{P}$ by specifying its action on a generic wave function:

$$\hat{P}\psi(x) = \psi(-x). \tag{4.49}$$

Exercise 4.4.1   Show that the operator $\hat{P}$ is Hermitian.

The two solutions above are interpreted as right movers and left movers, respectively. A parity transformation exchanges the two solutions as expected:

$$\hat{P}\psi_{k,+}(x) = \psi_{k,+}(-x) = \psi_{k,-}(x), \tag{4.50}$$

where the first equality stems from the definition of the parity operator itself.

Note that a more intuitive definition of the parity operator can be obtained as follows. Since the parity operator implements a reflection around the origin of the real axis $x = 0$, its action on the eigenstate of position must be

$$\hat{P}|x\rangle = |-x\rangle. \tag{4.51}$$

From Eq. (4.51) it follows that

$$\hat{P}^2 = 1, \tag{4.52}$$

and therefore its eigenvalues are ± 1, and the operator is Hermitian. Then we have

$$\langle x|\hat{P}|\psi\rangle = \left(\langle\psi|\hat{P}|x\rangle\right)^*, \tag{4.53}$$

$$= \left(\langle\psi|-x\rangle\right)^*, \tag{4.54}$$

$$= \langle -x|\psi\rangle, \tag{4.55}$$

which is exactly Eq. (4.49).

Wave Packets Revisited

If the system is prepared in the state $\psi_{k,+}$ at time $t = 0$, then the time evolution results in a plane wave propagating with momentum $p = \hbar k$:

$$\Psi(x, 0) = \psi_{k,+}(x) \quad \Longrightarrow \quad \Psi(x, t) = e^{\frac{i}{\hbar}(px - Et)}, \tag{4.56}$$

where $E = p^2/(2m)$ as above. The solution corresponding to a plane wave is not normalisable though. Physical states are made up of superpositions of plane waves, called wave packets, as seen in Section 3.10.3.

Let us consider a quantum state at time $t = 0$ such that

$$|W(0)\rangle = |W\rangle = \int dp\, g(p)\, |p\rangle, \tag{4.57}$$

with

$$g(p) = \frac{1}{(2\pi\sigma^2)^{1/4}} \exp\left[-\frac{1}{4\sigma^2}(p - p_0)^2\right]. \tag{4.58}$$

According to Eq. (4.35), the state vector at time t is

$$|W(t)\rangle = \int dp\, g(p)\, e^{-iE_p t/\hbar}\, |p\rangle, \tag{4.59}$$

where, for a free particle:

$$E_p = p^2/(2m). \tag{4.60}$$

Using the explicit expression for $g(p)$, Eq. (4.58), we can rewrite the terms in the argument of the exponential:

$$-\frac{1}{4\sigma^2}(p - p_0)^2 - i\frac{t\,p^2}{\hbar\,2m} = -\left(\frac{1}{4\sigma^2} + \frac{it}{2m\hbar}\right)(p - p_0)^2 - \frac{it p_0}{\hbar m}p + \frac{it\,p_0^2}{\hbar\,2m}. \tag{4.61}$$

Following the reasoning that led to Eq. (3.113), the wave function in position space at time t is

$$\langle x|W(t)\rangle = e^{itp_0^2/(\hbar 2m)} \int dp \frac{1}{(2\pi\sigma^2)^{1/4}} \exp\left[-\frac{1}{4\tilde{\sigma}^2}(p - p_0)^2\right] \frac{1}{(2\pi\hbar)^{1/2}} \exp\left[i\frac{p\tilde{x}}{\hbar}\right], \tag{4.62}$$

where

$$\frac{1}{4\tilde{\sigma}^2} = \frac{1}{4\sigma^2} + \frac{it}{2m\hbar}, \quad \tilde{x} = x - \frac{tp_0}{m}. \tag{4.63}$$

Note that $\tilde{\sigma}^2$ is now a complex number. The integral over p can be computed as before by completing the square in the exponential, shifting the integration variable and performing the Gaussian integral. The result reads

$$\psi_W(x, t) = \langle x|W(t)\rangle = K e^{-itp_0^2/(\hbar 2m)} \exp\left[ip_0 x/\hbar\right] \exp\left[-\frac{\tilde{\sigma}^2\tilde{x}^2}{\hbar^2}\right]. \tag{4.64}$$

The prefactor K can conveniently be rewritten as

$$K = \left(\frac{\tilde{\sigma}^2 + \tilde{\sigma}^{2*}}{\pi\hbar^2}\right)^{1/4} \left(\frac{\tilde{\sigma}^2}{\tilde{\sigma}^{2*}}\right)^{1/4}, \tag{4.65}$$

$$= \left(\frac{2|\tilde{\sigma}^2|^2}{\pi\sigma^2\hbar^2}\right)^{1/4} e^{i\varphi}, \tag{4.66}$$

where in the second line we have emphasised that the first factor is a real number and the second one is pure complex phase.

The probability density of finding the system at position x and at time t is given by

$$\left|\psi_W(x,t)\right|^2 \propto \exp\left[-\frac{2\left|\tilde{\sigma}^2\right|^2}{\hbar^2\sigma^2}\tilde{x}^2\right] ; \qquad (4.67)$$

this is a Gaussian distribution centred at tp_0/m with width $\hbar^2\sigma^2/\left|\tilde{\sigma}^2\right|^2$. We see that the centre of the distribution, i.e. the average value of x, moves with velocity p_0/m, while the width of the distribution increases as $|t|$ increases. The latter phenomenon is known as **spreading** of the wave packet.

4.5 Some Properties of One-Dimensional Potentials

4.5.1 General Restrictions on $\Psi(x,t)$

We see from the form of Eq. (4.43) that its solutions must satisfy the following properties:

1. $\Psi(x,t)$ must be a single-valued function of x and t.
2. $\Psi(x,t)$ must be a continuous function of x and t.
3. $\frac{\partial\Psi}{\partial x}(x,t)$ must be a continuous function of x.

There is an exception to this set of rules.

Restriction 3 does not apply when the potential energy function $V(x)$ has infinite discontinuities, e.g. the 'particle-in-a-box' problem, for which

$$V(x) = \begin{cases} \infty & \text{if } x > a \text{ or } x < 0, \\ 0 & \text{otherwise.} \end{cases}$$

We shall see later a detailed description of this system.

4.5.2 Energy Expectation Value

Let $|\psi\rangle$ be a quantum state of a one-dimensional system, whose dynamics is specified by the potential V, then we have

$$\langle\psi|\hat{H}|\psi\rangle = \int dx\, \psi(x)^*\hat{H}\psi(x), \qquad (4.68)$$

$$= \int dx\, \psi(x)^* \left\{-\frac{\hbar^2}{2m}\frac{d^2}{dx^2} + V(x)\right\}\psi(x), \qquad (4.69)$$

$$= \int dx\, \left\{\frac{\hbar^2}{2m}\left|\psi'(x)\right|^2 + V(x)\left|\psi(x)\right|^2\right\}. \qquad (4.70)$$

The first term in the integral above is clearly positive, and is the expectation value of the kinetic energy. Altogether we have

$$\langle \psi | \hat{H} | \psi \rangle \geq V_{\min} \int dx \, |\psi(x)|^2 = V_{\min}. \tag{4.71}$$

Hence we have shown that the expectation value of the energy is always larger than the minimum of the potential, for *any* state $|\psi\rangle$. In particular, this implies that all the eigenvalues of the Hamiltonian are larger than $V_{\min}$.

4.5.3 Ehrenfest's Theorem

Let us consider a generic state evolving in time, $|\Psi(t)\rangle$. We denote the expectation value at time t as

$$\langle m\hat{X} \rangle_t = \langle \Psi(t) | m\hat{X} | \Psi(t) \rangle, \tag{4.72}$$

and hence

$$\frac{d}{dt} \langle m\hat{X} \rangle_t = m \frac{1}{i\hbar} \langle [\hat{X}, \hat{H}] \rangle_t, \tag{4.73}$$

$$= \frac{m}{i\hbar} \left(\langle \left[\hat{X}, \frac{\hat{P}^2}{2m} \right] \rangle_t + \langle [\hat{X}, V(\hat{X})] \rangle_t \right), \tag{4.74}$$

$$= \langle \hat{P} \rangle_t. \tag{4.75}$$

This result is known as **Ehrenfest's theorem**: the expectation values satisfy the *classical* relation between the derivative of the position and the momentum.

Exercise 4.5.1   Show that we also have

$$\frac{d}{dt} \langle \hat{P} \rangle_t = -\langle \frac{d}{dx} V(\hat{X}) \rangle_t.$$

4.5.4 Degeneracy

The discrete energy levels of a one-dimensional potential are *not* degenerate.

Proof. Let us assume the contrary, namely that there is an energy value E for which we have two degenerate eigenstates ψ_1 and ψ_2. The two eigenfunctions satisfy the same equation:

$$\psi_1''(x) = -\frac{2m}{\hbar^2} (E - V(x)) \, \psi_1(x), \tag{4.76}$$

$$\psi_2''(x) = -\frac{2m}{\hbar^2} (E - V(x)) \, \psi_2(x). \tag{4.77}$$

Multiplying the first equation by ψ_2 and the second one by ψ_1, and subtracting them, we obtain

$$\psi_2(x)\psi_1''(x) - \psi_1(x)\psi_2''(x) = 0. \tag{4.78}$$

Integrating with respect to x yields

$$\psi_1'(x)\psi_2(x) - \psi_2'(x)\psi_1(x) = A, \tag{4.79}$$

where A is a constant independent of x. Since the eigenstates must be normalisable, we deduce that $\lim_{x\to\pm\infty}\psi_1(x) = \lim_{x\to\pm\infty}\psi_2(x) = 0$, and hence $A = 0$, i.e.

$$\psi_1'(x)\psi_2(x) - \psi_2'(x)\psi_1(x) = 0. \tag{4.80}$$

Integrating again with respect to x:

$$\log\psi_1(x) = \log\psi_2(x) + \text{const}, \tag{4.81}$$

i.e. ψ_1 and ψ_2 are proportional and therefore describe the *same* eigenstate, which contradicts the initial statement.

4.5.5 Nodes

The wave function of the nth level has $(n-1)$ zeros, and therefore $\psi_0(x)$ has no nodes.

4.6 Probability Current

The probability of finding the system at time t in an interval $[a, b]$ is given by the integral

$$P_{[a,b]}(t) = \int_a^b dx \, |\Psi(x,t)|^2. \tag{4.82}$$

Its derivative with respect to time is given by

$$\frac{d}{dt}P_{[a,b]}(t) = \int_a^b dx \left[\left(\frac{\partial}{\partial t}\Psi(x,t)^*\right)\Psi(x,t) + \Psi(x,t)^*\left(\frac{\partial}{\partial t}\Psi(x,t)^*\right)\right]. \tag{4.83}$$

Using Schrödinger's equation for the time evolution of the wave function, and writing the Hamiltonian as a differential operator acting in position space, we find

$$\frac{d}{dt}P_{[a,b]}(t) = \frac{1}{i\hbar}\int_a^b dx \, \frac{\partial}{\partial x}\left[\left(\frac{\hbar^2}{2m}\frac{\partial}{\partial x}\Psi(x,t)^*\right)\Psi(x,t) + \Psi(x,t)^*\left(-\frac{\hbar^2}{2m}\frac{\partial}{\partial x}\Psi(x,t)\right)\right]$$

$$= -\int_a^b dx \, \frac{\partial}{\partial x}j(x,t), \tag{4.84}$$

$$= j(a,t) - j(b,t), \tag{4.85}$$

where we have introduced the **probability current**

$$j(x,t) = \frac{i\hbar}{2m}\left[\left(\frac{\partial}{\partial x}\Psi(x,t)^*\right)\Psi(x,t) - \Psi(x,t)^*\left(\frac{\partial}{\partial x}\Psi(x,t)\right)\right]. \tag{4.86}$$

Equation (4.85) allows us to interpret $j(x,t)$ as the probability flux at time t and position x per unit time. Furthermore, Eq. (4.84) leads to the continuity equation

$$\frac{\partial}{\partial t} |\Psi(x,t)|^2 + \frac{\partial}{\partial x} j(x,t) = 0. \tag{4.87}$$

Example 4.2 Consider the time evolution of a free plane wave normalised as in Eq. (3.88):

$$\Psi(x,t) = \frac{1}{\sqrt{2\pi\hbar}} \exp\left[\frac{i}{\hbar}(px - Et)\right], \tag{4.88}$$

with $E = p^2/(2m)$. The corresponding probability current is

$$j(x,t) = \left(\frac{1}{2\pi\hbar}\right) v, \tag{4.89}$$

where $v = p/m$ is the classical velocity of the particle. Note that this is a constant that does not depend either on t or on x. We will generalise this result in one of the problems at the end of the chapter. In this particular case the probability flux at x has a simple physical interpretation. It is given by the probability density of finding the particle at x – namely $1/(2\pi\hbar)$ – multiplied by the classical velocity v.

Example 4.3 Consider now a free particle in a state that is a superposition of eigenstates of momentum with eigenvalues $\pm\hbar k$:

$$\psi(x) = Ae^{ikx} + Be^{-ikx}. \tag{4.90}$$

We have suppressed the time evolution in the equation above since it is just a trivial phase factor that does not enter into the expression for the current. We invite the reader to verify that the probability current in this case is

$$j(x) = \frac{\hbar k}{m}\left[|A|^2 - |B|^2\right]. \tag{4.91}$$

The cross-terms cancel and the flux is given by the probability density of particles moving to the right multiplied by their classical velocity *minus* the probability density of particles moving to the left multiplied by their classical velocity. Note that the overall normalisation is not specified here. What really matters is the ratio $|A|^2/|B|^2$.

Summary

- The time evolution of the state vector is determined by the Schrödinger equation; the change of the state vector in an infinitesimal time interval is given by the action of the Hamiltonian operator on the state. (4.1)

- The eigenvalue problem for the Hamiltonian operator, which is also known as the time-independent Schrödinger equation, yields the possible outcomes of a measurement of the energy of the system; the corresponding eigenstates are known as stationary states. (4.2)
- Physical characterisation of the stationary states. (4.2)
- The evolution of a state over a finite time is obtained by acting on the state with the time evolution operator. The stationary states yield a convenient basis to express the time evolution operator and compute the time evolution of a generic state. (4.3)
- General properties of the solutions of Schrödinger's equation for the wave function of a one-dimensional system. (4.4, 4.5)
- Definition of the probability current and its physical interpretation. (4.6)

Problems

4.1   Consider an operator $\hat{f}$ such that

$$\left[\hat{f}, \hat{H}\right] = 0.$$

Show that the expectation value $\langle\Psi(t)|\hat{f}|\Psi(t)\rangle$ is a constant for any state $|\Psi(t)\rangle$.

Note that the operator does not have an explicit dependence on time, and therefore the time dependence is entirely due to the fact that the state vector evolves according to the Schrödinger equation.

4.2   Let us consider an operator $\hat{O}(t)$ corresponding to some observable, where we have considered now the possibility of having a parametric dependence on time t. The expectation value of $\hat{O}(t)$ will evolve with time:

$$\langle\hat{O}(t)\rangle = \langle\Psi(t)|\hat{O}(t)|\Psi(t)\rangle.$$

The time evolution is due to *both* the parametric dependence on t, and the fact that the wave function evolves in time.

Show that the expectation value in Eq. (4.2) evolves according to

$$\frac{d}{dt}\langle\hat{O}(t)\rangle = \langle\Psi(t)|\left(\frac{d}{dt}\hat{O}(t) + \frac{1}{i\hbar}\left[\hat{O}, \hat{H}\right]\right)|\Psi(t)\rangle.$$

4.3   The time evolution operator $\hat{U}(t) = \exp\left[-i\hat{H}t/\hbar\right]$ is defined via the Taylor expansion of the exponential:

$$\hat{U}(t) = \sum_{n=0}^{\infty} \frac{1}{n!}\left(\frac{-i}{\hbar}\right)^n \hat{H}^n t^n.$$

If $|\psi\rangle$ is an eigenstate of $\hat{H}$ with eigenvalue E, show that

$$\hat{U}(t)|\psi\rangle = e^{-iEt/\hbar}|\psi\rangle.$$

4.4   Consider the two-state system described in Problem 3.4. At $t = 0$ the system is in the
state

$$|\psi(0)\rangle = \frac{1}{\sqrt{2}}|1\rangle + \frac{i}{\sqrt{2}}|2\rangle.$$

Let us assume that the states $|1\rangle$ and $|2\rangle$ are eigenstates of some operator $\hat{O}$ associated
with an observable O, with respective eigenvalues O_1 and O_2. Determine the
probability $P_1(t)$ that a measurement of observable O at time t yields O_1. Make a
sketch of the function $P_1(t)$.

4.5   Consider a particle in an infinite potential well:

$$V(x) = \begin{cases} 0, & \text{for } |x| < a, \\ \infty, & \text{otherwise.} \end{cases}$$

Physically this potential confines the particle to the region $|x| < a$, and therefore
its wave function must vanish identically outside this region. Hence the solution of
Schrödinger's equation must satisfy the boundary condition

$$\psi(-a) = \psi(a) = 0.$$

Verify that the wave function

$$\psi(x,t) = \begin{cases} A \sin \left[\frac{\pi x}{a} \right] e^{-iEt/\hbar}, & \text{if } -a < x < a, \\ 0, & \text{if } |x| > a, \end{cases}$$

is a solution to Schrödinger's equation. Calculate the energy of this state and the
probability density $P(x)$ to find the particle at a given x. Does $P(x)$ differ from the
corresponding result in classical mechanics?

4.6   A box containing a particle is divided into a left and a right compartment by a thin
partition (see Fig. 4.1). Suppose that the amplitude for the particle being on the left
side of the box is $\psi_1(t)$ and the amplitude for being on the right side of the box is

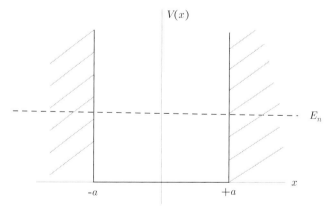

Infinite well potential of size $2a$, as described in Problem 4.6

$\psi_2(t)$. We will neglect the spatial dependence of the wave functions inside the two halves of the box. Suppose that the particle can tunnel through the partition, and that the rate of change of the amplitude on the right is $K/(i\hbar)$ times the amplitude on the left, where K is real:

$$i\hbar\frac{d}{dt}\psi_2(t) = K\psi_1(t).$$

What is the equation of motion for ψ_1? Write the Hamiltonian for this system.

4.7 Show that the eigenfunctions of the time-independent Schrödinger equation are unchanged if the potential is shifted by an arbitrary constant V_0. What does happen to the eigenvalues? What is the physical interpretation of this result? Does it have a classical analogue?

4.8 Show that, for a stationary state, the current

$$j(x) = \frac{\hbar}{2mi}\left(\Psi(x,t)^*\left(\frac{\partial}{\partial x}\Psi(x,t)\right) - \left(\frac{\partial}{\partial x}\Psi(x,t)^*\right)\Psi(x,t)\right)$$

is independent of both x and t.

4.9 Show that for symmetric potentials, for which $V(-x) = V(x)$, you can always construct energy eigenfunctions that have definite parity, i.e. that are either even or odd functions of x.
Hint: Make the substitution $x \to -x$ in Schrödinger's equation.

4.10 Let us consider a complete set of eigenstates of the energy for a one-dimensional quantum system. We assume that the spectrum is discrete and that the eigenstates are described by their wave functions in position space:

$$\hat{H}u_n(x) = E_nu_n(x), \quad n \text{ integer}.$$

Write down the expansion of a generic state at time $t = 0$, $\Psi(x,0)$ in the basis of the energy eigenstates. Write down an explicit expression for the coefficients of the expansion.
Deduce an expression for the wave function $\Psi(x,t)$ at time t.
Show that

$$\Psi(x,t) = \int dx'\left[\sum_n e^{-iE_nt/\hbar}u_n(x)u_n(x')^*\right]\Psi(x',0).$$

Verify that

$$\Psi(x,t) = \int dx'\, K(xt,x't')\Psi(x',t'),$$

where

$$K(xt,x't') = \sum_n e^{-iE_n(t-t')/\hbar}u_n(x)u_n(x')^*.$$

4.11 Consider a free particle in one dimension. Write the Hamiltonian for this system using the momentum operator $\hat{P}$. The eigenstates of the momentum operator with eigenvalue k are denoted by $|k\rangle$, and the state of the system at $t = 0$ is given by the vector

$$|\psi\rangle = \int \frac{dk}{2\pi} g(k) |k\rangle.$$

Find the state of the system at time t, $|\psi(t)\rangle$.

Compute the expectation value of the momentum at time t, $\langle\psi(t)|\hat{P}|\psi(t)\rangle$. How does it depend on t? Show that your answer is consistent with Ehrenfest's theorem. Compute

$$\frac{d}{dt}\langle\psi(t)|\hat{X}^2|\psi(t)\rangle \quad \text{and} \quad \frac{d}{dt}\langle\psi(t)|\hat{X}\hat{P} + \hat{P}\hat{X}|\psi(t)\rangle.$$

Integrate these equations, and find the time dependence of $\langle\psi(t)|\hat{X}^2|\psi(t)\rangle$.
Discuss under which conditions we have

$$(\Delta X)^2 = \frac{(\Delta P)^2}{m^2}t^2 + (\Delta X)_0^2.$$

Let us set $\hbar = 1$ and

$$g(k) = \left(\frac{1}{\pi\sigma^2}\right)^{1/4} \exp\left[-\frac{(k - k_0)^2}{2\sigma^2}\right].$$

Check that

$$\psi(x) = \int \frac{dk}{2\pi} g(k)e^{ikx}$$

is correctly normalised. Let $\psi(x,t)$ be the wave function of the state obtained by evolving $\psi(x)$ at time $t = 0$ until time t. Find the x dependence of $\psi(x,t)$, and discuss its physical interpretation.

Potentials

One-dimensional systems are an ideal playground to test the concepts introduced so far in this book. Solving a few examples explicitly and discussing their physical interpretation will allow us to get familiar with the description of quantum systems and their dynamics. In this chapter we will solve Schrödinger's equation for some simple one-dimensional potentials. Having found the solution of the mathematical equations, we will focus on how to extract physical information from these solutions. For these problems it is convenient to work in the position representation and describe the state of the system by its wave function.

The Schrödinger equation in a generic potential $V(x)$ is

$$i\hbar \frac{\partial}{\partial t} \Psi(x,t) = \left[-\frac{\hbar^2}{2m} \frac{\partial^2}{\partial x^2} + V(x) \right] \Psi(x,t), \tag{5.1}$$

and the corresponding time-independent Schrödinger equation is

$$\left[-\frac{\hbar^2}{2m} \frac{d^2}{dx^2} + V(x) \right] \psi(x) = E\psi(x). \tag{5.2}$$

The latter is a second-order differential equation for the function ψ. As discussed in the previous chapter, under some very general assumptions on the properties of the potential $V(x)$, the wave function ψ must be single-valued and continuous, and its first derivative with respect to x must also be continuous if the potential is not singular – see Section 4.4. In this chapter we will select some explicit forms for the potential and solve the corresponding eigenvalue problems. Despite their simplicity, some of these potentials will allow us to model important physical phenomena like tunnelling, which are characteristic of quantum mechanics.

5.1 Potential Step

The simplest case we are going to study is a **potential step**:

$$V(x) = \begin{cases} 0, & \text{for } x < 0, \\ V > 0, & \text{for } x > 0, \end{cases} \tag{5.3}$$

as shown in Fig. 5.1. Remember that the energy eigenvalues are always larger than the minimum of the potential, and hence in this case $E > 0$. There are two distinct cases that need to be analysed.

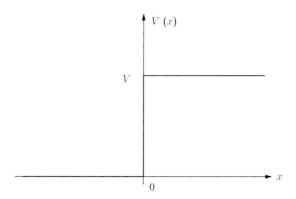

Potential step. The potential vanishes for $x < 0$ and takes some constant positive value V for $x > 0$.

Case 1

We shall start from the case where the total energy E is greater than V. We can then rewrite Eq. (5.2) as

$$\begin{cases} \psi''(x) = -(p/\hbar)^2 \psi(x), & \text{for } x < 0, \\ \psi''(x) = -(\bar{p}/\hbar)^2 \psi(x), & \text{for } x > 0, \end{cases} \tag{5.4}$$

where we have introduced the real numbers

$$p = \sqrt{2mE}, \quad \bar{p} = \sqrt{2m(E - V)} \tag{5.5}$$

and the double prime symbol denotes two derivatives with respect to the position x.

In classical mechanics, when a particle moving from left to right arrives at $x = 0$, its potential energy increases. In order for energy to be conserved, the kinetic energy of the particle has to decrease, i.e. the particle continues to move to the right, at a smaller velocity. The classical momenta of the particle to the left and to the right of the well are given respectively by p and $\bar{p}$. There is clearly no reflection when $E > V$ in classical mechanics.

The solutions of the eigenvalue equation for the energy of the quantum system are

$$\psi(x) = \begin{cases} Ae^{ipx/\hbar} + Be^{-ipx/\hbar}, & \text{for } x < 0, \\ Ce^{i\bar{p}x/\hbar} + De^{-i\bar{p}x/\hbar}, & \text{for } x > 0. \end{cases} \tag{5.6}$$

In each region they represent a superposition of plane waves travelling to the right and to the left.

We want to consider the case of an incident particle from the left ($x < 0$) that hits the barrier. Hence for $x > 0$ we do not expect to have a wave travelling to the left, and therefore $D = 0$.

We are now left with three unknown coefficients in Eq. (5.6), namely A, B and C. The coefficient A determines the overall normalisation of the wave function, or equivalently the flux of incoming particles from $-\infty$, while the other two can be determined by imposing the continuity of the wave function and its derivative at the origin:

$$\lim_{x \to 0^-} \psi(x) = \lim_{x \to 0^+} \psi(x), \tag{5.7}$$

$$\lim_{x \to 0^-} \psi'(x) = \lim_{x \to 0^+} \psi'(x). \tag{5.8}$$

You can readily check that

$$\text{Eq. (5.7)} \Rightarrow A + B = C, \tag{5.9}$$

$$\text{Eq. (5.8)} \Rightarrow p(A - B) = \bar{p}C. \tag{5.10}$$

The system can easily be solved:

$$B = \frac{p - \bar{p}}{p + \bar{p}} A, \tag{5.11}$$

$$C = \frac{2p}{p + \bar{p}} A. \tag{5.12}$$

Let us briefly discuss the physical content of this solution. To the right of the step, $x > 0$, we have a *transmitted* wave propagating to the right. As discussed in Section 4.6, the flux of particles moving to the right for $x > 0$ is $(\bar{p}/m)|C|^2$. To the left of the step, $x < 0$, we find a superposition of the incident wave and a *reflected* wave. The relative amplitudes for the particle to be in each of these states are respectively A and B. The fluxes of particles moving respectively to the right and to the left for $x < 0$ are $(p/m)|A|^2$ and $(p/m)|B|^2$. Taking the ratios of fluxes, we can define a **transmission coefficient**

$$T = \frac{\bar{p}|C|^2}{p|A|^2}, \tag{5.13}$$

and a **reflection coefficient**

$$R = \frac{p|B|^2}{p|A|^2} = \frac{|B|^2}{|A|^2}. \tag{5.14}$$

For the case of the potential step, this yields

$$T = \frac{4p\bar{p}}{(p + \bar{p})^2}, \tag{5.15}$$

$$R = \frac{(p - \bar{p})^2}{(p + \bar{p})^2}. \tag{5.16}$$

As expected on physical grounds,[1] we find

$$R + T = 1. \tag{5.17}$$

As already discussed in Section 4.6, it is only the ratio of the square of the amplitudes that matters, the overall normalisation $|A|^2$ simply fixes the overall incoming flux of particles.

We have found a phenomenon that has no classical analogue: there is a possibility for the particle to be reflected by the potential step even if its energy is greater than the height of the barrier.

---

[1] What is not transmitted must be reflected.

Case 2

We now turn to the case where $E < V$. In classical mechanics this situation corresponds to an incident particle that does not have enough kinetic energy to get past the potential barrier, and therefore is reflected at $x = 0$. Formally, the eigenvalue equations for the quantum system are the same as Eq. (5.4), but now $\bar{p}$ is a purely imaginary number. In order to have a probability density that is bounded for $x > 0$, we need to choose the solution that yields a decreasing exponential:

$$\bar{p} = i\sqrt{2m(V - E)} \equiv i\tilde{p}, \tag{5.18}$$

$$\psi(x) = Ce^{-\tilde{p}x/\hbar}. \tag{5.19}$$

Note that in this case

$$B = \frac{p - i\tilde{p}}{p + i\tilde{p}}A, \tag{5.20}$$

and therefore $|A|^2 = |B|^2$. The magnitudes of the incident and the reflected waves are the same. The two amplitudes are related by a simple phase shift:

$$\frac{p - i\tilde{p}}{p + i\tilde{p}} = -e^{2i\delta(E)}. \tag{5.21}$$

The phase shift can be expressed as a function of p and $\tilde{p}$:

$$\tilde{p} = p \cot \delta, \tag{5.22}$$

which shows that $\delta \to 0$ as $E \to 0$.

It is important to remark here that there is a nonvanishing probability of finding the particle in the classically forbidden region $x > 0$. However, this probability decays exponentially as x is increased. The typical range in which we can expect to find the particle is of the order of $\hbar/\tilde{p}$. It is straightforward to see that the probability current vanishes when the wave function is real. The physical interpretation is that there is no flux of particles for $x > 0$ in this case, and indeed the probability of finding the particle at x goes to zero as $x \to \infty$. In the region where $E > V(x)$, the wave function has an oscillatory behaviour, while in the classically forbidden region the amplitude of the wave function falls off exponentially.

5.2 Tunnelling

We have seen in the previous section that a quantum particle can access regions that are classically forbidden. This fact leads to a very important phenomenon called **tunnelling**, which is characteristic of quantum mechanics.

A step potential of height V and size a as shown in Fig. 5.2 yields a simple example of tunnelling that we can solve analytically.

We are looking for solutions that would not penetrate the barrier classically, hence $0 < E < V$. In this case, following the method used in the previous section, we can

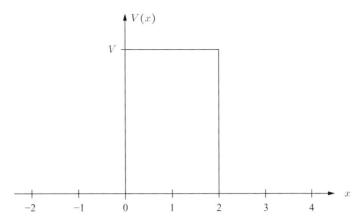

Fig. 5.2 Potential step of height V and size a ($a = 2$ in this example) that allows the system to tunnel from one side to the other.

write the most general function for the wave function representing a wave incident from the left:

$$\psi(x) = \begin{cases} Ae^{ipx/\hbar} + Be^{-ipx/\hbar}, & \text{for } x < 0, \\ Ce^{-\tilde{p}x/\hbar} + De^{\tilde{p}x/\hbar}, & \text{for } 0 < x < a, \\ AS(E)e^{ip(x-a)/\hbar}, & \text{for } x > a. \end{cases} \qquad (5.23)$$

Note that in this case we can keep the exponentially growing solution in the interval $0 < x < a$. Since this interval is finite, the wave function remains bounded. Once again the fact that we have an incident particle from the left is encoded in the fact that there is no left-propagating wave for $x > a$, i.e. no term of the form $e^{-ipx/\hbar}$. For $x > a$ there is no probability flux to the left for any value of x.

In this case we have a system with five unknown variables A, B, C, D and $S(E)$. We have to impose the continuity equations at $x = 0$ and $x = a$. The latter are four equations which enable us to determine the unknown coefficients up to an overall normalisation. In particular here we are interested in computing the transition amplitude $S(E)$, i.e. the ratio of the probability amplitude for the outgoing wave at $x > a$ to the probability amplitude for the incoming wave at $x < 0$. The computation is not difficult, but rather lengthy, and will be discussed in the problems at the end of the chapter. The solution for the transmission amplitude is

$$S(E) = \frac{2ip\tilde{p}}{(p^2 - \tilde{p}^2)\sinh(\frac{\tilde{p}a}{\hbar}) + 2ip\tilde{p}\cosh(\frac{\tilde{p}a}{\hbar})}. \qquad (5.24)$$

The transmissivity T is defined as the probability of the particle tunnelling through the barrier:

$$T = |S(E)|^2 = \left[1 + \frac{\sinh^2(\tilde{p}a/\hbar)}{4(E/V)(1 - E/V)}\right]^{-1}. \qquad (5.25)$$

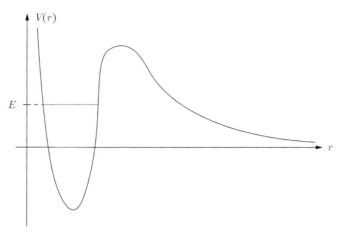

Fig. 5.3 Potential that describes the binding of an α particle in the nucleus. We have not tried to reproduce a realistic potential here, but rather focused on the fact that the shape of the potential allows for tunnelling phenomena.

This is a monotonically increasing function of E, as long as $E < V$. On the other hand, if the energy of the state is fixed, we see from Eq. (5.25) that T decreases exponentially with the size of the barrier a:

$$T \propto \exp\left[-2\sqrt{2m(V-E)}a/\hbar\right]. \tag{5.26}$$

Tunnelling is a phenomenon that has no classical analogue. Its experimental observation is one of the many strong confirmations that the quantum-mechanical framework yields correct predictions. Tunnelling has numerous applications in physics at the atomic and subatomic scale.

1. The α decay of nuclei can be modelled as a tunnelling process. The α particle inside the nucleus can be described as a quantum-mechanical system with the potential depicted in Fig. 5.3. A particle with energy E would be confined inside the nucleus – i.e. inside the dip in the potential – according to classical mechanics. However, in quantum mechanics, the particle has a finite probability of tunnelling to the right.
2. In solid-state physics, if we separate two metals with a thin insulator layer, and apply a voltage across the metals, then the insulator will act as a potential barrier. Any current observed in such an experiment is due to the tunnelling of the electrons across the potential barrier.
3. The scanning tunnelling microscope (STM) allows individual atoms at the surface of a metal to be imaged. The layer of air between the surface of the metal and the STM's needle can be seen as a potential barrier that the electrons need to tunnel through. Since the tunnelling probability depends exponentially on the distance between the STM's needle and the surface, one can determine such distance very precisely by measuring the current of electrons that are actually tunnelling.

We will come back to this idea in Section 17.5, where we discuss approximate methods for solving Schrödinger's equation.

5.3 Infinite Potential Well

Finally, let us examine another class of potentials that give rise to interesting phenomena, viz. the potential wells. We are going to start by solving the time-independent Schrödinger equation for the infinite potential well first. The solution for the finite well is deferred to the final section of this chapter.

In the examples that we have seen in the previous cases, the energy could take any real value from zero to infinity – these are examples of *continuous spectra*. In the case of potential wells, we will find, in addition, bound states with only discrete values of the energy – i.e. the constraints that we impose on the wave functions not only fix the amplitudes, as we discussed in the previous sections, but also restrict the possible values of the eigenvalues.

Physical Setting

The system we consider here is a single particle that evolves in the potential

$$V(x) = \begin{cases} 0, & \text{for } -\frac{a}{2} < x < \frac{a}{2}, \\ \infty, & \text{otherwise.} \end{cases} \tag{5.27}$$

The potential is sketched in Fig. 5.4. Note that the infinite potential well is completely specified by its width a. Therefore we expect the physics of this system to depend on a, and on the particle mass m. Once we have an explicit solution, we will come back to this point and focus on the dependence on a and m.

As discussed in the previous section, if the walls are of finite height $V > E$, the wave function outside the wall decays exponentially – $e^{-\tilde{p}\delta/\hbar}$, where δ is the distance from the edge of the potential well – and we need to impose continuity at the boundary. As the

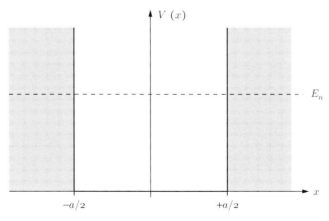

Fig. 5.4 Infinite potential well. The potential vanishes inside the interval $[-a/2, a/2]$ and goes to infinity otherwise. The region with $|x| > a/2$ is classically forbidden, and we can think of this potential as an example of a particle confined in a box of size a.

height of the potential barrier tends to infinity, we have that $\tilde{p} \to \infty$, and hence the wave function outside the potential $e^{-\tilde{p}|x|/\hbar} \to 0$. In other words, we expect the wave function to vanish outside the potential well, thereby implying that the probability of finding the particle outside the well is exactly zero: the particle is confined inside the container.

With that in mind, we want to find the stationary states of this system, i.e. we want to solve the time-independent Schrödinger equation.

Eigenvalue Equation

In order to find the stationary states we need to solve the eigenvalue problem for the Hamiltonian, which turns the physical problem above into a well-defined mathematical problem. We need to find the solutions of the free-particle equation

$$-\frac{\hbar^2}{2m}\frac{d^2}{dx^2}\psi(x) = E\psi(x),$$

(5.28)

with the boundary conditions

$$\psi(a/2) = \psi(-a/2) = 0.$$

(5.29)

All the information about the system is encoded in Eqs (5.28) and (5.29). In order to extract this information we need to solve the equation.

Solution of the Eigenvalue Problem

Setting $k = \sqrt{\frac{2mE}{\hbar^2}}$, Eq. (5.28) becomes the familiar equation

$$\psi''(x) = -k^2\psi(x).$$

(5.30)

The generic solution is a linear superposition of plane waves:

$$\psi(x) = Ae^{ikx} + Be^{-ikx}.$$

(5.31)

So far, the solution of the eigenvalue problem looks very similar to the solutions we discussed for the potential barriers. The peculiar features of the infinite well are encoded in the boundary conditions specified in Eq. (5.29). The latter can be written as

$$\psi(a/2) = Ae^{ika/2} + Be^{-ika/2} = 0,$$

(5.32)

$$\psi(-a/2) = Ae^{-ika/2} + Be^{ika/2} = 0.$$

(5.33)

We have here a system of two equations. One of the amplitudes will be fixed by overall normalisation of the wave function, therefore the two equations will constrain the other amplitude and the possible values of k. Indeed, we see that Eq. (5.32) yields

$$A = -Be^{-ika}$$

and replacing this result in Eq. (5.33), we obtain

$$\sin(ka) = 0,$$

(5.34)

where we see explicitly that the boundary conditions actually restrict the possible values of k, and hence the possible energy values. In order to satisfy Eq. (5.34), we need

$$ka = n\pi, \quad n \text{ integer.} \tag{5.35}$$

Hence, finally, the solutions of the time-independent Schrödinger equation in the infinite well are

$$u_n(x) = \begin{cases} \sqrt{\frac{2}{a}} \sin\left(\frac{n\pi x}{a}\right), & n \text{ even,} \\ \sqrt{\frac{2}{a}} \cos\left(\frac{n\pi x}{a}\right), & n \text{ odd,} \end{cases} \tag{5.36}$$

where n is an integer from 1 to ∞. We see from the explicit expression of the solutions that $n = 0$ is excluded because the wave function would vanish identically.

Physical Interpretation

We can now look at our solutions and discuss the physical properties that are embodied in the mathematical expressions.

Firstly, the possible values of the energy are

$$E_n = n^2 \frac{\pi^2 \hbar^2}{2ma^2}, \quad n = 1, 2, \dots ; \tag{5.37}$$

solutions are labelled by an integer: this is an example of a discrete spectrum. Note that the eigenfunctions u_n are superpositions of eigenstates of the momentum operator; the allowed values of the momentum are

$$p_n = \pm n \frac{\pi \hbar}{a}. \tag{5.38}$$

Note that the lowest value of the energy is greater than zero, and dictated by the size of the box. This is a consequence of the uncertainty principle. The particle is confined inside a box of size a. Hence the maximum uncertainty in the determination of its position is a, which implies that the minimum uncertainty in the value of the momentum is $\Delta p \geq \hbar/(2a)$. As a consequence, the particle has kinetic energy

$$E \sim (\Delta p)^2/(2m). \tag{5.39}$$

The kinetic energy that stems from the uncertainty principle is called **zero point energy**. A striking manifestation of zero point energy is the existence of liquid helium all the way down to $T = 0$. In order to have solid helium, we need to localise the particles in a lattice. This localisation entails a zero point energy, which becomes large for light particles, as shown by Eq. (5.39). In the case of helium the zero point energy is so large that the interatomic forces can no longer constrain the particles in a lattice, and the system remains in a liquid state.

5.4 Symmetry under Parity

By inspecting the energy eigenstates in Eq. (5.36) we see that

$$u_n(-x) = u_n(x), \quad n \text{ odd}, \tag{5.40}$$

$$u_n(-x) = -u_n(x), \quad n \text{ even}. \tag{5.41}$$

Under the parity operation defined in Eq. (4.49), the eigenfunctions $u_n(x)$ are multiplied by a constant $(-1)^{n+1}$, i.e. they are eigenstates of the parity operator $\mathcal{P}$, with eigenvalue $(-1)^{n+1}$. The eigenvalue is called the parity of the state. States with parity $(+1)$ are called *even*, states with parity (-1) are called *odd*.

More generally, for a potential that is symmetric under parity, $V(x) = V(-x)$, you can readily prove that

$$\hat{\mathcal{P}}\hat{H}\psi(x) = \hat{H}\hat{\mathcal{P}}\psi(x), \tag{5.42}$$

for any wave function ψ, which is the same as saying that $\left[\hat{\mathcal{P}}, \hat{H}\right] = 0$. The proof is left as an exercise for the interested reader.

As a consequence of the commutation relation, we know that there must be a basis made of common eigenvectors of $\hat{H}$ and $\hat{\mathcal{P}}$. Let us briefly recall the proof of this statement for this particular example. For each eigenstate of the Hamiltonian u_n, we can construct another eigenstate of $\hat{H}$ with the *same* eigenvalue by acting with $\hat{\mathcal{P}}$ on it:

$$\begin{aligned} \hat{H}\hat{\mathcal{P}}u_n(x) &= \hat{\mathcal{P}}\hat{H}u_n(x) \\ &= t\hat{h}\left(E_n u_n(x)\right) \\ &= E_n\hat{\mathcal{P}}u_n(x). \end{aligned} \tag{5.43}$$

We can then construct the states

$$\left(1 \pm \hat{\mathcal{P}}\right)u_n(x), \tag{5.44}$$

which are simultaneous eigenstates of $\hat{H}$ and $\hat{\mathcal{P}}$. Notice that the result above does not necessarily imply that the energy levels are doubly degenerate, since one of the linear combinations defined in Eq. (5.44) could vanish. If the Hamiltonian is symmetric under parity, i.e. if $[\hat{\mathcal{P}}, \hat{H}] = 0$, then there exists a set of eigenfunctions of $\hat{H}$ that are also eigenstates of $\hat{\mathcal{P}}$. Or, equivalently, we can find a complete basis of eigenstates of $\hat{H}$ made up of **either even or odd** functions of the position x.

5.5 Finite Potential Well

The final example that we are going to discuss is the finite potential well, which corresponds to the potential

$$V(x) = \begin{cases} -V_0, & \text{for } |x| < a/2, \\ 0, & \text{otherwise}. \end{cases} \tag{5.45}$$

Once again we have a potential that is symmetric under parity, $V(-x) = V(x)$. Therefore we can choose to look for solutions of the energy eigenvalue equation that are also parity eigenstates.

As discussed previously, energy eigenvalues must be larger than $-V_0$. This means that we can have states with $-V_0 < E < 0$; they are called **bound states** – the wave function for these states decays exponentially at large $|x|$, so that the probability of finding the particle outside the well becomes rapidly very small. On the other hand, states with $E > 0$ correspond to incident plane waves that are distorted by the potential. We will concentrate on the bound states here.

The Schrödinger equation reads

$$\psi'' = \begin{cases} -\frac{2m}{\hbar^2} E\psi, \\ -\frac{2m}{\hbar^2}(E + V_0)\psi. \end{cases} \tag{5.46}$$

The main difference between the infinite and the finite well comes from the fact that the wave function does not have to vanish outside the classically allowed region $|x| < a/2$. As we discussed before, the wave function for $|x| > a/2$ will decay exponentially.

For $-V_0 < E < 0$, the even parity solutions are of the form

$$\psi(x) = \begin{cases} A\cos(px/\hbar), & |x| < a/2, \\ Ce^{-\bar{p}x/\hbar}, & x > a/2, \\ Ce^{\bar{p}x/\hbar}, & x < -a/2, \end{cases} \tag{5.47}$$

while the odd parity solutions are

$$\psi(x) = \begin{cases} A\sin(px/\hbar), & |x| < a/2, \\ Ce^{-\bar{p}x/\hbar}, & x > a/2, \\ -Ce^{\bar{p}x/\hbar}, & x < -a/2. \end{cases} \tag{5.48}$$

As usual, we have introduced the momenta

$$p = \sqrt{2m(E + V_0)}, \quad \bar{p} = \sqrt{-2mE}. \tag{5.49}$$

Remember that we are looking for bound states, and hence $-V_0 < E < 0$ so the arguments of the square root in Eq. (5.49) are both positive. Note that the symmetry properties of the solutions have already been taken into account in the chosen parametrisations of the solutions.

Once again the values of E cannot be arbitrary, they are determined when we impose the continuity condition. Imposing continuity of the wave function and its derivative in the even-parity sector, we obtain

$$A\cos\left(\frac{pa}{2\hbar}\right) = Ce^{-\bar{p}a/(2\hbar)},$$

$$-\frac{p}{\hbar}A\sin\left(\frac{pa}{2\hbar}\right) = -\frac{\bar{p}}{\hbar}Ce^{-\bar{p}a/(2\hbar)}.$$

Dividing the bottom equation by the top one, we obtain the quantisation condition for the energy levels:

$$p\tan\left(\frac{pa}{2\hbar}\right) = \bar{p}. \tag{5.50}$$

Similarly in the odd-parity sector, we obtain

$$p \cot\left(\frac{pa}{2\hbar}\right) = -\bar{p}. \tag{5.51}$$

The graphical solution of these equations is discussed in the problems at the end of the chapter.

The main physical feature of this system is that there is always at least one even bound state. For very small V_0, i.e. for a shallow well, you can show that

$$E = -\frac{mV_0^2 a^2}{2\hbar^2}. \tag{5.52}$$

Summary

- Solution of the time-independent Schrödinger equation for a potential step, consequences of the continuity conditions for the solution of the equations. Definition of the transmission and reflection coefficients and their physical interpretation. (5.1)
- A simple extension of the potential step is the top-hat potential. The solution of the eigenvalue problem in this case shows that particles that would be reflected in classical mechanics can propagate through the potential. Discussion of the rate at which particles can tunnel. (5.2)
- We have discussed the detailed solution of the infinite potential well. This is an important example: the continuity condition in this case implies a quantisation condition leading to a discrete energy spectrum, another distinctive feature of quantum mechanics. The allowed energy levels arise as a direct consequence of the mathematical properties of the wave function. (5.3)
- We have introduced the idea of zero point energy, and its connection with the uncertainty principle. (5.3)
- Potentials symmetric under parity. We have established some properties of the eigenfunctions of the Hamiltonian for this type of potential. (5.4)
- Finite potential well. Using symmetry under parity allows us to simplify the equations leading to the eigenfunctions. This is a first important example of how the symmetries of a problem can be used to classify the solutions. The latter is a very powerful idea that finds applications in many different realms. (5.5)

Problems

5.1 Imposing the continuity of the wave function and its derivative at $x = 0$ and $x = a$, check the results for the transmission amplitude for the finite potential step, Eq. (5.24), in Section 5.2.

5.2 The continuity equations for the finite well yield

$$p \tan\left(\frac{pa}{2\hbar}\right) = \bar{p},$$

$$p \cot\left(\frac{pa}{2\hbar}\right) = -\bar{p},$$

respectively in the even- and odd-parity sectors. Let us define new variables:

$$\xi = pa/(2\hbar),$$
$$\eta = \bar{p}a/(2\hbar).$$

Check that the new variables are dimensionless.
Check that the continuity equations become

$$\eta = \xi \tan (\xi),$$
$$\eta = -\xi \cot (\xi),$$

and that

$$\xi^2 + \eta^2 = \frac{ma^2 V_0}{2\hbar^2} \equiv R^2.$$

Find a graphical solution to this set of equations in the (ξ, η) plane. Discuss the number of solutions as R is varied. Discuss the limit when $a \to 0$, $V_0 \to \infty$ and $aV_0 = g$ is kept fixed.

5.3 Let us consider now the following potential:

$$V(x) = -g\delta(x), g > 0.$$

This is an attractive potential, and we are going to look for bound states, i.e. states with $E < 0$.

The continuity condition for the wave function reads

$$\lim_{x \to 0^-} \psi(x) = \lim_{x \to 0^+} \psi(x).$$

Write down the Schrödinger equation, and integrate it over dx in the interval $[-\varepsilon, \varepsilon]$. Taking the limit $\varepsilon \to 0$, show that

$$\lim_{x \to 0^+} \psi'(x) - \lim_{x \to 0^-} \psi'(x) = -\frac{2mg}{\hbar^2} \psi(0).$$

Normalisable solutions have the form

$$\psi(x) = \begin{cases} e^{-\kappa x}, & \text{for } x > 0, \\ e^{\kappa x}, & \text{for } x < 0, \end{cases}$$

with $\kappa = \sqrt{-2mE}/\hbar$. Show that the continuity equation for the derivative yields

$$E = -\frac{mg^2}{2\hbar^2}.$$

6 Harmonic Oscillator

We shall now look at the solutions of Schrödinger's equation for the quantum harmonic oscillator. In this chapter we will focus on the one-dimensional case, which can be seen as another example of a one-dimensional potential. Unlike the infinite potential well, the potential for the harmonic oscillator is finite for all finite values of x, and only diverges when $x \to \pm\infty$. As we will see, the solution of the time-independent Schrödinger equation leads to a quantised spectrum. Stationary states correspond to quanta of energy. An important new methodology is introduced in this chapter: the time-independent Schrödinger equation will be solved using creation and annihilation operators, a technique that we will use again later in this book. Creation and annihilation operators yield a simple and physical interpretation of the energy eigenstates, and are used widely, e.g. in relativistic theories to build the space of physical states in quantum field theories. We will also solve the eigenvalue equation as a differential equation for the wave function. It is important to compare and understand the relation between these two different methods. The harmonic oscillator is ubiquitous in physics: every system close to the minimum of the potential is described by a quadratic potential, i.e. it behaves like a harmonic oscillator. The techniques and results presented in this chapter are applied widely to a variety of physical systems.

6.1 The Harmonic Oscillator in Classical Mechanics

Classically, the equation of motion for a particle of mass m, subject to a restoring force proportional to the displacement, x, is

$$m\ddot{x} = -k\,x, \tag{6.1}$$

where the constant of proportionality, k, is usually called the spring constant. This has oscillatory solutions of angular frequency $\omega = \sqrt{k/m}$. Energy conservation can be obtained by writing

$$\ddot{x} = v\frac{\mathrm{d}v}{\mathrm{d}x}, \tag{6.2}$$

and integrating the equation of motion above to give

$$\tfrac{1}{2}mv^2 + \tfrac{1}{2}kx^2 = E, \tag{6.3}$$

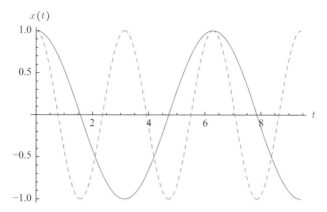

Fig. 6.1 Classical trajectory of the harmonic oscillator for $\omega = 1$ (solid curve) and $\omega = 2$ (dashed curve). The curves represent the position of the system as a function of time.

where E, the total energy of the system, is a constant. The first term on the left-hand side is the kinetic energy of the system, while the potential energy function, V, is given by

$$V(x) = \tfrac{1}{2}kx^2 = \tfrac{1}{2}m\omega^2 x^2. \tag{6.4}$$

It is the quadratic dependence on the position that characterises the harmonic motion. The evolution of the position of the oscillator as a function of time $x(t)$ is shown in Fig. 6.1 for two different values of ω. The total energy of the oscillator is $E = \tfrac{1}{2}m\omega^2 L^2$, where L is the maximum amplitude of the oscillations. The larger the amplitude of the oscillations, the larger the energy of the system. In the case depicted in Fig. 6.1, the amplitude $L = 1$ was used for both curves.

6.2 The Quantum Harmonic Oscillator

By analogy with the classical case, the quantum version of the **harmonic oscillator** is a one-dimensional system (for now), whose potential is a quadratic function of its coordinate. The time-independent Schrödinger equation is

$$\hat{H}|n\rangle = E_n|n\rangle, \tag{6.5}$$

where we denote by $|n\rangle$ the eigenstates of the Hamiltonian

$$\hat{H} = \frac{\hat{P}^2}{2m} + \tfrac{1}{2}m\omega^2 \hat{X}^2. \tag{6.6}$$

The notation used above suggests that the spectrum is discrete and the eigenstates of the Hamiltonian are labelled by some integer n. We will see below that this is indeed the case.

Taking the scalar product of Eq. (6.5) with eigenstates of the position operator $|x\rangle$, we can recast the time-independent Schrödinger equation as a differential equation for the wave functions of the eigenstates:

$$\left[-\frac{\hbar^2}{2m} \frac{d^2}{dx^2} + \tfrac{1}{2}m\omega^2 x^2 \right] u_n(x) = E_n u_n(x), \tag{6.7}$$

where the wave function is defined as usual:

$$u_n(x) = \langle x|n\rangle.$$

We will discuss the detailed solution of this differential equation at the very end of the chapter. Before dwelling on the details of such derivations, we will develop an elegant algebraic method to find the eigenfunctions. For now it suffices to say that acceptable solutions only arise for certain values of the total energy, i.e. the energy is *quantised*. The **energy eigenvalues** are

$$E_n = (n + \tfrac{1}{2})\hbar\omega, \quad n = 0, 1, 2, 3, \ldots, \infty. \tag{6.8}$$

Note that $\hbar\omega$ has got the correct dimensions, i.e. it is an energy. The state of lowest energy, or *ground state*, is labelled $n = 0$ and has an energy $\tfrac{1}{2}\hbar\omega$ rather than zero. The corresponding **eigenfunctions** may be written as

$$u_n(x) = C_n \exp\{-\alpha^2 x^2/2\} H_n(\alpha x), \tag{6.9}$$

where C_n is a normalisation constant, $\alpha^2 \equiv m\omega/\hbar$ and $H_n(\alpha x)$ are polynomials of degree n, known as *Hermite polynomials*. Once again using dimensional analysis, you can show that α has dimensions of inverse length, so that the product αx is dimensionless. The Hermite polynomials satisfy the orthogonality relation

$$\int_{-\infty}^{\infty} \exp\{-s^2\} H_m(s) H_n(s)\, ds = 2^n \sqrt{\pi}\, n!\; \delta_{mn}, \tag{6.10}$$

implying that the energy eigenfunctions are orthogonal (orthonormal with an appropriate choice of the constant, C_n).

The first few Hermite polynomials are

$$H_0(s) = 1, \tag{6.11}$$

$$H_1(s) = 2s, \tag{6.12}$$

$$H_2(s) = 4s^2 - 2, \tag{6.13}$$

$$H_3(s) = 8s^3 - 12s. \tag{6.14}$$

Note that the polynomials, and hence the energy eigenfunctions, are alternately even and odd functions of x, i.e. they are of either even or odd parity. Acting with the parity operator introduced in Eq. (4.49), you can readily check that

$$\hat{\mathcal{P}}\, u_n(x) = (-1)^n\, u_n(x), \tag{6.15}$$

the parity of the state u_n is $(-1)^n$. This is consistent with the discussion at the end of Chapter 5, since the potential for the harmonic oscillator is symmetric under reflections:

$$V(x) = V(-x). \tag{6.16}$$

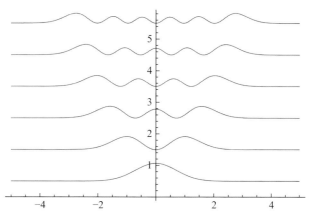

Fig. 6.2 Probability of finding the oscillator at x for the first five levels of the harmonic oscillator for $\alpha = 1$. Each curve is shifted upwards by $n + \frac{1}{2}$ to improve the readability of the plot.

The probability density to find the oscillator at a given position x is $|u_n(x)|^2$. The probability densities for the first five levels are shown in Fig. 6.2. It is clear from the shape of the probability densities in the figure that the probability of finding the system at larger values of x increases as the energy of the state is increased, i.e. for larger values of n. In order to quantify the last sentence we can compute the average position, and the fluctuations of the position, when the system is in the nth stationary state. The expectation value of the position of the quantum harmonic oscillator is given by

$$\langle x \rangle_n = \int_{-\infty}^{+\infty} dx \, x \left| C_n \exp\{-\alpha^2 x^2/2\} \, H_n(\alpha x) \right|^2 , \tag{6.17}$$

where the suffix n is there to remind us that the expectation value is computed in the state $u_n(x)$. You can easily convince yourself that $\langle x \rangle = 0$, because the integrand is always an *odd* function of x. The vanishing of the mean value of the position is what we would have naively expected, since the system is symmetric under parity. The quantum oscillator is equally likely to be to the left or to the right of the origin at $x = 0$. Note that here we are *not* considering the time evolution of the oscillator, but rather the probability density of finding the system at position x when the state of the system is described by one of the stationary states. The average in Eq. (6.17), and the average discussed in the next paragraph, is not an average over time, it is the average over multiple measurements performed after preparing the system always in the same state u_n.

It is more instructive to compute the mean square deviation $\langle x^2 \rangle$, which measures the variance of the probability distribution of the outcome of a measurement of the position of the system. You can compute the integrals explicitly for the first few Hermite polynomials, or simply look up the answer in a table of integrals:

$$\langle x^2 \rangle_n = \left(n + \frac{1}{2} \right) \frac{1}{\alpha^2} . \tag{6.18}$$

Once again this is the quantitative expression of some intuitive understanding: the larger the energy, the larger the variance. Observations of a large value of the position of the system become more probable as the energy of the system increases, more precisely:

$$E_n = m\omega^2 \langle x^2 \rangle_n, \tag{6.19}$$

which is the quantum version of the relation between the energy and the size of the oscillations of the harmonic oscillator. As you can see in Eq. (6.19), we found a relation between observables: the energy of the system and its position. In a stationary state, measurements of the energy will yield the eigenvalue E_n with probability 1. Measurements of the position will yield a distribution of values – remember that the results of experiments are stochastic variables – with mean value equal to 0 and variance proportional to the energy of the state. This is the kind of intuition that you may want to develop in quantum mechanics, in order to make contact between the abstract formalism and physical reality.

In the following sections we shall proceed to solve the time-independent Schrödinger equation. The solution will involve a number of steps:

1. we will rewrite the Hamiltonian in a factorised form;
2. we will then introduce and discuss creation and annihilation operators;
3. finally, we will be able to construct explicitly the solutions.

6.3 Factorising the Hamiltonian

The Hamiltonian for the harmonic oscillator is

$$\hat{H} = \frac{\hat{P}^2}{2m} + \tfrac{1}{2}m\omega^2 \hat{X}^2. \tag{6.20}$$

Let us factor out $\hbar\omega$, and rewrite the Hamiltonian as

$$\hat{H} = \hbar\omega \left[\frac{\hat{P}^2}{2m\hbar\omega} + \frac{m\omega}{2\hbar} \hat{X}^2 \right]. \tag{6.21}$$

Checking the dimensions of the constants, you can readily verify that

$$[\hbar\omega] = \text{energy}, \quad [2m\omega\hbar] = \text{momentum}^2, \quad \left[\frac{2\hbar}{m\omega} \right] = \text{length}^2. \tag{6.22}$$

Introducing the dimensionless quantities

$$\hat{\xi} = \sqrt{\frac{m\omega}{2\hbar}} \hat{X}, \tag{6.23}$$

$$\hat{\eta} = \frac{\hat{P}}{\sqrt{2m\hbar\omega}}, \tag{6.24}$$

the Hamiltonian becomes

$$\hat{H} = \hbar\omega \left[\hat{\eta}^2 + \hat{\xi}^2 \right]. \tag{6.25}$$

The operators $\hat{\xi}$ and $\hat{\eta}$ are simply the position and the momentum operators rescaled by some real constants; therefore both of them are Hermitian. Their commutation relation can easily be computed using the canonical commutation relations

$$\left[\hat{\xi}, \hat{\eta}\right] = \frac{1}{2\hbar} \left[\hat{X}, \hat{P}\right] = \frac{i}{2}. \tag{6.26}$$

If $\hat{\xi}$ and $\hat{\eta}$ were commuting variables, we would be tempted to factorise the Hamiltonian as

$$\hat{H} = \hbar\omega \left(\hat{\xi} + i\hat{\eta}\right) \left(\hat{\xi} - i\hat{\eta}\right). \tag{6.27}$$

However, we must be careful here, because the operators do not commute. So let us introduce

$$\begin{cases} \hat{a} = \hat{\xi} + i\hat{\eta}, \\ \hat{a}^\dagger = \hat{\xi} - i\hat{\eta}; \end{cases} \tag{6.28}$$

the expressions for $\hat{a}$ and $\hat{a}^\dagger$ in terms of $\hat{X}$ and $\hat{P}$ are

$$\hat{a} = \sqrt{\frac{m\omega}{2\hbar}} \, \hat{X} + \frac{i}{\sqrt{2m\omega\hbar}} \, \hat{P}, \tag{6.29}$$

$$\hat{a}^\dagger = \sqrt{\frac{m\omega}{2\hbar}} \, \hat{X} - \frac{i}{\sqrt{2m\omega\hbar}} \, \hat{P}. \tag{6.30}$$

We can then compute

$$\hat{a}\hat{a}^\dagger = \hat{\xi}^2 + i \left[\hat{\eta}, \hat{\xi}\right] + \hat{\eta}^2, \tag{6.31}$$

$$\hat{a}^\dagger\hat{a} = \hat{\xi}^2 - i \left[\hat{\eta}, \hat{\xi}\right] + \hat{\eta}^2. \tag{6.32}$$

Summing the two equations above, we get a factorised expression for the Hamiltonian:

$$\hat{H} = \frac{\hbar\omega}{2} \left(\hat{a}\hat{a}^\dagger + \hat{a}^\dagger\hat{a}\right). \tag{6.33}$$

Subtracting the same two equations yields the commutation relation between $\hat{a}$ and $\hat{a}^\dagger$:

$$\left[\hat{a}, \hat{a}^\dagger\right] = 1. \tag{6.34}$$

This commutation relation plays an important role in the rest of this chapter. Using the commutation relation, we can derive an alternative, and more useful, expression for $\hat{H}$:

$$\hat{H} = \left(\hat{a}^\dagger\hat{a} + \tfrac{1}{2}\right)\hbar\omega. \tag{6.35}$$

6.4 Creation and Annihilation

We are now going to find the eigenvalues of $\hat{H}$ using the operators $\hat{a}$ and $\hat{a}^\dagger$. We begin by computing the commutators $[\hat{H}, \hat{a}]$ and $[\hat{H}, \hat{a}^\dagger]$:

$$[\hat{H}, \hat{a}] = [\left(\hat{a}^\dagger\hat{a} + \tfrac{1}{2}\right)\hbar\omega, \hat{a}] = \hbar\omega[\hat{a}^\dagger\hat{a}, \hat{a}], \quad \text{since} \quad [\tfrac{1}{2}, \hat{a}] = 0. \tag{6.36}$$

Now

$$[\hat{a}^\dagger \hat{a}, \hat{a}] = [\hat{a}^\dagger, \hat{a}]\hat{a} = -\hat{a}, \tag{6.37}$$

so that we obtain

$$[\hat{H}, \hat{a}] = -\hbar\omega\hat{a}. \tag{6.38}$$

Similarly

$$[\hat{H}, \hat{a}^\dagger] = [\left(\hat{a}^\dagger \hat{a} + \tfrac{1}{2}\right)\hbar\omega, \hat{a}^\dagger] = \hbar\omega[\hat{a}^\dagger \hat{a}, \hat{a}^\dagger], \quad \text{since} \quad [\tfrac{1}{2}, \hat{a}^\dagger] = 0, \tag{6.39}$$

and

$$[\hat{a}^\dagger \hat{a}, \hat{a}^\dagger] = \hat{a}^\dagger[\hat{a}, \hat{a}^\dagger] = \hat{a}^\dagger, \tag{6.40}$$

and therefore

$$[\hat{H}, \hat{a}^\dagger] = \hbar\omega\hat{a}^\dagger. \tag{6.41}$$

Using these relations, we can compute

$$\hat{H}\left(\hat{a}|n\rangle\right) = \hat{a}\hat{H}|n\rangle + \left[\hat{H}, \hat{a}\right]|n\rangle, \tag{6.42}$$

$$= E_n\hat{a}|n\rangle - \hbar\omega\hat{a}|n\rangle, \tag{6.43}$$

$$= (E_n - \hbar\omega)\left(\hat{a}|n\rangle\right). \tag{6.44}$$

We have found an eigenvalue equation: it states that $\hat{a}|n\rangle$ is an eigenfunction of $\hat{H}$ belonging to the eigenvalue $(E_n - \hbar\omega)$, unless $\hat{a}|n\rangle \equiv 0$. We say that the operator $\hat{a}$ is a **lowering operator**; its action on an energy eigenstate is to turn it into another energy eigenstate of lower energy. It is also called an **annihilation operator**, because it removes one quantum of energy $\hbar\omega$ from the system.

Similarly, it is straightforward to show that

$$\hat{H}\hat{a}^\dagger|n\rangle = (E_n + \hbar\omega)\hat{a}^\dagger|n\rangle, \tag{6.45}$$

which says that $\hat{a}^\dagger|n\rangle$ is an eigenfunction of $\hat{H}$ belonging to the eigenvalue $(E_n + \hbar\omega)$, unless $\hat{a}^\dagger|n\rangle \equiv 0$. We say that the operator $\hat{a}^\dagger$ is a **raising operator**; its action on an energy eigenstate is to turn it into another energy eigenstate of higher energy. It is also called a **creation operator**, because it adds one quantum of energy $\hbar\omega$ to the system.

We can summarise these results by denoting the states of energy $E_n \pm \hbar\omega$ by $|n \pm 1\rangle$ and writing

$$\hat{a}|n\rangle = c_n|n-1\rangle \quad \text{and} \quad \hat{a}^\dagger|n\rangle = d_n|n+1\rangle, \tag{6.46}$$

where c_n and d_n are constants of proportionality (NOT eigenvalues!) and

$$\hat{H}|n-1\rangle = E_{n-1}|n-1\rangle = (E_n - \hbar\omega)|n-1\rangle, \tag{6.47}$$

$$\hat{H}|n+1\rangle = E_{n+1}|n+1\rangle = (E_n + \hbar\omega)|n+1\rangle. \tag{6.48}$$

6.5 Eigensystem

6.5.1 Eigenvalues

It should be clear that repeated applications of the lowering operator, $\hat{a}$, generate states of successively lower energy ad infinitum *unless there is a state of lowest energy*; application of the operator to such a state must yield the null vector.

Does such a state exist? The answer is yes because the Hamiltonian can only have positive eigenvalues, since the minimum of the potential is $V_{\min} = 0$ as discussed in the general case in Chapter 4. Alternatively, you can consider the expectation value of $\hat{H}$ in an arbitrary state $|\Psi\rangle$:

$$\langle \hat{H} \rangle = \left\langle \frac{\hat{P}^2}{2m} \right\rangle + \left\langle \tfrac{1}{2} m \omega^2 \hat{X}^2 \right\rangle, \tag{6.49}$$

and both terms on the right-hand side are non-negative. Thus there cannot be any states of negative energy.[1]

We denote the state of lowest energy, or ground state, by $|0\rangle$. Since there cannot be a state of lower energy, we must have

$$\hat{a} |0\rangle = |\emptyset\rangle, \tag{6.50}$$

where we introduced $|\emptyset\rangle$ to denote the null vector in the space of physical states, so that it does not get confused with the ground state $|0\rangle$. Applying the Hamiltonian to this state we see that

$$\hat{H} |0\rangle = \hbar\omega \, (\hat{a}^\dagger \hat{a} + \tfrac{1}{2}) \, |0\rangle = \tfrac{1}{2}\hbar\omega \, |0\rangle \equiv E_0 \, |0\rangle. \tag{6.51}$$

Thus we have found the ground-state energy: $E_0 = \tfrac{1}{2}\hbar\omega$. Application of the raising operator to the ground state generates the state $|1\rangle$ with energy $E_1 = \tfrac{3}{2}\hbar\omega$, whilst n applications of the raising operator generate the state $|n\rangle$ with energy $(n + \tfrac{1}{2})\hbar\omega$, so that

$$E_n = (n + \tfrac{1}{2})\hbar\omega, \qquad n = 0, 1, 2, 3, \ldots, \tag{6.52}$$

which is the result we stated at the beginning of this chapter for the energy eigenvalues of the one-dimensional oscillator!

6.5.2 Normalisation of Eigenstates

Requiring that both $|n\rangle$ and $|n-1\rangle$ be normalised enables us to determine the constant of proportionality c_n. Consider

$$\langle n|\hat{a}^\dagger \hat{a}|n\rangle = c_n \langle n|\hat{a}^\dagger|n-1\rangle \quad \text{from property of } \hat{a}, \tag{6.53}$$

$$= c_n \langle n-1|\hat{a}|n\rangle^* \quad \text{from definition of } \dagger, \tag{6.54}$$

$$= c_n c_n^* \langle n-1|n-1\rangle^* \quad \text{from property of } \hat{a}, \tag{6.55}$$

$$= |c_n|^2 \quad \text{since } \langle n-1|n-1\rangle^* = 1. \tag{6.56}$$

---

[1] This reasoning is identical to what we used in the general case, earlier in this book. It is applied here to the special case of a harmonic potential.

We can evaluate the left-hand side if we note that $\hat{a}^\dagger \hat{a} = \left(\hat{H}/\hbar\omega\right) - \frac{1}{2}$, giving

$$\langle n|\hat{a}^\dagger \hat{a}|n\rangle = n\langle n|n\rangle = n. \tag{6.57}$$

Thus $|c_n|^2 = n$ and if we choose the phase so that c_n is real, we can write

$$c_n = \sqrt{n}. \tag{6.58}$$

A similar calculation shows that

$$\langle n|\hat{a}\hat{a}^\dagger|n\rangle = |d_n|^2 = (n+1), \tag{6.59}$$

so that if we again choose the phase so that d_n is real, we obtain

$$d_n = \sqrt{n+1}. \tag{6.60}$$

In summary then, we have

$$\hat{a}\,|n\rangle = \sqrt{n}\,|n-1\rangle \quad \text{and} \quad \hat{a}^\dagger\,|n\rangle = \sqrt{n+1}\,|n+1\rangle. \tag{6.61}$$

6.5.3 Wave Functions

Finally let us show that we can reproduce the analytic expression for the eigenfunctions of the energy. The ground state is defined by the relation

$$\hat{a}|0\rangle = |\emptyset\rangle. \tag{6.62}$$

We can rewrite the equation above as a differential operator acting on the wave function of the ground state $u_0(x)$:

$$\hat{a}u_0(x) = \left[\sqrt{\frac{m\omega}{2\hbar}}\,\hat{X} + \frac{i}{\sqrt{2m\omega\hbar}}\,\hat{P}\right]u_0(x), \tag{6.63}$$

$$= \left[\sqrt{\frac{m\omega}{2\hbar}}\,x + \sqrt{\frac{\hbar}{2m\omega}}\,\frac{d}{dx}\right]u_0(x), \tag{6.64}$$

$$= 0. \tag{6.65}$$

Hence

$$\frac{d}{dx}u_0(x) = -\sqrt{\frac{2m\omega}{\hbar}}\sqrt{\frac{m\omega}{2\hbar}}\,xu_0(x), \tag{6.66}$$

$$= -\frac{m\omega}{\hbar}\,xu_0(x), \tag{6.67}$$

$$= -\alpha^2 xu_0(x), \tag{6.68}$$

where $\alpha^2 = m\omega/\hbar$. The solution of the equation above is the Gaussian that we have already seen in Section 6.2:

$$u_0(x) = C_0 \exp[-\alpha^2 x^2/2]. \tag{6.69}$$

Every other eigenfunction is obtained by repeatedly applying the creation operator $\hat{a}^\dagger$ to the ground state:

$$u_n(x) = \frac{1}{\sqrt{n!}}\left(\hat{a}^\dagger\right)^n u_0(x). \tag{6.70}$$

Remember that $\hat{a}^\dagger$ is represented as a differential operator acting on wave functions.

Exercise 6.5.1 Check that you can reproduce the wave functions for the first and second excited states of the harmonic oscillator, including the correct normalisation.

6.6 Brute Force Solution

Finally let us discuss the direct solution of the time-independent Schrödinger equation cast as a differential equation for the wave function. The method can be applied more generally to second-order equations, and indeed we will use it again to find the energy spectrum of the hydrogen atom. The equation that we are trying to solve for the harmonic oscillator is

$$\left[-\frac{\hbar^2}{2m}\frac{d^2}{dx^2} + \tfrac{1}{2}m\omega^2 x^2\right]\psi(x) = E\psi(x). \tag{6.71}$$

It is often useful, when solving mathematical equations, to work with dimensionless variables – after all, mathematicians don't care about metres, Joules, seconds, etc. Following our earlier convention we introduce the dimensionless variables

$$\xi = \sqrt{\frac{m\omega}{\hbar}}\,x, \quad \epsilon = E/(\hbar\omega), \quad \phi(\xi) = \sqrt{\frac{\hbar}{m\omega}}\psi(x). \tag{6.72}$$

Note that we have omitted a factor of $1/\sqrt{2}$ in the definition of ξ in order to get a more symmetric differential equation below, and we have rescaled the wave function by a factor $\sqrt{\frac{\hbar}{m\omega}}$, so that it is also dimensionless.[2] The eigenvalue equation becomes

$$\left[\frac{d^2}{d\xi^2} - \left(\xi^2 - 2\epsilon\right)\right]\phi(\xi) = 0. \tag{6.73}$$

In order to proceed, note that

$$\lim_{x\to\pm\infty}\frac{d^2}{d\xi^2}\phi(\xi) = \xi^2\phi(\xi), \tag{6.74}$$

with corrections that are suppressed as $1/\xi^2$. The Gaussian function $e^{-\xi^2/2}$ satisfies the same asymptotic behaviour, albeit with a different subleading correction. Therefore we are going to look for a solution of the form

$$\phi(\xi) = h(\xi)e^{-\xi^2/2}. \tag{6.75}$$

---

[2] Remember that the wave function of a quantum system on a line has dimensions of length$^{-1/2}$, so that its modulus squared is a probability density.

Substituting this ansatz in Eq. (6.73) yields an equation for $h(\xi)$:

$$\left[\frac{d^2}{d\xi^2} - 2\xi \frac{d}{d\xi} + (2\epsilon - 1) \right] h(\xi) = 0. \tag{6.76}$$

We are going to look for a solution by expanding $h(\xi)$ in a Taylor series around $\xi = 0$:

$$h(\xi) = \sum_{m=0}^{\infty} a_{2m} \xi^{2m+p}. \tag{6.77}$$

In this expansion we assume $a_0 \neq 0$, and we used the fact that the potential is symmetric under parity and therefore we can look for solutions that are either even or odd in ξ. In Eq. (6.77), p even (respectively odd) corresponds to an even (respectively odd) solution.

Equation (6.76) leads to a recursion relation for the coefficients of the Taylor expansion:

$$(2m + p + 2)(2m + p + 1)a_{2m+2} = (4m + 2p - 2\epsilon + 1)a_{2m}, \tag{6.78}$$

while for the lowest power of ξ we have

$$p(p-1)a_0 \xi^{p-2} = 0, \tag{6.79}$$

and therefore $p = 0, 1$, since a_0 does not vanish. The recursion relation implies that, for $m \to \infty$:

$$\frac{a_{2m+2}}{a_{2m}} \sim \frac{1}{m}. \tag{6.80}$$

In turn, Eq. (6.80) can be used to show that $h(\xi)$ grows like $e^{\lambda \xi^2}$ with $0 < \lambda < 1$, and therefore the divergence of $h(\xi)$ can overcome the Gaussian that we factored out in the ansatz: $\phi(\xi)$ is unbounded for $\xi \to \pm\infty$ and the solution is not normalisable.

The only way we can have a normalisable solution is if there exists a value m_0 such that

$$(4m_0 + 2p - 2\epsilon + 1) = 0. \tag{6.81}$$

In this case $a_{2m} = 0$ for all $m > m_0$, and $h(\xi)$ is a simple polynomial. Setting $2m_0 + p = n$, the quantisation condition, Eq. (6.81), can be written

$$\epsilon = \epsilon_n = n + \tfrac{1}{2}, \tag{6.82}$$

which yields the energy levels that we found above using creation and annihilation operators.

Summary

- We started this chapter with a brief summary of harmonic motion in classical mechanics, which we want to contrast with the quantum solution. We should keep in mind in particular the relation between the size of the oscillations and the energy of the classical system, and compare it later on with the quantum result. (6.1)

- Before discussing the different techniques to solve the eigenvalue problem for the Hamiltonian, we focus on the statement of the problem, the form of the eigenvalue equation and its solutions. In the first place the solutions are just given and their properties are discussed. (6.2)
- We develop the algebraic solution of the problem, starting from finding a factorised expression for the Hamiltonian. (6.3)
- From the factorised expression we can then define creation and annihilation operators and discuss their properties. (6.4)
- Using the creation and annihilation operators, we proceed to the construction of the eigenvalues and eigenvectors. A simple calculation allows us to determine the normalisation of the eigenstates. (6.5)
- Having found the normalised eigenvectors, we derive the corresponding wave functions, by expanding the eigenvectors in a basis made up of eigenvectors of the position operator. (6.5.3)
- Finally we compute directly the solutions of the eigenvalue equation as a differential equation, and show that we recover the quantisation condition and the same solutions that were found using creation and annihilation operators. (6.6)

Problems

6.1 Write the position operator $\hat{X}$ as a function of the raising and lowering operators $\hat{a}$ and $\hat{a}^{\dagger}$.

Deduce the matrix element of $\hat{X}$ between two eigenstates of the Hamiltonian: $\langle n|\hat{X}|m\rangle$.

Repeat the same computations for the momentum operator $\hat{P}$.

6.2 Consider a one-dimensional harmonic oscillator, consisting of a particle of mass m in a quadratic potential characterised by the frequency ω. Assume that the particle has a charge q and is placed in a uniform electric field $\mathcal{E}$ in the x direction. The Hamiltonian can be written as

$$\hat{H}(\mathcal{E}) = \frac{\hat{P}^2}{2m} + \tfrac{1}{2}m\omega^2\hat{X}^2 - q\mathcal{E}\hat{X}.$$

Write the eigenvalue equation for $\hat{H}(\mathcal{E})$.

Replacing the variable x by a new variable t:

$$t = x - \frac{q\mathcal{E}}{m\omega^2},$$

find the eigenvalues E'_n and the eigenfunctions $u'_n(x)$ that solve the equation above.

6.3 The contribution of the electron to the electric dipole moment is given by the matrix element of the operator:

$$\hat{D} = q\hat{X}.$$

In the absence of an electric field, show that

$$\langle \hat{D} \rangle = q\langle n|\hat{X}|n\rangle = 0.$$

Let us assume that the electric field is turned on. Compute the mean dipole moment:

$$\langle \hat{D} \rangle' = q\langle n'|\hat{X}|n'\rangle,$$

where $|n'\rangle$ are the kets associated with $u'_n(x)$, the wave functions found in the presence of an electric field in the question above.

Discuss the physical interpretation of this result.

6.4 Let us consider a one-dimensional harmonic oscillator. By recalling the action of the lowering operator a on the ground state, find the first-order differential equation that is satisfied by the wave function $u_0(x)$.

Using the creation operator, build the wave function for the first two excited states.

6.5 Let us consider the one-dimensional harmonic oscillator. Express $\hat{X}^2$ and $\hat{P}^2$ in terms of a and $a^\dagger$, and show that for an energy eigenstate:

$$\langle n|\frac{\hat{P}^2}{2m}|n\rangle = \langle n|\tfrac{1}{2}m\omega^2\hat{X}^2|n\rangle = \tfrac{1}{2}\langle n|\hat{H}|n\rangle.$$

Show that this result remains true when the oscillator is not in an eigenstate of the Hamiltonian, provided a time average over one classical oscillator period is taken. The time average is defined as

$$\bar{f} = \frac{1}{T}\int_0^T dt\, f(t).$$

6.6 Consider a one-dimensional harmonic oscillator defined by the potential

$$V(x) = \tfrac{1}{2}m\omega^2 x^2.$$

At $t = 0$ the state of the oscillator is

$$\Psi(x,0) = \sum_{n=0}^{\infty} c_n u_n(x),$$

where $u_n(x)$ are the eigenfunctions of the Hamiltonian.

Write down the equations satisfied by the wave functions $u_n(x)$. Recall the possible values of the energy E_n, and the corresponding eigenfunctions.

Write the wave function describing the system at time t.

Deduce the mean value of any observable A at time t:

$$\langle \Psi(t)|\hat{A}|\Psi(t)\rangle$$

as a function of the matrix elements

$$A_{mn} = \langle u_m|\hat{A}|u_n\rangle.$$

Compare with the time dependence of the mean value of A in a stationary state, i.e. in the case where only one of the coefficients c_n is different from zero.

Let us now consider the case where $c_0 = \cos \eta$, $c_1 = \sin \eta$ and $c_n = 0$ for $n > 1$. You should use this initial state in all subsequent questions.

What are the possible outcomes of a measurement of the energy? What are the probabilities of each outcome?

Consider the observables X and P. Using the explicit expressions for the eigenfunctions $u_0(x)$ and $u_1(x)$ in Eq. (6.9), determine which are the matrix elements X_{mn} and P_{mn} that do not vanish for $m, n = 0, 1$. (You don't need to compute the integrals, just identify those that do not vanish!)

Deduce that the mean values $\langle \Psi(t) | \hat{X} | \Psi(t) \rangle$ and $\langle \Psi(t) | \hat{P} | \Psi(t) \rangle$ are sinusoidal functions of time with angular frequency ω.

Verify that this result is consistent with Ehrenfest's theorem.

6.7 Consider a one-dimensional harmonic oscillator, of mass m and frequency ω. At time $t_0 = 0$, the system is in the state $|\psi(0)\rangle$ characterised by the following properties:

(a) a measurement of the energy yields $E_1 = 3/2\hbar\omega$ with probability 3/4, and $E_2 = 5/2\hbar\omega$ with probability 1/4;
(b) the mean value of the position in the state $|\psi(0)\rangle$ vanishes;
(c) the mean value of the momentum is positive.

The eigenstates of the Hamiltonian are identified by the kets $|n\rangle$:

$$\hat{H}|n\rangle = E_n|n\rangle.$$

Recall the possible values for the energy of the harmonic oscillator, and identify E_1 and E_2. Explain why the state of the system can be written as

$$|\psi(0)\rangle = \tfrac{1}{2}\left[\sqrt{3}e^{i\phi_1}|1\rangle + e^{i\phi_2}|2\rangle\right].$$

Write down the mean value of $\hat{X}$ at time 0 as a function of ϕ_1 and ϕ_2. You may use the following result:

$$\langle 1|\hat{X}|2\rangle = \sqrt{\frac{\hbar}{m\omega}}.$$

Deduce a condition on ϕ_1 and ϕ_2.

Starting from the matrix element of $\left[\hat{P}, \hat{H}\right]$, and using the canonical commutation relation between $\hat{X}$ and $\hat{P}$, show that

$$\langle 1|\hat{P}|2\rangle = -im\omega\langle 1|\hat{X}|2\rangle.$$

Compute the mean value of the momentum at $t = 0$:

$$\langle \psi(0)|\hat{P}|\psi(0)\rangle.$$

Deduce a second relation between ϕ_1 and ϕ_2.

Find the solutions for ϕ_1 and ϕ_2, and determine the state of the system at time t. What is the probability of finding E_2 if the energy is measured at time t? What is the mean value of the position at time t?

7 Systems in Three Spatial Dimensions

In previous chapters we have introduced the basic concepts of quantum mechanics, and studied them for some simple, yet relevant, one-dimensional systems. In this chapter we take another step towards the description of real physical phenomena and generalise the concepts introduced so far to systems that evolve in more than one spatial dimension. The generalisation is straightforward and it will give us the opportunity to review some of the key ideas about physical states, observables and time evolution. In the process, we will encounter and highlight new features that were not present for one-dimensional systems. Once again let us emphasise that three-dimensional in this context refers to the dimension of the physical space in which the system is defined, and *not* to the dimensionality of the Hilbert space of states; the latter clearly will depend on the type of system that we consider. The three-dimensional formulation will allow us to discuss more realistic examples of physical systems. It will be clear as we progress through this chapter that everything we discuss can be generalised to an arbitrary number of dimensions. In some physical applications, where a quantum system is confined to a plane, a two-dimensional formulation will be useful. More generally, it is instructive to think about problems in arbitrary numbers of dimensions. In this respect, it is fundamental to be able to work with vectors, tensors, indices, and all that. Problems and examples in this chapter should help develop some confidence in using an index notation to deal with linear algebra.

7.1 Quantum States

We discuss here the case of three spatial dimensions; positions in space are specified by a three-dimensional real vector $\underline{r}$. Having chosen a set of basis vectors, the position vector $\underline{r}$ can be represented for example by three Cartesian coordinates (x, y, z). Systems in two spatial dimensions can be treated in the same way, by simply dropping the dependence on the third coordinate z.

As in the one-dimensional case, quantum states are described by a state vector $|\psi\rangle$ that belongs to a Hilbert space $\mathcal{H}$. This state vector can be described by a wave function. Remember that the wave function can be interpreted as the coordinates of the state vector $|\psi\rangle$ in a basis made of eigenstates of the position operator

$$|\psi\rangle = \int d^3r \, |\underline{r}\rangle\langle\underline{r}|\psi\rangle, \tag{7.1}$$

where the wave function, $\psi(\underline{r}) = \langle \underline{r}|\psi \rangle$, is now a function of the position $\underline{r}$ of the system. You can compare Eq. (7.1) with Eq. (2.143) for the analogous expression for a system in one dimension. The spectrum of the position operator is continuous and therefore we need to integrate over the entire three-dimensional volume, while in Eq. (2.143) we were integrating along a one-dimensional line only.

The modulus square of the wave function has the same probabilistic interpretation discussed for one-dimensional systems. $|\Psi(\underline{r})|^2 \, d^3r$ is the probability that a measurement of the position of the particle yields a result in the infinitesimal volume element d^3r at $\underline{r}$ with $d^3r = dx \, dy \, dz$ in Cartesian coordinates or $r^2 \sin\theta \, dr \, d\theta \, d\phi$ in spherical polars. Thus $|\Psi(\underline{r})|^2$ is a *probability per unit volume*, and has dimensions of length^{-3}. The normalisation condition becomes

$$\int_{\text{all space}} d^3r \, |\Psi(\underline{r})|^2 = 1. \tag{7.2}$$

The physical meaning of this equation should be clear by now: the probability of finding the system *somewhere* in space must be equal to one.

Note that if the system under consideration is a two-dimensional system, then the position vector $\underline{r}$ is a two-dimensional Euclidean vector, and all the integrals are computed over two-dimensional surfaces.

For a time-dependent vector, we include the time dependence explicitly in the ket and denote the state vector as $|\Psi(t)\rangle$. The wave function associated with the state depends on both spatial coordinates and time:

$$\Psi(\underline{r}, t) = \langle \underline{r}|\Psi(t)\rangle. \tag{7.3}$$

The scalar product of two states now involves an integral over the whole volume, viz.

$$\langle \Phi(t')|\Psi(t)\rangle = \int d^3r \, \Phi(\underline{r}, t')^* \Psi(\underline{r}, t). \tag{7.4}$$

Note that in this example we have considered state vectors at different times. The kets can change as a function of time. The integral in Eq. (7.4) only runs over the spatial coordinates. The normalisation condition in Dirac's notation is

$$\langle \Psi(t)|\Psi(t)\rangle = 1. \tag{7.5}$$

When using Dirac's notation, the condition is written exactly in the same form in any number of dimensions. However, the details of the scalar product denoted by the bra and ket notation are different as we vary the number of dimensions – the integral in the scalar product extends over the whole spatial volume.

7.2 Observables

7.2.1 General Considerations

As far as observables are concerned, there is no conceptual difference between the systems on a one-dimensional line that we discussed earlier, and systems that are defined in higher-dimensional spaces. Observables are in one-to-one correspondence with linear Hermitian operators acting on the state vectors, and the postulates that we formulated earlier in this book remain valid.

Let us summarise the main features of the correspondence:

- The eigenvalues of the operator $\hat{O}$ yield the possible outcomes of a measurement of the observable O.
- The orthogonality relations that we proved for the eigenvectors of a Hermitian operator are still true.
- Likewise, the set of eigenvectors of a Hermitian operator

$$\hat{O}|\psi_k\rangle = O_k|\psi_k\rangle \tag{7.6}$$

are a *complete* set and therefore any state vector can be expanded using these eigenvectors as basis:

$$|\psi\rangle = \sum_k c_k|\psi_k\rangle. \tag{7.7}$$

The eigenstates $|\psi_k\rangle$ are the basis vectors, and the coefficients c_k are the coordinates of the vector $|\psi\rangle$ in that basis. As we discussed earlier, the vector space in quantum mechanics is a complex vector space, and the coefficients c_k are complex. Keeping the basis vectors $|\psi_k\rangle$ fixed and changing the state vector $|\psi\rangle$, the coefficients c_k will change.

Example

We can think of the operator $\hat{O}$ acting directly on the wave function, with the usual identification

$$\hat{O}\psi(\underline{r}) = \left(\hat{O}\psi\right)(\underline{r}) = \langle\underline{r}|\hat{O}|\psi\rangle, \tag{7.8}$$

and we can rewrite the results above using wave functions.

The eigenvalue equation for a Hermitian operator $\hat{O}$ in a three-dimensional system can be written in terms of the wave function of the eigenstates as

$$\hat{O}\psi_k(\underline{r}) = O_k\psi_k(\underline{r}). \tag{7.9}$$

Note that in this case the eigenfunctions depend on the vector $\underline{r}$. This is the only peculiarity of the three-dimensional case. The eigenvalues O_k are the possible outcomes of measuring O, the eigenfunctions $\psi_k(\underline{r})$ describe quantum states where a measurement of O would yield O_k with probability one.

The completeness of the eigenfunctions is expressed by the fact that

$$\psi(\underline{r}) = \sum_k c_k\psi_k(\underline{r}), \tag{7.10}$$

for any generic quantum state $\psi(\underline{r})$. As before, $|c_k|^2$ yields the probability of getting the value O_k upon measuring O in the state ψ. Using the orthonormality of the eigenstates, the coefficients c_k are obtained from the scalar products

$$c_k = \langle\psi_k|\psi\rangle, \tag{7.11}$$

where now the scalar product requires the evaluation of a three-dimensional integral, according to Eq. (7.4).

Using Dirac's notation, the definition of the Hermitian conjugate is formally unchanged:

$$\langle \phi | \hat{O}^\dagger | \psi \rangle = \left(\langle \psi | \hat{O} | \phi \rangle \right)^* . \tag{7.12}$$

The relation above can be rewritten using wave functions and integrals:

$$\int d^3r \, \phi(\underline{r})^* \hat{O}^\dagger \psi(\underline{r}) = \left(\int d^3r \, \psi(\underline{r})^* \hat{O} \phi(\underline{r}) \right)^* . \tag{7.13}$$

Once again, remember that now we need to integrate over three-dimensional space, d^3r is the infinitesimal integration volume.

7.2.2 Position

The position operator in three spatial dimensions is a vector of operators

$$\underline{\hat{X}} = \left(\hat{X}, \hat{Y}, \hat{Z} \right), \tag{7.14}$$

where each operator is associated with one Cartesian coordinate. The eigenstates of position satisfy

$$\hat{X}_k | \underline{r} \rangle = x_k | \underline{r} \rangle, \quad k = 1, 2, 3. \tag{7.15}$$

Their action on the wave function is

$$\hat{X} \psi(\underline{r}) = x \, \psi(\underline{r}), \tag{7.16}$$

$$\hat{Y} \psi(\underline{r}) = y \, \psi(\underline{r}), \tag{7.17}$$

$$\hat{Z} \psi(\underline{r}) = z \, \psi(\underline{r}), \tag{7.18}$$

where $\underline{r} = (x, y, z)$. We can combine these three equations using indices

$$\hat{X}_k \psi(\underline{r}) = x_k \psi(\underline{r}), \quad k = 1, 2, 3, \tag{7.19}$$

where $x_1 = x, x_2 = y, x_3 = z$. And once again, we can practice the use of Dirac's notation and write

$$\langle \underline{r} | \hat{X}_k | \psi \rangle = x_k \langle \underline{r} | \psi \rangle. \tag{7.20}$$

As a result, the components of the position operator commute between them:

$$\left[\hat{X}_k, \hat{X}_l \right] = 0, \quad k, l = 1, 2, 3. \tag{7.21}$$

7.2.3 Momentum

It is interesting to consider the generalisation of the momentum operator defined in Eq. (3.85) for the one-dimensional system. The momentum is a three-dimensional vector $\underline{P}$ which can be represented by its three components in a Cartesian reference frame (P_x, P_y, P_z). Following our discussion for a system in one spatial dimension, we define the momentum operator by defining its action on the wave function; with each component of the momentum we associate a Hermitian operator:

$$\hat{P}_x \rightarrow -i\hbar \frac{\partial}{\partial x}, \tag{7.22}$$

$$\hat{P}_y \rightarrow -i\hbar \frac{\partial}{\partial y}, \tag{7.23}$$

$$\hat{P}_z \rightarrow -i\hbar \frac{\partial}{\partial z}. \tag{7.24}$$

Hence for instance:[1]

$$\hat{P}_x \psi(\underline{r}) = -i\hbar \left. \frac{\partial \psi}{\partial x} \right|_{\underline{r}}. \tag{7.25}$$

A concise notation for the momentum operator is:

$$\underline{\hat{P}} \rightarrow -i\hbar \underline{\nabla}. \tag{7.26}$$

We shall use the notation x_i, with $i = 1, 2, 3$, to denote the three components x, y, z, respectively. Similarly $\frac{\partial}{\partial x_i}$ will be used to indicate $\frac{\partial}{\partial x}, \frac{\partial}{\partial y}, \frac{\partial}{\partial z}$, respectively. Using a simple property of partial derivatives, we have

$$\hat{P}_i \hat{P}_j \psi(\underline{r}) = -\hbar^2 \frac{\partial^2}{\partial x_i \partial x_j} \psi(\underline{r}) = -\hbar^2 \frac{\partial^2}{\partial x_j \partial x_i} \psi(\underline{r}) = \hat{P}_j \hat{P}_i \psi(\underline{r}). \tag{7.27}$$

Different components of the momentum commute with each other, i.e. they are compatible observables.

Following the derivation that led to Eq. (3.98), you can readily prove the canonical commutation relations in three dimensions:

$$\left[\hat{X}_i, \hat{P}_j \right] = i\hbar \delta_{ij}. \tag{7.28}$$

Thus $\hat{X}$ *does not* commute with $\hat{P}_x$, but *does* commute with $\hat{P}_y$ and $\hat{P}_z$. These commutation relations lead to Heisenberg uncertainty relations involving the components of the position and the momentum of the particle:

$$\Delta X_i \cdot \Delta P_j \geq \frac{\hbar}{2} \delta_{ij}. \tag{7.29}$$

7.3 Dynamics

Similarly to the case of one-dimensional systems, the dynamics is determined by the Schrödinger equation, which can be written in the position representation as

$$i\hbar \frac{\partial}{\partial t} \Psi(\underline{r}, t) = \hat{H} \Psi(\underline{r}, t). \tag{7.30}$$

---

[1] Note that in the previous equations we used an arrow to denote that the momentum operator can be associated with the partial derivative with respect to the Cartesian coordinates only in the position representation. When the operator acts on the wave function, as in Eq. (7.25), we are explicitly using the position representation and it is perfectly legal to use an equals sign.

In three dimensions the Hamiltonian for the system can be expressed as

$$\hat{H} = \hat{T} + \hat{V}$$

$$= -\frac{\hbar^2}{2m}\nabla^2 + V(\underline{\hat{X}}), \tag{7.31}$$

where

$$\nabla^2 = \frac{\partial^2}{\partial x^2} + \frac{\partial^2}{\partial y^2} + \frac{\partial^2}{\partial z^2} \tag{7.32}$$

is the Laplace operator. Thus the time evolution of a quantum state is found by solving the differential equation

$$i\hbar\frac{\partial}{\partial t}\Psi(\underline{r},t) = \left\{-\frac{\hbar^2}{2m}\nabla^2 + V(\underline{r})\right\}\Psi(\underline{r},t). \tag{7.33}$$

As in the one-dimensional case, we are going to solve the time-dependent Schrödinger equation by expanding the solution in eigenstates of the Hamiltonian. The latter are obtained by solving the eigenvalue problem for $\hat{H}$, i.e. the time-independent Schrödinger equation:

$$\hat{H}\psi_E(\underline{r}) = \left\{-\frac{\hbar^2}{2m}\nabla^2 + V(\underline{r})\right\}\psi_E(\underline{r}) = E\psi_E(\underline{r}). \tag{7.34}$$

There are many different ways to solve the time-independent Schrödinger equation, depending on the system under consideration. In the rest of this chapter we will explain how to use separation of variables and we are going to apply this technique to the case of the three-dimensional harmonic oscillator. Further examples will be considered in later chapters.

7.4 Separation of Variables: Three-Dimensional Harmonic Oscillator

Separation of variables is a useful technique for solving a time-independent Schrödinger equation when the Hamiltonian is the sum of terms, each depending on a subset of variables. As an illustration, we are going to discuss the solution of the time-independent Schrödinger equation for the isotropic harmonic oscillator in three dimensions, for which the potential is

$$V(\underline{r}) = \tfrac{1}{2}m\omega^2 r^2 = \tfrac{1}{2}m\omega^2\left(x^2 + y^2 + z^2\right). \tag{7.35}$$

The Hamiltonian of the system is the sum of three terms:

$$\hat{H} = \hat{H}_x + \hat{H}_y + \hat{H}_z, \tag{7.36}$$

where, as suggested by the notation, the first term on the right-hand side depends only on x, the second term only on y and the last one only on z.

The eigenvalue problem can be separated in Cartesian coordinates by writing the wave function as the product of three independent functions of x, y and z, respectively:

$$\psi_E(\underline{r}) = X(x)\,Y(y)\,Z(z). \tag{7.37}$$

Using this ansatz we can rewrite the time-independent Schrödinger equation as the sum of three terms. The term in the first bracket only depends on x, the second one only depends on y and the third one only depends on z:

$$\left\{ -\frac{\hbar^2}{2m} \frac{1}{X} \frac{d^2 X}{dx^2} + \frac{1}{2} m\omega^2 x^2 \right\} + \left\{ -\frac{\hbar^2}{2m} \frac{1}{Y} \frac{d^2 Y}{dy^2} + \frac{1}{2} m\omega^2 y^2 \right\}$$
$$+ \left\{ -\frac{\hbar^2}{2m} \frac{1}{Z} \frac{d^2 Z}{dz^2} + \frac{1}{2} m\omega^2 z^2 \right\} = E. \tag{7.38}$$

Each term in braces must be equal to a constant, so we can write

$$E_{n_x} + E_{n_y} + E_{n_z} = E \tag{7.39}$$

with, for example:

$$\left\{ -\frac{\hbar^2}{2m} \frac{d^2 X}{dx^2} + \frac{1}{2} m\omega^2 x^2 \, X \right\} = E_{n_x} X, \tag{7.40}$$

which is the time-independent Schrödinger equation for a one-dimensional oscillator problem for which we know the solution:

$$E_{n_x} = \left(n_x + \frac{1}{2} \right) \hbar\omega, \quad n_x = 0, 1, 2, 3 \ldots, \tag{7.41}$$
$$X(x) = u_{n_x}(x) = C_{n_x}, \exp(-\alpha^2 x^2 / 2) \, H_{n_x}(\alpha x). \tag{7.42}$$

Similar solutions are found for $Y(y)$ and $Z(z)$. Thus, the eigenvalues for the three-dimensional harmonic oscillator are

$$E_n = \left(n_x + n_y + n_z + \frac{3}{2} \right) \hbar\omega, \quad n_x, n_y, n_z = 0, 1, 2, 3 \ldots, \tag{7.43}$$
$$\equiv \left(n + \frac{3}{2} \right) \hbar\omega, \quad n = 0, 1, 2, 3 \ldots. \tag{7.44}$$

The corresponding eigenfunctions are given by the product of the respective solutions of the one-dimensional problem:

$$\psi_{n_x,n_y,n_z}(\underline{r}) = u_{n_x}(x) \, u_{n_y}(y) \, u_{n_z}(z). \tag{7.45}$$

Creation and Annihilation Operators

Note that the algebraic solution that we developed for the one-dimensional case can be extended to the three-dimensional system by defining creation and annihilation operators in each direction:

$$\hat{a}_k = \sqrt{\frac{m\omega}{2\hbar}} \, \hat{X}_k + \frac{i}{\sqrt{2m\omega\hbar}} \, \hat{P}_k, \tag{7.46}$$

$$\hat{a}_k^\dagger = \sqrt{\frac{m\omega}{2\hbar}} \, \hat{X}_k - \frac{i}{\sqrt{2m\omega\hbar}} \, \hat{P}_k, \tag{7.47}$$

where $k = 1, 2, 3$ in three dimensions.[2] Using the canonical commutation relations, Eq. (7.28), we obtain

$$\left[\hat{a}_k, \hat{a}_l^\dagger\right] = \delta_{kl},$$ (7.48)

and hence the Hamiltonian can be written as

$$\hat{H} = \hbar\omega \left(\hat{a}_x^\dagger \hat{a}_x + \hat{a}_y^\dagger \hat{a}_y + \hat{a}_z^\dagger \hat{a}_z + \frac{3}{2}\right).$$ (7.49)

The wave functions found in Eq. (7.45) are obtained as

$$\psi_{n_x, n_y, n_z}(\underline{r}) = \langle \underline{r} | n_x n_y n_z \rangle,$$ (7.50)

where

$$|n_x n_y n_z\rangle = \frac{1}{\sqrt{n_x! \, n_y! \, n_z!}} \left(\hat{a}_x^\dagger\right)^{n_x} \left(\hat{a}_y^\dagger\right)^{n_y} \left(\hat{a}_z^\dagger\right)^{n_z} |0\rangle,$$ (7.51)

and $|0\rangle$ denotes the vacuum state:

$$\hat{a}_k |0\rangle = |\emptyset\rangle, \quad k = 1, 2, 3.$$ (7.52)

7.5 Degeneracy

We see in the previous example a new feature, that we had not encountered in the one-dimensional case. The three-dimensional harmonic oscillator displays **degeneracy**, i.e. more than one eigenfunction corresponds to the *same* eigenvalue. Equation (7.43) shows that a given value of E_n can arise in more than one way. For example:

n	n_x	n_y	n_z	g_n
0	0	0	0	1
	1	0	0	
1	0	1	0	3
	0	0	1	

g_n is the number of ways that a given value of n, and hence of the energy E_n, can arise, and is called the *degree of degeneracy*. We say, for example, that the $n = 1$ level is three-fold degenerate, meaning that there are three distinct quantum states of the same energy, corresponding to the eigenfunctions

$$u_1(x) \cdot u_0(y) \cdot u_0(z), \quad u_0(x) \cdot u_1(y) \cdot u_0(z), \quad u_0(x) \cdot u_0(y) \cdot u_1(z).$$ (7.53)

---

[2] In an arbitrary number of dimensions d, the index k spans them all, $k = 1, 2, \ldots, d$.

Summary

- We start the chapter with a brief description of the quantum states of a three-dimensional quantum system as vectors of a Hilbert space. It is important to notice how the use of Dirac notation makes the transition from one-dimensional to three-dimensional systems almost seamless. (7.1)
- Following the general principles introduced in previous chapters, observables are implemented as Hermitian operators. (7.2.1)
- Position and momentum operators are generalised to an arbitrary number of dimensions and obey canonical commutation relations. (7.2.2, 7.2.3)
- The time evolution of a three-dimensional quantum system is given by the Schrödinger equation. The stationary states are the eigenstates of the Hamiltonian and have the same properties that were discussed in the one-dimensional case. (7.3)
- For specific forms of the Hamiltonian, the three-dimensional time-independent Schrödinger equation can be simplified using a technique known as separation of variables. This is a very useful technique, so make sure that you understand the general idea and the conditions under which it can be used. (7.4)
- Solving the time-independent Schrödinger equation, we encounter the first examples of degenerate eigenvalues. (7.5)

Problems

7.1   Find the energy eigenstates for a particle in a two-dimensional square infinite potential well of size $a \times a$.

Discuss the degeneracy of the first two energy levels.

7.2   Consider a two-dimensional system. The position of the particle can be described equivalently by the Cartesian coordinates (x, y), or the polar ones (r, θ). Sketch a two-dimensional plane, and identify the Cartesian and polar coordinates of a point P. Find the relation between the Cartesian and the polar coordinates.

The gradient operator in two-dimensional polar coordinates is

$$\underline{\nabla} = \hat{r}\frac{\partial}{\partial r} + \hat{\theta}\frac{1}{r}\frac{\partial}{\partial \theta}. \tag{7.54}$$

Use this expression to compute the Laplacian in polar coordinates.

7.3   Write down the TISE for the two-dimensional harmonic oscillator of frequency ω. Find the eigenvalues, eigenfunctions and their degeneracy by separating the Cartesian coordinates x, y.

Use the result in the problem above to write the equation for the radial wave function of a stationary state of the two-dimensional harmonic oscillator.

7.4   Use raising and lowering operators to solve the two-dimensional harmonic oscillator. Discuss the energy levels and their degeneracy.

7.5 Consider two harmonic oscillators, 1 and 2, and assume their potential energy to be

$$V_0(\hat{X}_1, \hat{X}_2) = \tfrac{1}{2}m\omega^2(\hat{X}_1 - a)^2 + \tfrac{1}{2}m\omega^2(\hat{X}_2 + a)^2.$$

Factorise the Hamiltonian, and define the creation and annihilation operators. Using the latter, find the eigenstates and eigenvalues of the Hamiltonian.

Consider now the case where the two oscillators are coupled by an extra term in the potential:

$$\Delta V = \lambda m\omega^2(\hat{X}_1 - \hat{X}_2)^2.$$

Introduce the observables

$$\hat{X}_G = \tfrac{1}{2}(\hat{X}_1 + \hat{X}_2), \quad \hat{P}_G = \hat{P}_1 + \hat{P}_2, \quad \hat{X}_R = (\hat{X}_1 - \hat{X}_2), \quad \hat{P}_R = \tfrac{1}{2}(\hat{P}_1 - \hat{P}_2).$$

Compute the commutation relations between these new operators.

Keeping in mind the commutation relations computed above, rewrite the Hamiltonian in terms of the new operators. (Two new frequencies $\omega_G = \omega$ and $\omega_R = \omega\sqrt{1 + 4\lambda}$ should appear.) Define new creation and annihilation operators, and find the eigenvalues and eigenvectors of the coupled oscillators.

8 Angular Momentum

Now that we have developed the tools to describe three-dimensional systems, we are ready to introduce into our quantum-mechanical framework the concept of **angular momentum**. Recall that in classical mechanics angular momentum is defined as the vector product of position and momentum:

$$\underline{L} = \underline{r} \times \underline{p}. \tag{8.1}$$

The angular momentum is itself a three-dimensional *vector*. The three Cartesian components of the angular momentum are

$$L_x = y\,p_z - z\,p_y, \qquad L_y = z\,p_x - x\,p_z, \qquad L_z = x\,p_y - y\,p_x. \tag{8.2}$$

Following the general principles of quantum mechanics, this chapter will define the Hermitian operators that can be associated with the components of the angular momentum. We will study their properties and find a set of eigenfunctions, which provide a basis to expand the angular dependence of three-dimensional wave functions. We will find that the properties of the angular momentum of quantum systems are quite different from the ones we are familiar with in classical mechanics. We will highlight these peculiar features and try to develop some intuition of how angular momentum works for a quantum system.

8.1 Angular Momentum Operator

For a quantum system the angular momentum is an observable, which can be measured by some appropriate experimental apparatus. The state of the system determines the probability distribution of the outcomes of repeated measurements. According to the postulates that we have spelled out in previous chapters, we need to associate with each observable a Hermitian operator and therefore we need to define one Hermitian operator for each component of the angular momentum. We have already defined the operators $\hat{\underline{X}}$ and $\hat{\underline{P}}$ associated respectively with the position and the momentum of a particle. Using the expression for the angular momentum in classical mechanics as a guide, we define the vector of operators

$$\hat{\underline{L}} = \hat{\underline{X}} \times \hat{\underline{P}}, \tag{8.3}$$

which we associate with angular momentum. Using the canonical commutation relations, Eq. (7.28), we can easily prove that

$$[\hat{L}_x, \hat{L}_y] = i\hbar\hat{L}_z, \qquad [\hat{L}_y, \hat{L}_z] = i\hbar\hat{L}_x, \qquad [\hat{L}_z, \hat{L}_x] = i\hbar\hat{L}_y. \tag{8.4}$$

Exercise 8.1.1 Check that you can derive the commutation relations above.

Once again, it is useful to get familiar with the more compact notation:

$$\left[\hat{L}_i, \hat{L}_j\right] = i\hbar\,\varepsilon_{ijk}\hat{L}_k. \tag{8.5}$$

On the right-hand side of Eq. (8.5), the factor of $\hbar$ guarantees that the equation is correct on dimensional grounds. The factor of i makes the operator on the right-hand side anti-Hermitian, as expected since on the left-hand side we have a commutator of Hermitian operators. Finally, the Levi–Civita tensor is antisymmetric in the indices i and j, which matches the symmetry properties of the left-hand side under the exchange of these indices. Note that, as the indices i and j take the values $1, 2, 3$, Eq. (8.5) is a compact notation that summarises nine separate equations.

As usual when dealing with the spatial properties of quantum states, we can choose a basis made up of eigenstates of the position operator, $|\underline{r}\rangle$, and represent the state vector $|\psi\rangle$ as a wave function, $\psi(\underline{r}) = \langle\underline{r}|\psi\rangle$, and operators as differential operators acting on the wave function. We have already seen that the momentum operator in this basis is represented as $\underline{\hat{P}} = -i\hbar\underline{\nabla}$. Equation (8.3) yields explicit expressions for the components of the angular momentum as differential operators:

$$\hat{L}_x \to -i\hbar\left(y\frac{\partial}{\partial z} - z\frac{\partial}{\partial y}\right), \ \hat{L}_y \to -i\hbar\left(z\frac{\partial}{\partial x} - x\frac{\partial}{\partial z}\right), \ \hat{L}_z \to -i\hbar\left(x\frac{\partial}{\partial y} - y\frac{\partial}{\partial x}\right). \tag{8.6}$$

Once again we used an arrow to emphasise that the angular momentum operator is represented by that specific differential operator in the position representation. Equation (8.6) can be economically rewritten as

$$\hat{L}_i = -i\hbar\,\varepsilon_{ijk}\,x_j\,\frac{\partial}{\partial x_k}, \tag{8.7}$$

where we have to sum over the repeated indices j and k.

Mathematical Aside: Levi–Civita and Indices
In Eq. (8.7) we use the following convention:

$$x_1 = x, \quad x_2 = y, \quad x_3 = z \tag{8.8}$$

to denote the three components of the position vector. The same convention is also used for the partial derivatives:

$$\frac{\partial}{\partial x_1} = \frac{\partial}{\partial x}, \quad \frac{\partial}{\partial x_2} = \frac{\partial}{\partial y}, \quad \frac{\partial}{\partial x_3} = \frac{\partial}{\partial z}. \tag{8.9}$$

In general, the components of a vector $\underline{V}$ can be labelled as

$$V_1 = V_x, \quad V_2 = V_y, \quad V_3 = V_z. \tag{8.10}$$

The symbol ε_{ijk} denotes the totally antisymmetric unit tensor:

$$\varepsilon_{123} = \varepsilon_{231} = \varepsilon_{312} = 1, \quad \text{cyclic indices,} \tag{8.11}$$

$$\varepsilon_{213} = \varepsilon_{132} = \varepsilon_{321} = -1, \quad \text{anticyclic indices.} \tag{8.12}$$

Out of 27 components, only the six above are actually different from zero. Check that you are comfortable with Eq. (8.7).

The following relations are useful:

$$\varepsilon_{ikl}\varepsilon_{imn} = \delta_{km}\delta_{ln} - \delta_{kn}\delta_{lm}, \tag{8.13}$$

$$\varepsilon_{ikl}\varepsilon_{ikm} = 2\delta_{lm}, \tag{8.14}$$

$$\varepsilon_{ikl}\varepsilon_{ikl} = 6. \tag{8.15}$$

Example

Instead of using the canonical commutation relations, we can derive the commutation relations between the components $\hat{L}_i$ using their representation as differential operators:

$$\hat{L}_x\hat{L}_y\,\psi(x,y,z) = -\hbar^2\left(y\frac{\partial}{\partial z} - z\frac{\partial}{\partial y}\right)\left(z\frac{\partial}{\partial x} - x\frac{\partial}{\partial z}\right)\psi(x,y,z)$$

$$= -\hbar^2\left\{y\frac{\partial}{\partial x} + yz\frac{\partial^2}{\partial z\partial x} - yx\frac{\partial^2}{\partial z^2} - z^2\frac{\partial^2}{\partial y\partial x} + zx\frac{\partial^2}{\partial y\partial z}\right\}\psi(x,y,z), \tag{8.16}$$

whilst

$$\hat{L}_y\hat{L}_x\,\psi(x,y,z) = -\hbar^2\left(z\frac{\partial}{\partial x} - x\frac{\partial}{\partial z}\right)\left(y\frac{\partial}{\partial z} - z\frac{\partial}{\partial y}\right)\psi(x,y,z)$$

$$= -\hbar^2\left\{zy\frac{\partial^2}{\partial x\partial z} - z^2\frac{\partial^2}{\partial x\partial y} - xy\frac{\partial^2}{\partial z^2} + xz\frac{\partial^2}{\partial z\partial y} + x\frac{\partial}{\partial y}\right\}\psi(x,y,z). \tag{8.17}$$

Noting the usual properties of partial derivatives

$$\frac{\partial^2}{\partial x\partial z} = \frac{\partial^2}{\partial z\partial x}, \quad \text{etc.} \tag{8.18}$$

we obtain on subtraction the desired result:

$$\left[\hat{L}_x, \hat{L}_y\right]\psi(x,y,z) = \hbar^2\left(x\frac{\partial}{\partial y} - y\frac{\partial}{\partial x}\right)\psi(x,y,z) = i\hbar\hat{L}_z\,\psi(x,y,z). \tag{8.19}$$

Angular momentum plays a central role in discussing *central potentials*, i.e. potentials that only depend on the radial coordinate r. For that purpose, it will prove useful to have expressions for the operators $\hat{L}_x$, $\hat{L}_y$ and $\hat{L}_z$ in spherical polar coordinates. Using the expression for the Cartesian coordinates as functions of the spherical ones, and the chain rule for the derivative, yields

$$\hat{L}_x \rightarrow i\hbar \left(\sin\phi \frac{\partial}{\partial\theta} + \cot\theta \, \cos\phi \frac{\partial}{\partial\phi} \right), \tag{8.20}$$

$$\hat{L}_y \rightarrow i\hbar \left(-\cos\phi \frac{\partial}{\partial\theta} + \cot\theta \, \sin\phi \frac{\partial}{\partial\phi} \right), \tag{8.21}$$

$$\hat{L}_z \rightarrow -i\hbar \frac{\partial}{\partial\phi}. \tag{8.22}$$

The expressions for the components of angular momentum in spherical coordinates only involve the angles θ and ϕ.

Note once again that the Cartesian components of the angular momentum **do not** commute with each other. Following our previous discussion on compatible observables, this means that the components are **not** compatible observables. We cannot measure, for instance, L_x and L_y simultaneously, and we do not have a basis of common eigenfunctions of the two operators. Physically, this also implies that measuring one component of the angular momentum modifies the probability of finding a given result for the other two.

8.2 Squared Norm of the Angular Momentum

Let us now introduce an operator that represents the square of the magnitude of the angular momentum:

$$\hat{L}^2 = \hat{L}_x^2 + \hat{L}_y^2 + \hat{L}_z^2 = \sum_{i=1}^{3} \hat{L}_i^2. \tag{8.23}$$

We will often refer to this quantity as the **angular momentum squared**. The operator can be represented as a differential operator acting on the wave function in spherical polar coordinates:

$$\hat{L}^2 \rightarrow -\hbar^2 \left[\frac{1}{\sin\theta} \frac{\partial}{\partial\theta} \left(\sin\theta \frac{\partial}{\partial\theta} \right) + \frac{1}{\sin^2\theta} \frac{\partial^2}{\partial\phi^2} \right]. \tag{8.24}$$

It is important to realise that this observable *is compatible with any of the Cartesian components of the angular momentum*:

$$[\hat{L}^2, \hat{L}_x] = [\hat{L}^2, \hat{L}_y] = [\hat{L}^2, \hat{L}_z] = 0. \tag{8.25}$$

Sample Proof

It is useful to derive one of these relations in full detail so that the reader can get familiar with manipulations that involve the angular momentum operators. Consider for instance the commutator $[\hat{L}^2, \hat{L}_z]$:

$$[\hat{L}^2, \hat{L}_z] = [\hat{L}_x^2 + \hat{L}_y^2 + \hat{L}_z^2, \hat{L}_z] \quad \text{from the definition of } \hat{L}^2$$

$$= [\hat{L}_x^2, \hat{L}_z] + [\hat{L}_y^2, \hat{L}_z] + [\hat{L}_z^2, \hat{L}_z]$$

$$= [\hat{L}_x^2, \hat{L}_z] + [\hat{L}_y^2, \hat{L}_z] \quad \text{since } \hat{L}_z \text{ commutes with itself}$$

$$= \hat{L}_x[\hat{L}_x, \hat{L}_z] + [\hat{L}_x, \hat{L}_z]\hat{L}_x + \hat{L}_y[\hat{L}_y, \hat{L}_z] + [\hat{L}_y, \hat{L}_z]\hat{L}_y.$$

We can use the commutation relation $[\hat{L}_x, \hat{L}_z] = -i\hbar\hat{L}_y$ to rewrite the first two terms on the right-hand side as

$$-i\hbar\left(\hat{L}_x\hat{L}_y + \hat{L}_y\hat{L}_x\right).$$

In a similar way, we can use $[\hat{L}_y, \hat{L}_z] = i\hbar\hat{L}_x$ to rewrite the second two terms as

$$i\hbar\left(\hat{L}_y\hat{L}_x + \hat{L}_x\hat{L}_y\right).$$

Thus, on substituting we find that

$$[\hat{L}^2, \hat{L}_z] = -i\hbar\hat{L}_x\hat{L}_y - i\hbar\hat{L}_y\hat{L}_x + i\hbar\hat{L}_y\hat{L}_x + i\hbar\hat{L}_x\hat{L}_y = 0.$$

8.3 Eigenfunctions

The compatibility theorem tells us that $\hat{L}^2$ and $\hat{L}_z$ thus have *simultaneous eigenfunctions*. The expressions in Eqs. (8.22) and (8.24) show that these eigenfunctions depend only on the angles θ and ϕ of the spherical coordinates. The eigenfunctions turn out to be the **spherical harmonics**

$$Y_\ell^m(\theta, \phi) = (-1)^m \left[\frac{2\ell + 1}{4\pi}\frac{(\ell - m)!}{(\ell + m)!}\right]^{1/2} P_\ell^m(\cos\theta) \exp\left(im\phi\right), \qquad (8.26)$$

where $\ell = 0, 1, 2, 3, \ldots$ and $P_\ell^m(\cos\theta)$ are known as the **associated Legendre polynomials**. Some examples of spherical harmonics will be given below, while the construction of the generic solution is discussed using algebraic methods later in this chapter.

The eigenvalue equation for $\hat{L}^2$ is

$$\hat{L}^2 Y_\ell^m(\theta, \phi) = \hbar^2\ell(\ell + 1) Y_\ell^m(\theta, \phi). \qquad (8.27)$$

The eigenvalue $\hbar^2\ell(\ell + 1)$ is *degenerate*; there exist $(2\ell + 1)$ eigenfunctions corresponding to a given ℓ and they are distinguished by the label m which can take any of the $(2\ell + 1)$ values

$$m = \ell, \ell - 1, \ldots, -\ell. \qquad (8.28)$$

In fact it is easy to show that m labels the eigenvalues of $\hat{L}_z$. Since

$$Y_\ell^m(\theta, \phi) \sim \exp\left(im\phi\right), \qquad (8.29)$$

we obtain directly that

$$\hat{L}_z Y_\ell^m(\theta, \phi) = -i\hbar\frac{\partial}{\partial\phi} Y_\ell^m(\theta, \phi) = m\hbar Y_\ell^m(\theta, \phi), \qquad (8.30)$$

confirming that the spherical harmonics are also eigenfunctions of $\hat{L}_z$ with eigenvalues $\hbar m$. Note that in both cases the eigenvalues have the right physical dimensions to be values of the square of the angular momentum and of one of its components, respectively. The eigenvalues of $\hat{L}^2$ and $\hat{L}_z$ are sufficient to identify uniquely one eigenfunction. The set made up of these two observables is a CSCO for this problem.

The explicit solution of the eigenvalue equations for $\hat{L}^2$ and $\hat{L}_z$ is discussed in detail below using algebraic methods that are reminiscent of the ones introduced to solve the harmonic oscillator.

Mathematical Aside: The Spherical Harmonics

A few examples of spherical harmonics are

$$Y_0^0(\theta, \phi) = \frac{1}{\sqrt{4\pi}},$$

$$Y_1^0(\theta, \phi) = \sqrt{\frac{3}{4\pi}}\cos\theta,$$

$$Y_1^1(\theta, \phi) = -\sqrt{\frac{3}{8\pi}}\sin\theta \exp\left(i\phi\right),$$

$$Y_1^{-1}(\theta, \phi) = \sqrt{\frac{3}{8\pi}}\sin\theta \exp\left(-i\phi\right).$$

You can plug these expressions into the eigenvalue equations, act with the partial derivatives and check that indeed they yield the expected eigenvalues.

8.4 Physical Interpretation

We have arrived at the important conclusion that *angular momentum is quantised*. The square of the magnitude of the angular momentum can only assume one of the discrete set of values

$$\hbar^2 \ell(\ell + 1), \qquad \ell = 0, 1, 2, \dots \tag{8.31}$$

and the z-component of the angular momentum can only assume one of the discrete set of values

$$\hbar m, \qquad m = \ell, \ell - 1, \dots, -\ell \tag{8.32}$$

for a given value of ℓ. The integers ℓ and m are called the **angular momentum quantum number** and the **magnetic quantum number**, respectively.

Let us end this section with some jargon: we refer to a particle in a state with angular momentum quantum number ℓ as *having angular momentum ℓ*, rather than saying, more clumsily but accurately, that it has angular momentum of magnitude $\sqrt{\ell(\ell + 1)}\,\hbar$.

8.5 Algebraic Solution of the Eigenvalue Equations

In the sections above we have introduced the angular momentum starting from the classical expression

$$\underline{L} = \underline{r} \times \underline{p}, \tag{8.33}$$

and have defined a quantum-mechanical operator by replacing $\underline{r}$ and $\underline{p}$ with the corresponding operators: Eq. (8.33) then defines a triplet of operators that can be represented as differential operators acting on the wave functions. The eigenvalue equations for $\hat{L}^2$ and $\hat{L}_z$ can be written as differential equations, whose solutions yield the eigenvalues and the eigenfunctions of the angular momentum that we described above.

In this section we are going to follow a different approach, namely we will derive the quantisation of angular momentum directly from the commutation relations of the components of $\underline{\hat{L}}$. This approach is completely generic and does not rely on the specific realisation of the angular momentum as a differential operator. To emphasise this point, and to be able to use the same results when we introduce spin later in this book, we will use in this context the symbol $\underline{\hat{J}}$ to denote the angular momentum.

Commutation Relations

Remember that $\underline{\hat{J}}$ is a vector, i.e. it is a triplet of operators. In Cartesian coordinates $\underline{\hat{J}} = (\hat{J}_x, \hat{J}_y, \hat{J}_z)$, and the commutation relations of its components are

$$[\hat{J}_x, \hat{J}_y] = i\hbar \hat{J}_z, \quad [\hat{J}_y, \hat{J}_z] = i\hbar \hat{J}_x, \quad [\hat{J}_z, \hat{J}_x] = i\hbar \hat{J}_y. \tag{8.34}$$

We will consider these equations as the defining properties of operators that can be associated with angular momentum. They are the only assumptions that we are going to use in the derivations below, and do not require the operators to be represented as differential operators. The square of the angular momentum is represented by the operator

$$\hat{J}^2 = \hat{J}_x^2 + \hat{J}_y^2 + \hat{J}_z^2, \tag{8.35}$$

with the property that

$$[\hat{J}^2, \hat{J}_x] = [\hat{J}^2, \hat{J}_y] = [\hat{J}^2, \hat{J}_z] = 0. \tag{8.36}$$

The compatibility theorem tells us that, for example, the operators $\hat{J}^2$ and $\hat{J}_z$ have simultaneous eigenstates. We denote these common eigenstates by $|\lambda, m\rangle$, so that

$$\hat{J}^2 |\lambda, m\rangle = \hbar^2 \lambda |\lambda, m\rangle, \tag{8.37}$$

$$\hat{J}_z |\lambda, m\rangle = \hbar m |\lambda, m\rangle, \tag{8.38}$$

which means that the eigenvalues of $\hat{J}^2$ are denoted by $\hbar^2 \lambda$, whilst those of $\hat{J}_z$ are denoted by $\hbar m$.

Raising and Lowering Operators

We now address the problem of finding the allowed values of λ and m. In the spirit of what we did for the harmonic oscillator, we introduce **raising** and **lowering** operators:

$$\hat{J}_\pm = \hat{J}_x \pm i\hat{J}_y, \tag{8.39}$$

and calculate the commutators with $\hat{J}^2$ and $\hat{J}_z$. Since $\hat{J}^2$ commutes with both $\hat{J}_x$ and $\hat{J}_y$, we get immediately that

$$[\hat{J}^2, \hat{J}_\pm] = 0, \tag{8.40}$$

whilst

$$[\hat{J}_z, \hat{J}_+] = [\hat{J}_z, \hat{J}_x] + i[\hat{J}_z, \hat{J}_y] = i\hbar\,\hat{J}_y + i(-i\hbar\,\hat{J}_x) = \hbar\,\hat{J}_+ \tag{8.41}$$

and

$$[\hat{J}_z, \hat{J}_-] = [\hat{J}_z, \hat{J}_x] - i[\hat{J}_z, \hat{J}_y] = i\hbar\,\hat{J}_y - i(-i\hbar\,\hat{J}_x) = -\hbar\,\hat{J}_-. \tag{8.42}$$

The structure of these commutation relations is similar to the one we obtained when solving the TISE for the harmonic oscillator: we can show that $\hat{J}_+$ and $\hat{J}_-$ act as raising and lowering operators for the z-component of the angular momentum. Consider the action of the commutator in Eq. (8.41) on an eigenstate $|\lambda, m\rangle$:

$$[\hat{J}_z, \hat{J}_+]\,|\lambda, m\rangle = \left(\hat{J}_z\hat{J}_+ - \hat{J}_+\hat{J}_z\right)|\lambda, m\rangle = \hbar\,\hat{J}_+\,|\lambda, m\rangle\,; \tag{8.43}$$

we can use the eigenvalue equation $\hat{J}_z\,|\lambda, m\rangle = m\hbar\,|\lambda, m\rangle$ to rewrite[1] this as

$$\hat{J}_z\left(\hat{J}_+\,|\lambda, m\rangle\right) = (m+1)\hbar\left(\hat{J}_+\,|\lambda, m\rangle\right), \tag{8.44}$$

which says that $\hat{J}_+\,|\lambda, m\rangle$ is also an eigenstate of $\hat{J}_z$ but with eigenvalue $\hbar\,(m+1)$, unless $\hat{J}_+\,|\lambda, m\rangle = 0$. Thus the operator $\hat{J}_+$ acts as a raising operator for the z-component of angular momentum.

Similarly, the second commutator can be used to show that

$$\hat{J}_z\hat{J}_-\,|\lambda, m\rangle = (m-1)\hbar\,\hat{J}_-\,|\lambda, m\rangle, \tag{8.45}$$

which says that $\hat{J}_-\,|\lambda, m\rangle$ is also an eigenstate of $\hat{J}_z$ but with eigenvalue $\hbar\,(m-1)$, unless $\hat{J}_-\,|\lambda, m\rangle = 0$. Thus the operator $\hat{J}_-$ acts as a lowering operator for the z-component of angular momentum.

Notice that since $[\hat{J}^2, \hat{J}_\pm] = 0$, we also have

$$\hat{J}^2\left(\hat{J}_\pm\,|\lambda, m\rangle\right) = \hat{J}_\pm\left(\hat{J}^2\,|\lambda, m\rangle\right) = \hbar^2\lambda\left(\hat{J}_\pm\,|\lambda, m\rangle\right), \tag{8.46}$$

so that the states generated by the action of $\hat{J}_\pm$ are still eigenstates of $\hat{J}^2$ belonging to the same eigenvalue $\hbar^2\lambda$. Thus we can write

$$\hat{J}_+\,|\lambda, m\rangle = c_+\hbar\,|\lambda, m+1\rangle,$$
$$\hat{J}_-\,|\lambda, m\rangle = c_-\hbar\,|\lambda, m-1\rangle,$$

where $c_\pm$ are (dimensionless) constants of proportionality.

---

[1] We use the parentheses here to emphasise the interpretation of the equation.

Bounds on the Magnetic Quantum Number

For a given value of λ, we can show that $m^2 \leq \lambda$, so that m must have both a *maximum value*, $m_{\max}$ and a *minimum value*, $m_{\min}$.

Proof. Starting from the definitions introduced so far, we can write

$$\left(\hat{J}^2 - \hat{J}_z^2\right) |\lambda, m\rangle = \left(\hat{J}_x^2 + \hat{J}_y^2\right) |\lambda, m\rangle, \qquad \text{implying that}$$

$$\left(\lambda - m^2\right) \hbar^2 |\lambda, m\rangle = \left(\hat{J}_x^2 + \hat{J}_y^2\right) |\lambda, m\rangle.$$

Taking the scalar product with $\langle \lambda, m|$ yields

$$\left(\lambda - m^2\right) \hbar^2 = \langle \hat{J}_x^2 + \hat{J}_y^2 \rangle \geq 0, \tag{8.47}$$

so that

$$\lambda - m^2 \geq 0 \qquad \text{or} \qquad -\sqrt{\lambda} \leq m \leq \sqrt{\lambda}. \tag{8.48}$$

Hence the spectrum of $\hat{J}_z$ is bounded above and below, for a given λ. We can deduce that there must be two states such that

$$\hat{J}_+ |\lambda, m_{\max}\rangle = 0 \quad \text{and} \quad \hat{J}_- |\lambda, m_{\min}\rangle = 0. \tag{8.49}$$

To proceed further, we need a couple of identities for $\hat{J}^2$, which follow from the definitions of $\hat{J}_\pm$. Noting that

$$\hat{J}_+\hat{J}_- = \hat{J}_x^2 + \hat{J}_y^2 + i\hat{J}_y\hat{J}_x - i\hat{J}_x\hat{J}_y = \hat{J}_x^2 + \hat{J}_y^2 + \hbar\hat{J}_z,$$

$$\hat{J}_-\hat{J}_+ = \hat{J}_x^2 + \hat{J}_y^2 - i\hat{J}_y\hat{J}_x + i\hat{J}_x\hat{J}_y = \hat{J}_x^2 + \hat{J}_y^2 - \hbar\hat{J}_z,$$

we can write

$$\hat{J}^2 = \hat{J}_+\hat{J}_- - \hbar\hat{J}_z + \hat{J}_z^2 \quad \text{or, equally,} \quad \hat{J}^2 = \hat{J}_-\hat{J}_+ + \hbar\hat{J}_z + \hat{J}_z^2. \tag{8.50}$$

Applying the first of these to the state of minimum m, we find

$$\hat{J}^2 |\lambda, m_{\min}\rangle = \left(\hat{J}_+\hat{J}_- - \hbar\hat{J}_z + \hat{J}_z^2\right) |\lambda, m_{\min}\rangle$$

$$= \hbar^2 \left(-m_{\min} + m_{\min}^2\right) |\lambda, m_{\min}\rangle, \quad \text{since } \hat{J}_- |\lambda, m_{\min}\rangle = 0$$

$$= \hbar^2 m_{\min}(m_{\min} - 1) |\lambda, m_{\min}\rangle$$

$$= \hbar^2 \lambda |\lambda, m_{\min}\rangle.$$

Thus we deduce that

$$\lambda = m_{\min}(m_{\min} - 1).$$

Similarly, using the second of the two identities for $\hat{J}^2$:

$$\hat{J}^2 |\lambda, m_{\max}\rangle = \left(\hat{J}_-\hat{J}_+ + \hbar\hat{J}_z + \hat{J}_z^2\right) |\lambda, m_{\max}\rangle$$

$$= \hbar^2 \left(m_{\max} + m_{\max}^2\right) |\lambda, m_{\max}\rangle, \quad \text{since } \hat{J}_+ |\lambda, m_{\max}\rangle = 0$$

$$= \hbar^2 m_{\max}(m_{\max} + 1) |\lambda, m_{\max}\rangle$$

$$= \hbar^2 \lambda |\lambda, m_{\max}\rangle,$$

and hence we obtain a second expression for λ:

$$\lambda = m_{max}(m_{max} + 1).$$

Usually, m_{max} is denoted by j and so we write this as

$$\lambda = j(j + 1) = m_{min}(m_{min} - 1).$$

This is a quadratic equation for m_{min}:

$$m_{min}^2 - m_{min} - j^2 - j = 0,$$

which can be factorised as

$$(m_{min} + j)(m_{min} - j - 1) = 0,$$

and we see that, since $m_{min} \leq j$ by definition, the only acceptable root is

$$m_{min} = -j.$$

Now since m_{max} and m_{min} must differ by some integer,[2] k, say, we can write

$$m_{max} - m_{min} = k, \quad k = 0, 1, 2, 3, \ldots,$$

or

$$j - (-j) = 2j = k, \quad k = 0, 1, 2, 3, \ldots.$$

Hence the allowed values of j are

$$j = 0, \tfrac{1}{2}, 1, \tfrac{3}{2}, 2, \ldots \tag{8.51}$$

For a given value of j, we see that m ranges over the values

$$-j, -j + 1, \ldots, j - 1, j, \tag{8.52}$$

a total of $(2j + 1)$ values.

8.6 Nomenclature

It is useful at this stage to summarise the main results obtained above.

- The eigenvalues of $\hat{J}^2$ are $j(j + 1)\hbar^2$, where j is one of the allowed values

$$j = 0, \tfrac{1}{2}, 1, \tfrac{3}{2}, 2, \ldots. \tag{8.53}$$

- Since $\lambda = j(j + 1)$, we can equally well label the simultaneous eigenstates of $\hat{J}^2$ and $\hat{J}_z$ by j rather than by λ, so that

$$\hat{J}^2 |j, m\rangle = \hbar^2 j(j + 1) |j, m\rangle, \tag{8.54}$$

$$\hat{J}_z |j, m\rangle = \hbar m |j, m\rangle. \tag{8.55}$$

---

[2] Since we can go from one to the other in steps of one using raising/lowering operators.

- For a given value of j, there are $(2j + 1)$ possible eigenvalues of $\hat{J}_z$, denoted $m\hbar$, where m runs from j to $-j$ in integer steps.
- The set of $(2j+1)$ states $\{|j, m\rangle\}$ is called a **multiplet** of states with angular momentum j.

8.7 Normalisation

We need to determine the constants of proportionality $c_\pm$ that appear in the equations

$$\hat{J}_\pm|j, m\rangle = \hbar\, c_\pm|j, m \pm 1\rangle. \tag{8.56}$$

We will perform this calculation in full detail, in order to illustrate some common manipulations involving angular momenta. Let us consider

$$
\begin{aligned}
\langle j, m|\hat{J}_-\hat{J}_+|j, m\rangle &= c_+\hbar \langle j, m|\hat{J}_-|j, m + 1\rangle \\
&= c_+\hbar \langle j, m + 1|(\hat{J}_-)^\dagger|j, m\rangle^*, \quad \text{from the definition of } \dagger \\
&= c_+\hbar \langle j, m + 1|\hat{J}_+|j, m\rangle \quad \text{since } (\hat{J}_-)^\dagger = \hat{J}_+ \\
&= c_+ c_+^*\hbar^2 \langle j, m + 1|j, m + 1\rangle \\
&= |c_+|^2\hbar^2, \quad \text{from orthonormality.}
\end{aligned}
$$

But we can evaluate the left-hand side by making use of the identity Eq. (8.50):

$$\hat{J}_-\hat{J}_+ = \hat{J}^2 - \hat{J}_z^{\,2} - \hbar\hat{J}_z,$$

yielding

$$
\begin{aligned}
\langle j, m|\hat{J}_-\hat{J}_+|j, m\rangle &= \langle j, m|\hat{J}^2 - \hat{J}_z^{\,2} - \hbar\hat{J}_z|j, m\rangle \\
&= \hbar^2 \langle j, m|j(j + 1) - m^2 - m|j, m\rangle \\
&= \hbar^2 \left[j(j + 1) - m(m + 1)\right], \quad \text{from orthonormality.}
\end{aligned}
$$

Thus we obtain

$$|c_+|^2 = j(j + 1) - m(m + 1). \tag{8.57}$$

In similar fashion:

$$
\begin{aligned}
\langle j, m|\hat{J}_+\hat{J}_-|j, m\rangle &= \hbar\, c_- \langle j, m|\hat{J}_+|j, m - 1\rangle \\
&= \hbar\, c_- \langle j, m - 1|(\hat{J}_+)^\dagger|j, m\rangle^*, \quad \text{from the definition of } \dagger \\
&= \hbar\, c_- \langle j, m - 1|\hat{J}_-|j, m\rangle, \quad \text{since } (\hat{J}_+)^\dagger = \hat{J}_- \\
&= \hbar^2\, c_- c_-^* \langle j, m - 1|j, m - 1\rangle \\
&= \hbar^2\, |c_-|^2, \quad \text{from orthonormality,}
\end{aligned}
$$

together with the other identity

$$\hat{J}_+\hat{J}_- = \hat{J}^2 - \hat{J}_z^{\,2} + \hbar\hat{J}_z,$$

yields

$$|c_-|^2 = j(j+1) - m(m-1).$$

Condon–Shortley Phase Convention

Taking $c_\pm$ to be real and positive gives

$$c_\pm = \sqrt{j(j+1) - m(m \pm 1)}. \tag{8.58}$$

This is sometimes referred to as the Condon–Shortley phase convention.

8.8 Matrix Representations

For a given j, the quantities $\langle j, m' | \hat{J}_z | j, m \rangle$ are known as the *matrix elements* of $\hat{J}_z$. We can calculate what they are as follows:

$$\langle j, m' | \hat{J}_z | j, m \rangle = m\hbar \langle j, m' | j, m \rangle = m\hbar \, \delta_{m',m}, \tag{8.59}$$

where we have used the orthonormality properties of the basis. Why matrix elements? As we have already seen, we can regard the labels m' and m as *labelling the rows and columns, respectively, of a matrix*. Given that m' and m can each take $(2j + 1)$ values, the matrix in question is $(2j + 1) \times (2j + 1)$. Similarly, the matrix elements of the raising and lowering operators are given by

$$\langle j, m' | \hat{J}_\pm | j, m \rangle = c_\pm \hbar \langle j, m' | j, m \pm 1 \rangle = \sqrt{j(j+1) - m(m \pm 1)} \, \hbar \, \delta_{m', m \pm 1}.$$

Some of these matrix representations are worked out explicitly in the problems at the end of this chapter.

8.9 Wave Functions

Finally, for the case of orbital angular momentum, we can go back to the expressions of the operators $\hat{L}_k$ as differential operators acting on wave functions and use the results above to derive the expression of the spherical harmonics.

First, let us recall the expression for the raising and lowering operators

$$\hat{L}_\pm = \hbar e^{\pm i\phi} \left[\pm \frac{\partial}{\partial \theta} + i \cot \theta \frac{\partial}{\partial \phi} \right], \tag{8.60}$$

and for the third component of the angular momentum

$$\hat{L}_z = -i\hbar \frac{\partial}{\partial \phi}. \tag{8.61}$$

State with Highest m

As discussed earlier, the state with the maximum eigenvalue of L_z satisfies

$$\hat{L}_+ Y_\ell^\ell(\theta, \phi) = 0 \quad \text{and} \quad \hat{L}_z Y_\ell^\ell(\theta, \phi) = \hbar \ell Y_\ell^\ell(\theta, \phi). \tag{8.62}$$

The second equation above suggests that we should look for single-valued solutions of the form

$$Y_\ell^\ell(\theta, \phi) = F_\ell^\ell(\theta) e^{i\ell\phi}, \tag{8.63}$$

where ℓ is a non-negative integer. We then obtain

$$\left[\frac{d}{d\theta} - \ell \cot\theta \right] F_\ell^\ell(\theta) = 0. \tag{8.64}$$

The solution of Eq. (8.64) is

$$F_\ell^\ell(\theta) = c_\ell (\sin\theta)^\ell. \tag{8.65}$$

The computation of the normalisation c_ℓ is done by imposing

$$\int d\cos\theta d\phi \left| Y_\ell^\ell(\theta, \phi) \right|^2 = 1. \tag{8.66}$$

The computation of the integral is left as an exercise, which yields

$$c_\ell = \frac{(-)^\ell}{2^\ell \ell!} \sqrt{\frac{(2\ell+1)!}{4\pi}}, \tag{8.67}$$

where the overall phase of c_ℓ is a matter of convention.

Other States

The other states are obtained from Y_ℓ^ℓ by acting repeatedly with L_-, keeping in mind that

$$\hat{L}_\pm Y_\ell^m(\theta, \phi) = \hbar \sqrt{\ell(\ell+1) - m(m\pm 1)} Y_\ell^{m\pm 1}(\theta, \phi). \tag{8.68}$$

In particular, we have

$$Y_\ell^m(\theta, \phi) = \sqrt{\frac{(\ell+m)!}{(2\ell)!\,(\ell-m)!}} \left(\frac{\hat{L}_-}{\hbar} \right)^{\ell-m} Y_\ell^m(\theta, \phi). \tag{8.69}$$

The interested reader can check that Eq. (8.69) yields the spherical harmonics that we gave explicitly, earlier in this chapter.

Summary

- The first logical step in our discussion is the definition of the operators associated with the Cartesian components of the angular momentum, and the derivation of the commutation relations between them. The fact that the components of the angular momentum do not commute with each other is the basis of the quantum properties of angular momentum. (8.1)

- Having defined the Cartesian components, we can consider the operator associated with the angular momentum squared. We then analyse the compatibility of the angular momentum squared with the components of the angular momentum vector and deduce that $\left\{\hat{L}^2, \hat{L}_z\right\}$ form a set of compatible observables. (8.2)
- We write down and analyse the equations that yield the simultaneous eigenfunctions of $\hat{L}^2$ and $\hat{L}_z$. (8.3)
- We discuss their solution using algebraic methods, which leads to the introduction of raising and lowering operators for $\hat{L}_z$. It is useful to compare the operators introduced here to the creation and annihilation operators introduced for the solution of the harmonic oscillator. The main result of this analysis is that angular momentum in quantum mechanics is quantised; only very specific eigenvalues are possible. (8.5)
- We found an explicit expression for the normalised eigenfunctions. (8.7)
- Working at fixed values of ℓ, we find a matrix representation of the angular momentum operators. (8.8)
- We end this chapter by looking at the explicit expression for the wave functions associated with the eigenvalues of $\hat{L}^2$ and $\hat{L}_z$. (8.9)

Problems

8.1 Compute the commutation relations of the momentum operator $\hat{\underline{P}}$ and the angular momentum $\hat{\underline{L}}$.

8.2 Using the canonical commutation relations, compute

$$\left[\hat{L}_i, \hat{L}_j\right].\tag{8.70}$$

8.3 The gradient operator in spherical polar coordinates is $\vec{\nabla} = \hat{r}\partial/\partial r + \hat{\theta}\partial/r\partial\theta + \hat{\phi}\partial/r\sin\theta\partial\phi$.

Derive the following entries in the table (e.g. $\partial\hat{r}/\partial\theta = \hat{\theta}$):

	$\hat{r}$	$\hat{\theta}$	$\hat{\phi}$
$\partial/\partial r$	0	0	0
$\partial/\partial\theta$	$\hat{\theta}$	$-\hat{r}$	0
$\partial/\partial\phi$	$\hat{\phi}\sin\theta$	$\hat{\phi}\cos\theta$	$-(\hat{r}\sin\theta + \hat{\theta}\cos\theta)$

Calculate ∇^2 in spherical polar coordinates.
Calculate L_x, L_y and L_z in spherical polar coordinates.
Calculate L^2 in spherical polar coordinates.

8.4 A particle has a wave function $u(x, y, z) = A z \exp[-b(x^2 + y^2 + z^2)]$, where b is a constant.

Show that this wave function is an eigenfunction of $\hat{L}^2$ and of $\hat{L}_z$ and find the corresponding eigenvalues.

Hint: Use the spherical polar expressions for $\hat{L}^2$ and $\hat{L}_z$, and write the wave function in spherical polars.

Can you identify the physical system for which this is an energy eigenstate?

8.5 The wave function of a particle is known to have the form

$$u(r, \theta, \phi) = A\, R(r)\, f(\theta)\, \cos 2\phi,$$

where f is an unknown function of θ. What can be predicted about the results of measuring:

- the z-component of angular momentum;
- the square of the angular momentum?

Hint: Note that $\cos 2\phi = \{\exp(2i\phi) + \exp(-2i\phi)\}/2$.

   Answer the same questions for the wave function

$$u(r, \theta, \phi) = A\, R(r)\, f(\theta)\, \cos^2 \phi.$$

8.6 Compute the commutation relations of the position operator $\hat{\underline{R}}$ and the angular momentum $\hat{\underline{L}}$. Deduce the commutation relations of $\hat{\underline{R}}^2$ with the angular momentum $\hat{\underline{L}}$.

8.7 Let us introduce the operator

$$\hat{P}_r \psi = -i\hbar \frac{1}{r} \frac{\partial}{\partial r}(r\psi).$$

Show that the Hamiltonian can be written as

$$\hat{H} = \frac{1}{2m}\left(\hat{P}_r^2 + \frac{\hat{L}^2}{r^2}\right).$$

8.8 The space of functions with $\ell = 1$ is a three-dimensional subspace of the space of functions of the two angular variables, $f(\theta, \phi)$. A basis for this subspace is given by the three spherical harmonics $Y_1^m(\theta, \phi)$ for $m = -1, 0, 1$. Explicitly construct the three 3×3 matrices that represent L_x, L_y and L_z:

$$(L_i)_{m,m'} = \langle \ell = 1, m | \hat{L}_i | \ell = 1, m' \rangle = \int \sin\theta\, d\theta\, d\phi\, Y_1^m(\theta, \phi)^* \hat{L}_i Y_1^{m'}(\theta, \phi).$$

Compute explicitly using the matrix representation the commutator of $\hat{L}_1$ with $\hat{L}_2$.
   Compute the expectation value of $\hat{L}_x$ in the state

$$\psi(\underline{r}) = F(r)\left[\frac{\sqrt{2}}{\sqrt{5}}Y_1^1(\theta, \phi) - \frac{\sqrt{3}}{\sqrt{5}}Y_1^0(\theta, \phi)\right],$$

where $F(r)$ denotes the radial dependence of the wave function.
   What are the possible outcomes if we measure L_z in this state?

8.9 The parity operator changes the sign of the position vector of a particle: $\underline{r} \rightarrow -\underline{r}$. Express this geometrical transformation in spherical coordinates. Check the properties of the first few spherical harmonics under parity.

8.10 Let us consider a particle in three dimensions. The operator associated with the z-component of the angular momentum in spherical polar coordinates is

$$\hat{L}_z = -i\hbar \frac{\partial}{\partial \phi}.$$

Find the eigenfunctions of $\hat{L}_z$. Explain why the eigenvalues are of the form $\hbar m$ with integer m. Check the dimensions on the left-hand side and the right-hand side of the equation above.

Write $\hat{L}_z$ using Cartesian coordinates. Show that

$$\psi_1(x, y, z) = f(r)(x + iy), \quad \psi_2(x, y, z) = f(r)(x - iy),$$

where $f(r)$ is a generic function, are eigenstates of $\hat{L}_z$ and find the corresponding eigenvalues. Check that you obtain the same results using the expression for L_z in spherical coordinates given in the question above.

Show that

$$\phi_1(x, y, z) = f(r)(z + ix), \quad \phi_2(x, y, z) = f(r)(z - ix)$$

are eigenfunctions of L_y, and find again the corresponding eigenvalues.

Let the time evolution of the system be determined by the Hamiltonian

$$\hat{H} = \mu B \hat{L}_y,$$

and let the system at time $t = 0$ be described by a wave function

$$\psi(x, y, z) = C \exp(-r/a)x.$$

Find the state of the system at time t.

What is the expectation value of $\hat{L}_z$ in the state ψ?

9

Spin

Using the commutation relations for the components of the angular momentum, we have found that the allowed eigenvalues for $\hat{J}^2$ are $\hbar^2 j(j+1)$, where $j = 0, \frac{1}{2}, 1\frac{3}{2}, \ldots$. For each value of j, the eigenvalues of J_z are $\hbar m$, with $m = -j, -j+1, \ldots, j-1, j$.

Compared with the solutions of the eigensystem discussed in the previous chapter, we see that we have found more solutions than there are in Eq. (8.27). In this chapter we are going to discuss the reason why the orbital angular momentum can only take integer values and then we are going to see that the idea of angular momentum can be extended in a way that allows half-integer values too. Note that the orbital angular momentum is expressed as a function of the position and momentum operators, i.e. it probes the spatial distribution of the quantum system. When working in position space, Eq. (8.27) is a partial differential equation in θ and ϕ. The solutions of this equation are the spherical harmonics $Y_\ell^m(\theta, \phi)$, Eq. (8.26). The explicit expression for the spherical harmonics shows that the ϕ dependence is simply

$$Y_\ell^m(\theta, \phi) \propto \exp(im\phi), \tag{9.1}$$

as expected, since the spherical harmonics are *also* eigenfunctions of $L_z = -i\hbar\frac{\partial}{\partial\phi}$. Since we required that wave functions must be single-valued, the spherical harmonics must be periodic in ϕ with period 2π:

$$Y_\ell^m(\theta, \phi) = Y_\ell^m(\theta, \phi + 2\pi). \tag{9.2}$$

Equation (9.2) requires m to be an integer, and hence ℓ must also be an integer.

However, as soon as we consider the angular momentum as an intrinsic property of the system, no longer related to the spatial dependence of its wave function, half-integer values of j become possible. This intrinsic angular momentum is called **spin**.[1] An obvious question at this point is: Is there any evidence in nature of systems with non-integer angular momentum?

9.1 The Stern–Gerlach Experiment

This experiment, performed for the first time in 1922 by Stern and Gerlach, provided evidence that the component of the spin along a given direction, S_z, is quantised and its spectrum allows only two values $\pm\hbar/2$, and therefore corresponds to a state with total

---

[1] When dealing with spin we denote the spin angular momentum vector by $\underline{S}$ and its square by S^2.

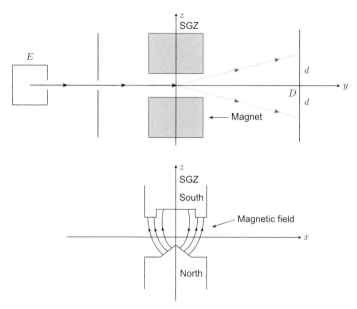

A beam of atoms, each with $j = \frac{1}{2}$, leave the emitter (E) with random spin orientation. They travel along the y-axis passing through a Stern–Gerlach magnet oriented along the z-direction (SGZ). The beam is split by SGZ into only two possible directions, arriving at the detector at D a distance d either above or below the y-axis.

angular momentum $\frac{1}{2}$. It is instructive to discuss the Stern–Gerlach experiment and its interpretation according to the postulates of quantum mechanics.

The Stern–Gerlach experiment is schematically represented in Fig. 9.1. A beam of atoms travels along the y-direction and is deflected by a variable magnetic field $\underline{B}$ in the z-direction. The atoms travel in a potential

$$V = \gamma \underline{S} \cdot \underline{B}, \tag{9.3}$$

where γ, the gyromagnetic ratio, is the proportionality coefficient that relates the magnetic moment of the atom to its spin. The deflection of the atoms, d, either above or below the y-axis, measured by the detector at D is proportional to the force acting on the system, i.e. the gradient of the potential V:

$$d \propto \gamma S_z \frac{\partial B_z}{\partial z}. \tag{9.4}$$

The important point here is that the deflection is proportional to the z-component of the spin of the system, all other factors in Eq. (9.4) being fixed in the experiment. The Stern–Gerlach experiment is a measurement of the z-component of the spin of the atoms.

As usual in quantum mechanics, we prepare the system in the same identical state, perform the experiment multiple times and study the probability distribution of the outcomes of the measurement. The probability distribution for the deflection, and therefore for the value of S_z, is peaked around two values, symmetrical with respect to the y-axis, which we interpret as a quantised spectrum with two possible eigenvalues, $S_z = \pm\hbar/2$, and therefore the atoms must be systems with spin $S = \frac{1}{2}$. We will come back to the Stern–Gerlach experiment in later chapters.

In order to understand the physical meaning of states with $S = \frac{1}{2}$, we will now investigate their properties in more detail.

9.2 Physical States

As far as the spin degrees of freedom are concerned, the system is a two-state system, with a basis made up of simultaneous eigenvectors of $\hat{S}^2$ and $\hat{S}_z$:

$$\left\{ |S = \tfrac{1}{2}, S_z = \tfrac{1}{2}\rangle, |S = \tfrac{1}{2}, S_z = -\tfrac{1}{2}\rangle \right\}. \tag{9.5}$$

Remember that with our conventions, we have

$$\hat{S}^2 |S = \tfrac{1}{2}, S_z = +\tfrac{1}{2}\rangle = \frac{3}{4}\hbar^2 |S = \tfrac{1}{2}, S_z = +\tfrac{1}{2}\rangle,$$

$$\hat{S}_z |S = \tfrac{1}{2}, S_z = +\tfrac{1}{2}\rangle = +\tfrac{1}{2}\hbar |S = \tfrac{1}{2}, S_z = +\tfrac{1}{2}\rangle,$$

$$\hat{S}^2 |S = \tfrac{1}{2}, S_z = -\tfrac{1}{2}\rangle = \frac{3}{4}\hbar^2 |S = \tfrac{1}{2}, S_z = -\tfrac{1}{2}\rangle,$$

$$\hat{S}_z |S = \tfrac{1}{2}, S_z = -\tfrac{1}{2}\rangle = -\tfrac{1}{2}\hbar |S = \tfrac{1}{2}, S_z = -\tfrac{1}{2}\rangle.$$

It is economical to introduce a shorthand notation for these basis vectors:

$$|S = \tfrac{1}{2}, S_z = +\tfrac{1}{2}\rangle = |+\tfrac{1}{2}\rangle = |\uparrow\rangle, \tag{9.6}$$

$$|S = \tfrac{1}{2}, S_z = -\tfrac{1}{2}\rangle = |-\tfrac{1}{2}\rangle = |\downarrow\rangle. \tag{9.7}$$

A generic state of a spin-$\frac{1}{2}$ system can be written as a linear combination of the basis vectors:

$$|\psi\rangle = \psi_1 |\uparrow\rangle + \psi_2 |\downarrow\rangle, \tag{9.8}$$

where $\psi_1, \psi_2 \in \mathbb{C}$. As usual, having chosen a basis, the vector can be represented by its coordinates:

$$|\psi\rangle \rightarrow \begin{pmatrix} \psi_1 \\ \psi_2 \end{pmatrix}, \tag{9.9}$$

and we use the arrow once again to stress the fact that $|\psi\rangle$ is represented by a complex vector *in a given basis*. It is important to emphasise once again that spin is not related to the spatial degrees of freedom of the system. The spin states are vectors in a two-dimensional Hilbert space $\mathcal{H}_{\text{spin}}$, while the states that correspond to the motion of the particle in space are vectors in an infinite-dimensional vector space $\mathcal{H}_{\text{space}}$, which are described for example by a wave function. If we are interested in describing both the spatial distribution and the spin degrees of freedom of the system, then we need to consider states in the tensor product

$$\mathcal{H} = \mathcal{H}_{\text{space}} \otimes \mathcal{H}_{\text{spin}}. \tag{9.10}$$

A basis for $\mathcal{H}$, made up of the tensor product of eigenstates of position and eigenstates of $\hat{S}_z$, is given by

$$\left\{ |\underline{r}, s_z\rangle = |\underline{r}\rangle |s_z\rangle, \underline{r} \in \mathbb{R}^3, s_z = \pm \tfrac{1}{2} \right\}, \tag{9.11}$$

and a generic state can be expanded as

$$|\Psi\rangle = \int d^3 r' \sum_{s_z} |\underline{r}', s_z\rangle \langle \underline{r}', s_z |\Psi\rangle, \tag{9.12}$$

$$= \int d^3 r' \left(\psi_1(\underline{r}') |\underline{r}'\rangle | \uparrow\rangle + \psi_2(\underline{r}') |\underline{r}'\rangle | \downarrow\rangle \right), \tag{9.13}$$

where in the second line we have written explicitly the sum over the two values of s_z and used the notation with the arrows introduced above for eigenstates of $\hat{S}_z$. Clearly, in Eq. (9.13)

$$\psi_1(\underline{r}') = \langle \underline{r}', \uparrow |\Psi\rangle, \tag{9.14}$$

$$\psi_2(\underline{r}') = \langle \underline{r}', \downarrow |\Psi\rangle. \tag{9.15}$$

Starting from Eq. (9.13), taking the scalar product with an eigenstate of position and using the orthonormality of position eigenstates yields

$$\langle \underline{r}|\Psi\rangle = \psi_1(\underline{r}) | \uparrow\rangle + \psi_2(\underline{r}) | \downarrow\rangle. \tag{9.16}$$

Once again we can represent $\langle \underline{r}|\Psi\rangle$ as a column vector in the specific basis in $\mathcal{H}_{\text{spin}}$:

$$\langle \underline{r}|\Psi\rangle \longrightarrow \begin{pmatrix} \psi_1(\underline{r}) \\ \psi_2(\underline{r}) \end{pmatrix}. \tag{9.17}$$

The physical interpretation of these components is the standard one, i.e. the coordinates in a given basis are probability amplitudes. Hence

1. $\left| \psi_1(\underline{r}) \right|^2$ yields the probability density of finding the system at position $\underline{r}$ with spin $s_z = +\tfrac{1}{2}$.
2. $\left| \psi_2(\underline{r}) \right|^2$ yields the probability density of finding the system at position $\underline{r}$ with spin $s_z = -\tfrac{1}{2}$.
3. The probability of finding the system anywhere in space with spin $s_z = +\tfrac{1}{2}$ is

$$P_\uparrow = \int d^3 r \left| \psi_1(\underline{r}) \right|^2. \tag{9.18}$$

4. The probability density of finding the system at position $\underline{r}$ with any value of s_z is

$$P(\underline{r}) = \left| \psi_1(\underline{r}) \right|^2 + \left| \psi_2(\underline{r}) \right|^2. \tag{9.19}$$

9.3 Matrix Representation

For $s = \tfrac{1}{2}$, let us compute the matrix elements $\langle \tfrac{1}{2}, m' | \hat{S}_i | \tfrac{1}{2}, m \rangle$; the possible values for m' and m are $m' = \tfrac{1}{2}$ or $-\tfrac{1}{2}$ and $m = \tfrac{1}{2}$ or $-\tfrac{1}{2}$. If we choose the convention that the row and column labels start with the largest value of s_z and decrease, so that the first row (column) is labelled $m'(m) = \frac{1}{2}$ and the second row (column) is labelled $m'(m) = -\frac{1}{2}$, we find the following 2×2 matrix representing $\hat{S}_z$:

$$\hat{S}_z \rightarrow \frac{\hbar}{2} \begin{pmatrix} 1 & 0 \\ 0 & -1 \end{pmatrix}.$$

We say that this matrix *represents* the operator $\hat{S}_z$ in the $s = \frac{1}{2}$ multiplet.

In order to compute the matrix representations of $\hat{S}_x$ and $\hat{S}_y$, it is easier to start from the raising and lowering operators. The only non-zero element for the matrix representing $\hat{S}_+$ is when $m' = \frac{1}{2}$ and $m = -\frac{1}{2}$, for which

$$c_+ = \sqrt{\tfrac{1}{2}(\tfrac{1}{2} + 1) + \tfrac{1}{2}(\tfrac{1}{2})} = 1,$$

and hence the matrix is

$$\hat{S}_+ \longrightarrow \hbar \begin{pmatrix} 0 & 1 \\ 0 & 0 \end{pmatrix},$$

whilst the only non-zero element for the matrix representing $\hat{S}_-$ is when $m' = -\frac{1}{2}$ and $m = \frac{1}{2}$, for which

$$c_- = \sqrt{\tfrac{1}{2}(\tfrac{1}{2} + 1) + \tfrac{1}{2}(\tfrac{1}{2})} = 1,$$

also giving

$$\hat{S}_- \longrightarrow \hbar \begin{pmatrix} 0 & 0 \\ 1 & 0 \end{pmatrix}.$$

From these two matrices it is easy to construct the matrices representing $\hat{S}_x$ and $\hat{S}_y$, since

$$\hat{S}_x = \tfrac{1}{2}(\hat{S}_+ + \hat{S}_-),$$

$$\hat{S}_y = \frac{1}{2i}(\hat{S}_+ - \hat{S}_-).$$

Thus

$$\hat{S}_x \longrightarrow \frac{\hbar}{2} \begin{pmatrix} 0 & 1 \\ 1 & 0 \end{pmatrix}, \quad \hat{S}_y \longrightarrow \frac{\hbar}{2} \begin{pmatrix} 0 & -i \\ i & 0 \end{pmatrix}.$$

You can readily verify that these 2×2 matrices satisfy the angular momentum commutation relations from which we started. We say, therefore, that they *provide a matrix representation* of the angular momentum operators.

The set of three 2×2 matrices which appear above in the matrix representations of $\hat{S}_x$, $\hat{S}_y$ and $\hat{S}_z$ are known as the **Pauli spin matrices** and are usually denoted as follows:

$$\sigma_x = \begin{pmatrix} 0 & 1 \\ 1 & 0 \end{pmatrix}; \quad \sigma_y = \begin{pmatrix} 0 & -i \\ i & 0 \end{pmatrix}; \quad \sigma_z = \begin{pmatrix} 1 & 0 \\ 0 & -1 \end{pmatrix}.$$

Collectively, we can write

$$\underline{\hat{S}} \longrightarrow \tfrac{1}{2}\hbar \, \underline{\sigma},$$

meaning $\hat{S}_x \longrightarrow \frac{1}{2}\hbar \sigma_x$, etc. Often we will just write $=$ instead of $\longrightarrow$, but you should remember that this is just one possible choice for representing the operators, dictated by the choice of basis we made above in $\mathcal{H}_{\text{spin}}$.

The Pauli matrices are Hermitian, as expected since they are associated with an observable, and have the following property, which you can easily verify:

$$\sigma_x^2 = \sigma_y^2 = \sigma_z^2 = 1,$$

where 1 denotes the unit 2×2 matrix.

9.4 Eigenvectors

We can now rederive the eigenvectors using the matrix representation introduced above. It is trivial to show that the matrix σ_z has eigenvectors which are just two-component column vectors:

$$\begin{pmatrix} 1 & 0 \\ 0 & -1 \end{pmatrix} \begin{pmatrix} 1 \\ 0 \end{pmatrix} = \begin{pmatrix} 1 \\ 0 \end{pmatrix}, \quad \begin{pmatrix} 1 & 0 \\ 0 & -1 \end{pmatrix} \begin{pmatrix} 0 \\ 1 \end{pmatrix} = -\begin{pmatrix} 0 \\ 1 \end{pmatrix},$$

so that the eigenvalue equations for $\hat{S}_z$ are

$$\frac{1}{2}\hbar \begin{pmatrix} 1 & 0 \\ 0 & -1 \end{pmatrix} \begin{pmatrix} 1 \\ 0 \end{pmatrix} = \frac{1}{2}\hbar \begin{pmatrix} 1 \\ 0 \end{pmatrix}, \quad \frac{1}{2}\hbar \begin{pmatrix} 1 & 0 \\ 0 & -1 \end{pmatrix} \begin{pmatrix} 0 \\ 1 \end{pmatrix} = -\frac{1}{2}\hbar \begin{pmatrix} 0 \\ 1 \end{pmatrix}.$$

We see explicitly that $\hat{S}_z$, has eigenvalues $\pm\frac{1}{2}\hbar$, as it should for a system with $s = \frac{1}{2}$, and that the basis vectors are the eigenstates of $\hat{S}_z$.

Furthermore, if we construct the matrix representing $\hat{S}^2$, we see that it has these same two column vectors as eigenvectors with a common eigenvalue $s(s+1)\hbar^2 \equiv \frac{3}{4}\hbar^2$:

$$\hat{S}^2 \equiv \hat{S}_x^2 + \hat{S}_y^2 + \hat{S}_z^2 = \frac{1}{4}\hbar^2 \left[\begin{pmatrix} 0 & 1 \\ 1 & 0 \end{pmatrix}^2 + \begin{pmatrix} 0 & -i \\ i & 0 \end{pmatrix}^2 + \begin{pmatrix} 1 & 0 \\ 0 & -1 \end{pmatrix}^2 \right] = \frac{3}{4}\hbar^2 \begin{pmatrix} 1 & 0 \\ 0 & 1 \end{pmatrix}$$

and

$$\frac{3}{4}\hbar^2 \begin{pmatrix} 1 & 0 \\ 0 & 1 \end{pmatrix} \begin{pmatrix} 1 \\ 0 \end{pmatrix} = \frac{3}{4}\hbar^2 \begin{pmatrix} 1 \\ 0 \end{pmatrix}, \quad \frac{3}{4}\hbar^2 \begin{pmatrix} 1 & 0 \\ 0 & 1 \end{pmatrix} \begin{pmatrix} 0 \\ 1 \end{pmatrix} = \frac{3}{4}\hbar^2 \begin{pmatrix} 0 \\ 1 \end{pmatrix}.$$

We can thus identify the two column vectors with the two simultaneous eigenstates of the operators $\hat{S}^2$ and $\hat{S}_z$:

$$|s = \tfrac{1}{2}, m = \tfrac{1}{2}\rangle \longrightarrow \begin{pmatrix} 1 \\ 0 \end{pmatrix}, \quad |s = \tfrac{1}{2}, m = -\tfrac{1}{2}\rangle \longrightarrow \begin{pmatrix} 0 \\ 1 \end{pmatrix}.$$

As already discussed previously, an arbitrary state $|\psi\rangle$ with $s = \frac{1}{2}$ may be represented as a linear combination of these two states since they span the two-dimensional space of two-component column vectors:

$$|\psi\rangle = \sum_{m=-\frac{1}{2}}^{\frac{1}{2}} c_m \, |s = \tfrac{1}{2}, m\rangle \qquad (9.20)$$

is represented by

$$\begin{pmatrix} c_{\frac{1}{2}} \\ c_{-\frac{1}{2}} \end{pmatrix} = c_{\frac{1}{2}} \begin{pmatrix} 1 \\ 0 \end{pmatrix} + c_{-\frac{1}{2}} \begin{pmatrix} 0 \\ 1 \end{pmatrix}.$$

9.5 Scalar Products

The Dirac kets we have seen are represented by column vectors of rank $(2s + 1)$; the corresponding conjugates, or Dirac bras, are represented by row vectors of the same rank. The rule for scalar products is that if

$$|\psi\rangle \longrightarrow \begin{pmatrix} \psi_1 \\ \psi_2 \end{pmatrix}, \quad \text{then} \quad \langle\psi| \longrightarrow \begin{pmatrix} \psi_1^* & \psi_2^* \end{pmatrix}.$$

The scalar product of two states $|\psi\rangle$ and $|\phi\rangle$ is then defined to be

$$\langle\phi|\psi\rangle \equiv \begin{pmatrix} \phi_1^* & \phi_2^* \end{pmatrix} \begin{pmatrix} \psi_1 \\ \psi_2 \end{pmatrix} = \phi_1^*\psi_1 + \phi_2^*\psi_2. \qquad (9.21)$$

Thus, for example, for a normalised state

$$\langle\psi|\psi\rangle = \psi_1^*\psi_1 + \psi_2^*\psi_2 = |\psi_1|^2 + |\psi_2|^2 = 1.$$

The orthonormality property of the eigenvectors is also obvious and may be used to project out the coefficients c_m in the expansion of the arbitrary state $|\psi\rangle$ in Eq. (9.20).

9.6 Eigenvectors of $\hat{S}_x$

So far we have concentrated on the eigenvalues and eigenstates of $\hat{S}_z$, but what of the other Cartesian components of angular momentum? It is clear that, since there is nothing special about the z-direction ('space is isotropic'!), we should also expect that measuring say the x-component of the angular momentum for a system with $s = \tfrac{1}{2}$ can only yield either $\pm\tfrac{1}{2}\hbar$. Let us verify this. The matrix representing $\hat{S}_x$ is $\tfrac{1}{2}\hbar\,\sigma_x$, so we need to find the eigenvalues and eigenvectors of the 2×2 matrix σ_x. Let us write

$$\sigma_x \chi = \rho \chi \quad \text{with} \quad \chi = \begin{pmatrix} \chi_1 \\ \chi_2 \end{pmatrix},$$

where ρ denotes an eigenvalue and χ the corresponding eigenvector. We find the eigenvalues by rewriting this as

$$\left(\sigma_x - \rho\,1\right)\chi = 0,$$

where 1 denotes the unit 2×2 matrix. This is a pair of simultaneous equations for χ_1 and χ_2, which only have a non-trivial solution if the determinant of the 2×2 coefficient matrix on the left-hand side is singular. The condition for this is

$$\det\left(\sigma_x - \rho\,1\right) = \begin{vmatrix} -\rho & 1 \\ 1 & -\rho \end{vmatrix} = 0,$$

which yields

$$\rho^2 - 1 = 0 \quad \text{implying that} \quad \rho = \pm 1.$$

The eigenvalues of $\hat{S}_x$ are thus $\pm\frac{1}{2}\hbar$ as anticipated. *More generally, the eigenvalues of $\hat{S}_x$ are written $m_x\hbar$*, where we have factored out the dimensionful unit of angular momentum $\hbar$ and m_x is a dimensionless number. In this case, where $s = \frac{1}{2}$, we have $m_x = \pm\frac{1}{2}$. Let us now find the eigenvectors corresponding to the two eigenvalues.

Case $\rho = 1$

The equation for the eigenvectors becomes

$$\begin{pmatrix} 0 & 1 \\ 1 & 0 \end{pmatrix}\begin{pmatrix} \chi_1 \\ \chi_2 \end{pmatrix} = \begin{pmatrix} \chi_1 \\ \chi_2 \end{pmatrix} \quad \Rightarrow \quad \begin{pmatrix} \chi_2 \\ \chi_1 \end{pmatrix} = \begin{pmatrix} \chi_1 \\ \chi_2 \end{pmatrix} \quad \Rightarrow \quad \chi_2 = \chi_1.$$

We can pick any two-component column matrix which satisfies this condition. In particular, a suitably normalised eigenvector which represents the state with $s = \frac{1}{2}$ and $m_x = \frac{1}{2}$ is

$$|s = \tfrac{1}{2}, m_x = \tfrac{1}{2}\rangle \longrightarrow \frac{1}{\sqrt{2}}\begin{pmatrix} 1 \\ 1 \end{pmatrix}.$$

Case $\rho = -1$

In close analogy with the computation above, let us write

$$\begin{pmatrix} 0 & 1 \\ 1 & 0 \end{pmatrix}\begin{pmatrix} \chi_1 \\ \chi_2 \end{pmatrix} = -\begin{pmatrix} \chi_1 \\ \chi_2 \end{pmatrix} \quad \Rightarrow \quad \begin{pmatrix} \chi_2 \\ \chi_1 \end{pmatrix} = -\begin{pmatrix} \chi_1 \\ \chi_2 \end{pmatrix} \quad \Rightarrow \quad \chi_2 = -\chi_1.$$

Thus a suitably normalised eigenvector which represents the state with $s = \frac{1}{2}$ and $m_x = -\frac{1}{2}$ is

$$|s = \tfrac{1}{2}, m_x = -\tfrac{1}{2}\rangle \longrightarrow \frac{1}{\sqrt{2}}\begin{pmatrix} 1 \\ -1 \end{pmatrix}.$$

Comments

Let us briefly comment on the solutions found above.

- The eigenvectors corresponding to $m_x = \frac{1}{2}$ and $m_x = -\frac{1}{2}$ are orthogonal, as they must be:

$$\frac{1}{\sqrt{2}} \begin{pmatrix} 1 & 1 \end{pmatrix} \frac{1}{\sqrt{2}} \begin{pmatrix} 1 \\ -1 \end{pmatrix} = 0.$$

- The eigenvectors of $\hat{S}_x$ are expressible as linear combinations of the eigenvectors of $\hat{S}_z$:

$$\frac{1}{\sqrt{2}} \begin{pmatrix} 1 \\ 1 \end{pmatrix} = \frac{1}{\sqrt{2}} \begin{pmatrix} 1 \\ 0 \end{pmatrix} + \frac{1}{\sqrt{2}} \begin{pmatrix} 0 \\ 1 \end{pmatrix}$$

and

$$\frac{1}{\sqrt{2}} \begin{pmatrix} 1 \\ -1 \end{pmatrix} = \frac{1}{\sqrt{2}} \begin{pmatrix} 1 \\ 0 \end{pmatrix} - \frac{1}{\sqrt{2}} \begin{pmatrix} 0 \\ 1 \end{pmatrix}.$$

Thus if a system with $s = \frac{1}{2}$ is in an eigenstate of $\hat{S}_x$, for example with $m_x = \frac{1}{2}$, then the probability that a measurement of $\hat{S}_z$ yields the result $m = \frac{1}{2}$ is $|\frac{1}{\sqrt{2}}|^2 = \frac{1}{2}$.

What has emerged from this analysis is that we can consider systems with $s = \frac{1}{2}$ as having *intrinsic* angular momentum, which has nothing to do with the orbital motion of the particle about some point. It is a property of the system in its own rest frame, and can be understood as a new degree of freedom of the particle. Hence the quantum numbers describing the state of the system must also include an index m labelling the values of these degrees of freedom. For the case $s = \frac{1}{2}$, m can take two values, and therefore the wave functions have two components as discussed above. For the general case of spin s, the wave functions have $2s + 1$ components. This intrinsic angular momentum is known as *spin* and does not have any classical analogue. Electrons, protons, neutrons and many of the more unstable particles have spin $\frac{1}{2}$.

9.7 The Stern–Gerlach Experiment Reloaded

We can now revisit the result of the original Stern–Gerlach experiment, which was conducted with a beam of silver atoms and found two emergent beams, corresponding to $s = \frac{1}{2}$. We are going to consider here a more elaborate experiment involving not one but several Stern–Gerlach magnets, which we use to make *successive* measurements of various components of angular momentum.

We assume that we can neglect any interaction between the particles in the beam. The two beams emerging from the first magnet have $m = \frac{1}{2}$ and $m = -\frac{1}{2}$, respectively, but only the former is allowed to proceed to the second magnet. Thus, *we know that each particle entering the second magnet is in the state* with $m = \frac{1}{2}$, represented by the column vector

$$|s = \tfrac{1}{2}, m = \tfrac{1}{2}\rangle \rightarrow \begin{pmatrix} 1 \\ 0 \end{pmatrix}.$$

The second magnet, which has its mean field aligned with the x-direction, measures the x-component of angular momentum. We can predict the outcome by expanding the state $|s = \frac{1}{2}, m = \frac{1}{2}\rangle$ in eigenstates of $\hat{S}_x$ and finding the probability amplitudes for the two possible outcomes, $m_x = \frac{1}{2}$ and $m_x = -\frac{1}{2}$. Thus

$$\begin{pmatrix} 1 \\ 0 \end{pmatrix} = a \frac{1}{\sqrt{2}} \begin{pmatrix} 1 \\ 1 \end{pmatrix} + b \frac{1}{\sqrt{2}} \begin{pmatrix} 1 \\ -1 \end{pmatrix}.$$

We find the amplitudes a and b by orthogonal projection in the usual way:

$$a = \frac{1}{\sqrt{2}} \begin{pmatrix} 1 & 1 \end{pmatrix} \begin{pmatrix} 1 \\ 0 \end{pmatrix} = \frac{1}{\sqrt{2}},$$

$$b = \frac{1}{\sqrt{2}} \begin{pmatrix} 1 & -1 \end{pmatrix} \begin{pmatrix} 1 \\ 0 \end{pmatrix} = \frac{1}{\sqrt{2}}.$$

We have then for the desired probabilities:

$$\text{probability of getting } m_x = \tfrac{1}{2} \text{ is } |a|^2 = \tfrac{1}{2},$$
$$\text{probability of getting } m_x = -\tfrac{1}{2} \text{ is } |b|^2 = \tfrac{1}{2}.$$

Since each particle is therefore equally likely to be found with $m_x = \frac{1}{2}$ or $m_x = -\frac{1}{2}$, equal numbers, on average, go into each of the two emergent beams and so the two beams will have *equal intensity*.

Regeneration

What happens if we select only those particles with $m_x = \frac{1}{2}$ emerging from the second magnet and allow them to impinge on a third magnet whose mean field is aligned with the z-direction? This is the situation illustrated in Fig. 9.2. We are, in effect, remeasuring $\hat{S}_z$ by means of the third apparatus. We know that the state of particles entering the third magnet is $|s = \frac{1}{2}, m_x = \frac{1}{2}\rangle$, and we can expand this state in terms of the complete set of eigenstates of $\hat{S}_z$. The expansion coefficients will be the probability amplitudes required to compute the probabilities of getting the two possible outcomes $m = \frac{1}{2}$ and $m = -\frac{1}{2}$ when we measure $\hat{S}_z$ for each particle:

$$\frac{1}{\sqrt{2}} \begin{pmatrix} 1 \\ 1 \end{pmatrix} = \frac{1}{\sqrt{2}} \begin{pmatrix} 1 \\ 0 \end{pmatrix} + \frac{1}{\sqrt{2}} \begin{pmatrix} 0 \\ 1 \end{pmatrix}.$$

Fig. 9.2   A beam of atoms, each with $s = \frac{1}{2}$, travelling along the y-axis passes through a sequence of Stern–Gerlach magnets whose mean fields are oriented along either the z-direction (SGZ) or the x-direction (SGX). The shaded boxes represent absorbers.

We see that the desired amplitudes are both $\frac{1}{\sqrt{2}}$, so giving equal probabilities for the two outcomes. The remarkable feature of this result is that the probability of getting $m = -\frac{1}{2}$ is non-zero *despite our having eliminated the beam with $m = -\frac{1}{2}$ which emerged from the first magnet!* This phenomenon is referred to as *regeneration*. It has arisen here because the second measurement, of the x-component of angular momentum, was *incompatible with the first measurement*, of the z-component.

General Remarks

More generally, if the second apparatus is aligned so that its mean field lies not in the x-direction, but in the x–z plane at an angle θ to the z-axis, then it measures the component of angular momentum not along the x-direction but along the direction of a unit vector

$$\underline{n} = \sin\theta\,\underline{e}_x + \cos\theta\,\underline{e}_z,$$

where $\underline{e}_x$ and $\underline{e}_z$ are the usual Cartesian unit vectors in the x- and z-directions, respectively. The relevant eigenstates are then those of the matrix

$$\underline{\sigma}.\underline{n} = \sigma_x\,\sin\theta + \sigma_z\,\cos\theta = \begin{pmatrix} \cos\theta & \sin\theta \\ \sin\theta & -\cos\theta \end{pmatrix}.$$

We shall investigate some of the properties of spin along a generic direction $\underline{n}$ in the problems at the end of the chapter.

We will come back to discussing Stern–Gerlach in more detail when we introduce entanglement later in this book.

Summary

- The Stern–Gerlach experiment. Experimental evidence for half-integer spin in nature. (9.1)
- We describe the Hilbert space of physical states with spin $\frac{1}{2}$. After choosing a complete set of commuting observables and the corresponding basis, we can expand arbitrary states in that basis and discuss the physical interpretation of the coefficients in the expansion. Spin and orbital properties of a quantum system. (9.2)
- Following what we did in the case of the orbital angular momentum, we can construct a matrix representation of the angular momentum operators for $s = \frac{1}{2}$. (9.3)
- We study the eigenvectors and eigenvalues of the spin component in arbitrary directions. We look at explicit examples in order to get familiar with the Hilbert space of physical states of a system with spin $\frac{1}{2}$. (9.4–9.6)
- Having developed an understanding of the states that are associated with the spin degrees of freedom, we can revisit the Stern–Gerlach experiment and look at some peculiar physical properties. (9.7)

Problems

9.1 The Hamiltonian that describes the interaction of a static spin-$\frac{1}{2}$ particle with an external magnetic field, $\underline{B}$, is

$$\hat{H} = -\hat{\underline{\mu}}.\underline{B},$$

where the magnetic moment operator, $\hat{\underline{\mu}}$, is related to the spin operator, $\hat{\underline{s}}$, by $\hat{\underline{\mu}} = \gamma\,\hat{\underline{s}}$, with γ the gyromagnetic ratio. Use the Pauli representation of the spin vector $\hat{\underline{s}}$ to find the energy eigenvalues in a static uniform magnetic field in the z-direction, $\underline{B}_0 = (0, 0, B_0)$.

9.2 Construct the matrix $\underline{\sigma} \cdot \underline{e}$, where $\underline{e}$ is a unit vector with Cartesian components e_x, e_y, e_z. $\underline{\sigma}$ has Cartesian components which are the Pauli matrices, $\sigma_x,\ \sigma_y,\ \sigma_z$, so that

$$\underline{\sigma} \cdot \underline{e} = e_x\sigma_x + e_y\sigma_y + e_z\sigma_z.$$

Show that the eigenvalues of $\underline{\sigma} \cdot \underline{e}$ are ± 1 and hence deduce that a measurement of the component of spin along the direction of $\underline{e}$, of a spin-$\frac{1}{2}$ particle, can only yield the result $\frac{1}{2}\hbar$ or $-\frac{1}{2}\hbar$. Obtain the *normalised* eigenvectors of $\underline{\sigma}.\underline{e}$ corresponding to each of the eigenvalues and verify your answers by considering the special case $e_x = 1$, $e_y = e_z = 0$. Remember that the normalisation condition for a two-component column matrix with entries ψ_1 and ψ_2 is

$$|\psi_1|^2 + |\psi_2|^2 = 1.$$

9.3 A beam of spin-$\frac{1}{2}$ particles is sent through a Stern–Gerlach apparatus which divides the incident beam into two spatially separated beams having $m = \pm\frac{1}{2}$, respectively. The beam with $m = -\frac{1}{2}$ is removed, whilst the beam with $m = \frac{1}{2}$ is allowed to impinge on a second Stern–Gerlach apparatus whose mean field is also perpendicular to the beam direction, but inclined at an angle θ with respect to that of the first apparatus. Calculate the relative intensities of the two emergent beams.

9.4 Construct the four-dimensional representation of the components of the intrinsic angular momentum $\hat{S}_x, \hat{S}_y, \hat{S}_z$ acting in the space of states with spin $s = 3/2$. Write explicitly the matrix corresponding to the operator $\hat{S}^2$; verify that you obtain the expected result.

9.5 The state of a particle of spin $\frac{1}{2}$ in one dimension is described by a two-component wave function in the basis of eigenstates of $\hat{S}_z$:

$$\Psi(x) = \left(\begin{array}{c} \psi_1(x) \\ \psi_2(x) \end{array} \right).$$

Let us consider a Hamiltonian with a potential that depends on the spin of the particle:

$$\hat{H} = \frac{1}{2m}\left[\hat{P}^2 + W(\hat{X})^2 + 2\hat{S}_z W'(\hat{X}) \right],$$

where $\hat{P} = -i\hbar\frac{d}{dx}$ is the momentum operator, $\hat{X}$ is the position operator, $\hat{S}_z$ is the operator associated with the z-component of the spin and $W(x)$ is a real function of the position x, such that $|W| \to \infty$ for $x \to \pm\infty$ and $W'(x) = \frac{d}{dx}W(x)$. The spin operator is $\hat{S}_i = \frac{\hbar}{2}\sigma_i$, where σ_i are the Pauli matrices:

$$\sigma_1 = \begin{pmatrix} 0 & 1 \\ 1 & 0 \end{pmatrix}, \quad \sigma_2 = \begin{pmatrix} 0 & -i \\ i & 0 \end{pmatrix}, \quad \sigma_3 = \begin{pmatrix} 1 & 0 \\ 0 & -1 \end{pmatrix}.$$

Compute the action of each term in the Hamiltonian on the wave function specified above. In particular, show that

$$2\hat{S}_z W'(\hat{X})\Psi(x) = \begin{pmatrix} \hbar W'(x)\psi_1(x) \\ -\hbar W'(x)\psi_2(x) \end{pmatrix}.$$

Show that

$$\hat{Q}_1^2 = \hat{Q}_2^2 = \hat{H},$$

where

$$\hat{Q}_1 = \frac{1}{\sqrt{2m}}\left[\sigma_1\hat{P} + \sigma_2 W(\hat{X})\right], \quad \hat{Q}_2 = \frac{1}{\sqrt{2m}}\left[\sigma_2\hat{P} - \sigma_1 W(\hat{X})\right]$$

and σ_1, σ_2 are the Pauli matrices.
  Prove that for any state $|\Psi\rangle$:

$$\langle\Psi|\hat{H}|\Psi\rangle \geq 0.$$

Show that the energy of the ground state vanishes if its wave function satisfies

$$\left[\hat{P} + iW(\hat{X})\sigma_3\right]\Psi_0(x) = 0.$$

Are $\hat{H}$ and $\hat{S}_z$ compatible observables?
  Let us denote by $\Psi^{(n,\pm)}(x)$ the simultaneous eigenstates of $\hat{H}$ and $\hat{S}_z$, with respective eigenvalues E_n and $\pm\hbar/2$. Show that

$$\Psi^{(n,+)}(x) = \begin{pmatrix} \psi_1^{(n,+)}(x) \\ 0 \end{pmatrix}.$$

What conditions need to be satisfied for the ground-state wave function $\Psi^{(0,+)}$ with $E_0 = 0$ to be normalisable?
  Prove that $[\hat{Q}_1, \hat{H}] = 0$.
  What can you deduce about the degeneracy of the energy eigenvalues?

10

Addition of Angular Momenta

Building on what we have discussed in the previous two chapters, we now turn to the problem of dealing with the **addition of two angular momenta**. For example, we might wish to consider an electron which has both an intrinsic spin and some orbital angular momentum, as in a real hydrogen atom. Or we might have a system of two electrons and wish to know what possible values the total spin of the system can take. Classically, the angular momentum is a vector quantity, and the total angular momentum is simply $\underline{J} = \underline{J}^{(1)} + \underline{J}^{(2)}$. The maximum and minimum values that $\underline{J}$ can take correspond to the case where either $\underline{J}^{(1)}$ and $\underline{J}^{(2)}$ are parallel, so that the magnitude of $\underline{J}$ is $|\underline{J}^{(1)}| + |\underline{J}^{(2)}|$, or antiparallel, when it has magnitude $\left| |\underline{J}^{(1)}| - |\underline{J}^{(2)}| \right|$. Once again the situation is different in the quantum-mechanical case; this should not come as a surprise at this stage, since we have seen that the angular momentum in quantum mechanics obeys very specific properties. This chapter discusses the addition of angular momenta for a quantum system. Superscript (1) and (2) identify the two angular momenta that are being added throughout this chapter.

10.1 Total Angular Momentum Operator

In the quantum case, the total angular momentum is represented by the operator

$$\hat{\underline{J}} \equiv \hat{\underline{J}}^{(1)} + \hat{\underline{J}}^{(2)}. \tag{10.1}$$

They can be the orbital angular momenta of two quantum systems, or the orbital angular momentum and the spin of a single quantum system. The only assumption here is that $\hat{\underline{J}}_1$ and $\hat{\underline{J}}_2$ are *independent angular momenta*, meaning each satisfies the usual angular momentum commutation relations

$$[\hat{J}_x^{(n)}, \hat{J}_y^{(n)}] = i\hbar\, \hat{J}_z^{(n)}, \quad \text{etc. and} \quad [\hat{J}^{(n)\,2}, \hat{J}_i^{(n)}] = 0, \quad \text{etc.,} \tag{10.2}$$

where $n = 1, 2$ labels the individual angular momenta, $i = x, y, z$, and etc. stands for cyclic permutations. The operators $\hat{J}_k^{(1)}$ act on state vectors that belong to some Hilbert space $\mathcal{H}_1$, while the operators $\hat{J}_k^{(2)}$ act on state vectors in some other Hilbert space $\mathcal{H}_2$. The total angular momentum $\hat{\underline{J}}$ acts in the product space

$$\mathcal{H} = \mathcal{H}_1 \otimes \mathcal{H}_2,$$

according to the rules discussed in Chapter 2. *As a consequence, any component of $\underline{\hat{J}}^{(1)}$ commutes with any component of $\underline{\hat{J}}^{(2)}$:*

$$[\hat{J}_i^{(1)}, \hat{J}_k^{(2)}] = 0, \quad i, k = x, y, z, \tag{10.3}$$

so that the two angular momenta are compatible observables. It follows that the four operators $\hat{J}^{(1)2}, \hat{J}_z^{(1)}, \hat{J}^{(2)2}, \hat{J}_z^{(2)}$ are *mutually commuting* and so must admit a set of common eigenvectors that form a basis of $\mathcal{H}$. This common eigenbasis is known as the **uncoupled basis** and is denoted $\{|j_1, m_1, j_2, m_2\rangle\}$ in Dirac notation, with the numbers in the ket identifying the eigenvalues of the four operators. For each value of the pair j_1, j_2 there are $(2j_1 + 1) \times (2j_2 + 1)$ vectors in the basis. The vectors satisfy

$$\hat{J}^{(1)2} |j_1, m_1, j_2, m_2\rangle = j_1(j_1 + 1)\hbar^2 |j_1, m_1, j_2, m_2\rangle, \tag{10.4}$$

$$\hat{J}_z^{(1)} |j_1, m_1, j_2, m_2\rangle = m_1\hbar |j_1, m_1, j_2, m_2\rangle, \tag{10.5}$$

$$\hat{J}^{(2)2} |j_1, m_1, j_2, m_2\rangle = j_2(j_2 + 1)\hbar^2 |j_1, m_1, j_2, m_2\rangle, \tag{10.6}$$

$$\hat{J}_z^{(2)} |j_1, m_1, j_2, m_2\rangle = m_2\hbar |j_1, m_1, j_2, m_2\rangle. \tag{10.7}$$

The reader can readily check that the total angular momentum operators satisfy the usual commutation relations

$$[\hat{J}_x, \hat{J}_y] = i\hbar \hat{J}_z, \quad \text{etc. and} \quad [\hat{J}^2, \hat{J}_i] = 0. \tag{10.8}$$

As for any angular momentum operator then, $\hat{J}^2$ has eigenvalues $j(j + 1)\hbar^2$ whilst the operator corresponding to the z-component of the total angular momentum has eigenvalues $m\hbar$ with m running between j and $-j$ in integer steps for a given j.

10.2 Addition Theorem

The question which then arises is, given two angular momenta, corresponding to angular momentum quantum numbers j_1 and j_2, respectively, what are the allowed values of the total angular momentum quantum number j? The answer is provided by the **angular momentum addition theorem**.

Theorem 10.1   *The allowed values of the total angular momentum quantum number j, given two angular momenta corresponding to quantum numbers j_1 and j_2, are*

$$j = j_1 + j_2, j_1 + j_2 - 1, \ldots, |j_1 - j_2| \tag{10.9}$$

and for each of these values of j, m takes on the $(2j + 1)$ values

$$m = j, j - 1, \ldots, -j. \tag{10.10}$$

The proof of this theorem is beyond the scope of this book, and is deferred to more advanced texts.

Example 10.1 If we have a particle with orbital angular momentum quantum number $\ell = 1$ and spin $s = \frac{1}{2}$, the possible values for the total angular momentum quantum number are $j = 3/2, \frac{1}{2}$.

Example 10.2 If we have two particles with orbital angular momentum quantum numbers $\ell_1 = 1$ and $\ell_2 = 1$, then the total orbital angular momentum quantum number can be $\ell = 2, 1, 0$.

Example 10.3 If we have two particles with spin $s_1 = s_2 = \frac{1}{2}$, the total spin of the system can be $s = 1, 0$. This case is discussed in detail at the end of the chapter.

It is easy to show that $\hat{J}^2$ commutes with $\hat{J}^{(1)\,2}$ and $\hat{J}^{(2)\,2}$ but *not* with $\hat{J}_z^{(1)}$ or $\hat{J}_z^{(2)}$ by writing

$$\hat{J}^2 = (\underline{\hat{J}}^{(1)} + \underline{\hat{J}}^{(2)})^2 = \{\hat{J}^{(1)\,2} + \hat{J}^{(2)\,2} + 2\underline{\hat{J}}^{(1)} \cdot \underline{\hat{J}}^{(2)}\}. \tag{10.11}$$

The dot product in the last term on the right-hand side contains the x- and y-components of the two angular momenta which do not commute with the respective z-components. The operator $\hat{J}_z$ commutes with $\hat{J}^{(1),2}$ and $\hat{J}^{(2),2}$, so the set of four operators $\hat{J}^2, \hat{J}_z, \hat{J}_1^2, \hat{J}_2^2$ are *also* a set of mutually commuting operators with a common eigenbasis known as the **coupled basis**.

Exercise 10.2.1 Using the commutation relations for the components of the angular momenta $\underline{\hat{J}}, \underline{\hat{J}}^{(1)}$ and $\underline{\hat{J}}^{(2)}$, check that you can derive all the commutation relations stated so far.

Exercise 10.2.2 For given values of j_1 and j_2, check that the number of vectors in both the coupled and uncoupled bases is $(2j_1 + 1) \times (2j_2 + 1)$. Clearly the number of vectors has to be the same in the two bases, since they are bases of the same Hilbert space.

The vectors of the coupled basis are denoted $\{|j, m, j_1, j_2\rangle\}$ and satisfy

$$\hat{J}^2 |j, m, j_1, j_2\rangle = j(j + 1)\hbar^2 |j, m, j_1, j_2\rangle, \tag{10.12}$$

$$\hat{J}_z |j, m, j_1, j_2\rangle = m\hbar |j, m, j_1, j_2\rangle, \tag{10.13}$$

$$\hat{J}^{(1)\,2} |j, m, j_1, j_2\rangle = j_1(j_1 + 1)\hbar^2 |j, m, j_1, j_2\rangle, \tag{10.14}$$

$$\hat{J}^{(2)\,2} |j, m, j_1, j_2\rangle = j_2(j_2 + 1)\hbar^2 |j, m, j_1, j_2\rangle. \tag{10.15}$$

These are states of *definite total angular momentum and definite z-component of total angular momentum* but *not* in general states with definite $J_z^{(1)}$ or $J_z^{(2)}$. In fact, they are expressible as linear combinations of the states of the uncoupled basis, with coefficients known as **Clebsch–Gordan coefficients**, some of which you can find tabulated below. For instance, using the completeness of the states of the uncoupled basis

$$\sum_{m_1, m_2} |j_1, m_1, j_2, m_2\rangle\langle j_1, m_1, j_2, m_2| = 1 \tag{10.16}$$

yields

$$|j, m, j_1, j_2\rangle = \sum_{m_1, m_2} |j_1, m_1, j_2, m_2\rangle\langle j_1, m_1, j_2, m_2|j, m, j_1, j_2\rangle, \tag{10.17}$$

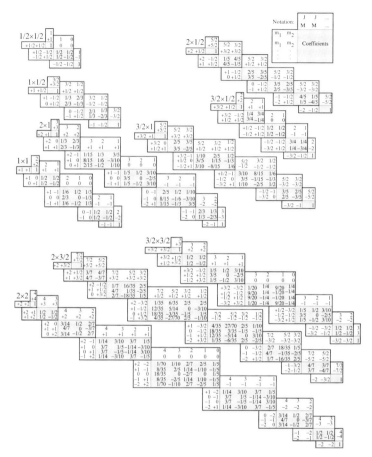

Fig. 10.1   Tables of Clebsch–Gordan coefficients. The notation is explained in the upper right corner of the figure. The figure is taken from P. A. Zyla et al. (Particle Data Group), *Prog. Theor. Exp. Phys.* 2020, 083C01 (2020).

where the Clebsch–Gordan coefficients appear explicitly as the scalar products on the right-hand side of the equation.

Mathematical Aside: Clebsch–Gordan coefficients

The Clebsch–Gordan coefficients that appear for the smaller values of the angular momenta are summarised in Fig. 10.1. Note that a square-root sign is to be understood over every coefficient, e.g. for $-8/15$ read $-\sqrt{8/15}$.

10.3   Example: Two Spin-$\frac{1}{2}$ Particles

We consider the important case of two spin-$\frac{1}{2}$ particles for which the spin quantum numbers are $s_1 = \frac{1}{2}$ and $s_2 = \frac{1}{2}$, respectively. According to the theorem, the total spin quantum number s takes on the values $s_1 + s_2 \equiv 1$ and $|s_1 - s_2| \equiv 0$ only.

Thus two electrons can have a total spin of 1 or 0 only: these states of definite total spin are referred to as *triplet* and *singlet* states, respectively, because in the former there are three possible values of the spin magnetic quantum number, $m_s = 1, 0, -1$, whereas in the latter there is only one such value, $m_s = 0$.

Using the notation introduced in the previous chapter, the states of the uncoupled basis can be written as

$$|\uparrow\uparrow\rangle, \quad |\uparrow\downarrow\rangle, \quad |\downarrow\uparrow\rangle, \quad |\downarrow\downarrow\rangle, \tag{10.18}$$

where the first and second arrows in the ket refer to electrons 1 and 2, respectively. The operators $\hat{S}_1^2$ and $\hat{S}_{1z}$ act only on the first quantum number, i.e. the eigenvalue of $\hat{S}_{1z}$ represented by the first arrow, whilst $\hat{S}_2^2$ and $\hat{S}_{2z}$ act only on the second quantum number.

It should be clear that since $|\uparrow\uparrow\rangle$ has $m_{s_1} = \frac{1}{2}$ and $m_{s_2} = \frac{1}{2}$, it must have $m_s = 1$, that is, the total z-component of spin $\hbar$, and *can therefore only be* $s = 1$ *and not* $s = 0$. This is an example of what is known as a *stretched state*: it has the maximum possible value of the z-component of total angular momentum (spin) and must therefore be a member of both the coupled and uncoupled basis:

$$|\uparrow\uparrow\rangle = |s = 1, m_s = 1, s_1 = \tfrac{1}{2}, s_2 = \tfrac{1}{2}\rangle. \tag{10.19}$$

A similar argument shows that $|\downarrow\downarrow\rangle$ has $m_s = -1$ and thus also can only be $s = 1$, so that

$$|\downarrow\downarrow\rangle = |s = 1, m_s = -1, s_1 = \tfrac{1}{2}, s_2 = \tfrac{1}{2}\rangle. \tag{10.20}$$

The remaining two states of the coupled basis are, however, non-trivial linear combinations of the two remaining states of the uncoupled basis:

$$|s = 1, m_s = 0, s_1 = \tfrac{1}{2}, s_2 = \tfrac{1}{2}\rangle = \frac{1}{\sqrt{2}}[|\uparrow\downarrow\rangle + |\downarrow\uparrow\rangle], \tag{10.21}$$

$$|s = 0, m_s = 0, s_1 = \tfrac{1}{2}, s_2 = \tfrac{1}{2}\rangle = \frac{1}{\sqrt{2}}[|\uparrow\downarrow\rangle - |\downarrow\uparrow\rangle]. \tag{10.22}$$

Proof. We apply the lowering operator, $\hat{S}_-$, for the z-component of total spin to the stretched state:

$$\hat{S}_-|\uparrow\uparrow\rangle = (\hat{S}_{1-} + \hat{S}_{2-})|\uparrow\uparrow\rangle. \tag{10.23}$$

The left-hand side we write as

$$\hat{S}_-|s = 1, m_s = 1, s_1 = \tfrac{1}{2}, s_2 = \tfrac{1}{2}\rangle = \sqrt{1(1+1) - 1(1-1)}\,\hbar$$
$$\times |s = 1, m_s = 0, s_1 = \tfrac{1}{2}, s_2 = \tfrac{1}{2}\rangle, \tag{10.24}$$
$$= \sqrt{2}\,\hbar\,|s = 1, m_s = 0, s_1 = \tfrac{1}{2}, s_2 = \tfrac{1}{2}\rangle, \tag{10.25}$$

using the usual properties of the lowering operator, whilst we write the right-hand side as

$$(\hat{S}_{1-} + \hat{S}_{2-})|s_1 = \tfrac{1}{2}, m_{s_1} = \tfrac{1}{2}, s_2 = \tfrac{1}{2}, m_{s_2} = \tfrac{1}{2}\rangle,$$

and note that

$$\hat{S}_{1-}|s_1 = \tfrac{1}{2}, m_{s_1} = \tfrac{1}{2}, s_2 = \tfrac{1}{2}, m_{s_2} = \tfrac{1}{2}\rangle$$
$$= \sqrt{\tfrac{1}{2}(\tfrac{1}{2}+1) - \tfrac{1}{2}(\tfrac{1}{2}-1)}\,\hbar\,|s_1 = \tfrac{1}{2}, m_{s_1} = -\tfrac{1}{2}, s_2 = \tfrac{1}{2}, m_{s_2} = \tfrac{1}{2}\rangle$$
$$= \hbar\,|s_1 = \tfrac{1}{2}, m_{s_1} = -\tfrac{1}{2}, s_2 = \tfrac{1}{2}, m_{s_2} = \tfrac{1}{2}\rangle$$
$$= \hbar\,|\downarrow\uparrow\rangle,$$

whilst

$$\hat{S}_{2-}|s_1 = \tfrac{1}{2}, m_{s_1} = \tfrac{1}{2}, s_2 = \tfrac{1}{2}, m_{s_2} = \tfrac{1}{2}\rangle$$
$$= \sqrt{\tfrac{1}{2}(\tfrac{1}{2}+1) - \tfrac{1}{2}(\tfrac{1}{2}-1)}\,\hbar\,|s_1 = \tfrac{1}{2}, m_{s_1} = \tfrac{1}{2}, s_2 = \tfrac{1}{2}, m_{s_2} = -\tfrac{1}{2}\rangle$$
$$= \hbar\,|s_1 = \tfrac{1}{2}, m_{s_1} = \tfrac{1}{2}, s_2 = \tfrac{1}{2}, m_{s_2} = -\tfrac{1}{2}\rangle$$
$$= \hbar\,|\uparrow\downarrow\rangle.$$

Equating the two sides then yields the stated result:

$$|s = 1, m_s = 0, s_1 = \tfrac{1}{2}, s_2 = \tfrac{1}{2}\rangle = \tfrac{1}{\sqrt{2}}[|\uparrow\downarrow\rangle + |\downarrow\uparrow\rangle]. \qquad (10.26)$$

The remaining member of the coupled basis must be a linear combination of $|\uparrow\downarrow\rangle$ and $|\downarrow\uparrow\rangle$ orthogonal to this, which we can take to be

$$|s = 0, m_s = 0, s_1 = \tfrac{1}{2}, s_2 = \tfrac{1}{2}\rangle = \tfrac{1}{\sqrt{2}}[|\uparrow\downarrow\rangle - |\downarrow\uparrow\rangle]. \qquad (10.27)$$

Exercise 10.3.1   Check that the Clebsch–Gordan coefficients obtained by this brute force calculation coincide with the ones in Fig. 10.1.

Summary

- We have introduced the angular momentum operator for a system of two particles: as one would naively expect, the total angular momentum operator is simply the sum of the operators associated with each particle. The sum acts in the tensor space of the Hilbert spaces of states of particle 1 and particle 2. (10.1)
- We have discussed the choice of complete sets of commuting observables, and the corresponding coupled and uncoupled bases. (10.1, 10.2)
- Addition theorem for angular momenta is stated without proof. Make sure that you understand how to apply it to practical examples. (10.2)
- Example: coupled and uncoupled bases for a system of two particles of spin $\tfrac{1}{2}$. (10.3)

Problems

10.1 What are the allowed values of the total angular momentum quantum number, j, for a particle with spin $s = \frac{1}{2}$ and orbital angular momentum $\ell = 2$? If two spinless particles each have orbital angular momentum $\ell = 1$, what are the allowed values of the total orbital angular momentum?

10.2 Given a system of two non-interacting particles with orbital angular momentum $\ell = 1$, $|m_1, m_2\rangle$ denotes a state where particles 1 and 2 have L_z components $m_1\hbar$ and $m_2\hbar$, respectively.

Construct the operators $\hat{L}^2$ and $\hat{L}_z$ for the system in terms of the operators $\hat{L}_z^{(1)}$, $\hat{L}_z^{(2)}$, $\hat{L}_\pm^{(1)}$ and $\hat{L}_\pm^{(2)}$.

Normalise the following wave functions. Are these states eigenstates of $\hat{L}^2$ and $\hat{L}_z$? If yes, calculate the eigenvalues:

$$|1, 1\rangle, \quad |-1, -1\rangle, \quad |1, 0\rangle + |1, 1\rangle, \quad |1, 0\rangle + |0, 1\rangle, \quad |1, 0\rangle - |0, 1\rangle,$$
$$|1, -1\rangle + 2|0, 0\rangle + |-1, 1\rangle, \quad |1, -1\rangle - |-1, 1\rangle.$$

10.3 For given values of j_1 and j_2, the number of states in the uncoupled basis $|j_1 m_1, j_2 m_2\rangle$ is $(2j_1 + 1)(2j_2 + 1)$. The number of states in the coupled basis $|JM, j_1, j_2\rangle$ should be the same, since the two bases are linearly related. A partial proof of the angular momentum addition theorem consists of showing that this is the case provided that the allowed values of j run from $|j_1 - j_2|$ in integer steps to $j_1 + j_2$. You are invited to prove, therefore, that

$$\sum_{j=|j_1-j_2|}^{j_1+j_2} (2j + 1) = (2j_1 + 1)(2j_2 + 1).$$

10.4 Verify that the states

$$|s = 1, m_s = 1\rangle = |\uparrow\uparrow\rangle,$$
$$|s = 1, m_s = 0\rangle = \frac{1}{\sqrt{2}}\{|\uparrow\downarrow\rangle + |\downarrow\uparrow\rangle\},$$
$$|s = 1, m_s = -1\rangle = |\downarrow\downarrow\rangle,$$
$$|s = 0, m_s = 0\rangle = \frac{1}{\sqrt{2}}\{|\uparrow\downarrow\rangle - |\downarrow\uparrow\rangle\}$$

of the coupled basis are indeed eigenstates of the operator $\hat{S}^2$ with eigenvalues $s(s + 1)\hbar^2$ by using the identity

$$\hat{S}^2 = \hat{S}^{(1)\,2} + \hat{S}^{(2)\,2} + 2\hat{S}_z^{(1)}\hat{S}_z^{(2)} + \hat{S}_+^{(1)}\hat{S}_-^{(2)} + \hat{S}_-^{(1)}\hat{S}_+^{(2)}.$$

10.5 Using the table of Clebsch–Gordan coefficients in Fig. 10.1, write the vector $|3, 1, \frac{3}{2}, \frac{3}{2}\rangle$ as a linear combination of vectors of the relevant uncoupled basis.

10.6 The stationary states of the harmonic oscillator can be constructed by acting on the ground state $|0\rangle$ with a creation operator $\hat{a}^\dagger$. The eigenstates satisfy the eigenvalue equation

$$\hat{H}|n\rangle = \hbar\omega \left(n + \tfrac{1}{2}\right)|n\rangle,$$

where $\hat{H} = \hbar\omega(\hat{a}^\dagger\hat{a} + \tfrac{1}{2})$ and the creation and annihilation operators satisfy the commutation relation $[\hat{a}, \hat{a}^\dagger] = 1$. The eigenstates of the energy, which we denote by $|n\rangle$, are also eigenstates of the 'number' operator $N = \hat{a}^\dagger\hat{a}$, with eigenvalue n.

Let us now consider two independent harmonic oscillators – one described by the operators $\hat{a}_+$ and $\hat{a}_+^\dagger$, the other described by the operators $\hat{a}_-$ and $\hat{a}_-^\dagger$. The fact that the oscillators are independent translates into the fact that all '+' operators commute with all '−' operators.

Explain why it is possible to find a basis of simultaneous eigenvalues of $\hat{H}_+$ and $\hat{H}_-$, where

$$\hat{H}_+ = \hbar\omega \left(\hat{a}_+^\dagger\hat{a}_+ + \tfrac{1}{2}\right),$$
$$\hat{H}_- = \hbar\omega \left(\hat{a}_-^\dagger\hat{a}_- + \tfrac{1}{2}\right).$$

Let the ground state $|0\rangle$ satisfy $\hat{a}_+|0\rangle = 0$ and $\hat{a}_-|0\rangle = 0$. Show that the states

$$|n_+, n_-\rangle = \frac{\left(\hat{a}_+^\dagger\right)^{n_+}}{\sqrt{n_+!}} \frac{\left(\hat{a}_-^\dagger\right)^{n_-}}{\sqrt{n_-!}}|0\rangle$$

are normalised simultaneous eigenstates of $\hat{H}_+$ and $\hat{H}_-$, and find the corresponding eigenvalues.

Let us define the operators

$$\hat{J}_x = \tfrac{1}{2}\hbar(\hat{a}_+^\dagger\hat{a}_- + \hat{a}_-^\dagger\hat{a}_+),$$
$$\hat{J}_y = \frac{1}{2i}\hbar(\hat{a}_+^\dagger\hat{a}_- - \hat{a}_-^\dagger\hat{a}_+),$$
$$\hat{J}_z = \tfrac{1}{2}\hbar(\hat{a}_+^\dagger\hat{a}_+ - \hat{a}_-^\dagger\hat{a}_-).$$

Show that $[\hat{J}_x, \hat{J}_y] = i\hbar\hat{J}_z$.

We assume that the components of $\hat{J}_i$ satisfy the commutation relations

$$[\hat{J}_i, \hat{J}_j] = i\epsilon_{ijk}\hbar\hat{J}_k.$$

Show that

$$\hat{J}^2 = \hat{J}_x^2 + \hat{J}_y^2 + \hat{J}_z^2 = \hbar^2 \frac{\hat{N}}{2}\left(\frac{\hat{N}}{2} + 1\right),$$

where $\hat{N} = \hat{a}_+^\dagger\hat{a}_+ + \hat{a}_-^\dagger\hat{a}_-$.

Show that the states $|n_+, n_-\rangle$ are simultaneous eigenstates of $\hat{J}^2$ and $\hat{J}_z$, and find the corresponding eigenvalues.

For a fixed value of $n_+ + n_-$, find the possible values of the eigenvalues of $\hat{J}_z$. Find an expression for the raising operator $\hat{J}_+$ in terms of the creation and annihilation operators, and discuss the physical interpretation.

11

Central Potentials

We are now ready to study a generic class of three-dimensional physical systems. They are the systems that evolve in a central potential, i.e. a potential energy that depends only on the distance r from the origin: $V(\underline{r}) = V(r)$. If we use spherical coordinates to parametrise our three-dimensional space, a central potential does *not* depend on the angular variables θ and ϕ. In other words, these are systems that are spherically symmetric, i.e. invariant under rotations. We will see in this chapter that working in spherical coordinates allows us to reduce the time-independent Schrödinger equation for the three-dimensional system into an equation for the radial part of the wave function, i.e. an eigenvalue problem for a one-dimensional system defined on the positive real semi-axis. An example of central potential is the Coulomb potential between electrically charged particles that we will study in detail when we discuss the hydrogen atom.

11.1 Stationary States

As usual, the dynamics of the system is encoded in the solutions of the time-independent Schrödinger equation. Working in the position representation, we can write the eigenvalue problem for the Hamiltonian operator as a differential equation for the wave functions of the stationary states:

$$\left[-\frac{\hbar^2}{2\mu} \nabla^2 + V(r) \right] u(\underline{r}) = E u(\underline{r}). \tag{11.1}$$

We have denoted the mass of the particle by μ in order to avoid confusion with the magnetic quantum number m, which will appear in the solutions below. It is clearly convenient to use spherical coordinates and write the Laplacian as

$$\nabla^2 = \frac{1}{r^2} \frac{\partial}{\partial r} \left(r^2 \frac{\partial}{\partial r} \right) + \frac{1}{r^2 \sin \theta} \frac{\partial}{\partial \theta} \left(\sin \theta \frac{\partial}{\partial \theta} \right) + \frac{1}{r^2 \sin^2 \theta} \frac{\partial^2}{\partial \phi^2}. \tag{11.2}$$

The key observation to solve the eigensystem in Eq. (11.1) is that we can use Eq. (8.24) and rewrite the Laplacian as

$$\nabla^2 = \frac{1}{r^2} \frac{\partial}{\partial r} \left(r^2 \frac{\partial}{\partial r} \right) - \frac{1}{\hbar^2 r^2} \hat{L}^2, \tag{11.3}$$

and thus the eigenvalue equation becomes

$$\frac{\hbar^2}{2\mu}\left[-\frac{1}{r^2}\frac{\partial}{\partial r}\left(r^2\frac{\partial}{\partial r}\right)+\frac{1}{\hbar^2 r^2}\hat{L}^2\right]u(r,\theta,\phi) = (E - V(r))\,u(r,\theta,\phi). \qquad (11.4)$$

Separation of Variables

We have already seen that the eigenfunctions of the operator $\hat{L}^2$ are the spherical harmonics $Y_\ell^m(\theta,\phi)$. Therefore it makes sense to look for a solution of the time-independent Schrödinger equation by separating the solution into

$$u(r,\theta,\phi) = R_\ell(r)Y_\ell^m(\theta,\phi). \qquad (11.5)$$

We have written the general solution $u(\underline{r})$ as the product of a radial function $R_\ell(r)$, which depends only on the radius r, times the spherical harmonics. The latter encode all the angular dependence of the solutions $u(\underline{r})$.

Using Eq. (8.27), we can rewrite Eq. (11.4) as an ordinary differential equation for the function $R_\ell(r)$:

$$\frac{1}{r^2}\frac{d}{dr}\left(r^2\frac{dR_\ell}{dr}\right) - \frac{\ell(\ell+1)}{r^2}R_\ell + \frac{2\mu}{\hbar^2}\left[E - V(r)\right]R_\ell = 0. \qquad (11.6)$$

Note that the magnetic quantum number m does not enter into the equation for the radial wave function R_ℓ. The dependence of the radial function on the angular momentum ℓ is emphasised by the suffix. Each energy level will have a $(2\ell+1)$-fold degeneracy, identified by the value of m.

Equation for the Radial Wave Function

Let us now discuss the solution for the radial part of the equation. The radial equation is simplified by the substitution

$$R_\ell(r) = \frac{\chi_\ell(r)}{r}; \qquad (11.7)$$

inserting Eq. (11.7) into Eq. (11.6) yields

$$\frac{d^2\chi_\ell}{dr^2} + \left[\frac{2\mu}{\hbar^2}(E - V(r)) - \frac{\ell(\ell+1)}{r^2}\right]\chi_\ell(r) = 0. \qquad (11.8)$$

We have obtained a one-dimensional eigenvalue problem. Equation (11.8) is the time-independent Schrödinger equation for a one-dimensional system in the potential:

$$V_\ell(r) = V(r) + \frac{\hbar^2}{2\mu}\frac{\ell(\ell+1)}{r^2}. \qquad (11.9)$$

This is the sum of the potential $V(r)$ and a 'centrifugal' term

$$\frac{1}{2\mu}\frac{\hat{L}^2}{r^2}. \qquad (11.10)$$

Boundary Condition for the Radial Wave Function

We know that the solution of the Schrödinger equation, which we call $u(r, \theta, \phi)$ in this section, must be finite. Therefore the radial part $R_\ell(r)$ has to be finite everywhere including the origin. For this to happen we need

$$\chi_\ell(0) = 0. \tag{11.11}$$

Actually this condition turns out to be true also for a potential energy that diverges as $r \to 0$.

Hence the dynamics of a quantum system in a central potential can be reduced to the dynamics of a one-dimensional system in the region $r > 0$ – remember that r is the radial coordinate and hence is always positive.

The normalisation condition for the radial wave function is

$$\int_0^\infty |R_\ell(r)|^2 \, r^2 dr = \int_0^\infty |\chi_\ell(r)|^2 \, dr = 1. \tag{11.12}$$

11.2 Physical Interpretation

The solution of the one-dimensional problem in Eq. (11.8) only depends on the value of the energy E. Given that the angular part is given by the spherical harmonics Y_ℓ^m, we obtain that the three-dimensional wave function is entirely determined by the values of E, ℓ, m. The eigenstates can be written as

$$u_{n\ell m}(\underline{r}) = R_{n\ell}(r) Y_\ell^m(\theta, \phi), \tag{11.13}$$

where we have added an index n that labels the allowed values of the energy E.[1] As already noted above, the magnetic quantum number does not enter into the radial equation, and therefore the solution $R_{n\ell}(r)$ only depends on the two indices n and ℓ.

Let us discuss the behaviour of the solution $R_{n\ell}$ near the origin for a potential such that

$$\lim_{r \to 0} \left[r^2 V(r) \right] = 0, \tag{11.14}$$

i.e. for a potential that diverges in the origin *less* than $1/r^2$. As a consequence, the short-distance behaviour of the effective one-dimensional potential $V_\ell(r)$ is dominated by the centrifugal term. We shall look for a solution of the form

$$R_{n\ell}(r) = \text{const} \cdot r^s. \tag{11.15}$$

Substituting Eq. (11.15) into Eq. (11.6) and neglecting terms that vanish in the limit $r \to 0$, we find

$$s(s + 1) = \ell(\ell + 1). \tag{11.16}$$

---

[1] The notation here suggests that we are going to deal with a discrete spectrum indexed by integer numbers. If the energy spectrum is continuous, the index n can be replaced by the actual eigenvalue E.

The latter has two solutions for s:

$$s = \ell, \quad \text{or } s = -(\ell + 1). \tag{11.17}$$

Clearly the solution with $s = -(\ell + 1)$ is divergent at the origin $r = 0$ and therefore does not satisfy the boundary condition for the radial wave function. Hence, close to the origin, the solutions with angular momentum ℓ are proportional to r^ℓ.

The probability density for a particle to be at a distance r from the origin is given by

$$|R|^2 r^2 \simeq r^{2(\ell+1)}. \tag{11.18}$$

The larger the angular momentum, the more rapidly the probability goes to zero at $r = 0$. This is the precise quantum realisation of a centrifugal potential. The larger the value of ℓ, the stronger the potential, and the less probable it becomes to observe the system close to the origin.

11.3 Quantum Rotator

We can conclude this chapter with a simple example of physical relevance: the quantum rotator. The quantum rotator is a quantum system made up of two particles of mass m_1 and m_2 separated by a fixed distance r_e. It is a simple but effective description of the rotational degrees of freedom of a diatomic molecule. By solving the time-independent Schrödinger equation we are going to find the rotational energy levels, i.e. the energy levels of the molecule in the limit where we neglect the vibrational energy.

Let us choose the centre of mass of the molecule as the origin of the reference frame. The state of the system is completely specified by two angles θ and ϕ that specify the orientation of the axis of the molecule with respect to the axes of the reference frame, as illustrated in Fig. 11.1.

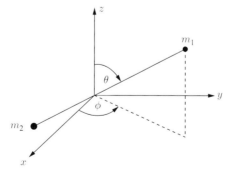

Fig. 11.1 Rotator in three-dimensional space. The length of the axis between the two atoms is fixed to be r_e. The origin of the reference frame is chosen to coincide with the centre of mass of the system. The direction of the axis of the rotator is completely specified by the two angles θ and ϕ.

The distance from the origin to the first mass is denoted by r_1, and similarly the distance from the origin to the second mass is r_2. Since we are neglecting the vibrational degrees of freedom, both r_1 and r_2 are constant, and clearly $r_1 + r_2 = r_e$. Since we have chosen the origin of the reference frame to be the centre of mass of the system, we also have

$$m_1 r_1 = m_2 r_2, \qquad (11.19)$$

and therefore

$$r_1/m_2 = r_2/m_1 = r_e/(m_1 + m_2). \qquad (11.20)$$

The moment of inertia of this system is simply

$$I = m_1 r_1^2 + m_2 r_2^2 \equiv \mu r_e^2, \qquad (11.21)$$

where we have introduced the reduced mass

$$\frac{1}{\mu} = \frac{1}{m_1} + \frac{1}{m_2}. \qquad (11.22)$$

In classical mechanics the angular momentum of the system is given by

$$|\underline{L}| = I \omega_R, \qquad (11.23)$$

where ω_R is the angular velocity of the system. The energy can be expressed as

$$H = \tfrac{1}{2} I \omega_R^2 = \frac{L^2}{2I} = \frac{L^2}{2\mu r_e^2}. \qquad (11.24)$$

The Hamiltonian for the quantum rotator is defined starting from Eq. (11.24), and promoting L^2 to be the operator associated with the square of the angular momentum:

$$\hat{H} = \frac{\hat{L}^2}{2\mu r_e^2}. \qquad (11.25)$$

The reduced mass μ and the radius of the molecule r_e are constants that define the physical system under study: different diatomic molecules have different reduced masses, or sizes.

Note that the wave function of the system does not depend on r since we are neglecting the vibrations of the molecule. So the states of the quantum system are described by a wave function $\psi(\theta, \phi)$ that depends only on the angular variables.

The Hamiltonian for the diatomic molecule is proportional to $\hat{L}^2$; we have already computed the eigenvalues and the eigenfunctions of $\hat{L}^2$ when we discussed the angular momentum. The stationary states are the spherical harmonics

$$\hat{H} Y_\ell^m(\theta, \phi) = \frac{\ell(\ell+1)\hbar^2}{2\mu r_e^2} Y_\ell^m(\theta, \phi). \qquad (11.26)$$

The constant $B = \hbar/(4\pi \mu r_e^2)$ is usually called the 'rotational constant'. It has the dimensions of a frequency. The energy levels are therefore

$$E_\ell = B h \ell(\ell+1). \qquad (11.27)$$

This simple model yields a good quantitative description of the spectrum of diatomic molecules.

11.4 Central Square Well

We can now consider the problem of a central square well, i.e. the solution of the time-independent Schrödinger equation for a constant central potential for $r \in \mathcal{I}$, where $\mathcal{I}$ is some interval on the positive real axis. The eigenvalue problem can be written as

$$-\nabla^2 u(\underline{r}) = k^2 u(\underline{r}), \tag{11.28}$$

where we have introduced

$$k^2 = \frac{2\mu(E - V_0)}{\hbar^2}. \tag{11.29}$$

Note that k has dimensions of length^{-1} and is real for $E > V_0$. Using the substitution in Eq. (11.7) yields

$$\chi_\ell''(r) + \left[k^2 - \frac{\ell(\ell+1)}{r^2} \right] \chi_\ell(r) = 0. \tag{11.30}$$

It is useful at this stage to introduce the dimensionless variable

$$\rho = kr, \quad \text{and} \quad \eta_\ell(\rho) = \chi_\ell(r), \tag{11.31}$$

so that we can rewrite Eq. (11.30) as

$$\eta_\ell''(\rho) + \left[1 - \frac{\ell(\ell+1)}{\rho^2} \right] \eta_\ell(\rho) = 0. \tag{11.32}$$

It is instructive to discuss the solution of Eq. (11.32) in detail, as it allows us to introduce some interesting mathematical tools. Defining

$$\eta_\ell(\rho) = \rho^{\ell+1} \xi_\ell(\rho) \tag{11.33}$$

allows us to rewrite the equation above as

$$\rho \xi_\ell''(\rho) + 2(\ell+1)\xi_\ell'(\rho) + \rho \xi_\ell(\rho) = 0. \tag{11.34}$$

In this form, the differential equation can be solved using a general technique known as the **Laplace method**.

Mathematical Aside: The Laplace method

Differential equations of the form

$$F(y, x) = (a_n + b_n x) \frac{d^n y}{dx} + (a_{n-1} + b_{n-1} x) \frac{d^{n-1} y}{dx^{n-1}} + \cdots + (a_0 + b_0 x) y$$

$$= 0 \tag{11.35}$$

admit solutions of the form

$$y(x) = \int_\Gamma dz \, Z(z) e^{zx}, \tag{11.36}$$

where z is a complex integration variable and Γ is a path in the complex plane to be determined as part of the solution.[2]

Substituting Eq. (11.36) allows Eq. (11.35) to be written as

$$F(y, x) = \int_\Gamma dz\, e^{zx}\, Z(z)\, [P(z) + Q(z)x], \tag{11.37}$$

where

$$P(z) = \sum_{k=0}^{n} a_k z^k, \quad Q(z) = \sum_{k=0}^{n} b_k z^k. \tag{11.38}$$

Choosing

$$Z(z) = \frac{1}{Q(z)} \exp \int^z dw\, \frac{P(w)}{Q(w)} \tag{11.39}$$

leads to

$$F(y, x) = \int_\Gamma d\, (Z(z)Q(z)e^{zx}) = Z(z)Q(z)e^{zx}\Big|_{z_1}^{z_2}, \tag{11.40}$$

where z_1 and z_2 are the ends of the path Γ. In order to solve the equation $F(y, x) = 0$, we need to find a path such that the function ZQe^{zx} takes the same value at both ends.

Using Laplace's method to solve Eq. (11.34), we obtain

$$\xi_\ell(\rho) \propto (-)^\ell \left(\frac{1}{\rho} \frac{d}{d\rho} \right)^\ell \frac{\sin \rho}{\rho}. \tag{11.41}$$

Substituting back into the equations above yields for the radial part of the wave function

$$R_\ell(r) \propto (kr)^\ell \xi_\ell(kr) = j_\ell(kr), \tag{11.42}$$

which is known as the **spherical Bessel function**. The asymptotic behaviour for large distances r is

$$R_\ell(r) \sim \frac{\sin(kr - \ell\pi/2)}{kr}. \tag{11.43}$$

Mathematical Aside: Spherical Bessel Functions

Let us discuss here the details of the solution using Laplace's method. Applying the general formulae to the specific case of Eq. (11.34), we readily obtain

$$Z(z) = \frac{1}{z^2 + 1} \exp \int dw\, \frac{2(\ell + 1)w}{w^2 + 1} = (1 + z^2)^\ell. \tag{11.44}$$

It is easy to guarantee that ZQe^{zx} vanishes at the ends of Γ by choosing the path to go from $-i$ to i along the imaginary axis. Hence

---

[2] The explanation of the Laplace method presented here is taken from E. Onofri and C. Destri, *Istituzioni di Fisica Teorica*, Carocci Editore, Rome, 1998.

$$\xi_\ell(\rho) = \kappa \int_{-i}^{i} dz\, e^{z\rho} (1 + z^2)^\ell, \tag{11.45}$$

$$= C \int_{-1}^{1} dt\, e^{it\rho}(1 - t^2)^\ell, \tag{11.46}$$

where C is a normalisation constant that can be fixed a posteriori. The recursion relation

$$\frac{d}{d\rho} \xi_\ell(\rho) = -\frac{\rho}{2(\ell + 1)} \xi_{\ell+1}(\rho) \tag{11.47}$$

is obtained by integrating by parts, while the initial value

$$\xi_0(\rho) = \frac{\sin \rho}{\rho} \tag{11.48}$$

is computed explicitly from Eq. (11.46).

The recursion relation leads to Eq. (11.41). The leading asymptotic behaviour for large ρ is obtained when all derivatives in Eq. (11.41) act on the sine function, which yields

$$\xi_\ell(\rho) \sim (-1)^\ell \frac{1}{\rho^{\ell+1}} \left(\frac{d}{d\rho}\right)^\ell \sin \rho \sim \frac{\sin(\rho - \ell\pi/2)}{\rho^{\ell+1}}. \tag{11.49}$$

The **spherical Bessel function** is defined as

$$j_\ell(\rho) = \rho^\ell \xi_\ell(\rho). \tag{11.50}$$

A second linear independent solution is obtained from the same recursion, starting from

$$n_0(\rho) = -\cos(\rho)/\rho. \tag{11.51}$$

Being singular for $\rho \to 0$, these solutions do not satisfy the boundary condition for the radial wave function.

It is useful to introduce the linear combinations

$$h_\ell^{(+)}(\rho) = n_\ell(\rho) + i j_\ell(\rho) \quad \text{and} \quad h_\ell^{(-)}(\rho) = n_\ell(\rho) - i j_\ell(\rho). \tag{11.52}$$

For completeness, we report the *small* distance behaviour of the spherical Bessel functions:

$$j_\ell(\rho) \sim \frac{\rho^\ell}{(2\ell + 1)!!}, \tag{11.53}$$

$$n_\ell(\rho) \sim -\frac{(2\ell - 1)!!}{\rho^{\ell+1}}. \tag{11.54}$$

As we discussed in the case of one-dimensional potentials, the solutions above for the case $E > V_0$ remain valid when $E < V_0$ provided k is replaced by $i\kappa$, where

$$\kappa^2 = \frac{2\mu(V_0 - E)}{\hbar^2}. \tag{11.55}$$

The only solution that is normalisable at large distance is $h_\ell^{(+)}(\rho)$.

Summary

- We started the chapter by writing down the time-independent Schrödinger equation for a system in a central potential. Make sure that you understand what is specific to the case of the central potential. (11.1)
- Using separation of variables, we reduced the equation for the radial dependence of the wave function to a one-dimensional problem. (11.1)
- We have derived the form of the effective potential, which includes a centrifugal term. The finiteness of the three-dimensional wave function at the origin implies a boundary condition for the solution of the one-dimensional problem. (11.1)
- We presented a discussion of the solutions for the stationary states and their generic properties. (11.2)
- The quantum rotator is a simple example of a central potential. We have discussed the formulation of the problem, its physical interpretation and the energy spectrum that we obtain from the eigenvalue equation for the Hamiltonian. (11.3)
- Another example of a central potential is the central square well. We have discussed the definition of the problem and the solution of the TISE. The solutions of the eigenvalue problem in this case are the spherical Bessel functions. (11.4)

Problems

11.1 Compute the energy levels of the quantum rotator discussed in Section 11.3. Discuss their degeneracy, and compute the distance between successive levels.

Let $\hat{Z} = r_e \cos\theta$ be the operator associated with the projection of the molecule axis along the z-direction. Using the fact that

$$\cos\theta\, Y_\ell^m(\theta, \phi) = \sqrt{\frac{\ell^2 - m^2}{4\ell^2 - 1}} Y_{\ell-1}^m(\theta, \phi) + \sqrt{\frac{(\ell + 1)^2 - m^2}{4(\ell + 1)^2 - 1}} Y_{\ell+1}^m(\theta, \phi),$$

show that

$$\langle \ell', m'|Z|\ell, m\rangle = r_e \delta_{m',m} \left[\delta_{\ell',\ell-1} \sqrt{\frac{\ell^2 - m^2}{4\ell^2 - 1}} + \delta_{\ell',\ell+1} \sqrt{\frac{(\ell + 1)^2 - m^2}{4(\ell + 1)^2 - 1}} \right].$$

Deduce the time evolution of $\langle Z(t)\rangle = \langle \Psi(t)|\hat{Z}|\Psi(t)\rangle$, where $\Psi(\theta, \phi, t)$ describes the state of the system at time t.

11.2 The parity operator changes the sign of the position vector of a particle: $\underline{r} \to -\underline{r}$. Express this geometrical transformation in spherical coordinates. Check the properties of the first few spherical harmonics under parity.

11.3 Find the stationary states in a spherical well of depth V_0 and size a. Discuss the cases $-V_0 < E < 0$ and $E > 0$ separately. Find explicitly a quantisation condition for a system in s wave, i.e. for $\ell = 0$.

11.4 The angular momentum operator in two dimensions can be defined as

$$\hat{L} = \hat{X}\hat{P}_y - \hat{Y}\hat{P}_x.$$

Rewrite the operator $\hat{L}$ as a differential operator in polar coordinates. Show that the eigenvalues of L are $\hbar m$, where m is an integer, and find the eigenfunctions.

11.5 Write down the time-independent Schrödinger equation for a particle in a two-dimensional circular infinite well of radius R. Separate the variables by writing the wave function as

$$\psi(r, \theta) = R(r)\Phi(\theta).$$

Find the differential equation for the radial part of the wave function $R_m(r)$ for each value of the angular momentum m.

The solution of the radial equation is given by the Bessel function $J_{|m|}(kr)$, where k is related to the energy E by $E = \hbar^2 k^2/(2m)$. Write down the boundary condition for this problem. Let us denote by $a_{n,m}$ the nth zero of $J_{|m|}$. Find the energy levels as functions of the zeros of the Bessel functions.

For $m = 0$ the first zero of the Bessel function $J_0(z)$ occurs for $z = 2.405$. Deduce the value of the energy for the ground state of the system. Compare this with the ground-state energy of the two-dimensional infinite square potential well of size $L = \sqrt{\pi}R$.

11.6 Consider a system made up of two atoms of masses m_1 and m_2. The vibrations of this system can be described by a one-dimensional Schrödinger equation for a system of reduced mass μ, evolving in a potential $V(r)$ that depends on the relative distance r of the two atoms.

The potential $V(r)$ is repulsive at short distances, has a minimum $-V_0$ at some value r_e and goes to zero when $r \to \infty$. Sketch the potential, describe the physical meaning of r_e for a classical system and write the time-independent Schrödinger equation for this system.

In order to study the small deviations around r_e, we introduce the variable $z = r - r_e$ and expand the potential:

$$V(r) = -V_0 + \tfrac{1}{2}V''(r_e)z^2 + O(z^3).$$

Explain the origin of the expansion above. What can you say about $V''(r_e)$? Compute the dimensions of the parameter

$$\omega = \sqrt{\frac{V''(r_e)}{\mu}}.$$

Write the time-independent Schrödinger equation for small oscillations, using ω and neglecting terms that are cubic or higher order in z.

State the energy spectrum of the system in this approximation. What can be learned about the potential V by measuring the energy eigenvalues?

Let us denote by $|n\rangle$ the nth energy eigenstate. Compute the typical size of the fluctuations $d_n = \sqrt{\langle n|\hat{z}^2|n\rangle}$ of the position. You can use the fact that for the harmonic oscillator:

$$\langle n|\hat{V}_{\text{ho}}|n\rangle = \tfrac{1}{2}\langle n|\hat{H}_{\text{ho}}|n\rangle.$$

Deduce a bound on the number of energy levels that can be accurately described by the quadratic approximation introduced above.

Consider the next *two* terms in the Taylor expansion of the potential around r_e. The wave function for the ground state of the harmonic oscillator is given by

$$u_0(x) = C_0 \exp(-\alpha^2 x^2/2), \quad \alpha^2 = \hbar\omega.$$

Compute the shift of the energy level of the ground state due to these new terms at first order in perturbation theory. You may find the following integrals useful:

$$\int_{-\infty}^{+\infty} dx \, \exp(-\alpha^2 x^2) = \left(\frac{\pi}{\alpha^2}\right)^{1/2}, \quad \int_{-\infty}^{+\infty} dx \, \exp(-\alpha^2 x^2) \, x^4 = \frac{3}{4}\frac{\sqrt{\pi}}{\alpha^5}.$$

Note: In order to answer this last part of the problem, you need to study the chapter on time-independent perturbation theory.

11.7 Consider a free particle in three dimensions; its dynamics is determined by the Hamiltonian

$$\hat{H} = \frac{\hat{P}^2}{2\mu},$$

where $\hat{P}^2 = \hat{P}_x^2 + \hat{P}_y^2 + \hat{P}_z^2$ and μ denotes the mass of the particle.

We can define two sets of compatible observables:

$$S_1 = \left\{\hat{H}, \hat{P}_x, \hat{P}_y, \hat{P}_z,\right\},$$

$$S_2 = \left\{\hat{H}, \hat{L}^2, \hat{L}_z\right\}.$$

Check that the observables in S_1 are compatible, and find their simultaneous eigenfunctions. Is the spectrum discrete or continuous? Write the normalisation condition for the eigenfunctions.

Consider a generic state described by the wave function $\psi(\underline{x})$. Write $\psi(\underline{x})$ as a superposition of the eigenstates of the observables in S_1. Show that the coefficients of such a superposition are given by the Fourier transform $\tilde{\psi}(\underline{p})$ of the wave function. What is the probabilistic interpretation of $\tilde{\psi}(\underline{p})$?

Check that the observables in S_2 are compatible. Write the set of equations obeyed by their simultaneous eigenfunctions. In particular, find the equation for the radial part of the wave function. Find explicitly the solution for the case $\ell = 0$, specifying carefully the boundary conditions.

Consider now the case $\ell \neq 0$. Using the ansatz

$$\chi_\ell(r) = r^{\ell+1}\eta_\ell(r),$$

find the differential equation obeyed by $\eta_\ell(r)$.

Check that

$$\eta_\ell = K_\ell \left(\frac{1}{r}\frac{d}{dr}\right)^\ell \frac{\sin(kr)}{kr},$$

where $k = \sqrt{2\mu E/\hbar^2}$ is a solution of the differential equation that you found in the first part of this question.

11.8 A generic state represented by the wave function $\psi(\underline{x})$ can be expanded in the eigenstates with defined angular momentum discussed above. Write such an expansion for a plane wave travelling along the z-direction with momentum $p = \hbar k$ in terms of unknown coefficients $c(k')_{lm}$. Show that $c(k')_{lm}$ are non-zero only if $k' = k$ and $m = 0$.

11.9 The three-dimensional isotropic harmonic oscillator of mass μ can be studied as a system in a central potential

$$V(r) = \tfrac{1}{2}\mu\omega^2 r^2,$$

where r is the distance from the origin in three-dimensional Euclidean space. The square of the angular momentum, $\hat{L}^2$, and the z-component, $\hat{L}_z$, commute with the Hamiltonian, and therefore we can find simultaneous eigenvectors of these three operators that we denote by $|n\ell m\rangle$, where ℓ and m are the angular momentum and magnetic quantum number, respectively. In the position representation these states are represented by the wave functions

$$\langle \underline{r}|n\ell m\rangle = \frac{\eta_{n\ell}(r)}{r} Y_\ell^m(\theta, \phi).$$

The function $\eta_{n\ell}$ is the solution of the one-dimensional TISE in an effective potential

$$\left[-\frac{\hbar^2}{2\mu}\frac{d^2}{dr^2} + \frac{\hbar^2\ell(\ell+1)}{2\mu r^2} + \tfrac{1}{2}\mu\omega^2 r^2 \right] \eta_{n\ell}(r) = E_n \eta_{n\ell}(r),$$

We know from the solution using separation of variables that the energy eigenvalues are

$$E_n = (n + 3/2)\,\hbar\omega\,,$$

where n is an integer. Solving the one-dimensional problem for each value of ℓ shows that we can rewrite the eigenvalues as

$$E_n = (2n_r + \ell + 3/2)\,\hbar\omega\,,$$

where n_r is an integer that characterises the eigenfunctions of the radial equation, and

$$n = 2n_r + \ell\,.$$

We introduce the operators

$$\hat{A}_1 = \frac{1}{\sqrt{2}}\left(\hat{a}_x - i\hat{a}_y\right),$$

$$\hat{A}_0 = \hat{a}_z,$$

$$\hat{A}_{-1} = \frac{1}{\sqrt{2}}\left(\hat{a}_x + i\hat{a}_y\right),$$

where $\hat{a}_k$ are the annihilation operators in direction k introduced in Eq. (7.46).

Compute the commutation relations of the operators $\hat{A}_m$ and $\hat{A}_m^\dagger$ (nine in total). Compute their commutation relations with the number operators $\hat{N}_m = \hat{A}_m^\dagger \hat{A}_m$ and deduce that they can be interpreted as creation and annihilation operators.

Show that

$$\hat{H} = \left(\hat{N}_1 + \hat{N}_0 + \hat{N}_{-1} + \frac{3}{2} \right) \hbar \omega,$$

$$\hat{L}_z = \left(\hat{N}_1 - \hat{N}_{-1} \right) \hbar.$$

Deduce that the states

$$|n_1 n_0 n_{-1}\rangle = \frac{1}{\sqrt{n_1! \, n_0! \, n_{-1}!}} \hat{A}_1^{\dagger \, n_1} \hat{A}_0^{\dagger \, n_0} \hat{A}_{-1}^{\dagger \, n_{-1}} |000\rangle$$

are eigenvectors of $\hat{H}$ and $\hat{L}_z$, with eigenvalues

$$(n_1 + n_0 + n_{-1}) \hbar \omega,$$

$$(n_1 - n_{-1}) \hbar.$$

Consider $\mathcal{H}_n$, the subspace of eigenstates of the Hamiltonian with eigenvalue $E_n = (n + 3/2) \hbar \omega$. Show that the number of vectors $|n_1 n_0 n_{-1}\rangle$ spanning this subspace is

$$\tfrac{1}{2}(n + 1)(n + 2).$$

Show that m can span all integer values from $-n$ to n, and that the number of vectors $|n_1 n_0 n_{-1}\rangle$ corresponding to each value of m is

$|m|$	n	$n-1$	$n-2$	$n-3$	$\ldots$	$n-2s$	$n-2s-1$	$\ldots$
c_m	1	1	2	2	$\ldots$	$s+1$	$s+2$	$\ldots$

Let us denote by d_ℓ the number of subspaces $\mathcal{H}_{n\ell}$ with angular momentum ℓ that are included in $\mathcal{H}_n$. For each value of ℓ greater than $|m|$, each subspace $\mathcal{H}_{n\ell}$ contains exactly one eigenvector of $\hat{L}_z$ with eigenvalue m. Therefore we have

$$c_m = \sum_{\ell > |m|} d_\ell.$$

Show that

$$d_\ell = 1, \quad \text{for } \ell = n, n-2, \ldots,$$

$$d_\ell = 0, \quad \text{otherwise.}$$

Hence the possible values of ℓ are

$$\ell = n, n-2, \ldots, 0, \quad \text{if } n \text{ even,}$$

$$\ell = n, n-2, \ldots, 1, \quad \text{if } n \text{ odd.}$$

Check that in both cases the total number of basis vectors is

$$\tfrac{1}{2}(n + 1)(n + 2),$$

as obtained above.

12

Hydrogen Atom

We have finally developed all the tools that are necessary to study the **hydrogen atom** (H atom) from a quantum-mechanical perspective. In this chapter we present a non-relativistic formulation of the problem, where the interaction is modelled with a static Coulomb potential.

The H atom is a bound state of a proton and an electron. The masses of the two particles are, respectively,

$$m_p = 1.67 \times 10^{-24} \text{ g}, \tag{12.1}$$

$$m_e = 0.91 \times 10^{-27} \text{ g}. \tag{12.2}$$

They have opposite charges, e and $-e$, with

$$e = 4.80 \times 10^{-10} \text{ statC}. \tag{12.3}$$

The ratio of the two masses is

$$m_p/m_e = 1836.15267247(80). \tag{12.4}$$

The interaction between the two particles is due to electromagnetism; in a non-relativistic formulation we can therefore model the H atom as a particle of reduced mass m:

$$\frac{1}{\mu} = \frac{1}{m_p} + \frac{1}{m_e} \approx \frac{1}{0.995 m_e}, \tag{12.5}$$

in a Coulomb potential

$$V(r) = -\frac{e^2}{r}. \tag{12.6}$$

Given that the proton mass is much larger than the electron one, the reduced mass of the ep system is very close to the electron mass. The distance r that appears in the expression for the Coulomb potential is the distance between the electron and the proton. We can identify the origin of our reference frame with the position of the proton; the potential is clearly symmetric under rotations around the origin. The parameters that specify the physical system are the reduced mass μ and the electron charge e.[1] The H atom is an example of motion in a central force, and we are going to use the formalism developed in Chapter 11 to find the stationary states of this system.

---

[1] Note that we use cgs units in this book.

12.1 Stationary States

Following the conventions established in the previous chapter, the Hamiltonian for the system is

$$\hat{H} = -\frac{\hbar^2}{2\mu}\nabla^2 - \frac{e^2}{r}. \tag{12.7}$$

Working with a central potential, we are going to look for simultaneous eigenfunctions of the energy, the square of the angular momentum and the z-component of the angular momentum. The time-independent Schrödinger equation

$$\hat{H}\psi(\underline{r}) = E\psi(\underline{r}) \tag{12.8}$$

is solved as usual by separation of variables. Using for the solution $\psi(\underline{r})$ the ansatz

$$\psi_{n\ell m}(r, \theta, \phi) = \frac{\chi_{n\ell}(r)}{r}Y_\ell^m(\theta, \phi), \tag{12.9}$$

we obtain the radial one-dimensional equation

$$\left[-\frac{\hbar^2}{2\mu}\frac{d^2}{dr^2} + \frac{\hbar^2\ell(\ell+1)}{2\mu r^2} - \frac{e^2}{r}\right]\chi_{n\ell}(r) = E_{n\ell}\chi_{n\ell}(r). \tag{12.10}$$

Equation (12.10) is simply Eq. (11.6) for the specific case of the Coulomb potential. Note that here the notation is anticipating the fact that we are going to find a discrete spectrum, where the energy eigenvalues are labelled by the integer n. The effective potential in this case is

$$V_l(r) = -\frac{e^2}{r} + \frac{\hbar^2}{2\mu}\frac{\ell(\ell+1)}{r^2}. \tag{12.11}$$

The effective potential is sketched in Fig. 12.1. The boundary condition for the radial function $\chi_{n\ell}(r)$ is

$$\chi_{n\ell}(0) = 0. \tag{12.12}$$

Remember that, since we are looking for the bound states of the system, we are only interested in solutions with *negative* energy, i.e. $E_{n\ell} < 0$. Equation (12.10) suggests that the energy levels will depend on the total angular momentum ℓ.

It is useful to describe the solutions of the Schrödinger equation in terms of two physical quantities: a characteristic length

$$a_0 = \frac{\hbar^2}{\mu e^2} \approx 0.52 \text{ Å} \tag{12.13}$$

and energy

$$E_I = \frac{\mu e^4}{2\hbar^2} = \frac{e^2}{2a_0} \approx 13.6 \text{ eV}. \tag{12.14}$$

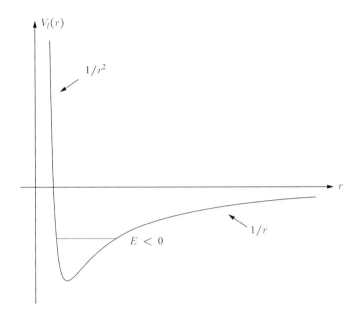

Fig. 12.1 Effective potential for the one-dimensional radial Schrödinger equation for a system with total angular momentum l. We look for solutions of the time-independent Schrödinger equation with negative energy E corresponding to bound states of the proton/electron system.

Having denoted by $E_{n\ell}$ the energy levels, we introduce the dimensionless variables

$$\rho = r/a_0, \quad \lambda_{n\ell} = \sqrt{-E_{n\ell}/E_I}. \tag{12.15}$$

Equation (12.10) can be written as

$$\left[-\frac{\hbar^2}{2\mu} \frac{1}{a_0^2} \frac{d^2}{d\rho^2} + \frac{\hbar^2 \ell(\ell+1)}{2\mu a_0^2 \rho^2} - \frac{e^2}{a_0 \rho} - E_{n\ell} \right] \chi_{n\ell}(a_0 \rho) = 0. \tag{12.16}$$

Introducing

$$u_{n\ell}(\rho) = \chi_{n\ell}(a_0 \rho), \tag{12.17}$$

Eq. (12.16) can be rewritten as

$$\left[\frac{d^2}{d\rho^2} - \frac{\ell(\ell+1)}{\rho^2} + \frac{2\mu e^2 a_0}{\hbar^2} \frac{1}{\rho} + \frac{2\mu a_0^2}{\hbar^2} E_{n,\ell} \right] u_{n\ell}(\rho) = 0, \tag{12.18}$$

and finally

$$\left[\frac{d^2}{d\rho^2} - \frac{\ell(\ell+1)}{\rho^2} + \frac{2}{\rho} - \lambda_{n\ell}^2 \right] u_{n\ell}(\rho) = 0. \tag{12.19}$$

We need to solve Eq. (12.19) in order to find $\lambda_{n\ell}$, and hence the eigenvalues of the Hamiltonian $E_{n\ell}$, and the corresponding radial wave functions $u_{n\ell}(r)$, while the angular part of the wave functions is given by the spherical harmonics.

12.2 Solution of the Radial Equation

In this section we shall discuss some technical details related to the solution of the radial Eq. (12.19). We wish to keep the mathematical details of the solution separate from the physical interpretation, which will be discussed in the next section.

In order to shine a light on the form of the solution, we can start by considering its limiting behaviours as $\rho \to 0$ and $\rho \to \infty$.

Let us first discuss the large-distance regime. As ρ is increased, both the centrifugal and Coulomb potentials tend to zero and become unimportant in Eq. (12.19), which becomes

$$\left[\frac{d^2}{d\rho^2} - \lambda_{n\ell}^2 \right] u_{n\ell}(\rho) = 0. \tag{12.20}$$

The solutions to this latter equation are simply

$$u_{n\ell}(\rho) = \exp(\pm \lambda_{n\ell} \rho), \tag{12.21}$$

and the solution that grows exponentially must be discarded because it yields a non-normalisable wave function.

When trying to solve the full equation, taking into account the centrifugal and the Coulomb potentials, it is convenient to factor out the leading behaviour at large distances; hence we will look for a complete solution of the form

$$u_{n\ell}(\rho) = e^{-\lambda_{n\ell}\rho} \eta_{n\ell}(\rho). \tag{12.22}$$

Equation (12.19) becomes

$$\eta_{n\ell}'' - 2\lambda_{n\ell} \eta_{n\ell}' + \left(-\frac{\ell(\ell+1)}{\rho^2} + \frac{2}{\rho} \right) \eta_{n\ell} = 0, \tag{12.23}$$

where the prime symbol denotes the differentiation with respect to ρ. The boundary condition for $u_{n\ell}$ translates into a boundary condition for $\eta_{n\ell}$, namely $\eta_{n\ell}(0) = 0$.

As $\rho \to 0$ we know from the earlier discussion that $u_{n\ell} \sim \rho^{\ell+1}$. Therefore we can look for a solution for $\eta_{n\ell}$ expanded as a power series in ρ:

$$\eta_{n\ell}(\rho) = \rho^{\ell+1} \sum_{q=0}^{\infty} c_q \rho^q. \tag{12.24}$$

The problem of finding $\eta_{n\ell}$, and hence $u_{n\ell}$, is therefore reduced to the problem of finding the coefficients c_q. Inserting this ansatz into Eq. (12.23), we obtain

$$\sum_q (q+l+1)(q+l)c_q \rho^{q+\ell-1} - 2\lambda_{n\ell}(q+\ell+1)c_q \rho^{q+\ell}$$
$$+ 2c_q \rho^{q+\ell} - \ell(\ell+1)c_q \rho^{q+\ell-1} = 0, \tag{12.25}$$

and hence

$$\sum_q q(q+2\ell+1)c_q \rho^{q+\ell-1} - 2\left[\lambda_{n\ell}(q+\ell+1) - 1 \right] c_q \rho^{q+\ell} = 0. \tag{12.26}$$

Shifting the summation index, $q \to q - 1$, in the second term of the sum above, we obtain

$$\sum_q \left[q(q + 2\ell + 1)c_q - 2 \left[\lambda_{n\ell}(q + \ell) - 1 \right] c_{q-1} \right] \rho^{q+\ell-1} = 0. \tag{12.27}$$

Since the last equality must hold for all values of ρ, we deduce

$$q(q + 2\ell + 1)c_q - 2 \left[\lambda_{n\ell}(q + l) - 1 \right] c_{q-1} = 0. \tag{12.28}$$

Equation (12.28) is a recursion relation between the coefficients of the Taylor expansion of $\eta_{n\ell}/\rho^{l+1}$. It is crucial to note that for large q:

$$\frac{c_q}{c_{q-1}} \overset{q \to \infty}{\sim} \frac{2\lambda_{n\ell}}{q}, \quad \text{i.e. } c_q \sim \frac{(2\lambda_{n\ell})^q}{q!}. \tag{12.29}$$

The asymptotic behaviour for c_q leads to a solution that grows exponentially at large distances:

$$\eta_{n\ell}(\rho) \sim \rho^{\ell+1} e^{2\lambda_{n\ell}\rho}, \tag{12.30}$$

which in turn yields a wave function $u_{n\ell}$ that is not normalisable.

Therefore we *must* have $c_q = 0$ for some finite value of q, that we will denote $q = n_r > 0$. According to Eq. (12.28), this can only happen if

$$\lambda_{n\ell} = \frac{1}{n_r + \ell} \equiv \frac{1}{n}. \tag{12.31}$$

Then the expansion in Eq. (12.24) only contains a finite number of terms, i.e. it is simply a polynomial in ρ of finite order $n_r - 1$.

We see from Eq. (12.31) that this condition implies that the energy eigenvalues are quantised. Remember that $\lambda_{n\ell}$ is related to the eigenvalues of the Hamiltonian via Eq. (12.15). Energy quantisation is a consequence of having required the wave function to be normalisable.

12.3 Physical Interpretation

The computation in the previous section shows that the eigenvalues of the Hamiltonian for the H atom are

$$E_{n\ell} = \frac{-E_I}{(n_r + \ell)^2}, \tag{12.32}$$

where ℓ is the angular momentum of the state, and $n_r > 0$ is an integer. We see that the value of the energy does not depend separately on n_r and ℓ, but only on their sum $n = n_r + \ell$. The integer n is called the **principal quantum number**; its value characterises the so-called *electron shells*. Note that the condition $n_r > 0$ translates into $\ell < n$. Therefore if $n = 1$, then the system *must* have $\ell = 0$; this is denoted as a $1s$ orbital. If $n = 2$, then ℓ can take the values 0 and 1, which correspond to $2s$ and $2p$ orbitals, respectively.

We can rewrite Eq. (12.14) as

$$E_I = \tfrac{1}{2}\alpha^2 \mu c^2,\tag{12.33}$$

where α is the **fine-structure constant**

$$\alpha = \frac{e^2}{\hbar c} \simeq \frac{1}{137}.\tag{12.34}$$

Note that $\mu \simeq m_e$, and therefore μc^2 is the rest energy of the electron. Equation (12.33) shows that the typical scale of the energy levels in the H atom is 10^{-4} smaller than the rest energy of the electron. This justifies the non-relativistic treatment of the H atom that we have used here. Clearly there *are* relativistic corrections; however, they are of small effect, typically of order α, and can be studied in perturbation theory.

Using Eq. (12.13), we can express the energy levels as

$$E_n = -\frac{e^2}{2n^2 a_0}, \quad n = 1, 2, \ldots.\tag{12.35}$$

This is the formula put forward by Bohr in 1913, before quantum mechanics was fully developed. The ionisation energy is the energy needed to extract the electron from the ground state and is given by E_I. This is known as the **Rydberg energy**. The electron in the H atom can go from one shell to a lower one by emitting a photon. The series of transitions from principal number $n \geq 2$ to $n = 1$ is called the *Lyman series*. The transitions are named by Greek letters: the transition from $n = 2$ to $n = 1$ is called Lyman-α, from 3 to 1 Lyman-β, etc. Likewise, the transitions from $n \geq 3$ to $n = 2$ form the *Balmer series*. These transitions, which were discovered before quantum mechanics was established, have been observed with great precision and provide strong experimental evidence in favour of quantum mechanics. However, only the full quantum-mechanical treatment that we discussed above yields the eigenvalues and degeneracies observed in nature. Equation (12.35) shows clearly that for a given principal quantum number n, ℓ can take the values $\ell = 0, 1, \ldots, n - 1$ (corresponding, respectively, to $n_r = n, n - 1, \ldots, 1$). Since for each value of ℓ we have a $(2\ell + 1)$ degeneracy, the total degeneracy of the level E_n is

$$\sum_{\ell=0}^{n-1}(2\ell + 1) = n(n - 1) + n = n^2.\tag{12.36}$$

The polynomials $\eta_{n\ell}$ are called the **associated Laguerre polynomials**. The full solution for the eigenfunctions of the energy is

$$\psi_{nlm}(\underline{r}) = R_{n\ell}(r)Y_\ell^m(\theta, \phi).\tag{12.37}$$

The first few radial functions are

$$R_{1,0} = 2a_0^{-3/2}e^{-r/a_0},\tag{12.38}$$

$$R_{2,0} = \frac{1}{2\sqrt{2}}a_0^{-3/2}\left(2 - \frac{r}{a_0}\right)e^{-r/(2a_0)},\tag{12.39}$$

$$R_{2,1} = \frac{1}{2\sqrt{6}}a_0^{-3/2}\frac{r}{a_0}\,e^{-r/(2a_0)}.\tag{12.40}$$

The exponential decay is common to all the eigenfunctions and allows the interpretation of a_0 as the typical spatial extension of the ground state, with higher energy levels spreading further. It is interesting to estimate the expectation value of the electron momentum. Since the typical size of the atom is a_0, we deduce from the Heisenberg uncertainty relation that $\sqrt{\langle p^2 \rangle} \sim \mu e^2/\hbar$. Thus we obtain

$$v \sim p/\mu \simeq e^2/\hbar = \alpha c \simeq \frac{1}{137} c \ll c. \tag{12.41}$$

We see *a posteriori* that the motion of the electron is non-relativistic, and hence our non-relativistic description is accurate. Relativistic corrections are expected to be $O(v/c) \simeq O(\alpha)$.

Finally, note that the treatment presented here can be applied to any hydrogen-like atom, i.e. an atom with an electron and a nucleus of charge Ze. Simply replace everywhere $e^2 \to Ze^2$.

This concludes our first analysis of the hydrogen atom, a cornerstone of our understanding of nature, and a triumph of quantum mechanics. Remember that one of the initial motivations for quantum mechanics was a quantitative understanding of the energy levels of the hydrogen atom and their multiplicities. Further details about this fundamental system will be explored in Part II of the book.

Mathematical Aside: Laguerre Polynomials
Laguerre polynomials are defined as

$$L_p^0(z) = e^z \frac{d^p}{dz^p} \left(e^{-z} z^p \right), \tag{12.42}$$

$$L_p^k(z) = (-)^k \frac{d^k}{dz^k} L_{p+k}^0(z), \tag{12.43}$$

where $k, p = 1, 2, \ldots$. They are polynomials of degree p with p nodes on the positive real semi-axis. They are the solution of Laplace's equation

$$\left[z \frac{d^2}{dz^2} + (k + 1 - z) \frac{d}{dz} + p \right] L_p^k(z) = 0, \tag{12.44}$$

and are orthonormal with respect to the scalar product

$$\int_0^\infty dz\, e^{-z}\, z^k L_p^k(z) L_q^k(z) = \frac{[(p + k)!\,]^3}{p!} \delta_{pq}. \tag{12.45}$$

The radial dependence of the eigenstates of the H atom can be written as

$$\frac{\eta_{n\ell}(r/a_0)}{r/a_0} \frac{1}{a_0} = a_0^{-3/2} N_{n\ell} F_{n\ell}(2r/a_0), \tag{12.46}$$

where

$$N_{n\ell} = \frac{2}{n^2} \sqrt{\frac{(n - \ell - 1)!}{[(n + \ell)!\,]^3}}, \tag{12.47}$$

$$F_{n\ell}(z) = z^\ell L_{n-\ell-1}^{2\ell+1}(z). \tag{12.48}$$

Summary

- We have begun the chapter by setting up the problem. Modelling the system so that we can define the Hamiltonian for the H atom is a fundamental first step in this chapter. (12.1)
- Separation of variables leads again to a one-dimensional equation for the radial wave function. We discussed the features of the effective potential for the one-dimensional problem and the boundary condition. (12.2)
- We discussed the detailed solution for the stationary states, which entails quantisation of the energy levels. This derivation is rather long and technical, but there are interesting physical insights at every step, so make sure you can follow the solution from beginning to end. (12.2)
- As usual, we do pay special attention to the physical interpretation of the mathematical results, i.e. we try to understand in detail the physics that is encoded in the mathematical details. (12.3)

Problems

12.1 *Muonic atom.* The muon is a particle with the same properties as the electron, except that its mass is 207 times heavier: $m_\mu/m_e = 207$. Its interactions with an atomic nucleus are essentially electromagnetic. A muon can be attracted by the Coulomb field of a nucleus and form a bound state, called a 'muonic atom'.

Describe the energy of the bound states of a muon in the Coulomb field of a heavy atom such as lead, with charge $Z = 82$, in the approximation where the nucleus is considered to be infinitely heavy.

Determine the Bohr radius for this system, and discuss the validity of the computation above, knowing that the radius of the lead nucleus is $\rho_0 \approx 8.5 \times 10^{-13}$ cm.

Assuming that the charge is distributed uniformly inside a sphere of radius ρ_0, the potential inside the atom is

$$V(r) = \frac{Ze^2}{2\rho_0^3} r^2 + C.$$

Find the constant C such that the potential is continuous at the boundary of the nucleus. Neglecting the Coulomb field at large r, find the energy of the ground state for this model.

12.2 Consider a hydrogen atom whose wave function at time $t = 0$ is a superposition of energy eigenfunctions:

$$\Psi(\underline{r}, t = 0) \equiv \psi(\underline{r}) = \frac{1}{\sqrt{78}} \left[2u_{100}(\underline{r}) - 7u_{200}(\underline{r}) + 5u_{322}(\underline{r}) \right].$$

Is this wave function an eigenfunction of the parity operator?

What is the probability of obtaining the result (i) E_1, (ii) E_2, (iii) E_3 on measuring the total energy?

What are the expectation values of the energy, of the square of the angular momentum and of the z-component of the angular momentum?

12.3 The normalised energy eigenfunction of the ground state of the hydrogen atom $(Z = 1)$ is

$$u_{100}(\underline{r}) = C \, \exp(-r/a_0),$$

where a_0 is the Bohr radius and C is a normalisation constant.

Calculate the normalisation constant, C; you may wish to note the useful integral

$$\int_0^\infty \exp(-br) \, r^n \, \mathrm{d}r = n! \, /b^{n+1}, \quad n > -1.$$

Determine the radial distribution function, $D_{10}(r)$, and sketch its behaviour; hence determine the most probable value of the radial coordinate, r, and the probability that the electron is within a sphere of radius a_0; recall that $Y_0^0(\theta, \phi) = 1/\sqrt{4\pi}$.

Calculate the expectation value of r, and the expectation value of the potential energy, $V(r)$.

Calculate the uncertainty, Δr, in r.

12.4 Starting from the time-independent Schrödinger equation for the H atom, set

$$r = \lambda z^2/2, \tag{12.49}$$

$$\frac{\chi(r)}{r} = \frac{F(z)}{z}. \tag{12.50}$$

Show that $F(z)$ obeys the radial equation of the two-dimensional harmonic oscillator discussed in Chapter 11.

Deduce the wave function for the ground state of the H atom.

(*This solution of the H atom is due to Schwinger.*)

13 Identical Particles

There are many systems in nature that are made up of several particles of the same species. These particles all have the same mass, charge and spin, and need to be treated as **identical particles**. For instance, the electrons in an atom are identical particles. Identical particles cannot be distinguished by measuring their intrinsic properties. While this is also true for classical particles, the laws of classical mechanics allow us to follow the trajectory of each individual particle, i.e. their time evolution in space. The trajectories uniquely identify each particle in classical mechanics, making identical particles distinguishable. This way of identifying the particles does not survive in quantum mechanics and we need to postulate that the exchange of identical particles leaves the state of the system unchanged. This simple concept has profound consequences that we will explore in this chapter, leading to the celebrated spin-statistics theorem and Pauli exclusion principle. A proof of the spin-statistics theorem is beyond the scope of this book, and we will simply assume its validity here. As you read this chapter, make sure that you understand its consequences and that you become familiar with the way it is applied to physical systems.

13.1 Permutation Symmetry

In quantum mechanics, the concept of trajectory does not exist and identical particles are truly *indistinguishable*. Let us consider for simplicity a system of two identical particles in three dimensions. The state of the system is described by a wave function:

$$\psi(\xi_1, \xi_2), \quad \xi = \{\underline{x}, \sigma\}, \tag{13.1}$$

where $\underline{x}$ is the position of the particle, and σ represents any other degree of freedom of the system, e.g. the z-component of the spin of the particle, if the latter is different from zero.

The state with the two particles exchanged is described by the wave function:

$$\psi(\xi_2, \xi_1) = \hat{P}_{12}\psi(\xi_1, \xi_2), \tag{13.2}$$

and we have introduced the **permutation operator** $\hat{P}_{12}$ that swaps the particles 1 and 2. If the two particles are identical, the two functions represent the *same* quantum state, and therefore

$$\psi(\xi_1, \xi_2) = e^{i\alpha}\psi(\xi_2, \xi_1). \tag{13.3}$$

Repeating the exchange of the two particles, we find

$$e^{2i\alpha} = 1 \implies e^{i\alpha} = \pm 1. \tag{13.4}$$

Hence the wave function of a system of two identical particles must be either symmetric or antisymmetric under the exchange of the two particles.

The Spin-Statistics Theorem

Systems of identical particles with integer spin ($s = 0, 1, 2, \ldots$), known as **bosons**, have wave functions which are symmetric under interchange of any pair of particle labels. The wave function is said to obey *Bose–Einstein* statistics.

Systems of identical particles with half-odd-integer spin ($s = \frac{1}{2}, \frac{3}{2}, \ldots$), known as **fermions**, have wave functions which are antisymmetric under interchange of any pair of particle labels. The wave function is said to obey *Fermi–Dirac* statistics.

This law was discovered by Wolfgang Pauli and is supported by experimental evidence. It can be derived in the framework of relativistic quantum mechanics as a consequence of the positivity of energy.

13.2 A First Look at Helium

In the simplest model of the helium atom, the Hamiltonian is

$$\hat{H} = \hat{H}_1 + \hat{H}_2 + \frac{e^2}{|\underline{r}_1 - \underline{r}_2|}, \tag{13.5}$$

where

$$\hat{H}_i = \frac{\hat{P}_i^2}{2\mu} - \frac{2e^2}{r_i}. \tag{13.6}$$

Note that it is *symmetric* under permutation of the indices 1 and 2 which label the two electrons. This must be the case if the two electrons are *identical* or *indistinguishable*: it cannot matter which particle we label 1 and which we label 2. This observation is quite general: the same argument holds for identical particles other than electrons and can be extended to systems of more than two identical particles.

Let us write the symmetry condition concisely as

$$\hat{H}(2, 1) = \hat{P}_{12}\hat{H}(1, 2)\hat{P}_{12} = \hat{H}(1, 2). \tag{13.7}$$

Suppose that

$$\hat{H}(1, 2)\psi(1, 2) = E\psi(1, 2), \tag{13.8}$$

then interchanging the labels 1 and 2 gives

$$\hat{H}(2, 1)\psi(2, 1) = E\psi(2, 1) \tag{13.9}$$

but using the symmetry property of $\hat{H}$ means that

$$\hat{H}(1,2)\psi(2,1) = E\psi(2,1),\qquad(13.10)$$

so we conclude that $\psi(1,2)$ and $\psi(2,1)$ are both eigenfunctions belonging to the *same* eigenvalue, E, as is any linear combination of $\psi(1,2)$ and $\psi(2,1)$. In particular, the normalised symmetric and antisymmetric combinations

$$\psi_\pm \equiv \frac{1}{\sqrt{2}}\{\psi(1,2) \pm \psi(2,1)\}\qquad(13.11)$$

are eigenfunctions belonging to the eigenvalue, E.

Using the particle interchange operator $\hat{P}_{12}$ that we introduced above, we have

$$\hat{P}_{12}\psi(1,2) = \psi(2,1),\qquad(13.12)$$

then the symmetric and antisymmetric combinations are eigenfunctions of $\hat{P}_{12}$ with eigenvalues ±1, respectively:

$$\hat{P}_{12}\psi_\pm = \pm\psi_\pm.\qquad(13.13)$$

Since $\psi_\pm$ are simultaneous eigenfunctions of $\hat{H}$ and $\hat{P}_{12}$, it follows that $[\hat{H}, \hat{P}_{12}] = 0$, and therefore that the symmetry of the wave function under $\hat{P}_{12}$ is a constant of motion.

13.3 Two-Electron Wave Function

In Chapter 9, we constructed the states of the coupled representation for two spin-$\frac{1}{2}$ electrons, the three triplet states

$$|1,1\rangle = |\uparrow\uparrow\rangle,$$

$$|1,0\rangle = \frac{1}{\sqrt{2}}\{|\uparrow\downarrow\rangle + |\downarrow\uparrow\rangle\},$$

$$|1,-1\rangle = |\downarrow\downarrow\rangle,$$

and the singlet state

$$|0,0\rangle = \frac{1}{\sqrt{2}}\{|\uparrow\downarrow\rangle - |\downarrow\uparrow\rangle\},\qquad(13.14)$$

where we have used a simplified notation for the states of the coupled basis:

$$|s, m_s\rangle = |s, m_s, s_1, s_2\rangle.$$

Notice that the triplet states are *symmetric* under interchange of the labels 1 and 2, whereas the singlet state is *antisymmetric*. If we are to satisfy the spin-statistics theorem, this has implications for the symmetry of the spatial wave functions that we combine with the spin functions to give the full wave function of the two-electron system. The two-electron wave function will have the general form

$$\langle \underline{r}_1, \underline{r}_2|\Psi(1,2)\rangle = \psi(\underline{r}_1, \underline{r}_2)|s, m_s\rangle.\qquad(13.15)$$

The symmetry properties of the various factors can be summarised as follows:

	symmetry of $\lvert s, m_s \rangle$	symmetry of ψ	symmetry of Ψ
$s = 0$ (singlet)	a	s	a
$s = 1$ (triplet)	s	a	a

Thus, the spatial wave function must be *antisymmetric* if the two electrons are in a *spin triplet state* but *symmetric* if they are in a *spin singlet state*.

13.4 More on the Helium Atom

Suppose for the moment that we neglect spin and also neglect the mutual Coulomb repulsion between the two electrons. That is, we treat the two electrons as moving independently in the Coulomb field of the nucleus. The Hamiltonian then reduces to

$$\hat{H} = \hat{H}_1 + \hat{H}_2 \qquad \text{where} \qquad \hat{H}_i = \frac{\hat{p}_i^2}{2\mu} - \frac{2e^2}{r_i}. \tag{13.16}$$

From the solution of the H atom, we already know what the eigenfunctions and eigenvalues for $\hat{H}_1$ and $\hat{H}_2$ are, namely

$$\hat{H}_1\, u_{n_1 \ell_1 m_1}(\underline{r}_1) = E_{n_1} u_{n_1 \ell_1 m_1}(\underline{r}_1), \tag{13.17}$$

$$\hat{H}_2\, u_{n_2 \ell_2 m_2}(\underline{r}_2) = E_{n_2} u_{n_2 \ell_2 m_2}(\underline{r}_2). \tag{13.18}$$

Using separation of variables, we can build eigenfunctions of $\hat{H}$ that are just products of the one-electron eigenfunctions:

$$
\begin{aligned}
\hat{H} u_{n_1 \ell_1 m_1}(\underline{r}_1)\, u_{n_2 \ell_2 m_2}(\underline{r}_2) &= (\hat{H}_1 + \hat{H}_2)\, u_{n_1 \ell_1 m_1}(\underline{r}_1)\, u_{n_2 \ell_2 m_2}(\underline{r}_2) \\
&= (E_{n_1} + E_{n_2})\, u_{n_1 \ell_1 m_1}(\underline{r}_1)\, u_{n_2 \ell_2 m_2}(\underline{r}_2) \\
&= E_n\, u_{n_1 \ell_1 m_1}(\underline{r}_1)\, u_{n_2 \ell_2 m_2}(\underline{r}_2).
\end{aligned}
\tag{13.19}
$$

Thus the energy eigenvalues are given by

$$E_n = E_{n_1} + E_{n_2} \qquad \text{where} \qquad E_{n_i} = -\frac{\mu}{2\hbar^2}\left(Ze^2\right)^2 \frac{1}{n_i^2}. \tag{13.20}$$

The Ground State

In this crude model the ground-state energy is just

$$E_{n=1} = E_{n_1=1} + E_{n_2=1} = 2\,E_{n_1=1}. \tag{13.21}$$

Setting $Z = 2$ in the Bohr formula thus yields for the ground-state energy

$$E_1 = 8 \times (-13.6 \text{ eV}) = -108.8 \text{ eV}, \tag{13.22}$$

to be compared with the experimentally measured value of -79.005 eV.

The ground-state spatial wave function has $n_1 = n_2 = 1$ and $\ell_1 = \ell_2 = m_1 = m_2 = 0$ and is thus

$$u_{100}(\underline{r}_1)\, u_{100}(\underline{r}_2). \tag{13.23}$$

Each electron is in a $1s$ state; we say that the **electronic configuration** is $(1s)^2$. If we now look at the spin degrees of freedom, we remember that the total wave function is a product of a spatial wave function and a spin vector of the correct symmetry. But the spatial wave function of the ground state determined above is *symmetric* and can thus *only be combined with a spin singlet spin function* to give an overall antisymmetric two-electron state vector:

$$\langle \underline{r}_1, \underline{r}_2 | \Psi(\text{ground state}) \rangle = u_{100}(\underline{r}_1)\, u_{100}(\underline{r}_2)\, |0, 0\rangle. \tag{13.24}$$

Notice that, since $\ell_1 = \ell_2 = 0$, the total orbital angular momentum quantum number of the ground-state configuration is $\ell = \ell_1 + \ell_2 = 0$. Thus the ground state has zero orbital and spin angular momentum, and hence zero total angular momentum.

The First Excited States

The first excited states correspond to one electron being excited to a $2s$ or $2p$ state, with the other remaining in a $1s$ state. The electronic configurations are denoted by $(1s)(2s)$, and $(1s)(2p)$, respectively. The (degenerate) energy eigenvalue can again be obtained from the Bohr formula with $Z = 2$:

$$E_{n=2} = E_{n_1=1} + E_{n_2=2} = E_{n_1=2} + E_{n_2=1} = 5 \times -13.6 \text{ eV} = -68.0 \text{ eV}. \tag{13.25}$$

In this case it is possible to construct spatial wave functions which are either *symmetric* or *antisymmetric*. The overall antisymmetric combinations are then

$$|\Psi(\text{singlet})\rangle = \frac{1}{\sqrt{2}} \{ u_{100}(\underline{r}_1)\, u_{2\ell m_\ell}(\underline{r}_2) + u_{2\ell m_\ell}(\underline{r}_1)\, u_{100}(\underline{r}_2) \} \, |0, 0\rangle, \tag{13.26}$$

$$|\Psi(\text{triplet})\rangle = \frac{1}{\sqrt{2}} \{ u_{100}(\underline{r}_1)\, u_{2\ell m_\ell}(\underline{r}_2) - u_{2\ell m_\ell}(\underline{r}_1)\, u_{100}(\underline{r}_2) \} \, |1, m_s\rangle. \tag{13.27}$$

13.5 Pauli Exclusion Principle

The results that we have just obtained for the independent particle approximation to the helium atom illustrate a more general result, related to the spin-statistics theorem and known as the **Pauli exclusion principle**, which states that two identical fermions cannot be in the same quantum state. For example:

- In the ground state, we see that although both electrons have $n = 1$ and $\ell = m_\ell = 0$, i.e. both are in a $1s$ state, they are in a spin singlet state, which means that if one electron is in the spin state α, the other must be in the spin state β: the two electrons cannot have an identical set of quantum numbers. If both were in the spin state α, the two-electron spin state would be a triplet state, which is ruled out by the spin-statistics theorem.

- In any excited state, both electrons can be in the spin state α, corresponding to the triplet state, but then the spatial wave function is forced to be antisymmetric, so that the quantum numbers n, ℓ, m_ℓ of the two electrons have to differ – otherwise the spatial wave function would vanish identically!

No such restriction applies to identical bosons; any number of identical bosons can occupy the same quantum state.

Summary

- We introduced the concept of indistinguishable particles in quantum mechanics and highlighted the differences from the classical description. As a consequence, we discussed permutation symmetry and the properties of the state vector under a permutation. These properties are summarised in the spin-statistics theorem. (13.1)
- We discussed the symmetry of the wave function, using a simplified model of the helium atom as an example. (13.2)
- We have considered the description of the state vector for a system of two-electron wave functions, taking into account both the orbital and spin degrees of freedom. (13.3)
- More on the helium atom. We have seen the relation between the symmetry of the orbital wave function under permutations and the spin state of the atom. (13.4)
- Last but not least, make sure that you understand how the antisymmetry of the state vector under permutations leads to the Pauli exclusion principle. (13.5)

Problems

13.1 Consider two identical particles with spin s. Show that all eigenstates of the total spin of the system $\hat{S}^2$ corresponding to the same eigenvalue $\hbar S(S + 1)$ have the same parity under permutations of the two particles. Find the relation between the quantum number S and the symmetry of the state.

Hint: This is the generalisation of the discussion of the permutation symmetry properties for a system of two spin-$\frac{1}{2}$ systems.

13.2 The stationary state of a quantum system of spin s is given by $\psi(\underline{r})$ and is independent of the spin degrees of freedom. Consider now two such systems together, with no interactions between the two systems. The orbital angular momentum of the systems is denoted by ℓ_1 and ℓ_2; count the number of linearly independent states for the cases where the two systems are bosons and fermions, respectively.

13.3 Let us denote by ξ_i, for $i = 1, 2, 3$, the (generalised) coordinates that describe particles 1, 2 and 3. The normalised wave functions that describe the states of each particle in isolation are denoted $\psi_{f_i}(\xi)$, where $f_i = 1, 2, 3$ are possible quantum numbers. Find a basis of the space of states of a system made up of three such particles under the assumption that they are bosons.

   Hint: Consider the cases $f_1 = f_2 = f_3$, $f_1 = f_2 \neq f_3$ and $f_1 \neq f_2 \neq f_3$ separately.

13.4 A quantum system is made up of three identical, non-interacting particles of spin 0. Each particle has angular momentum $\ell_i = 1$ for $i = 1, 2, 3$. Find a basis of the vector space of states of this system.

   Show that the system cannot be in a state with total angular momentum $L = 0$.

13.5 Find the spectrum of two identical bosons interacting via a quadratic potential

$$V(\underline{r}_1, \underline{r}_2) = \tfrac{1}{2}m\omega^2(\underline{r}_1 - \underline{r}_2)^2.$$

   How does the answer change for fermions?

Symmetries in Quantum Mechanics

Symmetries play a central role in the study of physical systems. They dictate the choice of the dynamical variables used to characterise the system, lead to conservation laws and set constraints on the evolution of the system. We shall see explicit examples of these features in this chapter, and set up the mathematical framework to be able to have a unified formalism for describing generic symmetries.

We shall see that group theory is the natural language to discuss symmetry. Both the solutions of classical equations of motion and the state vectors of quantum mechanics form linear vector spaces. Symmetries of physical systems imply the existence of distinctive regular patterns in these vector spaces. The exact structure of these patterns is entirely determined by the group-theoretical properties of the set of symmetry transformations and *is independent* of other dynamical details. In this chapter, we will work out some examples in detail in order to get familiar with the more abstract concepts. A full treatment of group theory applications to quantum mechanics would require a much longer description than the one we can present in this chapter. Here we aim to introduce selected important ideas. In particular, we want to make clear the role of momentum and angular momentum as generators of translations and rotations, respectively, and suggest the idea that these are really the defining properties of these observables.

14.1 Classical Symmetry

In order to introduce the basic ideas, we will start our discussion of symmetries from the more familiar case of classical mechanics, where we have an intuitive understanding of the concept of symmetry.[1]

Reference Frames and Transformation Laws

When studying a physical system, we can think of a **reference frame** as a set of rules which enable the observer to measure physical quantities. For instance, if we are studying a point-like particle, the reference frame is made up of a mapping of three-dimensional space, i.e. a choice of coordinates, a clock and a detector, so that at any time t we can identify the position $\underline{x}(t)$ of the system. The same phenomenon can be observed from

---

[1] This discussion follows closely the one presented in A. Di Giacomo, *Lezioni di Fisica Teorica*, Edizioni ETS, Pisa, 1992.

two different reference frames, which are unambiguously defined as soon as we know the mapping which relates the quantities measured in each reference frame. This mapping is called a **transformation law**. For instance, a system can be rotated by some angle θ according to the transformation law

$$t \mapsto t, \quad \underline{x} \mapsto \underline{x}' = R(\theta)\underline{x}, \tag{14.1}$$

where $R(\theta)$ is the operator implementing the rotation. Having chosen a reference frame, $R(\theta)$ can be represented as a matrix that relates the coordinates in the two reference frames.

Invariance under a Transformation

The evolution of a physical system is determined by the equations of motion, i.e. by mathematical equations that yield the time dependence of the coordinates of the system in a given reference frame.

Two reference frames are said to be **equivalent** for the study of a given physical system if:

1. *all* physical states that can be realised in one frame, can also be realised in the other frame, i.e. the mapping between the two frames is invertible;
2. we have the *same* equations of motion in both systems.

Consider a transformation law R; a physical system is said to be **invariant** under R if two reference frames related by this transformation are equivalent. The transformation R is called a **symmetry** of the system.

This concept is best illustrated by a concrete example. Let us consider a point particle whose evolution defines a trajectory from $\underline{x}(0)$ to $\underline{x}(t)$ in a given reference frame A. If we look at the same particle from the reference frame B, the evolution is given by the trajectory

$$R\underline{x}(0) \to R\underline{x}(t). \tag{14.2}$$

If the system is invariant under the transformation R, and we start a trajectory from the point $R\underline{x}(0)$ in A, then the evolution of the system in the reference frame A is the same as the evolution in the reference frame B. And therefore, in the reference frame A, the system evolves from $R\underline{x}(0)$ to $R\underline{x}(t)$; i.e. the time evolution commutes with the transformation R.

Let us now consider the constraints on the dynamics of a system that are implied by the invariance under the transformation R. The equation of motion for a point particle under the action of some force $\underline{f}$ is given by

$$m\underline{\ddot{x}} = \underline{f}(\underline{x}). \tag{14.3}$$

Starting from an initial state $(\underline{x}(0), \underline{\dot{x}}(0))$, the equations of motion determine the trajectory of the particle. If we observe the system from a transformed reference frame, the trajectory is given by $\underline{x}'(t) = R\underline{x}(t)$. In order to have invariance, the equation of motion needs to be the same in both reference frames, i.e.

$$m\underline{\ddot{x}}' = \underline{f}(\underline{x}'), \tag{14.4}$$

with the *same* function f appearing in both Eq. (14.3) and Eq. (14.4). Using the property that R is invertible, and assuming that R does not depend on time, we express $\underline{x}$ as a function of $\underline{x}'$, and rewrite Eq. (14.3) as

$$mR^{-1}\underline{\ddot{x}}' = \underline{f}(R^{-1}\underline{x}'), \tag{14.5}$$

and therefore

$$R\underline{f}(R^{-1}\underline{x}') = \underline{f}(\underline{x}') \quad \text{i.e.} \quad \underline{f}(R\underline{x}) = R\underline{f}(\underline{x}). \tag{14.6}$$

Equation (14.6) shows that the invariance under R implies a particular structure of the forces acting on the particle, and vice versa.

14.2 Quantum Symmetry

The invariance with respect to a transformation law R in quantum mechanics can be expressed in the following way:

1. The allowed states in both reference frames are the same; they are vectors in the same Hilbert space, i.e. the transformation law is a mapping from the Hilbert space of states $\mathcal{H}$ onto itself.
2. Starting from the same initial state, we have the same time evolution in both reference frames.

If the system is invariant under a change of reference frames, the transformation that implements the change of reference frame is called a *symmetry* of the system. Given two states $|\psi_1\rangle, |\psi_2\rangle \in \mathcal{H}$ in a given reference frame A, these states are mapped by the transformation R into $|\psi_1'\rangle, |\psi_2'\rangle \in \mathcal{H}$, respectively. These new states describe the system as seen from the new reference frame B. The transition from $|\psi_1\rangle$ to $|\psi_2\rangle$ in A is seen as a transition from $|\psi_1'\rangle$ to $|\psi_2'\rangle$ from the transformed reference frame. If the system is invariant under the transformation, the probability amplitude $\langle\psi_1|\psi_2\rangle$ should be the same as the probability amplitude $\langle\psi_1'|\psi_2'\rangle$. Symmetry transformations obey Wigner's theorem.

Theorem 14.1 *Given a transformation in a Hilbert space $\mathcal{H}$:*

$$|\psi\rangle \mapsto |\psi'\rangle, \tag{14.7}$$

such that

$$\left|\langle\psi_1'|\psi_2'\rangle\right|^2 = \left|\langle\psi_1|\psi_2\rangle\right|^2, \quad \forall\,|\psi_1\rangle, |\psi_2\rangle \in \mathcal{H}, \tag{14.8}$$

it is always possible to choose the complex phase of the quantum states in such a way that the transformation is realised on the vectors as either a unitary or an anti-unitary representation.

We shall not prove Wigner's theorem here, since we do not want to dwell for too long on the mathematical details. Nonetheless, the reader should be aware that this theorem plays a central role in describing the symmetry properties of quantum-mechanical systems.

Let us now analyse the implications of the invariance under a given transformation R, and its implementation in $\mathcal{H}$ that we denote $\hat{U}(R)$. As discussed above, a system is invariant under a given transformation only if the time evolution is the same in both reference frames. If we start from a state $|\psi\rangle$ and apply the unitary transformation $\hat{U}(R)$, we obtain the state $|\psi'\rangle$. The time evolution of $|\psi\rangle$ is dictated by the Hamiltonian:

$$i\hbar\frac{d}{dt}|\psi\rangle = \hat{H}|\psi\rangle. \tag{14.9}$$

Therefore the evolution of $|\psi'\rangle$ is

$$i\hbar\frac{d}{dt}|\psi'\rangle = i\hbar\frac{d}{dt}\hat{U}(R)|\psi\rangle = \hat{U}(R)\hat{H}|\psi\rangle$$
$$= \hat{U}(R)\hat{H}\left(\hat{U}(R)^\dagger\hat{U}(R)\right)|\psi\rangle = \hat{U}(R)\hat{H}\hat{U}(R)^\dagger|\psi'\rangle, \tag{14.10}$$

i.e. $\hat{H}' = \hat{U}(R)\hat{H}\hat{U}(R)^\dagger$. Since the system is invariant we must have $\hat{H}' = \hat{H}$, and hence R is a symmetry transformation of the system only if

$$\left[\hat{U}(R), \hat{H}\right] = 0; \tag{14.11}$$

the unitary operators implementing the symmetry transformations commute with the Hamiltonian.

Particle on a One-Dimensional Lattice

The main features of symmetry in quantum mechanics can be seen in a simple example. Let us consider a point particle (e.g. an electron) in a one-dimensional lattice with lattice spacing b. The dynamics of the system is dictated by the Hamiltonian:

$$\hat{H} = \hat{P}^2/(2m) + V(\hat{X}), \tag{14.12}$$

where m is the mass of the electron, $\hat{P}$ the operator associated with its momentum and $V(\hat{X})$ the operator associated with the potential. Because the electron is evolving in a lattice, the potential $V(x)$ is a periodic function of period b:

$$V(x + nb) = V(x), \forall n \in \mathbb{Z}. \tag{14.13}$$

The transformation law

$$x \mapsto x' = x + nb \tag{14.14}$$

is a symmetry of the system, since it leaves the potential unchanged. The electron appears to behave in exactly the same manner to two observers that are related by one of these translations.

We shall now translate this physical property into mathematical language. The state of the system is described by a state vector $|\psi\rangle$ in the Hilbert space $\mathcal{H}$. The transformed state after a translation by n steps of size b will be denoted $|\psi'\rangle$. We shall denote by $\hat{U}(n)$ the operator that relates $|\psi\rangle$ to $|\psi'\rangle$. $\hat{U}(n)$ is an operator mapping $\mathcal{H}$ onto itself:

$$\hat{U}(n): \quad \mathcal{H} \to \mathcal{H}, \quad |\psi\rangle \mapsto |\psi'\rangle = \hat{U}(n)|\psi\rangle. \tag{14.15}$$

According to Wigner's theorem, the operators $\{\hat{U}(n)\}$ form a unitary representation of the symmetry group of the Hamiltonian. Physical observables in quantum mechanics are represented by Hermitian operators. Under the transformation $\hat{U}(n)$, each operator is mapped into

$$\hat{O} \mapsto \hat{O}' = \hat{U}(n)\,\hat{O}\,\hat{U}(n)^{-1}. \tag{14.16}$$

Exercise 14.2.1   Show that the operator $\hat{O}'$ is Hermitian, and therefore can be associated with an observable.
Hint: The fact that $\hat{U}(n)$ is unitary plays an important role.

Let $|x\rangle$ be an eigenvector of the position operator $\hat{X}$ to the eigenvalue x, then

$$\hat{U}(n)|x\rangle = |x + nb\rangle. \tag{14.17}$$

Let us now consider the action of the operator $V(\hat{X})$ on the states $|x\rangle$:

$$V(\hat{X})|x\rangle = V(x)|x\rangle, \tag{14.18}$$

where we should remember that $V(\hat{X})$ is an operator acting on kets, while $V(x)$ is a number, i.e. the actual value of the potential at the position x. The operator $V(\hat{X})$ can be replaced by its eigenvalue because $|x\rangle$ is an eigenvector of the operator $\hat{X}$. Starting from Eq. (14.18), we obtain the following result:

$$\hat{U}(n)V(\hat{X})\hat{U}(n)^{-1}|x\rangle = \hat{U}(n)V(\hat{X})|x - nb\rangle, \tag{14.19}$$

$$= \hat{U}(n)V(x - nb)|x - nb\rangle, \tag{14.20}$$

$$= V(x - nb)|x\rangle. \tag{14.21}$$

Since the operators $\{|x\rangle\}$ form a basis in $\mathcal{H}$, we can deduce the following equality *between operators*:

$$\hat{U}(n)V(\hat{X})\hat{U}(n)^{-1} = V(\hat{X} - nb). \tag{14.22}$$

Exercise 14.2.2   Using the above results, prove that the quantum Hamiltonian (i.e. the operator $\hat{H}$ acting on the vector space of quantum states) is invariant under $\hat{U}(n)$: $\left[\hat{U}(n), \hat{H}\right]$.

Group Structure

For the case of discrete translations one can easily visualise that the following properties hold:

P1. $\hat{U}(n)\hat{U}(m) = \hat{U}(n + m)$, the successive application of two translations yields another translation.

P2. $\hat{U}(0) = 1$, where 1 indicates the identity operator in $\mathcal{H}$.

P3. $\hat{U}(-n) = \hat{U}(n)^{-1}$, i.e. each translation has an inverse, which is the translation in the opposite direction.

These three properties are exactly the defining properties of a **group**. Hence the set of translations by multiples of b is a group. It is easy to realise from the definitions discussed earlier in this chapter that the successive application of two symmetry transformations yields another symmetry of the system, and that any symmetry transformation is invertible. Therefore the set of symmetry transformations of a system must have the mathematical structure of a group.

14.3 Symmetry Groups in Physics

Group theory has important consequences in the study of quantum-physical systems. We list here some general properties, which we will illustrate in some explicit examples in the following sections.

1. Since the Hamiltonian commutes with all symmetry transformations, there exist bases that diagonalise $\hat{H}$ and some subset of the group transformations simultaneously.
2. The group-theoretical properties do not depend on the details of the physical system and can be found by general mathematical methods.
3. The physical interpretation of the mathematical results leads to valuable information on the states of the system.

We end this section by listing a number of important examples of symmetries that are commonly used in physics. While we will not have time to discuss each of these symmetries in this book, they play an important role in our current understanding of nature.

1. Translations in space: $\underline{x} \mapsto \underline{x} + \underline{a}$; in classical mechanics, momentum conservation follows from the invariance under space translations.
2. Rotations in three-dimensional space; this symmetry is associated with conservation of angular momentum in classical mechanics.
3. Parity transformation: $\underline{x} \mapsto -\underline{x}$; most interactions are symmetric under this transformation, with the notable exception of the so-called *weak interactions* in particle physics.
4. Permutation symmetry; quantum systems with identical particles are invariant under the exchange of these particles. This symmetry has important physical consequences, as we discussed in Chapter 13.
5. Lorentz transformations; these combine space and time transformations together and are the basis of Einstein's theory of special relativity.
6. Gauge invariance; electromagnetism is invariant under U(1) gauge transformations. Gauge invariance is a statement of the fact that the gauge potential contains more degrees of freedom than the physical ones, rather than a genuine symmetry of the theory. Here we will be content with calling this gauge symmetry. The generalisation of gauge symmetry to other groups is one of the guiding principles in constructing the standard model of particle physics.
7. Internal symmetries of strong interactions; this type of symmetry plays a crucial role in identifying the hadrons and their properties.

14.4 Group of Translations

Translations are defined by generalising the discrete transformations introduced in Eq. (14.14):

$$x \mapsto x + a, \quad a \in \mathbb{R}. \tag{14.23}$$

They are implemented as operators $\hat{U}(a)$ acting on the Hilbert space of states, which can be defined by specifying their action on the basis vector made up of eigenstates of position:

$$\hat{U}(a)\,|x\rangle = |x + a\rangle. \tag{14.24}$$

It can readily be shown that the translations defined in Eq. (14.24) satisfy the properties **P1**, **P2** and **P3** discussed in Section 14.2 and therefore form a group, whose elements are labelled by the continuous parameter a.

Working with a continuous transformation parameter allows us to consider infinitesimal transformations, which we can linearise in a neighbourhood of the identity:

$$\hat{U}(\epsilon) = 1 - i\epsilon\hat{T}, \tag{14.25}$$

where ϵ is the infinitesimal parameter and 1 denotes the identity operator that corresponds to $\epsilon = 0$. The factor of i is a convention; the operator $\hat{T}$ acting on a state vector implements the infinitesimal variation of that vector. Such an operator is called the **generator** of the transformations $\hat{U}(a)$.[2] In order to find the relation between the finite transformations and the generator, we can proceed as follows. Using a Taylor expansion we have

$$\hat{U}(a + \epsilon) = \hat{U}(a) + \epsilon\frac{d\hat{U}}{da}, \tag{14.26}$$

while using the group property **P1** yields

$$\hat{U}(a + \epsilon) = \hat{U}(\epsilon)\hat{U}(a), \tag{14.27}$$
$$= \hat{U}(a) - i\epsilon\hat{T}\,\hat{U}(a), \tag{14.28}$$

where we used Eq. (14.25) in going from the first line to the second one. Comparing Eqs (14.26) and (14.28) allows us to derive the relation

$$\frac{d\hat{U}(a)}{da} = -i\hat{T}\,\hat{U}(a), \tag{14.29}$$

and hence

$$\hat{U}(a) = \exp\left(-ia\hat{T}\right). \tag{14.30}$$

Note that if $\hat{U}(a)$ is unitary then $\hat{T}$ is Hermitian.

Let us now consider the action of $\hat{T}$ on the wave function. As usual, we write

$$\hat{T}\psi(x) = \langle x|\hat{T}|\psi\rangle. \tag{14.31}$$

---

[2] Note that here we use $\hat{U}(a)$ to denote the translation by a finite amount a, and $\hat{T}$ without an argument to denote the generator of the infinitesimal transformation.

Using the hermiticity of $\hat{T}$, we have

$$\langle x|\hat{T}|\psi\rangle = \left(\langle\psi|\hat{T}|x\rangle\right)^*, \tag{14.32}$$

$$= \left(\langle\psi|\frac{i}{\epsilon}\left[|x+\epsilon\rangle - |x\rangle\right]\right)^*, \tag{14.33}$$

$$= -\frac{i}{\epsilon}\left[\langle x+\epsilon| - \langle x|\right]|\psi\rangle, \tag{14.34}$$

$$= -i\frac{\psi(x+\epsilon) - \psi(x)}{\epsilon}, \tag{14.35}$$

$$= -i\frac{d}{dx}\psi(x). \tag{14.36}$$

By comparison with Eq. (3.100), we see that

$$\hat{T} = \hat{P}/\hbar. \tag{14.37}$$

We recover the result that we had anticipated in Section 3.10.1, namely that the momentum operator is the generator of translations (up to a factor of $\hbar$ that we should expect on dimensional grounds). In our opinion the derivation in this chapter is clearer since it starts by defining translations on eigenstates of the position operator. If the theory is invariant under translations, then

$$\left[\hat{P}, \hat{H}\right] = 0, \tag{14.38}$$

i.e. momentum is conserved as a consequence of invariance under translations. This is the first example of a very important property, namely that symmetries of the system are associated with conserved quantities and vice versa.

14.5 Group of Rotations

Before discussing rotation operators acting in the state space $\mathcal{H}$, we want to review some basic properties of geometrical rotations.

14.5.1 Rotations in Two Dimensions

In two dimensions rotations are uniquely defined by the angle of rotation. They preserve the length of a vector and the angle between vectors. For example, the image of a vector under a rotation by $\pi/3$ is represented in Fig. 14.1. Clearly the net result of two successive rotations is a rotation, the rotation by $\theta = 0$ is the identity and any rotation can be undone by rotating in the opposite direction. The set of all two-dimensional rotations forms a group, called SO(2), or U(1). The elements of the group are labelled by the angle of the rotation, $\theta \in [0, 2\pi)$. There is an infinite number of elements, denoted by a continuous parameter; groups where the elements are labelled by continuous parameters are called **continuous groups**. We will denote two-dimensional rotations by $\mathcal{R}(\theta)$. Note that the parameter labelling the rotations varies in a compact interval, the interval $[0, 2\pi)$

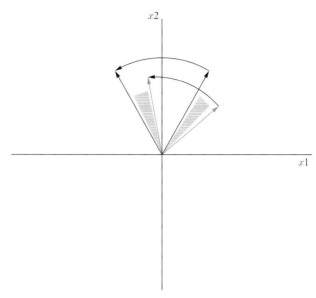

Fig. 14.1 Rotation of vectors by $\pi/3$. You can see from the picture that the length of the vectors, and the angle between them, are left unchanged.

in this case. Groups with parameters varying over compact intervals are called **compact groups**. The group of translations discussed above is also a continuous group, but not a compact one.

The action of rotations on real vectors in two dimensions defines a representation of the group.[3] Given a basis $\{\underline{e}_1, e_2\}$, a vector $\underline{r}$ is represented by two coordinates:

$$\underline{r} = x_1\underline{e}_1 + x_2\underline{e}_2. \tag{14.39}$$

The action of a rotation $\mathcal{R}(\theta)$ can be represented as a 2×2 matrix:

$$\begin{pmatrix} x_1 \\ x_2 \end{pmatrix} \mapsto \begin{pmatrix} x_1' \\ x_2' \end{pmatrix} = \begin{pmatrix} \cos\theta & -\sin\theta \\ \sin\theta & \cos\theta \end{pmatrix} \begin{pmatrix} x_1 \\ x_2 \end{pmatrix}. \tag{14.40}$$

Intuitively, two successive rotations by θ and ψ yield a rotation by $\theta + \psi$, and hence the order in which the rotations are performed does not matter. All group elements commute with each other, and therefore the group is called **Abelian**.

Exercise 14.5.1   Using the two-dimensional matrix representation defined above, check that

$$\mathcal{R}(\theta)\mathcal{R}(\theta') = \mathcal{R}(\theta + \theta'). \tag{14.41}$$

---

[3] At this point the reader will have noticed that we used several times the term 'representation'. The representation of a group has a precise mathematical definition. For the sake of brevity we will not develop these concepts in this book.

It is interesting to consider a one-dimensional complex representation of U(1). Given the coordinates (x_1, x_2) of a point in a two-dimensional space, we can define the complex number $z = x_1 + ix_2$. The transformation properties of z define a representation

$$z \mapsto z' = e^{i\theta} z. \tag{14.42}$$

With each rotation $R(\theta)$ we can associate a single complex number $D(\theta) = e^{i\theta}$, and we have

$$z' = D(\theta)z. \tag{14.43}$$

14.5.2 Rotations in Three Dimensions

Rotations in three dimensions are characterised by an axis (specified by a unit vector $\underline{u}$) and the angle of rotation θ ($0 \le \theta < 2\pi$). Hence a three-dimensional rotation is identified by three real parameters, and denoted by $\mathcal{R}_{\underline{u}}(\theta)$. The three real parameters can be chosen to be the components of a single vector:

$$\underline{\theta} = \theta\underline{u}, \tag{14.44}$$

whose norm is given by the angle θ, and whose direction defines the axis of the rotation.

As for the two-dimensional case, the set of three-dimensional rotations constitutes a group, called SO(3). However, SO(3) is not Abelian:

$$\mathcal{R}_{\underline{u}}(\theta)\mathcal{R}_{\underline{u}'}(\theta') \ne \mathcal{R}_{\underline{u}'}(\theta')\mathcal{R}_{\underline{u}}(\theta). \tag{14.45}$$

Rotation around a given axis defines subgroups of SO(3). Each of these subgroups is isomorphic to SO(2).

Infinitesimal Rotation

Since rotations are identified by a continuous rotation angle, we can consider rotations by infinitesimally small angles.

The action of an infinitesimal rotation on a vector is given by

$$\mathcal{R}_{\underline{u}}(d\theta)\underline{v} = \underline{v} + d\theta\, \underline{u} \times \underline{v}. \tag{14.46}$$

Every finite rotation can be decomposed as a product of infinitesimal ones:

$$\mathcal{R}_{\underline{u}}(\theta + d\theta) = \mathcal{R}_{\underline{u}}(\theta)\mathcal{R}_{\underline{u}}(d\theta) = \mathcal{R}_{\underline{u}}(d\theta)\mathcal{R}_{\underline{u}}(\theta). \tag{14.47}$$

Exercise 14.5.2   Show that

$$\mathcal{R}_{\mathbf{y}}(-d\theta')\mathcal{R}_{\mathbf{x}}(d\theta)\mathcal{R}_{\mathbf{y}}(d\theta')\mathcal{R}_{\mathbf{x}}(-d\theta) = \mathcal{R}_{\mathbf{z}}(d\theta d\theta'), \tag{14.48}$$

where $\mathcal{R}_{\mathbf{x}}$, $\mathcal{R}_{\mathbf{y}}$ and $\mathcal{R}_{\mathbf{z}}$ denote, respectively, rotations around the x, y and z-axes. Before you perform the explicit calculation, can you explain why the result has to be proportional to $d\theta d\theta'$?

14.6 Rotations in the Space of Quantum States

Let us consider a single particle in three-dimensional space. At any given time the state of the particle is described by a vector in a Hilbert space $|\psi\rangle \in \mathcal{H}$. The associated wave function is obtained by projecting the state vector onto the basis of eigenfunctions of the position operator:

$$\psi(\underline{r}) = \langle \underline{r} | \psi \rangle. \tag{14.49}$$

We can now rotate the system by a rotation $\mathcal{R}$, such that

$$\underline{r} \mapsto \underline{r}' = \mathcal{R}\underline{r}; \tag{14.50}$$

the state of the system after the rotation is described by a *different* vector $|\psi'\rangle \in \mathcal{H}$, and its associated wave function $\psi'(\underline{r}) = \langle \underline{r} | \psi' \rangle$. The transformation of the state is described by the action of an operator $\hat{R}$ acting in $\mathcal{H}$. It is natural to define the operator $\hat{R}$ by specifying its action on the elements of a basis made up of eigenvectors of the position operator:

$$\hat{R} |\underline{r}\rangle = |\mathcal{R}\underline{r}\rangle. \tag{14.51}$$

Note here the difference between $\mathcal{R}$, which is a rotation in the three-dimensional Euclidean space, and $\hat{R}$, which implements the corresponding transformation of the state vectors in $\mathcal{H}$. The two operators act in different vector spaces!

Having specified the action of $\hat{R}$ on a basis, we can deduce its action on a generic vector $|\psi\rangle$:

$$|\psi\rangle = \int d^3r \, |\underline{r}\rangle\langle \underline{r}|\psi\rangle, \tag{14.52}$$

and therefore

$$|\psi'\rangle = \hat{R}|\psi\rangle, \tag{14.53}$$

$$= \int d^3r \, \hat{R}|\underline{r}\rangle \, \langle \underline{r}|\psi\rangle, \tag{14.54}$$

$$= \int d^3r \, |\mathcal{R}\underline{r}\rangle \, \langle \underline{r}|\psi\rangle, \tag{14.55}$$

$$= \int d^3r' \, |\underline{r}'\rangle\langle \mathcal{R}^{-1}\underline{r}'|\psi\rangle. \tag{14.56}$$

Comparing this last equation with the expansion of $|\psi'\rangle$:

$$|\psi'\rangle = \int d^3r' \, |\underline{r}'\rangle\langle \underline{r}'|\psi'\rangle, \tag{14.57}$$

we see that the *value* of the initial wave function ψ at the point $\underline{r}$ is rotated to become the value of the function ψ' at the point $\underline{r}'$:

$$\psi'(\underline{r}') = \psi(\mathcal{R}^{-1}\underline{r}') \iff \psi'(\underline{r}') = \psi(\underline{r}). \tag{14.58}$$

Equation (14.58) can be rewritten as

$$\langle \underline{r} | \hat{R} | \psi \rangle = \langle \mathcal{R}^{-1} \underline{r} | \psi \rangle. \tag{14.59}$$

Exercise 14.6.1   Prove Eq. (14.59) using the unitarity of $\hat{R}$.

The operator $\hat{R}$ is called a **rotation operator**.

For an infinitesimal rotation angle $d\theta$, e.g. around the z-axis, the rotation operator can be expanded at first order in $d\theta$:

$$\hat{R}_z(d\theta) = 1 - id\theta \hat{T}_z + O(d\theta^2); \tag{14.60}$$

where we have denoted $\hat{T}_z$ the *generator* of rotations around the z-axis. Following the discussion in Section 14.4, a finite rotation can then be written as

$$\hat{R}_z(\theta) = \exp\left(-i\theta \hat{T}_z\right). \tag{14.61}$$

The generators of rotations around the other axes $\hat{T}_x, \hat{T}_y$ are defined in an analogous way.

Rotation Operators and Angular Momentum

Let us describe the vector $\underline{r}$ by its coordinates (x, y, z) in a given Cartesian basis, and let us consider the transformation of the wave function under a rotation by $d\theta$ around the z-axis. According to the discussion in the previous sections, we can write

$$\mathcal{R}_z^{-1}(d\theta) \begin{pmatrix} x \\ y \\ z \end{pmatrix} = \begin{pmatrix} x + y\, d\theta \\ y - x\, d\theta \\ z \end{pmatrix}, \tag{14.62}$$

and therefore

$$\psi'(x, y, z) = \psi(x + y\, d\theta, y - x\, d\theta, z). \tag{14.63}$$

Expanding at first order in $d\theta$ yields

$$\psi'(x, y, z) = \psi(x, y, z) - id\theta \frac{1}{\hbar}\left[x\left(-i\hbar\frac{\partial}{\partial y}\right) - y\left(-i\hbar\frac{\partial}{\partial x}\right)\right]\psi(x, y, z). \tag{14.64}$$

Inside the square bracket we recognise the expression for the z-component of the angular momentum in the R representation, $\hat{X}\hat{P}_y - \hat{Y}\hat{P}_x$. We have shown a very important result: *the angular momentum operator in quantum mechanics is the generator of rotations in the space of physical states*:

$$\hat{T}_z = \frac{1}{\hbar}\hat{L}_z. \tag{14.65}$$

The angular momentum of a state describes the transformation properties of a given system under rotations.

From Eq. (14.64) we can easily derive

$$\psi'(x, y, z) = \langle \underline{r} | \psi' \rangle = \langle \underline{r} | \left[1 - id\theta \frac{\hat{L}_z}{\hbar}\right] |\psi\rangle; \tag{14.66}$$

since $\{|\underline{r}\rangle\}$ is a complete basis in $\mathcal{H}$, we deduce

$$|\psi'\rangle = \hat{R}_z(d\theta)\,|\psi\rangle = \left[1 - id\theta\,\frac{\hat{L}_z}{\hbar}\right]|\psi\rangle. \tag{14.67}$$

The equation above is valid for arbitrary $|\psi\rangle$, and therefore we can write an identity between operators:

$$\hat{R}_z(d\theta) = 1 - id\theta\,\frac{L_z}{\hbar}. \tag{14.68}$$

Equation (14.48) induces the following relation between the rotation operators acting in $\mathcal{H}$:

$$\left[1 + id\theta'\frac{1}{\hbar}\hat{L}_y\right]\left[1 - id\theta\frac{1}{\hbar}\hat{L}_x\right]\left[1 - id\theta'\frac{1}{\hbar}\hat{L}_y\right]\left[1 + id\theta\frac{1}{\hbar}\hat{L}_x\right] = 1 - id\theta d\theta'\frac{1}{\hbar}L_z\,; \tag{14.69}$$

expanding the left-hand side, and comparing the coefficients of the $d\theta d\theta'$ term, we get the commutation relation of the components of angular momentum:

$$\left[\hat{L}_x, \hat{L}_y\right] = i\hat{L}_z. \tag{14.70}$$

Note that the commutation relations of angular momentum operators, which we obtained almost accidentally using the canonical commutation relations, are in fact a consequence of the non-Abelian structure of the group of geometrical rotations and encode the geometrical properties of rotations.

The full set of commutation relations between generators can be computed by a similar method, i.e. starting from geometrical properties of combinations of infinitesimal rotations. They can be summarised as

$$\left[\hat{L}_i, \hat{L}_j\right] = i\hbar\epsilon_{ijk}\hat{L}_k. \tag{14.71}$$

The corresponding finite rotation operator is obtained as usual by exponentiating the generator:

$$\hat{R}_z(\theta) = \exp\left[-i\theta\frac{1}{\hbar}\hat{L}_z\right]\,; \tag{14.72}$$

it can be generalised for a generic rotation around an axis $\underline{u}$:

$$\hat{R}_{\underline{u}}(\theta) = \exp\left[-i\theta\frac{1}{\hbar}\underline{u}\cdot\underline{\hat{L}}\right]. \tag{14.73}$$

Since the operators $\hat{L}_x, \hat{L}_y, \hat{L}_z$ do not commute:

$$\hat{R}_{\underline{u}}(\theta) \neq \exp\left[-i\theta\frac{1}{\hbar}u_x\hat{L}_x\right]\exp\left[-i\theta\frac{1}{\hbar}u_y\hat{L}_y\right]\exp\left[-i\theta\frac{1}{\hbar}u_z\hat{L}_z\right]. \tag{14.74}$$

Finally let us consider again rotations around the z-axis, and let us choose a basis in $\mathcal{H}$ composed of eigenvectors of $\hat{L}_z$, $\{|m, \tau\rangle\}$. The variable τ indicates all the other indices that are needed to specify the vectors of a given basis. Expanding a generic ket $|\psi\rangle$:

$$|\psi\rangle = \sum_{m,\tau} c_{m,\tau}|m, \tau\rangle, \tag{14.75}$$

where

$$\hat{L}_z|m, \tau\rangle = m|m, \tau\rangle. \tag{14.76}$$

Acting with a rotation operator on $|\psi\rangle$:

$$
\begin{aligned}
\hat{R}_z(\theta)|\psi\rangle &= \sum_{m,\tau} c_{m,\tau} \hat{R}_z(\theta)|m, \tau\rangle \\
&= \sum_{m,\tau} c_{m,\tau} e^{-im\theta}|m, \tau\rangle.
\end{aligned}
\tag{14.77}
$$

There are two physical properties encoded in the equation above that we should emphasise. First, we see that the values of the angular momentum determine the transformation properties of a system under rotations. Second, as we have already discussed, the eigenvalues of the orbital angular momentum component L_z are integer multiples of $\hbar$, and therefore

$$\hat{R}_z(2\pi) = 1. \tag{14.78}$$

The integer eigenvalues of $\hat{L}_z$ guarantee that the rotation operator corresponding to a rotation by 2π is the identity operator. This has to be the case if we are considering a system with no internal degrees of freedom: if the state of the system is uniquely determined by its position in space, then upon a rotation by 2π the particle *has* to come back to its initial state. In this case the angular momentum of the system is purely its orbital angular momentum. If the system has further degrees of freedom, i.e. spin, then a rotation by 2π does not necessarily bring the system back to its initial state – the spin degrees of freedom can be shuffled by the rotation – and therefore half-integer values of the (spin) angular momentum are allowed. Note that the intrinsic angular momentum can be defined as the generator of rotations in the space of the spin degrees of freedom. The *total* angular momentum is then the generator of rotations for a generic system.

14.7 Rotations Acting on Operators

Given the transformation properties of physical states, we can easily derive the transformation properties of the operators that act on them. Let us consider a state $|\psi\rangle$ and its image under a rotation $|\psi'\rangle = \hat{R}|\psi\rangle$. When acting on $|\psi\rangle$ with some operator $\hat{O}$ we obtain a state $|\phi\rangle = \hat{O}|\psi\rangle$. Under the rotation $\hat{R}$:

$$|\phi\rangle \mapsto |\phi'\rangle = \hat{R}|\phi\rangle, \tag{14.79}$$

and hence

$$
\begin{aligned}
|\phi'\rangle &= \hat{R}\hat{O}\,|\psi\rangle \\
&= (\hat{R}\hat{O}\hat{R}^{\dagger})\,(\hat{R}|\psi\rangle) \\
&= \hat{O}'|\psi'\rangle\,;
\end{aligned}
\tag{14.80}
$$

i.e. the operator in the rotated system is given by $\hat{O}' = \hat{R}\hat{O}\hat{R}^{\dagger}$.

Scalar Operators

A scalar operator is an operator which is invariant under rotations:

$$\hat{O}' = \hat{O}. \tag{14.81}$$

If we consider infinitesimal rotations, the condition above translates into

$$\left[\hat{O}, \hat{J}_k\right] = 0 \quad \text{for } k = 1, 2, 3. \tag{14.82}$$

Several scalars appear in physical problems; e.g. the angular momentum squared $\hat{J}^2$ is a scalar.

Vector Operators

A vector operator $\mathbf{V}$ is a set of three operators V_x, V_y, V_z which transform like the components of a geometric vector under rotations:

$$\begin{pmatrix} V_x \\ V_y \\ V_z \end{pmatrix} \mapsto \begin{pmatrix} V_x' \\ V_y' \\ V_z' \end{pmatrix} = R \begin{pmatrix} V_x \\ V_y \\ V_z \end{pmatrix} R^\dagger = \mathcal{R} \begin{pmatrix} V_x \\ V_y \\ V_z \end{pmatrix}, \tag{14.83}$$

where $\mathcal{R}$ is the 3×3 matrix describing the geometric rotation.

Exercise 14.7.1   Deduce the commutation relations of V_x with the three generators J_x, J_y, J_z.

Conservation of Angular Momentum

A system is invariant under rotations if its Hamiltonian is invariant under all the elements of the group. For this to be true it is necessary and sufficient that the Hamiltonian $\hat{H}$ commutes with all the generators of the rotation operators:

$$[\hat{H}, \hat{J}_k] = 0, \quad k = 1, 2, 3 \, ; \tag{14.84}$$

this property can be rephrased by saying that the Hamiltonian is a scalar operator. It implies that the angular momentum is conserved.

The operators $\hat{H}, \hat{J}^2, \hat{J}_z$ commute with each other and therefore can be diagonalised simultaneously. We can choose a basis for the vector space of physical states made up of eigenvalues of these three operators, which we denote $\{k, j, m\}$.

The degeneracy of the states and selection rules for the transition between these states are determined by their transformation properties under rotations.

14.8 Commutation Relations for a Generic Non-Abelian Group

The derivation we have just seen for the commutator of the generators of rotations can be generalised to any non-Abelian group.

Let us consider a generic continuous group G, whose elements are labelled by n real parameters $\alpha = (\alpha_1, \ldots, \alpha_n)$. We choose the parameters α in such a way that $g(0) = e$.

Let D be a representation of the group: with each element $g(\alpha) \in G$ we associate a linear operator $D(\alpha)$ acting on some vector space $\mathcal{H}$. For an infinitesimal transformation $D(d\alpha)$ we can expand linearly in $d\alpha$:

$$D(d\alpha) = \mathbb{1} + \sum_k d\alpha_k \left. \frac{\partial D}{\partial \alpha_k} \right|_0 + O(d\alpha_k^2). \tag{14.85}$$

The generators T_k are defined by identifying the expression above with

$$D(d\alpha) = \mathbb{1} - i \sum_k d\alpha_k T_k + O(d\alpha_k^2), \tag{14.86}$$

i.e.

$$T_k = -i \left. \frac{\partial D}{\partial \alpha_k} \right|_0 . \tag{14.87}$$

For a finite transformation we have

$$D(\alpha) = \lim_{N \to \infty} D(\alpha/N)^N = \lim_{N \to \infty} \left[1 - i \frac{\alpha_k T_k}{N} \right]^N = \exp[-i\alpha_k T_k]. \tag{14.88}$$

Consider now the group commutator

$$D(\alpha)D(\beta)D(\alpha)^{-1}D(\beta)^{-1}; \tag{14.89}$$

this is the product of four elements of the group, and therefore it is a member of the group. Hence there must be a vector of n real parameters γ such that:

$$D(\alpha)D(\beta)D(\alpha)^{-1}D(\beta)^{-1} = D(\gamma). \tag{14.90}$$

Clearly γ is a function of α and β.

If the group is Abelian the matrices commute, and the group commutator reduces to the identity, i.e. $\gamma(\alpha, \beta) = 0$, while for a non-Abelian group we have

$$\gamma(\alpha, \beta) \neq 0. \tag{14.91}$$

If we consider infinitesimal transformations, and expand $D(\alpha)$, $D(\beta)$ and $D(\gamma)$ at first order in their arguments, we find

$$D(\gamma) = \mathbb{1} - i\gamma_k T_k = \mathbb{1} + \beta_l \alpha_m [T_l, T_m]. \tag{14.92}$$

Let us concentrate now on $\gamma(\alpha, \beta)$. The following properties can easily be proved:

1. $\gamma(0, 0) = 0$;
2. if $\alpha = 0$, or $\beta = 0$, then $\gamma = 0$.

Therefore we conclude that γ must be a *quadratic* function of its arguments:

$$\gamma_t(\alpha, \beta) = -c_{rst}\alpha_r \beta_s, \tag{14.93}$$

where c_{rst} are real constants, and $r, s, t(= 1, \ldots, n)$ are the indices labelling the parameters of the transformation.

Using Eq. (14.93), we obtain

$$D(\gamma) = \mathbb{1} - ic_{rst}\alpha_r\beta_s T_t, \tag{14.94}$$

i.e.

$$[T_r, T_s] = ic_{rst}T_t. \tag{14.95}$$

The set of all real linear combinations of T_k is a vector space $\mathcal{G}$:

$$\mathcal{G} = \left\{ \sum_k c_k T_k, c_k \in \mathbb{R} \right\}. \tag{14.96}$$

The commutator $[,]$ defines a binary operation $\mathcal{G} \times \mathcal{G} \to \mathcal{G}$, such that

$$[X + Y, Z] = [X, Z] + [Y, Z], \tag{14.97}$$

$$[X, Y + Z] = [X, Y] + [X, Z], \tag{14.98}$$

$$[\alpha X, \beta Y] = \alpha\beta[X, Y]. \tag{14.99}$$

The vector space $\mathcal{G}$, equipped with the product law $[,]$, is called the **Lie algebra** of the group G. The coefficients c_{rst} which define the commutators of the generators are called **structure constants**.

Note that the structure constants were obtained starting from the parametrisation $\gamma(\alpha, \beta)$, defined in specific representations. It can be shown that they are actually independent of the representation.

Summary

- We introduced the idea of symmetry in a classical system. (14.1)
- We then discussed its 'natural' extension to the case of a quantum-mechanical system, which resulted in Wigner's theorem and (anti-)unitary representations of symmetry transformations. (14.2)
- The set of symmetry transformations of a physical system has the mathematical structure of a group. (14.2)
- We have examined a few examples of symmetry groups in physics and their importance in the contemporary description of natural phenomena. (14.3)
- We have analysed the structure of the group of translations. The momentum can be naturally defined as the generator of translations. (14.4)
- Groups of rotations in two and three dimensions. Abelian and non-Abelian structure of the two groups. (14.5)
- Rotations as unitary operators in the Hilbert space of states. We have introduced the possibility of defining the angular momentum as the generator of rotations. (14.6)
- We have discussed the transformations of operators under rotations, and defined scalar operators and vector operators. (14.7)
- We have considered the set of generators for a generic symmetry group and showed that they form an algebra, known as the Lie algebra of the group. (14.8)

Problems

14.1　Write down explicitly the 3×3 matrix which represents a rotation by the angle θ around the z-axis in three dimensions. Expand its elements for infinitesimal θ and deduce the generator of rotations around z for this three-dimensional representation.

14.2　Consider the set of 2×2 complex unitary matrices U, with $\det U = 1$. This set is a group under matrix multiplication called SU(2). Use the fact that $\det U = 1$ to prove that the generators are traceless. Since the matrices are unitary, the generators are also Hermitian.

Check that the matrices $\sigma_i/2$, where σ_i are the Pauli matrices, are a basis for the Lie algebra of SU(2). Compute explicitly the commutation relations of the SU(2) generators $\sigma_i/2$, and check that they satisfy the same algebra as the SO(3) generators.

14.3　Consider the quantum state described by the wave function

$$\psi(x) = \left(\frac{1}{2\pi\sigma^2}\right)^{1/4} \exp\left\{-\frac{x^2}{4\sigma^2}\right\}.$$

1. Compute $\hat{P}^2 \psi(x)$.
2. Write the wave function $\psi'(x)$ describing the same system shifted by a to the right along the x-axis. Show that

$$\psi'(x) = \exp\left[-i\hat{P}a/\hbar\right]\psi(x).$$

3. The angular momentum operator in two dimensions can be defined as:

$$\hat{L} = \hat{X}\hat{P}_y - \hat{Y}\hat{P}_x.$$

Rewrite the operator $\hat{L}$ as a differential operator in polar coordinates.
4. Consider a state described by a wave function $\phi(x, y)$. The wave function of the same system rotated by an angle ϵ is described by

$$\phi'(x, y) = \phi(x \cos \epsilon + y \sin \epsilon, -x \sin \epsilon + y \cos \epsilon).$$

For an infinitesimal rotation ϵ, show that

$$\phi'(x, y) = \phi(x, y) - \frac{i}{\hbar}\epsilon\hat{L}\psi(x, y).$$

5. Show that the eigenvalues of L are $\hbar m$, where m is an integer, and find the eigenfunctions.

PART II

APPLICATIONS

Quantum Entanglement

For observables like position and momentum, in quantum mechanics the quantum states in general do not give them an absolute existence. Their value in a particular system is generally only known once the measurement is made. Nevertheless, certain correlations can be present in a system. For a system that is made up of two or more parts that can be measured separately, such as at distinctly different spatial positions, the measurement of one part of the system may immediately imply what the measurement at another part of the system will be. This is a feature that can emerge in a quantum system which is entangled. Viewed on its own, quantum entanglement can look very mysterious. For spatially separated entangled parts of a system, this behaviour can even appear as a violation of causality. In the early phases in the development of quantum mechanics, entanglement took on a sort of mysterious reputation. However, if one carefully analyses a quantum entangled system, following all the rules of quantum mechanics, one finds no real mystery. Nevertheless, quantum entanglement leads to important questions and understandings of determinism and the nature of reality. And beyond these more lofty and sometimes philosophical concerns, exploiting the property of entanglement can lead to practical applications for areas such as computing, encryption and communication. This chapter will first explain in conceptual terms the meaning of quantum entanglement, formalise the idea and then examine the applications.

15.1 Hidden Variables and the Einstein–Podolsky–Rosen Paradox

Quantum mechanics asserts that the state of a system is fully specified by its state vector. The observables do not, in general, have definite values but are distributed according to a probability distribution. Unless the system is known to be in an eigenstate of the observable which we are measuring, all that we can predict is the probability with which a particular eigenvalue occurs when the measurement is repeated a large number of times on identically prepared systems. Quantum mechanics asserts that *making a measurement does not, in general, reveal a pre-existing value of the observable being measured*. This lack of determinism has troubled many physicists since the inception of quantum mechanics. Einstein, in particular, was dissatisfied with quantum mechanics as a fundamental theory of nature. His views are summed up in his famous statement, 'God does not play dice with the universe'. He seemed to believe that there was an underlying reality in which particles would have well-defined positions and speeds, and would evolve according to deterministic laws.

Attempts at a deterministic view are embodied in so-called **hidden variable** theories, which assert that there exist parameters which fully determine the values of observables, that quantum mechanics is an incomplete theory which does not involve these hidden variables and that the state vector simply reflects the probability distribution with respect to these parameters. The analogy that is sometimes drawn is to the relationship between classical mechanics and statistical mechanics. In the latter, a statistical description of physical properties, such as the motion of molecules in a gas, is all that is possible, even though the underlying microscopic behaviour of the motion of individual particles is in principle governed by the deterministic laws of Newtonian mechanics.

15.1.1 The Einstein–Podolsky–Rosen Paradox

In 1935, Einstein, Podolsky and Rosen published a famous paper which questioned the completeness of quantum theory, arguing that conventional quantum measurement theory leads to unreasonable conclusions. Their argument was later expressed in a somewhat simpler form by Bohm and we will look at it here.

Background

Firstly, let us recall some of the properties of spin-$\frac{1}{2}$ systems. If we measure the z-component of spin, we can only get one of the two possible values $\pm\frac{1}{2}\hbar$. This is illustrated in Fig. 15.1, where a beam of spin-$\frac{1}{2}$ particles enters a Stern–Gerlach magnet whose mean field is aligned with the z-axis, and as discussed in Section 9.1 two beams emerge, corresponding to particles with $m = \frac{1}{2}$ and $m = -\frac{1}{2}$, respectively.

In the standard representation, the emergent beams correspond to eigenstates of

$$\hat{s}_z \longrightarrow \tfrac{1}{2}\hbar \begin{pmatrix} 1 & 0 \\ 0 & -1 \end{pmatrix}, \tag{15.1}$$

which have the representation

$$|m = \tfrac{1}{2}\rangle \longrightarrow \begin{pmatrix} 1 \\ 0 \end{pmatrix}, \qquad |m = -\tfrac{1}{2}\rangle \longrightarrow \begin{pmatrix} 0 \\ 1 \end{pmatrix}. \tag{15.2}$$

Equally, if we measure the x-component of spin as shown in Fig. 15.2, we can only get one of the two possible values $\pm\frac{1}{2}\hbar$, which we denote by $m_x = \pm\frac{1}{2}$. Here, the emergent

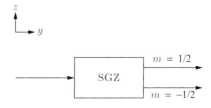

Fig. 15.1 Stern–Gerlach magnet with mean field aligned with the z-axis. The two emergent beams correspond to particles with $m = \frac{1}{2}$ and $m = -\frac{1}{2}$, respectively.

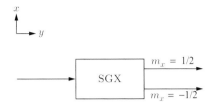

Fig. 15.2 Stern–Gerlach magnet with mean field aligned along the x-axis.

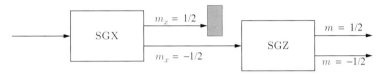

Fig. 15.3 Two Stern–Gerlach magnets with the first that the beam enters aligned along the x-axis and the next along the z-axis.

beams correspond to eigenstates of

$$\hat{s}_x \longrightarrow \tfrac{1}{2}\hbar \begin{pmatrix} 0 & 1 \\ 1 & 0 \end{pmatrix}, \tag{15.3}$$

which have the representation

$$|m_x = \tfrac{1}{2}\rangle \longrightarrow \frac{1}{\sqrt{2}} \begin{pmatrix} 1 \\ 1 \end{pmatrix}, \qquad |m_x = -\tfrac{1}{2}\rangle \longrightarrow \frac{1}{\sqrt{2}} \begin{pmatrix} 1 \\ -1 \end{pmatrix}. \tag{15.4}$$

These of course can be expressed as linear combinations of the eigenstates of $\hat{s}_z$:

$$\frac{1}{\sqrt{2}} \begin{pmatrix} 1 \\ 1 \end{pmatrix} = \frac{1}{\sqrt{2}} \begin{pmatrix} 1 \\ 0 \end{pmatrix} + \frac{1}{\sqrt{2}} \begin{pmatrix} 0 \\ 1 \end{pmatrix},$$

$$\frac{1}{\sqrt{2}} \begin{pmatrix} 1 \\ -1 \end{pmatrix} = \frac{1}{\sqrt{2}} \begin{pmatrix} 1 \\ 0 \end{pmatrix} - \frac{1}{\sqrt{2}} \begin{pmatrix} 0 \\ 1 \end{pmatrix}.$$

Thus if we prepare a beam of spin-$\frac{1}{2}$ particles in, say, the state $|m_x = -\frac{1}{2}\rangle$ by using an SGX magnet and beam stopper as shown in Fig. 15.3 and allow it to impinge on an SGZ magnet, the two emergent beams will have equal intensities. This is equivalent to the statement that, if we measure the z-component of spin of a particle in the state $|m_x = -\frac{1}{2}\rangle$, the probability of getting $\frac{1}{2}\hbar$ is $|1/\sqrt{2}|^2 = \frac{1}{2}$ and the probability of getting $-\frac{1}{2}\hbar$ is also $|1/\sqrt{2}|^2 = \frac{1}{2}$. More generally, we might consider measuring the component of spin in the direction of a unit vector $\underline{n} = (\sin\theta, 0, \cos\theta)$ in the x–z plane for which

$$\underline{\hat{s}} \cdot \underline{n} = \hat{s}_x \sin\theta + \hat{s}_z \cos\theta \longrightarrow \tfrac{1}{2}\hbar \begin{pmatrix} \cos\theta & \sin\theta \\ \sin\theta & -\cos\theta \end{pmatrix}. \tag{15.5}$$

The normalised eigenvectors are

$$|m_\theta = \tfrac{1}{2}\rangle \longrightarrow \begin{pmatrix} \cos\frac{1}{2}\theta \\ \sin\frac{1}{2}\theta \end{pmatrix}, \qquad |m_\theta = -\tfrac{1}{2}\rangle \longrightarrow \begin{pmatrix} \sin\frac{1}{2}\theta \\ -\cos\frac{1}{2}\theta \end{pmatrix}. \tag{15.6}$$

Fig. 15.4 EPR experiment with two spin-$\frac{1}{2}$ particles in an overall spin zero state with each going to one of the Stern–Gerlach magnets A or B.

The EPR Thought Experiment

Now consider a hypothetical process shown in Fig. 15.4 in which a spinless particle at rest decays into a pair of spin-$\frac{1}{2}$ particles, which, from momentum conservation, emerge back-to-back. Let us suppose that their individual spins are then analysed by means of two Stern–Gerlach magnets, A and B, which can each be oriented so that the component of spin in an arbitrary direction can be measured. Also, the experiment will be arranged such that A and B are sufficiently far apart so that information recorded at A, even travelling at the speed of light, could not reach B until after the measurement there is done and vice versa. Suppose that both magnets are oriented to measure the z-component of spin. Conservation of angular momentum tells us that, since the two spin-$\frac{1}{2}$ particles are produced in a spin singlet state ($S = 0$), the total z-component of spin of the two-particle system must be zero. Thus, if the result of measuring s_z at A is $\frac{1}{2}\hbar$, the result of measuring s_z at B *must be* $-\frac{1}{2}\hbar$, and vice versa.

In its own right this seems nothing special to quantum mechanics, since it is just conservation of angular momentum. In a classical physics system, if the particles measured at A and B were initially created in a combined zero angular momentum state, and the particle at A is measured along a given axis to have a certain angular momentum, then the particle at B measured along an axis pointing in the same direction will have its angular momentum in the opposite direction, just as found in the quantum-mechanical case. Nevertheless, in quantum mechanics the observer affects the state of the system from the act of making a measurement. So when the spin is measured at A along the z-axis, the particle spin is projected into one of the two states $\pm\frac{1}{2}\hbar$, yet somehow that measurement also then affects the state of the particle at B. What seems odd here is that since the measurement at A affected the state and projected it along one of the two directions only after the act of measurement, how did the particle at B know what happened? So for example if the measurement at A projects the particle into the $+\frac{1}{2}\hbar$ state, that act of measurement purely local to A then also projects the particle at B to be in the state $-\frac{1}{2}\hbar$.

The difference between the classical versus quantum measurement is, in the classical case, the angular momentum of the particles moving to A and B has absolute existence. If, for example, the angular momentum of the particle at A is measured in some other direction, not z, but maybe in the x-direction, whatever the projection of the angular momentum is along x, that is what will be measured. If the maximum value of the angular momentum happened to be fully along z, then in the classical case, the measurement along x would indicate zero angular momentum along it and likewise the measurement along the x-direction at B also would be zero. What is different in classical mechanics is the view that there is an absolute existence of those states, so that even before the measurement, the particles were in a specific state of angular momentum and the act of measurement had no effect on that state.

In contrast, in quantum mechanics, the measurement affects the system. Thus, if at A the measurement of the spin is done along the x-axis, the particle state vector collapses once again only on one of two possible outcomes, $\pm\frac{1}{2}\hbar$ this time along the x-axis. In fact, whatever axis one measures the spin at A on, the state vector for the particle spin collapses along that direction so that one will measure either one of these two values. The unusual thing, which EPR noted, is that not only does the measurement at A affect the outcome at A, but also at B, no matter how far away B might be. Recall that in quantum mechanics there is no predetermined value of the spin state of the particle at A. It was only due to the act of measurement at A that the particle state vector collapsed into one of two possible states. Yet somehow that measurement act and its outcome seemed to have been conveyed to B, since the measurement at B depended on what was measured at A. Because no matter in what direction $\underline{n}$ the spin is measured at A and B, whatever value is measured at A, the opposite value will be measured at B. We are driven to the conclusion that once we have measured the spin component in the direction $\underline{n}$ of the particle that enters A, we *know* the component of spin in that direction of the particle that enters B!

This raises some interesting questions:

- Is the fact that the outcome of the measurement at B is completely predictable a consequence of the measurement at A?
- If so, how did the information about what happened in the measurement at A get transmitted to the other particle?

Let us now suppose that the magnet at A is oriented to measure the x-component of spin. Again we know that there are only two possible outcomes, $m_x = \frac{1}{2}$ or $m_x = -\frac{1}{2}$. Suppose that we obtain the result $m_x = \frac{1}{2}$. The analysis above then tells us that, were we to measure the x-component of spin of the other particle, at magnet B, we are guaranteed to get the result $m_x = -\frac{1}{2}$. However, if magnet B measures the z-component of spin of the particle at B, we have equal probabilities of $\frac{1}{2}$ of getting the two possible outcomes, since the state with $m_x = -\frac{1}{2}$ is an equal superposition of states with $m = \pm\frac{1}{2}$, so that now the result at B is no longer predetermined! As another example, suppose A measures along the z-axis, so they can get $\pm\frac{1}{2}\hbar$ and suppose they measure $-\frac{1}{2}\hbar$. Then we know at B that the spin along the z-axis will be $\frac{1}{2}\hbar$. Suppose instead that B measures the spin along an axis at angle θ with respect to the z-axis in the x–z plane. Then we can decompose the $\frac{1}{2}\hbar$ state along the z-axis in terms of the basis states along the θ-axis, which gives

$$\begin{pmatrix} 1 \\ 0 \end{pmatrix} = \cos\tfrac{1}{2}\theta \begin{pmatrix} \cos\tfrac{1}{2}\theta \\ \sin\tfrac{1}{2}\theta \end{pmatrix} + \sin\tfrac{1}{2}\theta \begin{pmatrix} \sin\tfrac{1}{2}\theta \\ -\cos\tfrac{1}{2}\theta \end{pmatrix}$$

$$= \cos\tfrac{1}{2}\theta|m_\theta = \tfrac{1}{2}\rangle + \sin\tfrac{1}{2}\theta|m_\theta = -\tfrac{1}{2}\rangle. \tag{15.7}$$

Thus a measurement along the θ-axis would find spin-up with probability $\cos^2\left(\frac{1}{2}\theta\right)$ and spin-down with probability $\sin^2\left(\frac{1}{2}\theta\right)$. In all these cases the particle at B somehow 'knows' what the state of the particle A was when it was measured. How does the particle at B 'know' which way the magnet at A is pointing and the result that was found?

If one were to list the results of these experiments on the quantum system, one would conclude that there is a correlation between what A does and what B measures. For example, amongst the experiments, if one looked at the cases where the measurement axis for both A and B was the same, no matter what direction they point in, whatever is measured at A, the measurement at B is always found to be of the same magnitude but in the opposite direction. And for measurement directions at A and B that are not the same, there still appears some correlation between the two measurements. It seems that measuring the spin of one particle affects the result of measuring the spin of the other particle, *regardless of how far apart the two particles are when the measurements are made*. This would appear to imply action at a distance, or even transmission of information at superluminal speed, unless *any spin component of either particle has a definite value from the instant of decay of the parent particle* so that measurements simply reveal these pre-existing values. For Einstein, this latter possibility seemed much more reasonable. He thought that the real world consisted of components possessing objective properties which exist independently of any measurements by observers. Moreover, he thought that the result of a measurement of a property at location B cannot depend on an event at location A that is so far from B that information about the event at A, even travelling at the speed of light, could not have reached B until after the measurement has taken place. Theories meeting these requirements are called **realistic local theories**.

Here we will look at the contribution to this debate from the physicist John Bell, who derived a criterion which any realistic local theory should obey. The questions to be addressed are:

- Is quantum mechanics a realistic local theory?
- If not, can we make predictions about experiments which can distinguish between quantum mechanics and realistic local theories?

15.2 Bell's Inequality

We now turn to the question of whether or not quantum mechanics can be distinguished from a realistic local theory by suitable experimental measurements. Our treatment follows that of Wigner.[1]

Consider again the hypothetical process shown in Fig. 15.4, in which a spinless particle at rest decays into a pair of spin-$\frac{1}{2}$ particles, which, from momentum conservation, emerge back-to-back. Let us suppose that their individual spins are then analysed by means of two

---

[1] E. P. Wigner, 'On hidden variables and quantum mechanical probabilities', *Am. J. Phys.* **38** (1970) 1005, https://doi.org/10.1119/1.1976526. A shortened version may be found in the text *Quantum Mechanics* by F. Mandl (Wiley, 1992). An amusing non-technical account of Bell's inequality is in the article by David Mermin, 'Is the moon there when nobody looks? Reality and the quantum theory', *Physics Today* **38** (1985) 38, https://doi.org/10.1063/1.880968.

Stern–Gerlach magnets, A and B, which can each be oriented so that the component of spin in an arbitrary direction can be measured. In particular, we assume that we can orient the magnets A and B so that it is possible to measure the components of spin in three co-planar directions specified by unit vectors $\underline{n}_1$, $\underline{n}_2$ and $\underline{n}_3$.

In a realistic local theory, the individual spin-$\frac{1}{2}$ particles have definite values, each either $+\frac{1}{2}\hbar$ or $-\frac{1}{2}\hbar$, for the spin components in all three directions. Notice that we are *not* assuming that we can measure all three components simultaneously, but merely that if any one of them is measured by suitably orienting the SG magnet, the outcome is predictable with certainty. We will use an abbreviated notation ↑ and ↓ to denote the two possible outcomes. We can characterise the spin state of a given particle by, for example, $(\uparrow, \downarrow, \uparrow)$, meaning that a measurement of the spin component in the $\underline{n}_1$-direction is certain to produce the result ↑, that in the $\underline{n}_2$-direction the result ↓ and that in the $\underline{n}_3$-direction the result ↑. We can divide the results of a large number of measurements into groups specified by $(\sigma_1, \sigma_2, \sigma_3, \tau_1, \tau_2, \tau_3)$, where σ_i and τ_i denote the spin components in the direction $\underline{n}_i$ of the particles travelling to magnets A and B, respectively.

We will denote by $f(\sigma_1, \sigma_2, \sigma_3, \tau_1, \tau_2, \tau_3)$ the fraction of particle pairs belonging to the group $(\sigma_1, \sigma_2, \sigma_3, \tau_1, \tau_2, \tau_3)$. Since we are assuming that the spin-$\frac{1}{2}$ pair is produced in a singlet state, this tells us immediately that

$$f(\sigma_1, \sigma_2, \sigma_3, \tau_1, \tau_2, \tau_3) = 0 \qquad \text{unless} \quad \sigma_i = -\tau_i, \quad i = 1, 2, 3. \tag{15.8}$$

The locality requirement is satisfied, since the outcome of a measurement at A depends only on the values of σ_1, σ_2 and σ_3 and the orientation of the magnet A, but is independent of the orientation of the magnet B.

15.2.1 Spin Correlations

We can compute $P_{\uparrow\uparrow}(\underline{n}_i, \underline{n}_j)$, the probability that, for a given pair of spin-$\frac{1}{2}$ particles, a measurement of the spin component along $\underline{n}_i$ at A gives ↑ and a measurement of the spin component along $\underline{n}_j$ at B gives ↑. For example:

$$P_{\uparrow\uparrow}(\underline{n}_1, \underline{n}_2) = \sum_{\sigma_2 \sigma_3} \sum_{\tau_1 \tau_3} f(\uparrow, \sigma_2, \sigma_3; \tau_1, \uparrow, \tau_3) = f(\uparrow, \downarrow, \uparrow; \downarrow, \uparrow, \downarrow) + f(\uparrow, \downarrow, \downarrow; \downarrow, \uparrow, \uparrow), \tag{15.9}$$

where we have used the condition embodied in Eq. (15.8) to observe that the only non-zero terms in the sum are those for which $\sigma_1 = \uparrow, \tau_1 = \downarrow, \sigma_2 = \downarrow$, and $\tau_2 = \uparrow$. Similar arguments give

$$P_{\uparrow\uparrow}(\underline{n}_3, \underline{n}_2) = f(\uparrow, \downarrow, \uparrow; \downarrow, \uparrow, \downarrow) + f(\downarrow, \downarrow, \uparrow; \uparrow, \uparrow, \downarrow), \tag{15.10}$$

$$P_{\uparrow\uparrow}(\underline{n}_1, \underline{n}_3) = f(\uparrow, \uparrow, \downarrow; \downarrow, \downarrow, \uparrow) + f(\uparrow, \downarrow, \downarrow; \downarrow, \uparrow, \uparrow). \tag{15.11}$$

Since the fractions f are non-negative, we can add these last two results to obtain **Bell's inequality**:

$$P_{\uparrow\uparrow}(\underline{n}_1, \underline{n}_2) \leq P_{\uparrow\uparrow}(\underline{n}_3, \underline{n}_2) + P_{\uparrow\uparrow}(\underline{n}_1, \underline{n}_3). \tag{15.12}$$

We can now calculate the spin correlations given by quantum mechanics using the results that we obtained in the last section. In the spin singlet state the probability of getting the result $\downarrow$ when we measure the component of spin in the direction $\underline{n}_1$ at magnet A is $\frac{1}{2}$. If we get the result $\downarrow$, we know that a similar measurement at B of the spin component along the same direction $\underline{n}_1$ would yield the result $\uparrow$. Suppose $\underline{n}_1$ corresponds to the z-axis (we will call this the $\underline{k}$-direction) and at B we choose instead to measure the spin along the $\underline{n}_2$-direction, which is at an angle θ with respect to the z-axis. In that case we can examine what is the projection of a state, which is $\uparrow$ along the z-axis, when examined along the $\underline{n}_2$-axis, and this is simply Eq. (15.7). Thus, reading off the probability amplitudes, the probability of getting the result $\uparrow$ if we measure the $\underline{n}_2$-component of spin of the particle at B is $\cos^2(\frac{1}{2}\theta)$. As such, the overall probability of getting $\downarrow$ in the z-direction at A *and* $\uparrow$ in the $\underline{n}_2$-direction at B is

$$P_{\downarrow\uparrow}(\underline{k}, \underline{n}_2) = \tfrac{1}{2} \cos^2\left(\tfrac{1}{2}\theta\right). \tag{15.13}$$

Likewise, the probability of getting $\uparrow$ in the z-direction at A and $\uparrow$ in the $\underline{n}_2$-direction at B is

$$P_{\uparrow\uparrow}(\underline{k}, \underline{n}_2) = \tfrac{1}{2} \sin^2\left(\tfrac{1}{2}\theta\right). \tag{15.14}$$

This result just depends on the relative angle between the two directions from which spin is measured. Thus, writing this in a more general notation, if we denote two directions as $\underline{n}_i$ and $\underline{n}_j$ with angle between them θ_{ij}, then

$$P_{\uparrow\uparrow}(\underline{n}_i, \underline{n}_j) = \tfrac{1}{2} \sin^2\left(\tfrac{1}{2}\theta_{ij}\right),$$
$$P_{\downarrow\uparrow}(\underline{n}_i, \underline{n}_j) = \tfrac{1}{2} \cos^2\left(\tfrac{1}{2}\theta_{ij}\right). \tag{15.15}$$

For quantum mechanics to agree with the results for a realistic local theory, for example, the probabilities of finding $\uparrow$ in both directions must satisfy Bell's inequality in Eq. (15.12), so that

$$\sin^2\left(\tfrac{1}{2}\theta_{12}\right) \leq \sin^2\left(\tfrac{1}{2}\theta_{23}\right) + \sin^2\left(\tfrac{1}{2}\theta_{13}\right). \tag{15.16}$$

There clearly are geometries for which this inequality is violated. For example, recalling that $\underline{n}_1$, $\underline{n}_2$ and $\underline{n}_3$ are coplanar and supposing also that $\underline{n}_3$ bisects the angle between $\underline{n}_1$ and $\underline{n}_2$, then

$$\theta_{13} = \theta_{23} = \tfrac{1}{2}\theta_{12}. \tag{15.17}$$

Substituting into Bell's inequality gives

$$\sin^2(\theta_{13}) \leq 2\sin^2\left(\tfrac{1}{2}\theta_{13}\right), \tag{15.18}$$

which we can simplify to

$$\cos^2\left(\tfrac{1}{2}\theta_{13}\right) \leq \tfrac{1}{2}. \tag{15.19}$$

This is clearly *not* satisfied for $0 < \tfrac{1}{2}\theta_{13} < \pi/4$ and so we are driven to the conclusion that quantum mechanics violates Bell's inequality; at least some of its predictions are not consistent with it being a realistic local theory!

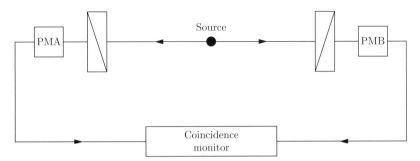

Fig. 15.5 Schematic of the Aspect *et al.* experimental apparatus. Here, PM are polarimeters that measure the linear polarisation of the photons.

15.2.2 Experimental Tests of Bell's Inequality

In the 1980s, A. Aspect, J. Dalibard and G. Roger carried out a remarkable series of experiments on an optical analogue of the thought experiment involving correlated spin-$\frac{1}{2}$ particles that we have been discussing.[2] For photons, a measurement of the linear polarisation is performed by polarisers at A and B in orientations $\underline{n}_A$ and $\underline{n}_B$, respectively (Fig. 15.5). A measurement along $\underline{n}_A$ yields the result V if the polarisation is parallel to $\underline{n}_A$ and H if perpendicular to $\underline{n}_A$, and so on. The *correlation coefficient* is defined to be

$$E(\underline{n}_A, \underline{n}_B) = P_{VV}(\underline{n}_A, \underline{n}_B) + P_{HH}(\underline{n}_A, \underline{n}_B) - P_{VH}(\underline{n}_A, \underline{n}_B) - P_{HV}(\underline{n}_A, \underline{n}_B). \qquad (15.20)$$

Aspect *et al.* considered the combination

$$S = E(\underline{n}_A, \underline{n}_B) - E(\underline{n}_A, \underline{n}'_B) + E(\underline{n}'_A, \underline{n}_B) + E(\underline{n}'_A, \underline{n}'_B), \qquad (15.21)$$

where $\underline{n}_A, \underline{n}'_A$ and $\underline{n}_B, \underline{n}'_B$ represent two different orientations of the polarisers at A and B, respectively. In a realistic local theory, a Bell inequality can be derived for this quantity:

$$-2 \leq S \leq 2. \qquad (15.22)$$

Quantum mechanics predicts that, for suitable orientations S can reach the values $\pm 2\sqrt{2}$, in clear contradiction to the Bell inequality.

[2] A. Aspect, J. Dalibard and G. Roger, 'Experimental realization of Einstein–Podolsky–Rosen–Bohm Gedanken-experiment: A new violation of Bell's inequalities', *Phys. Rev. Lett.* **49** (1982) 91–94, https://doi.org/10.1103/PhysRevLett.49.91 and A. Aspect, J. Dalibard and G. Roger, 'Experimental test of Bell's inequality using time-varying analyzers', *Phys. Rev. Lett.* **49** (1982) 1804–1807, http://dx.doi.org/10.1103/PhysRevLett.49.1804. The schematic in Fig. 15.5 and the sketched plot in Fig. 15.6 are based on these papers.

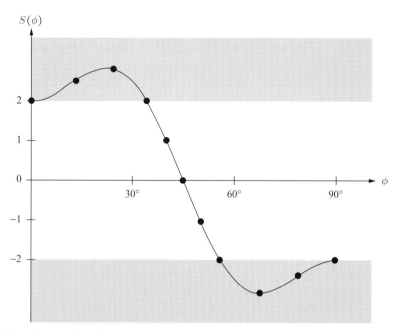

Fig. 15.6   A plot of S from Eq. (15.21) from the experiment of Aspect *et al.*

In the Aspect *et al.* experiment, the orientations were chosen so that

$$\underline{n}_A \cdot \underline{n}_B = \underline{n}'_A \cdot \underline{n}_B = \underline{n}'_A \cdot \underline{n}'_B = \cos\phi \quad \text{and} \quad \underline{n}_A \cdot \underline{n}'_B = \cos 3\phi, \tag{15.23}$$

for which Eq. (15.21) gives

$$S = 3\cos(2\phi) - \cos(6\phi). \tag{15.24}$$

It is left as a problem at the end of the chapter to compute this expression. The value of S was then determined for various values of ϕ between $0°$ and $90°$. The results are shown in Fig. 15.6, together with the predictions of quantum mechanics (solid curve). The shaded regions are those where Bell's inequality is violated. It is clear that the experimental results violate Bell's inequality and are in excellent agreement with quantum mechanics: Aspect *et al.* quote the value of S at $\phi = 22.5°$ where the greatest discrepancy with the inequality occurs,

$$S_{\text{expt}} = 2.697 \pm 0.015. \tag{15.25}$$

In a second paper, Aspect *et al.* reported further results in which the orientations of the polarisation analysers are selected only after the photon pair has left the source and too late for information travelling at the speed of light from one analyser to reach the other in time to affect the polarisation measurement there. Again, the results are in good agreement with quantum mechanics but violate Bell's inequality by five standard deviations! Note that the violation of Bell's inequality implies that no physical theory, even one that hypothetically may develop in the future and supersede quantum mechanics, can ever have a deterministic measurement outcome.

15.3 Characterising Entanglement

Quantum entanglement has a close association with quantum information theory. Thinking in terms of information content leads to a systematic formal structure for studying and quantifying entanglement. Moreover, some of the key applications for entanglement have been identified in this context, in particular quantum communication and quantum computing. It is important at a foundational level to understand the underlying ideas behind these subjects that have emerged from classical information theory and classical computing. In this section we will first review the basic ideas of classical information theory, which then lead to developing the ideas of quantum information theory. Then we will see how these concepts help to quantify and characterise quantum entanglement and more generally quantum correlations. In a following section on quantum computing, likewise we will start by reviewing the basics of classical computing.

15.3.1 Bit and Qubit

The formal treatment of quantum entanglement centres on the basic unit of information, called the qubit, which is the analogue to the bit from classical computing. In classical computing the two-state system is the basic building block. With the two possible states usually denoted as 0 or 1, a bit of information stores one of these numbers and represents the state of the classical system. Thus 0 would mean the system is in one state and 1 the other state. The quantum system differs from the classical system in that it can be in both states at the same time. The information characterising that for a two-state system is called a qubit and is expressed as a quantum state $|\psi\rangle$, which is a linear combination of the two possible states:

$$|\psi\rangle = a_\downarrow| \downarrow\rangle + a_\uparrow| \uparrow\rangle, \quad a_\downarrow, a_\uparrow \in C, \qquad (15.26)$$

where for quantum systems, adhering to our notation from the previous section, we will denote the two states of the system by $\uparrow$ and $\downarrow$, though later when discussing quantum computing they will be denoted respectively by 0 and 1. We will adhere throughout this chapter to the following convention in representing these two states:

$$| \uparrow\rangle \leftrightarrow |0\rangle \rightarrow \begin{pmatrix} 1 \\ 0 \end{pmatrix}, \quad | \downarrow\rangle \leftrightarrow |1\rangle \rightarrow \begin{pmatrix} 0 \\ 1 \end{pmatrix}. \qquad (15.27)$$

The complex numbers $a_\downarrow$ and $a_\uparrow$ in Eq. (15.26) are the probability amplitudes of being in the two states $| \downarrow\rangle$ and $| \uparrow\rangle$, respectively, and satisfy the normalisation condition $|a_\downarrow|^2 + |a_\uparrow|^2 = 1$. A common parametrisation of these amplitudes is in terms of polar angles in three-dimensional spherical coordinates:

$$|\psi\rangle = \cos\frac{\theta}{2}| \uparrow\rangle + e^{i\phi} \sin\frac{\theta}{2}| \downarrow\rangle, \qquad (15.28)$$

with the range $0 \leq \theta \leq \pi$ and $0 \leq \phi \leq 2\pi$. This is represented pictorially in Fig. 15.7 as the surface of a unit sphere, called the Bloch sphere, that has already been mentioned in

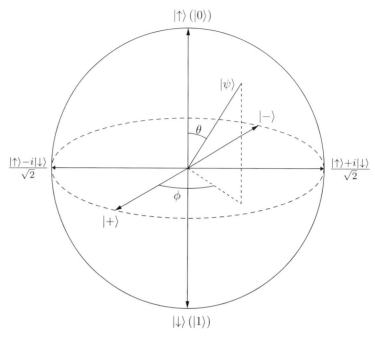

Fig. 15.7 Bloch sphere parametrised with spherical-polar coordinates.

Chapter 2. Here, another picture is drawn of this sphere with new notation for the states that will be relevant to our discussion in this chapter. Physically, the states $|\uparrow\rangle$ and $|\downarrow\rangle$ might characterise the two spin states of an electron or for that matter any two-state system. As an aside, as already alluded to above, note that in the field of quantum information the notation $|\uparrow\rangle$ and $|\downarrow\rangle$ is seldom used and it is more common to see the **computational basis** notation $|0\rangle$ and $|1\rangle$. Also in this field the spin states along the x-axis are often denoted $|+\rangle \equiv (|0\rangle + |1\rangle)/\sqrt{2}$ and $|-\rangle \equiv (|0\rangle - |1\rangle)/\sqrt{2}$.

Classical information about two bits would be contained in four possible states $00, 01, 10$ or 11. The quantum-mechanical analogue, or two-qubit state, is the linear combination of the following four states:

$$|\psi\rangle = a_{\downarrow\downarrow}|\downarrow\downarrow\rangle + a_{\downarrow\uparrow}|\downarrow\uparrow\rangle + a_{\uparrow\downarrow}|\uparrow\downarrow\rangle + a_{\uparrow\uparrow}|\uparrow\uparrow\rangle, \quad a_{ij} \in C, \tag{15.29}$$

with the normalisation condition $\sum_{ij}^{\downarrow\uparrow} |a_{ij}|^2 = 1$. Entanglement is a property that can only be discussed in the context of two systems, A and B. An **entangled state** between systems A and B is defined such that the state vector of the combined A and B systems cannot be written as a state vector just for system A tensor product with a state vector just for system B. A frequently used complete set of two-qubit states in which both qubits are entangled are the **Bell states**:

$$|\Phi^{\pm}\rangle = \frac{1}{\sqrt{2}}\left(|\uparrow\uparrow\rangle \pm |\downarrow\downarrow\rangle\right), \quad |\Psi^{\pm}\rangle = \frac{1}{\sqrt{2}}\left(|\uparrow\downarrow\rangle \pm |\downarrow\uparrow\rangle\right). \tag{15.30}$$

A Bell state is also referred to as one **ebit** of entanglement.

15.3.2 Information Theory Essentials – Shannon Entropy

Information theory is centred on quantifying information with the goal to devise efficient communication methods. There are many words here that need to be made precise. By information it is meant any collection of symbols and a set of rules on how to sequentially arrange them. For example, the information could be a language or mathematics. By communication it can mean either sending, receiving or storing the information. Finally, the word efficient has a precise meaning and goes to the main point of information theory. A form of information, such as a language, is itself a code. The goal of information theory is to take information expressed in one code, say a language, and see if another code can be found which expresses the exact same information but more compactly. An explicit example, and a very important one in practice, is to code information into binary numbers. In this case information theory is taking the original code, such as a statement in English, and expressing it in terms of some efficient binary code. There are a multitude of ways that this has been done. For example, a standard developed for this is the American Standard Code for Information Interchange, or ASCII. In ASCII the uppercase letter A is represented by the binary string 1000001, whereas lowercase a is represented by 1100001 and so forth. However, depending on what type of messages one is communicating, there could be more efficient, meaning shorter, binary strings for representing the letters and symbols of language.

The idea of communicating information more efficiently is not new. Within any language, one of the inherent goals of its users is to find ways to express ideas more efficiently and succinctly through better sentence structure or inventing single words that can express whole big ideas. The various means to manipulate a language, such as through slang or text messaging, are ways that ideas in a language might be conveyed more efficiently. Information theory is simply taking this basic idea and placing it in a quantitative framework, where one has a measure for assessing the efficiency with which information is being communicated.

The formal construct underlying Shannon information theory is as follows. Suppose there is a certain language (code) that is comprised of a set of N symbols $s_1 \ldots s_N$. A message X comprising M_X symbols formed from this language would have the generic form $X = s_i^1 s_j^2 s_k^3 s_l^4 \ldots s_t^{M_X}$, where the subscripts are integers all less than or equal to N associated with a particular symbol of this language. One can count how many times, M_i^X, the symbol s_i appears in this message, and thus its frequency $f_i^X = M_i^X / M_X$ in the string X. Shannon's theory is not focused on just a single message, but whole collections of messages that comprise basically the complete usage of that language. What Shannon theory does is abstract from this the idea that typical messages in a language utilise each of its symbols s_i with some probability p_i. In Shannon information theory the information content, or **Shannon information**, of a particular symbol s_i is defined as

$$I(p_i) \equiv -\log_2 p_i, \tag{15.31}$$

with the log expressed in base 2, so the information is in units of bits. The meaning of this quantity only makes sense when a large collection of symbols are being examined. In this context, I expresses either of these two similar interpretations, the uncertainty or surprise of that symbol in the message or the information gained as one reads the message and comes across the symbol s_i. Suppose, for example, the message is a series of words in some particular language and the symbols are the letters of that language. The higher the probability p_i of coming across s_i, according to Eq. (15.31), the lower is the information gained or the less it is surprising when one comes across this letter. That is because if a letter appears with high probability it means it occurs in many words of that language. Thus, coming across that letter in the message does not help very much in narrowing down what word it helps make. On the other hand, a letter s_j that is rare to find, so has a very low probability p_j, informs much more about the message, since the particular language would contain fewer words or arranged ideas with that letter, making it easier to narrow down the word. For example, in the English language, the letter a appears with a much higher probability than the letter z. If the first letter of a message contained the letter a, there would be many words that start with it and so one could not so easily narrow down what particular word is being sent. On the other hand, if the first letter we obtained was z, there are far fewer words that start with z, so we can better narrow down what word is being sent.

The above discussion is a qualitative explanation of Shannon information content, but let us also understand its quantitative meaning. First, let us look at the extreme case that the probability of a symbol s_1 is one, $p_1 = 1$, which from Eq. (15.31) means $I(p_1) = 0$. Suppose this is the case of a very peculiar language with just one letter. That means every symbol in the message is exactly the same. Anyone at the receiving end of the message will already know what symbol is being sent, since there is only one. As such, no new information is gained by the sending of this symbol, which is why $I = 0$. Next, suppose the probability of a particular symbol were 0.5, which from Eq. (15.31) means that the Shannon information content is 1 bit. Thus, in coming across this symbol in a string, one bit of information would be gained.

As an example that illustrates the meaning of the information function $I(p)$, consider the problem of trying to guess an unknown integer between 1 and 16. The probability of guessing the number is $p = 1/16$, so $I(1/16) = 4$ bits. The question is, what is the minimum number of yes/no guesses one needs in order to be certain to determine the number? The way to answer this is to recognise that the binary representation of a number up to 16 requires a string of up to four binary numbers. If one asked the four questions, is the binary number in the first position 1, then in the second, third and fourth positions that would be adequate to determine the number exactly, thus four bits, which is consistent with the information function.

For a string of N symbols, which have respective probability of appearing p_i, the **Shannon entropy** of that string is then the average of the information of all the symbols:

$$S_{\text{sh}} = -\sum_{i}^{N} p_i \log_2 p_i. \tag{15.32}$$

This, according to Shannon information theory, is the minimum bits per symbol needed to encode the string. This is the basic result of Shannon information theory, also called the Shannon source coding theory.

As an example, suppose one is working with a 'language' comprised of a set of symbols, and suppose one wishes to transmit strings of these symbols by first converting to a binary code. One naturally would want to seek the most efficient way to carry out this process, meaning create the shortest possible binary strings that contain all the information written in the original language. To apply Shannon information theory to answer this question, first one must determine the probability of occurrence of all the symbols in this language. This question demonstrates the empirical nature of Shannon information theory. There can be many ways to determine these probabilities. For example, should one take the probabilities as those found by looking at the entire body of literature in the language, and determine how often each symbol appears? Or should one consider all the strings of information one needs to send with the task at hand and determine from that set the frequency of each symbol? Perhaps one should just look at the single string one needs to send and determine the frequency of occurrence of each symbol. The latter clearly would serve for the single task at hand. However, that could mean for each string to be sent a new code would need to be devised and all receivers would need to be informed of this. Alternatively, if one universal code is established, it would be convenient in that all users would know it once and for all, but given strings of information may not have symbols appearing at the frequencies for which the code was initially determined. These are the sorts of practical issues one encounters in dealing with Shannon information theory.

As an explicit example, first suppose the language only contained two symbols, A and B. If p_A is the probability of symbol A appearing, then the probability of symbol B must be $1 - p_A$ and the Shannon entropy would be

$$S_{\mathrm{sh}} = -p_A \log_2 p_A - (1 - p_A) \log_2 (1 - p_A), \tag{15.33}$$

and is shown in Fig. 15.8. In the limits where either $p_A = 0$ or $p_A = 1$, the Shannon entropy is zero. These correspond to the cases where either all the symbols that appear in a message are respectively B or A. In both these cases, anyone receiving the message will already know beforehand what is in the message, so no new information is gained,

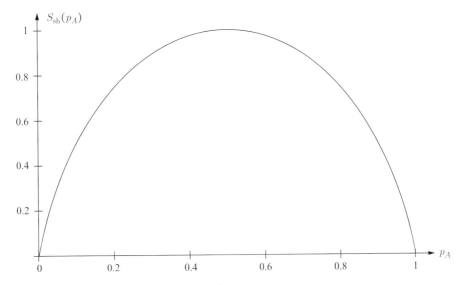

Fig. 15.8 Shannon entropy for a system with two symbols A and B.

thus it makes sense that the Shannon entropy is zero. On the other hand, if $p_A = 0.5$, $S = 1$ bit/symbol. The Shannon coding theory specifies that the most efficient message will involve 1 bit per symbol, so A could be represented by 1 and B by 0. This is the case of maximum randomness and so the coding theorem simply states that one cannot compress beyond each letter requiring a single bit.

As a more useful example, suppose we know that the four symbols in the language are A, B, C and D and they appear with respective probabilities 0.5, 0.25, 0.125 and 0.125. According to Eq. (15.32), the Shannon entropy of this language is

$$S_{\rm sh} = -0.5 \log_2 0.5 - 0.25 \log_2 0.25 - 0.125 \log_2 0.125 - 0.125 \log_2 0.125$$
$$= 1.75 \text{ bits/symbol.}$$

Thus the Shannon source coding theory specifies that the most efficient code of messages would involve 1.75 bits per symbol. Of course, a given symbol must be represented by a whole number of bits. However, if one takes a long string of such symbols and averages the total number of bits needed to code them, Shannon source coding theory states that the most efficient such coding will find the given rate $S_{\rm sh}$ and that one cannot do better.

As a representative example, suppose the following string needs to be transmitted:

$$AABABACD. \tag{15.34}$$

The basic idea is that the higher the probability for a symbol to appear, the smaller the associated bit length should be assigned to that symbol. Since A appears the most, one could start by considering that it is coded as a single binary number, so say 1. Then B is the next most frequent. One might say, why not then assign 0 to B? If these were the only two symbols in the language that would be fine. However, there is a problem once a third symbol exists. The symbol C needs to be assigned, so one might give it two bits, say 01. The problem now is, if these two numbers appear in a long code, it is ambiguous whether it represents C or AB. Coding by this approach would require what is called a **prefix**, in other words something in between each set of binary numbers representing a symbol, which indicates that the symbol is done and a different one is next. Having such prefixes will only lengthen the code, which is not what one is trying to achieve. By this reasoning one would conclude that A can be identified with a single binary number but B would require two binary numbers, say 01. Continuing with the same reasoning, a suitable coding of the remaining symbols could be 001 for C and 000 for D. If one examines the information $I(p_i)$ associated with the four symbols, one indeed finds that the binary length of each of the four symbols $ABCD$ should be respectively 1, 2, 3, 3. Moreover, with this particular choice, there would never be any ambiguity when reading a string of such binary numbers from left to right about which symbols in the original language they correspond to. Thus, the string in Eq. (15.34) would be coded as

$$11011011001000,$$

which is in total 14 binary numbers to represent 8 symbols in the original language, thus a rate of 1.75 bits/symbol. Based on Shannon source coding theory, this is the most efficient coding that is possible. There are of course many other ways to associate the original symbols with bits. For example, one could associate each symbol with two binary numbers, for example $ABCD$ associated respectively with 00, 01, 10 and 11. Then the original string would require 16 bits for coding the 8 symbols, thus requiring a rate higher than the minimum Shannon entropy. In fact, any other possible way to code the original symbols will have a Shannon entropy no lower than 1.75 bits/symbol.

This is not the full story. One could code the symbols in a very different way. Rather than coding the individual symbols one could look at sequences of symbols, such as AA, AB, etc. and determine how frequently they appear in the language and formulate a code based on that. Looking again at Eq. (15.34) and supposing this string is represented by sequences of two symbols, we find that BA appears with probability 0.5, AA with probability 0.25 and CD with probability 0.25. With this partitioning of the symbols, the Shannon entropy is

$$S_{\mathrm{sh}} = -0.5 \log_2 0.5 - 0.25 \log_2 0.25 - 0.25 \log_2 0.25 = 1.5 \text{ bits/symbol}.$$

As such, this approach would lead to even more compression of the initial symbols into binary numbers. One can proceed as before to now assign binary numbers to each of the three symbols in such a way as to not have any need for prefixes in the binary string representation of the information string. Similarly, one could imagine combining sequences of three symbols, four, etc. In fact, one could imagine a scheme that simply takes the entire string Eq. (15.34) and assigns a corresponding binary number to it, say 1. If this were the only string that ever appeared in this language, then it would have probability 1 and the Shannon information is $I = \log_2 1 = 0$. This means that such a message contains zero information and also it has zero Shannon entropy. The reason is that if a language had only one message, then no new information is ever exchanged. The receiver already knows what message is being sent, so no new information is gained by the message. If one wanted to associate the entire string of 8 symbols with a binary representation, it means one then needs to look at a large number of possible strings and determine the probability of occurrence of all such strings. This again is where the empirical nature of Shannon information theory enters.

This section has given a short overview of Shannon information theory and its application. The key general feature that can be extracted is that Shannon entropy is a measure of uncertainty. It quantifies how unpredictable a piece of information is. The more predictable the information, the lower is the Shannon entropy. The higher is the Shannon entropy, the more that information will tell us something that we were not anticipating.

15.3.3 Von Neumann Entropy

There are two types of states that are important in quantum information theory. The first is called the **pure state**, which is simply a quantum-mechanical state vector for a system. The state vector could be a single basis state or a linear superposition of basis states with respective probability amplitudes for each state in the expansion. Measuring observables from a pure state is probabilistic in the quantum-mechanical sense. In contrast, there is the **mixed state**, which is an ensemble of pure states, each with some classical probability ascribed to them. This can arise if one is not certain which pure state the system is in, but only have a probability for each pure state in the ensemble. Thus the probabilities associated with it are not intrinsic to nature, but rather of the statistical mechanical type due to our own lack of knowledge of how this system was prepared. A general way to represent a system accounting for both these possibilities is through the **density matrix**

$$\hat{\rho} = \sum_i p_i |\chi_i\rangle\langle\chi_i|, \tag{15.35}$$

where χ_i are quantum (pure) states that do not in general have to be mutually orthogonal and p_i is the probability for the pure state $|\chi_i\rangle$ within the ensemble.

Properties of the Density Matrix $\hat{\rho}$

1. **Hermiticity**: Follows on inspection of Eq. (15.35) that $\hat{\rho}^\dagger = \hat{\rho}$.
2. **Trace property** $\text{Tr}(\hat{\rho}) = 1$: Let $\{|\phi_i\rangle\}$ be a complete set of states. Then

$$\begin{aligned}
\text{Tr}(\hat{\rho}) &= \sum_{ij} \langle\phi_j|p_i|\chi_i\rangle\langle\chi_i|\phi_j\rangle \\
&= \sum_i p_i \sum_j \langle\chi_i|\phi_j\rangle\langle\phi_j|\chi_i\rangle \\
&= \sum_i p_i\langle\chi_i|\chi_i\rangle = \sum_i p_i = 1.
\end{aligned} \tag{15.36}$$

3. **Positive-semidefinite**: For an arbitrary vector $|\phi\rangle$, $\langle\phi|\hat{\rho}|\phi\rangle = \sum_i p_i\langle\phi|\chi_i\rangle\langle\chi_i|\phi\rangle = \sum_i p_i|\langle\phi|\chi_i\rangle|^2 \geq 0$.
4. **Eigenvalues** $0 \leq \alpha_i \leq 1$ **for all** i: Since $\hat{\rho}$ is Hermitian, it only has real eigenvalues $\hat{\rho}|\alpha_i\rangle = \alpha_i|\alpha_i\rangle$, where $\{|\alpha_i\rangle\}$ are the eigenstates of $\hat{\rho}$ with corresponding eigenvalues $\{\alpha_i\}$. Since $\hat{\rho}$ is positive-semidefinite every eigenvalue is non-negative, $\alpha_i \geq 0$ for all i. The trace Property 2 implies in the eigenstate basis, $\sum_i \alpha_i = 1$, thus each eigenvalue must satisfy $\alpha_i \leq 1$ for all i.
5. **Trace property for** $\hat{\rho}^2$, $\text{Tr}[\hat{\rho}^2] \leq 1$: In the eigenvector basis for $\hat{\rho}$, $\text{Tr}[\hat{\rho}^2] = \sum_i \alpha_i^2$. Since each α_i is non-negative, it implies $(\sum_i \alpha_i)^2 = 1 \geq \sum_i \alpha_i^2$.

Note as a special case, from Property 4, the density matrix for a pure state corresponds to when one eigenvalue is 1 and the rest are 0. Also observe $\hat{\rho}^2 = \sum_i \alpha_i^2|\alpha_i\rangle\langle\alpha_i|$, thus $\hat{\rho}^2 = \hat{\rho}$, a feature called idempotent, only for the density matrix of a pure state. As such, either condition $\text{Tr}(\hat{\rho}^2) = 1$ or $\hat{\rho}^2 = \hat{\rho}$ implies the other, and either is sufficient to imply that $\hat{\rho}$ represents a pure state.

An important feature of density matrices, useful in quantum information and quantum computation, is called **purification**. This means that for any density matrix one can find a suitable expanded Hilbert space in which it can be represented through a pure state. More precisely, for a given system A, with density matrix $\hat{\rho}_A$, it is possible to introduce another system R, and define a pure state for the joint system $|AR\rangle$, such that $\hat{\rho}_A = \text{Tr}_R[|AR\rangle\langle AR|]$ (here the notation is A and R are representative states in systems A and R, respectively). To prove this, recall that for any density matrix, since it is Hermitian, a suitable unitary transformation exists, which brings it to diagonal form. Alternatively stated, for any density matrix $\hat{\rho}_A$, a suitable basis of orthonormal states exists, where it is diagonal, $\hat{\rho}_A = \sum_i p_i|i_A\rangle\langle i_A|$. The purification process involves introducing a system R, which has the same space of states as A. Let the orthonormal basis states for R be denoted $|i_R\rangle$, which could be the same as the states $|i_A\rangle$ but does not necessarily have to be. One can use any complete set of orthonormal states for the system R. Define the pure state within the AR system $|AR\rangle \equiv \sum_i \sqrt{p_i}|i_A i_R\rangle$. Compute the reduced density matrix for system A by summing over all the R states:

$$\text{Tr}_R[|AR\rangle\langle AR|] = \sum_{ij} \sqrt{p_i p_j}|i_A\rangle\langle j_A|\text{Tr}[|i_R\rangle\langle j_R|]$$

$$= \sum_{ij} \sqrt{p_i p_j}|i_A\rangle\langle j_A|\delta_{ij}$$

$$= \sum_i p_i|i_A\rangle\langle i_A|$$

$$= \hat{\rho}_A. \tag{15.37}$$

The state $|AR\rangle$ is referred to as a purification of $\hat{\rho}_A$.

Properties of the von Neumann Entropy

The **von Neumann entropy** is defined as

$$S_{vN}(\hat{\rho}) = -\text{Tr}[\hat{\rho}\log_2\hat{\rho}]. \tag{15.38}$$

It is considered a suitable analogue in the quantum case to the Shannon entropy, Eq. (15.32), of the classical case, as a measure of the uncertainty. If one demands that $S_{vN}(\hat{\rho})$ is basis independent and coincides with the Shannon entropy for some choice of basis in which $\hat{\rho}$ is diagonal, then the von Neumann entropy, Eq. (15.38), is the only possibility. The von Neumann entropy contains several useful features for characterising information in a quantum system. To start with, we can examine the case of a mixed state where each state $|\phi_i\rangle$ appears with probability p_i and where all these states are mutually orthogonal. In this case the density matrix is diagonal in the $|\phi_i\rangle$ basis and the von Neumann entropy, Eq. (15.38), becomes exactly the same as the Shannon entropy, $S_{vN} = \sum_i p_i \log_2 p_i$. Another important case to consider is a pure state, where $\hat{\rho} = |\phi\rangle\langle\phi|$, for a quantum state ϕ, for which $S_{vN}(\hat{\rho}) = 0$. This is the quantum analogue to the classical case where there is only a single piece of information, so it is known with probability one, and thus has zero Shannon entropy, since a user already knows the information before it is sent.

Example 15.1 The von Neumann entropy differs from the Shannon entropy when there is a mixture of quantum states that are not all orthogonal. Suppose the ensemble consists of two spin states, one spin-up in the z-direction $|\uparrow\rangle$ and one spin-down in the y-direction $|y_\downarrow\rangle \equiv (|\downarrow\rangle + i|\uparrow\rangle)/\sqrt{2}$, and suppose both states in the mixture occur with probability 0.5. Then the density matrix for this mixed state is

$$\hat{\rho} = 0.5|\uparrow\rangle\langle\uparrow| + 0.5\left[\frac{1}{\sqrt{2}}(i|\uparrow\rangle + |\downarrow\rangle)\right]\left[\frac{1}{\sqrt{2}}(-i\langle\uparrow| + \langle\downarrow|)\right]$$

$$= \begin{pmatrix} 0.75 & i0.25 \\ -i0.25 & 0.25 \end{pmatrix}. \tag{15.39}$$

Solving for the eigenvalues of this matrix gives $\alpha = 0.146$ and 0.854, so the von Neumann entropy is

$$S_{\text{vN}} = -0.146 \log_2(0.146) - 0.854 \log_2(0.854) = 0.600 \text{ bits.}$$

Classically we would say that each of the two states in the mixture occur with probability 0.5, so that the Shannon entropy would be $S_{\text{sh}} = -0.5 \log_2 0.5 - 0.5 \log_2 0.5 = 1$ bit. Thus the von Neumann entropy is less than the Shannon entropy. In general, $S_{\text{vN}} \leq S_{\text{sh}}$, with equality only when the states in the mixed state density matrix are all orthogonal. Strictly speaking, the probabilities occurring in $\hat{\rho}$ should only be given a physical interpretation if the pure states appearing are all mutually orthogonal.

The von Neumann entropy has the following mathematical properties:

1. **Invariance under unitary transformation of density matrix.** Let $\hat{U}$ be a unitary operator. Then

$$\begin{aligned}
S_{\text{vN}}(\hat{U}\hat{\rho}\hat{U}^\dagger) &= -\text{Tr}[\hat{U}\hat{\rho}\hat{U}^\dagger \log_2 \hat{U}\hat{\rho}\hat{U}^\dagger] \\
&= -\text{Tr}[\hat{U}\hat{\rho} \log_2(\hat{\rho})\hat{U}^\dagger] \\
&= -\text{Tr}[\hat{\rho} \log_2 \hat{\rho}] = S_{\text{vN}}(\hat{\rho}).
\end{aligned} \tag{15.40}$$

2. **Positivity,** $S_{\text{vN}}(\hat{\rho}) \geq 0$. Since $\hat{\rho}$ is Hermitian, there is a unitary transformation that can bring it to diagonal form. Since all eigenvalues, α_j, of $\hat{\rho}$ are non-negative, it means each term in the trace $-\alpha_j \log_2 \alpha_j$ is non-negative and thus so is the sum. $S_{\text{vN}}(\hat{\rho})$ is zero only if one eigenvalue is 1 and the rest 0, and thus for a density matrix $\hat{\rho}$ of a pure state.

3. **Maximum** $S_{\text{vN}} \leq S_{\text{vN}}^{max} = \log_2 D$, **where** D **is the number of non-zero eigenvalues of the density matrix.** In the basis where $\hat{\rho}$ is diagonal, $S_{\text{vN}}(\hat{\rho}) = -\sum_i \alpha_i \log_2 \alpha_i$. By the property of the density matrix, we know $\sum_i \alpha_i = 1$. Thus by the properties of the $\log_2$ function, $S_{\text{vN}}(\hat{\rho})$ is maximum when all eigenvalues are the same. If D is the dimensionality of $\hat{\rho}$, then each eigenvalue must be $\alpha_i = 1/D$, and so $S_{\text{vN}}(\hat{\rho}) = -\sum_i^D \alpha_i \log_2 \alpha_i = \log_2 D$.

4. **Subadditivity** $S_{\text{vN}}(\hat{\rho}_{\text{AB}}) \leq S_{\text{vN}}(\hat{\rho}_{\text{A}}) + S_{\text{vN}}(\hat{\rho}_{\text{B}})$. The von Neumann entropy of the combined system of A and B is always less than or equal to the sum of the von Neumann entropies of the two individual systems A and B. Examine the simplest case, where the combined Hilbert space is simply the tensor product of the individual Hilbert spaces, $\mathcal{H}_{\text{AB}} = \mathcal{H}_{\text{A}} \otimes \mathcal{H}_{\text{B}}$, which means $\hat{\rho}_{\text{AB}} = \hat{\rho}_{\text{A}} \otimes \hat{\rho}_{\text{B}}$. If $\hat{U}_{\text{I}}$ is the unitary transformation that diagonalises $\hat{\rho}_{\text{I}}$, then $\hat{U}_{\text{A}} \otimes \hat{U}_{\text{B}}$ diagonalises $\hat{\rho}_{\text{A}} \otimes \hat{\rho}_{\text{B}}$. Going to this diagonal basis:

$$\begin{aligned}
S_{\text{vN}}(\hat{\rho}_{\text{AB}}) &= -\sum_{ij} p_i^{\text{A}} p_j^{\text{B}} \log_2(p_i^{\text{A}} p_j^{\text{B}}) \\
&= -\sum_{ij} p_i^{\text{A}} p_j^{\text{B}}[\log_2 p_i^{\text{A}} + \log_2 p_j^{\text{B}}] \\
&= -\sum_i p_i^{\text{A}} \log_2 p_i^{\text{A}} - \sum_j p_j^{\text{B}} \log_2 p_j^{\text{B}} \\
&= S_{\text{vN}}(\hat{\rho}_{\text{A}}) + S_{\text{vN}}(\hat{\rho}_{\text{B}}).
\end{aligned}$$

If the two Hilbert spaces are more correlated, then there will be fewer states available to the system. For example, if two particles are free they have more states accessible to them than if they are bound by some interaction between them. Thus when the Hilbert space $\mathcal{H}_{AB} \in \mathcal{H}_A \otimes \mathcal{H}_B$, in general $S_{vN}(\hat{\rho}_{AB}) \leq S_{vN}(\hat{\rho}_A) + S_{vN}(\hat{\rho}_B)$.

Using Klein's inequality, which is stated and proved in Appendix 15.A at the end of this chapter, a general proof follows. Let $\hat{\rho} = \hat{\rho}_{AB}$ and $\hat{\sigma} \equiv \hat{\rho}_A \otimes \hat{\rho}_B$. Then Klein's inequality states

$$-\mathrm{Tr}(\hat{\rho} \log_2 \sigma) = -\mathrm{Tr}[\hat{\rho}_{AB}(\log_2 \hat{\rho}_A + \log_2 \hat{\rho}_B)]$$
$$= -\mathrm{Tr}[\hat{\rho}_A \log_2 \hat{\rho}_A] - \mathrm{Tr}[\hat{\rho}_B \log_2 \hat{\rho}_B]$$
$$= S_{vN}(\hat{\rho}_A) + S_{vN}(\hat{\rho}_B). \tag{15.41}$$

From Klein's inequality it follows that $S_{vN}(\hat{\rho}_{AB}) \leq S_{vN}(\hat{\rho}_A) + S_{vN}(\hat{\rho}_B)$, where equality holds when $\hat{\rho}_{AB} = \hat{\rho}_A \otimes \hat{\rho}_B$.

5. **Concavity.** For $p_1, p_2, \ldots, p_n \geq 0$ such that $p_1 + p_2 + \cdots + p_n = 1$, $S_{vN}(p_1\hat{\rho}_1 + p_2\hat{\rho}_2 + \cdots + p_n\hat{\rho}_2) \geq p_1 S_{vN}(\hat{\rho}_1) + p_2 S_{vn}(\hat{\rho}_2) + \cdots + p_n S_{vN}(\hat{\rho}_n)$. The sum of density matrices $p_1\hat{\rho}_1 + p_2\hat{\rho}_2 + \cdots + p_n\hat{\rho}_n$ occurs when one has an ensemble of different density matrices mixed together. The von Neumann entropy is larger for this case, since one is more ignorant about how the state was prepared. To mathematically prove this property, let $\hat{\rho}_A \equiv \sum_j p_j\hat{\rho}_j$ and $\hat{\rho}_B \equiv \sum_j p_j|j\rangle\langle j|$ and the joint state be $\hat{\rho}_{AB} \equiv \sum_j p_j\hat{\rho}_j \otimes |j\rangle\langle j|$ for any orthonormal basis $\{|j\rangle\}$. Observe that $S_{vN}(\hat{\rho}_{AB}) = S_{sh}(\{p_i\}) + \sum_j p_j S_{vN}(\hat{\rho}_j)$. By applying subadditivity the concavity property is proved.

6. **Triangle inequality** $S_{vN}(\hat{\rho}_{AB}) \geq |S_{vN}(\hat{\rho}_A) - S_{vN}(\hat{\rho}_B)|$. When $\mathcal{H}_{AB} = \mathcal{H}_A \otimes \mathcal{H}_B$, from the discussion on subadditivity this inequality holds. To prove it in general, let R be a system that purifies systems A and B. Purification of, for example, system A means creating a pure state $|AR\rangle$ such that tracing over the states in R gives back $\hat{\rho}_A$, $\hat{\rho}_A = \mathrm{Tr}_R[|AR\rangle\langle AR|]$. As discussed earlier, purification can be done for any systems. Thus, since the system ABR by construction is a pure state, $S_{vN}(\hat{\rho}_{AR}) = S_{vN}(\hat{\rho}_B)$ and $S_{vN}(\hat{\rho}_R) = S_{vN}(\hat{\rho}_{AB})$. Subadditivity applied to systems A and R means $S_{vN}(\hat{\rho}_R) + S_{vN}(\hat{\rho}_A) \geq S_{vN}(\hat{\rho}_{AR})$. Combining these also means $S_{vN}(\hat{\rho}_{AB}) \geq S_{vN}(\hat{\rho}_B) - S_{vN}(\hat{\rho}_A)$. Noting the symmetry between the systems A and B, this also means $S_{vN}(\hat{\rho}_{AB}) \geq S_{vN}(\hat{\rho}_A) - S_{vN}(\hat{\rho}_B)$. Combining these gives the triangle inequality.

7. **Strong subadditivity** $S_{vN}(\hat{\rho}_{ABC}) + S_{vN}(\hat{\rho}_B) \leq S_{vN}(\hat{\rho}_{AB}) + S_{vN}(\hat{\rho}_{BC})$. The proof of this is involved and the reader is referred to Nielsen and Chuang in the References section at the end of this chapter.

Von Neumann Entropy and Information

In the quantum-mechanical case, information is encoded in the form of quantum states. It is an entirely different question how exactly one extracts useful information from a quantum state, and examples such as teleportation, superdense coding, and quantum computing, which we will discuss, are all procedures formulated to do just that. However, it is useful to quantify in abstract terms the information content of quantum-mechanical states and that is what quantum information theory does. As opposed to letters or symbols in the classical case, in the quantum-mechanical case quantum states take their place. Thus, suppose you have a collection of quantum-mechanical states at your disposal, $|i\rangle$, which need not be mutually orthogonal. For example, these could be spin states along a range of orientations. Suppose each of these states exists in the ensemble with probability p_i. The density matrix describing this mixed state is then

$$\hat{\rho} = \sum_i p_i |i\rangle\langle i|. \tag{15.42}$$

The collection of all possible messages of n states that can be sent would then be contained in the tensor product density matrix

$$\hat{\rho}^{\otimes n} = \hat{\rho} \otimes \hat{\rho} \otimes \cdots \otimes \hat{\rho}. \tag{15.43}$$

Naively, if we want to send n quantum states in the message, it requires a Hilbert space of dimension n. However, the quantum-mechanical analogue of the Shannon noiseless source coding theorem, called the Schumacher coding theorem,[3] tells us that the smallest dimension of the Hilbert space, $\mathcal{H}$, must be

$$\log_2(\dim \mathcal{H}) = nS_{\mathrm{vN}}(\hat{\rho}). \tag{15.44}$$

For example, suppose there were two orthogonal quantum states $|\uparrow\rangle$ and $|\downarrow\rangle$ with the density matrix

$$\hat{\rho} = 0.5 |\uparrow\rangle\langle\uparrow| + 0.5 |\downarrow\rangle\langle\downarrow|,$$

so in other words these are two mixed states which arise with probability $\frac{1}{2}$ each. The von Neumann entropy for this density matrix is $S_{\mathrm{vN}}(\hat{\rho}) = -0.5 \log_2 0.5 - 0.5 \log_2 0.5 = 1$. Thus each letter in this message requires one state in Hilbert space. This example is basically no different from the case of classical information theory. If one were to code this message with qubits, then one could just assign one state in Hilbert space for each of the two states, thus the $|\uparrow\rangle$ and $|\downarrow\rangle$ states of a qubit. Similarly one could look at cases where many orthogonal spin states were involved, such as $|\uparrow\uparrow\downarrow\downarrow\rangle, |\uparrow\downarrow\uparrow\downarrow\rangle$ etc. with some associated probability for each state. The Schumacher coding theorem would give no different result than the corresponding Shannon coding theorem, and one could go forth and devise compression schemes.

A difference arises when the set of mixed states are not all orthogonal. In that case, as shown in the example in Eq. (15.39), the von Neumann entropy for the system is less than the naive Shannon entropy. According to the Schumacher coding theorem, it implies that one can compress the quantum data even more than would be possible for classical data. The basic reason why this is possible is because if states are not orthogonal, then it means a given state can have partial projection on another state in the ensemble. As such, there is some overlap in information contained amongst these states. This sort of overlap has no classical counterpart. A thorough implementation of quantum compression is very detailed and will not be discussed here. Further details can be found in Preskill's lectures

[3] B. Schumacher, 'Quantum coding', *Phys. Rev. A* **51** (1995) 2738–2747, https://doi.org/10.1103/PhysRevA.51.2738.

and Nielsen and Chuang's book, with references to both at the end of this chapter. Our purpose in this section is only to illustrate the information content in quantum states and how to quantify that.

A few salient features about the compression of quantum information are worth noting here, since they highlight underlying quantum properties. For this let us return to the example in Eq. (15.39). Suppose one was communicating some quantum states and one of those states that was meant to be sent was $|\uparrow\rangle$, but instead the state $|y_\downarrow\rangle = (|\downarrow\rangle + i|\uparrow\rangle)/\sqrt{2}$ was sent in its place. One would be inclined to think that the state that was sent still contained some part of the actual information one wanted to send, since the state one wanted to send still appeared with probability $\frac{1}{2}$ in the state that was sent. The notion of fidelity between two quantum systems quantifies this idea. This is a measure of how close two quantum systems are. If the two systems under question are simply two pure states, $|\phi\rangle$ and $|\psi\rangle$, then the fidelity is just the norm of the overlap of the two states, $F(|\phi\rangle, |\psi\rangle) = |\langle\psi|\phi\rangle|$. More generally, the **fidelity** between a density matrix $\hat{\rho}$ and a pure state ϕ is defined as

$$F(|\phi\rangle, \hat{\rho}) \equiv \sqrt{\langle\phi|\hat{\rho}|\phi\rangle}. \tag{15.45}$$

Note that if $\hat{\rho}$ represents a pure state $|\psi\rangle\langle\psi|$, F reduces to the overlap between the two pure states. In quantum information theory one often uses fidelity to assess how close the information at hand is to the information one intended to utilise.

Example 15.2 Developing further the example of the density matrix in Eq. (15.39), its eigenstates are

$$|\uparrow'\rangle = \cos\frac{\pi}{8}|\uparrow\rangle - i\sin\frac{\pi}{8}|\downarrow\rangle,$$

$$|\downarrow'\rangle = \sin\frac{\pi}{8}|\uparrow\rangle + i\cos\frac{\pi}{8}|\downarrow\rangle,$$

with corresponding eigenvalues $\cos^2(\pi/8)$ for $|\uparrow'\rangle$ and $\sin^2(\pi/8)$ for $|\downarrow'\rangle$. The fidelity between the primed states and the original states $|\uparrow\rangle$ and $|y_\downarrow\rangle$ is $F(|\uparrow\rangle, |\uparrow'\rangle) = F(|y_\downarrow\rangle, |\uparrow'\rangle) = \cos(\pi/8) = 0.924$ and $F(|\uparrow\rangle, |\downarrow'\rangle) = F(|y_\downarrow\rangle, |\downarrow'\rangle) = \sin(\pi/8) = 0.383$. What this tells us is that even if we did not know which of the two unprimed states was sent, if one simply guessed the state to be $|\uparrow'\rangle$ then it would have considerable overlap with the actual state, whether $|\uparrow\rangle$ or $|y_\downarrow\rangle$, that was actually sent. Depending on the accuracy of the task one needed this information for, the state $|\uparrow'\rangle$ might still be a reasonable representation for the actual state one had in mind. Classically the situation would not be quite as good. Suppose two types of letters appearing with equal frequency were being communicated via a classical channel. If one of these letters were omitted, then simply by guessing, the receiver at the other end would only be right half the time, whereas communication in the quantum channel gives around a 92% chance of being correct. This is an underlying feature of the Schumacher coding theorem. Many different communication schemes have been devised to utilise this property to achieve accuracy in quantum communication.

15.3.4 Entanglement Entropy

One of the most useful properties of quantum mechanics that can be applied to quantum information is entanglement. The von Neumann entropy can be utilised in quantifying the information content of an entangled system. Entanglement of two systems, also called a bipartite system, means the state vector describing the state of the combined two systems cannot be written as a tensor product between the associated state vectors of the two separate systems. For example, in the case of a two-qubit system, it is said to be entangled if the state vector cannot be written as the tensor product of two one-qubit systems. In particular, the two-qubit system in Eq. (15.29) is entangled if no set of complex numbers $a_\downarrow, a_\uparrow, b_\downarrow$ and $b_\uparrow$ of the tensor product state

$$(a_\downarrow | \downarrow \rangle + a_\uparrow | \uparrow \rangle) \otimes (b_\downarrow | \downarrow \rangle + b_\uparrow | \uparrow \rangle) \tag{15.46}$$

produce that state.

For an entangled system, it is useful to have a quantitative measure of the degree of entanglement. The entanglement entropy is a mapping of the states of a quantum system into a single number. It is defined as follows. Suppose the total system consists of two subsystems A and B. The total Hilbert space of the combined system is the tensor product of the Hilbert spaces of the two subspaces $\mathcal{H} = \mathcal{H}_A \otimes \mathcal{H}_B$. Suppose $|\Psi\rangle$ is a pure state of the combined system. The question we wish to ask is whether this is an entangled state between subsystems A and B, and if so, to what degree is it entangled. Thus we wish to know to what degree subsystem A in this pure state is entangled, or likewise subsystem B. A commonly adopted measure of the entanglement entropy is given by the von Neumann entropy. Thus the **entanglement entropy of subsystem A** is given by

$$S_A = -\text{Tr}[\hat{\rho}_A \log_2 \hat{\rho}_A], \tag{15.47}$$

where $\hat{\rho}_A$ is the reduced density matrix of subsystem A, obtained by tracing out the states of system B as

$$\hat{\rho}_A = \text{Tr}_B[\hat{\rho}_{AB}]. \tag{15.48}$$

An explicit example of such a partial trace is given below. Here, $\hat{\rho}_{AB}$ is the density matrix for the combined state, which for a pure state $|\Psi\rangle$ is simply

$$\hat{\rho}_{AB} = |\Psi\rangle\langle\Psi|. \tag{15.49}$$

We have already said above that if the state $|\Psi\rangle$ is simply a tensor product of some state $|\Psi_A\rangle$ in $\mathcal{H}_A$ with some state $|\Psi_B\rangle$ in $\mathcal{H}_B$, then it is by definition not entangled. This can be checked from Eq. (15.47), using Eq. (15.48), $\hat{\rho}_A = |\Psi_A\rangle\langle\Psi_A|$. In general, a basis can be found where $|\Psi_A\rangle$ is an element of it. In this basis, thinking about $\hat{\rho}_A$ as a matrix, it will have entry 1 on the diagonal for $|\Psi_A\rangle$ and 0 everywhere else. Thus, from Eq. (15.47) it follows that $S = 0$, which means no entanglement.

Example 15.3 Consider a two-spin system, as discussed in Sections 15.1 and 15.2. Using the notation $| \uparrow \rangle$ and $| \downarrow \rangle$ for the two spin states (rather than the $|0\rangle$ and $|1\rangle$ notation), suppose the state of the system is

$$|\Psi\rangle = \frac{1}{\sqrt{2}} (|\uparrow\downarrow\rangle \pm |\downarrow\uparrow\rangle),$$

so $\hat{\rho}_{AB} = |\Psi\rangle\langle\Psi|$. To see this clearly, writing $\hat{\rho}_{AB}$ explicitly as the 4×4 matrix that it is, with entries along the row and columns being the basis states $|\uparrow\uparrow\rangle, |\uparrow\downarrow\rangle, |\downarrow\uparrow\rangle$ and $|\downarrow\downarrow\rangle$, respectively, gives

$$\hat{\rho}_{AB} = \begin{pmatrix} 0 & 0 & 0 & 0 \\ 0 & \frac{1}{2} & \pm\frac{1}{2} & 0 \\ 0 & \pm\frac{1}{2} & \frac{1}{2} & 0 \\ 0 & 0 & 0 & 0 \end{pmatrix}.$$

Then one can do the partial trace for example over the states of B as follows, to give the reduced density matrix for system A:

$$\hat{\rho}_A = \mathrm{Tr}_B(\hat{\rho}_{AB}) = \langle\uparrow_B |\hat{\rho}_{AB}| \uparrow_B\rangle + \langle\downarrow_B |\hat{\rho}_{AB}| \downarrow_B\rangle$$
$$= \tfrac{1}{2} (|\downarrow_A\rangle\langle\downarrow_A | + |\uparrow_A\rangle\langle\uparrow_A |).$$

In matrix form with row and columns ordered in the basis states $|\uparrow_A\rangle$ and $|\downarrow_A\rangle$, respectively:

$$\hat{\rho}_A = \begin{pmatrix} \frac{1}{2} & 0 \\ 0 & \frac{1}{2} \end{pmatrix}.$$

Thus the entanglement entropy of A is

$$S_A = -\mathrm{Tr}(\hat{\rho}_A \log_2 \hat{\rho}_A) = -\tfrac{1}{2} \log_2 \tfrac{1}{2} - \tfrac{1}{2} \log_2 \tfrac{1}{2} = \log_2 2 = 1.$$

Exercise 15.3.1 Practise the partial trace, by now doing this instead over the A states from $\hat{\rho}_{AB}$, and determine the reduced density matrix $\hat{\rho}_B$. Then compute the entanglement entropy of B, which will be the same as above.

15.3.5 Other Measures of Information

Since the early days of quantum mechanics up to almost the present, the role of measurement has been treated to a minimalist level, to the extent of making the subject functional. This has meant mainly accepting that an observer affects the measurement, the resulting collapse of the state vector/wave function idea and associated uncertainty relations amongst incompatible observables. It could be said that the subject of quantum mechanics has tried its best still to be viewed through classical eyes, and so therefore central to its description is identifying in what ways it is not like classical mechanics. What work has been done in further understanding the measurement process of quantum mechanics has sat at the fringes of the subject, and often even been considered philosophical. More recently there has been a growing shift in this way of thinking. A recognition has been taking hold that measurement can play a much more important role with utilitarian ends. This is an interesting topic, but beyond the scope of this book. For us it has to be sufficient that this point has been recognised here, and in this section a few more concepts related to measurement will be discussed in the context of correlations between two systems.

This will show some more ways in which measurement can be utilised to extract information and separate out what can be considered quantum behaviour.

Suppose there are two classical information systems A and B, each with their respective probability distribution for the symbols in that system. As discussed earlier, the respective Shannon entropy for each system

$$S_{sh}(J) = -\sum_{i_J} p_{i_J} \log_2 p_{i_J}, \tag{15.50}$$

for $J = A, B$, gives the average of the information in that system. One can also define the joint Shannon entropy of both systems:

$$S_{sh}(A, B) \equiv -\sum_i p_i^{AB} \log_2 p_i^{AB}, \tag{15.51}$$

where p_i^{AB} is the probability of finding a particular state i of the combined A and B systems. If A and B were two completely independent systems, thus having no correlation between them, then the states i would simply index all possible product states $i_A j_B$ of the two subsystems A and B with respective probability $p_{i_A}^A p_{i_B}^B$. However, if there were interactions between the two subsystems, in general the probabilities would not be of this simple product form. The **mutual information**

$$I_{sh}(A, B) = S_{sh}(A) + S_{sh}(B) - S_{sh}(A, B) \tag{15.52}$$

is a measure of the information content of both independent systems minus the information content of the combined system. When the two systems are completely independent, so that combined system probabilities are simply the product of all probabilities of the two individual systems, then $I_{sh}(A, B) = 0$.

Another useful measure of information when there are two systems is the conditional entropy of one system, when all states of the other system are summed over with their weighted probabilities. Thus the **conditional entropy** of A when B is known is

$$S_{sh}(A|B) \equiv \sum_{i_B} p_{i_B}^B S_{sh}(A|B = i_B), \tag{15.53}$$

where

$$S_{sh}(A|B = i_B) = -\sum_{i_A} p(A_{i_A}|B_{i_B}) \log_2 p(A_{i_A}|B_{i_B}). \tag{15.54}$$

The conditional entropy of A given above is the information content of A when B is measured in all its possible states. In classical probability theory $S_{sh}(A|B) = S_{sh}(A, B) - S_{sh}(B)$. As such, an equivalent expression of **mutual information** in classical probability theory is

$$J_{sh}(A : B) = S_{sh}(A) - S_{sh}(A|B). \tag{15.55}$$

For quantum information theory, the probability distributions for the systems are replaced by density matrices and the Shannon entropies are replaced by the von Neumann entropies. Thus the joint entropy would be $S_{vN}(\hat{\rho}_{AB})$, where $\hat{\rho}_{AB}$ is the density matrix of the combined A and B systems, and the **quantum mutual information** analogous to the classical case Eq. (15.52) is then

$$I_{vN}(A, B) = S_{vN}(\hat{\rho}_A) + S_{vN}(\hat{\rho}_B) - S_{vN}(\hat{\rho}_{AB}). \tag{15.56}$$

When $I_{vN}(A, B) \neq 0$, it is said that the systems A and B are correlated.

As conditional entropy involves measurement in the quantum-mechanical context, it brings in all the consequences associated with the collapse of the state vector. The evaluation of the conditional entropy in the quantum-mechanical case first requires a well-defined definition of measurement. The one most commonly used is called von Neumann measurement, or equivalently local projective measurement. For a system with density matrix $\hat{\rho}_{AB}$, suppose we make the measurement of B and it collapses into the state $|j_B\rangle$. The probability of obtaining this outcome j_B will be $p_{j_B} = \text{Tr}[\hat{\rho}_{AB|j_B}]$, where

$$\hat{\rho}_{AB|j_B} = \frac{1}{p_{j_B}}(I_A \otimes \hat{\Pi}_{j_B})\hat{\rho}_{AB}(I_A \otimes \hat{\Pi}_{j_B}). \tag{15.57}$$

Here $\{\hat{\Pi}_{j_B} \equiv |j_B\rangle\langle j_B|\}$ is part of a complete set of orthogonal projectors $\{\hat{\Pi}_{j_B}\}$ with properties $\sum_j \Pi_j = 1$, $\Pi_i\Pi_i = \Pi_i$ and $\Pi_i\Pi_j = 0$ for $i \neq j$. Then the conditional entropy in quantum mechanics for A when B is measured in all possible states is

$$S_{vN}(\hat{\rho}_{AB}|\{\hat{\Pi}_{j_B}\}) = \sum_{j_B} p_{j_B} S_{vN}(\hat{\rho}_{A|j_B}), \tag{15.58}$$

where

$$\hat{\rho}_{A|j_B} = \text{Tr}[\hat{\rho}_{AB|j_B}]. \tag{15.59}$$

Note that $S_{vN}(\hat{\rho}_{AB}|\{\hat{\Pi}_{j_B}\})$ will be dependent on the particular choice of projectors $\{\hat{\Pi}_{j_B}\}$. In other words, what is measured in system B will affect the conditional entropy for A. With these quantities now defined, the other form of mutual information Eq. (15.55) for the quantum-mechanical case can now be defined as

$$J_{vN}(\hat{\rho}_{AB}) = S_{vN}(\hat{\rho}_A) - S_{vN}(\hat{\rho}_{AB}|\{\hat{\Pi}_{j_B}\}). \tag{15.60}$$

If one varied this set of projectors until one obtained the maximum value for the conditional entropy Eq. (15.58), then the **quantum discord**[4] is defined as

$$D_{AB}(\hat{\rho}_{AB}) = I_{vN}(\hat{\rho}_{AB}) - \max_{\{\hat{\Pi}_{j_B}\}}[J_{vN}(\hat{\rho}_{AB})]. \tag{15.61}$$

In the classical case, both forms of mutual information are the same and therefore the quantum discord is zero. In quantum mechanics this is no longer the case, and the difference is called the discord. It is another measure of the quantumness of the correlation in a system. An alternative way to understand Eq. (15.61) is that the mutual information $I(A, B)$ accounts for both the quantum and classical corrections, whereas the second term is regarded as the classical correlations obtained from local measurements on system B. Thus the quantum discord is the difference between the total and the classical correlations.

---

[4] H. Ollivier and W. H. Zurek, 'Quantum discord: A measure of the quantumness of correlations', *Phys. Rev. Lett.* **88** (2001) 017901, https://doi.org/10.1103/PhysRevLett.88.017901; L. Henderson and V. Vedral, 'Classical, quantum and total correlations', *J. Phys. A* **34** (2001) 6899–6905, http://dx.doi.org/10.1088/0305-4470/34/35/315.

15.4   Quantum Communication

The entanglement of a quantum state can be utilised in various aspects of treating information. One key process associated with entanglement is quantum teleportation, which allows for the information about a quantum state to be transferred from one location to another. In this process, it should be carefully noted that matter is not transferred to a different location, only information. The central observation in this process is that if an identical particle is in a particular quantum state and that information is transferred to another location where a second identical particle of the same type is present, then this second particle can be placed in exactly the same quantum state as the first. In doing so, it is as if the material system in one location has been moved to another location. In this respect, quantum teleportation does not physically move particles, but rather it is a process of communication. In particular, as will be shown for quantum teleportation, to transport the information contained in a qubit from one location to another will require the combination of both a shared entanglement and a classical communication channel. As classical communication is required, it implies that information cannot be transferred any faster than the speed of light.

15.4.1   No-Cloning Theorem

This theorem states that it is impossible to create an identical copy of an arbitrary unknown quantum state.[5]

Let $|\chi\rangle$ be an unknown quantum state in Hilbert space $\mathcal{H}$, which we wish to make a duplicate copy of. The duplicate copy will sit in a tensor product Hilbert space $\mathcal{H} \otimes \mathcal{H}$. We will consider starting with some arbitrary, normalised, initial state $|j\rangle$. Thus initially we have in the Hilbert space $\mathcal{H} \otimes \mathcal{H}$ the tensor product state

$$|j\rangle \otimes |\chi\rangle = |j\rangle|\chi\rangle = |j\chi\rangle, \tag{15.62}$$

where the latter two expressions are shorthand notation that we will sometimes use. The question being asked is, can some process turn the state $|j\rangle$ into a duplicate state $|\chi\rangle$? There are two possible quantum operations which we can use on this composite state. First, we can perform an observation of this state. However, that will collapse the state, thus corrupting the information initially contained in it, so in general altering the state $|\chi\rangle$ we wanted to duplicate. So this option is not useful for our purposes here. Second, we can control the Hamiltonian for this system and evolve it with the unitary time evolution operator $\hat{U}(t)$. The statement of the no-cloning theorem is that there exists no unitary operator on $\mathcal{H} \otimes \mathcal{H}$ that can turn the state $|j\rangle$ into a given state $|\chi\rangle$ in $\mathcal{H}$.

---

[5] W. Wootters and W. Zurek, 'A single quantum cannot be cloned', *Nature* **299** (1982) 802–803, https://doi.org/10.1038/299802a0.

Proof: Suppose there exists a unitary operator $\hat{U}$ on $\mathcal{H} \otimes \mathcal{H}$ such that

$$\hat{U}|j\rangle|\chi\rangle = \exp(-i\Theta(j, \chi))|\chi\rangle|\chi\rangle,$$
$$\hat{U}|j\rangle|\phi\rangle = \exp(-i\Theta(j, \phi))|\phi\rangle|\phi\rangle, \tag{15.63}$$

where χ and ϕ are two arbitrary states in $\mathcal{H}$. Here $\Theta(j, \chi)$ is an angle which in general can be a function of the states on which $\hat{U}$ is acting. Then we have

$$\langle\chi|\phi\rangle = \langle\chi|\langle j|\hat{U}^{\dagger}\hat{U}|j\rangle|\phi\rangle = \exp[i(-\Theta(j, \phi) + \Theta(j, \chi))]\langle\chi|\langle\chi||\phi\rangle|\phi\rangle$$
$$= \exp[i(-\Theta(j, \phi) + \Theta(j, \chi))]\langle\chi|\phi\rangle^2. \tag{15.64}$$

Note in the first line above we simply inserted a factor of unity, $\hat{U}^{\dagger}\hat{U} = 1$. This equation leads to the condition $|\langle\chi|\phi\rangle| = |\langle\chi|\phi\rangle|^2$, which can only hold if $|\langle\chi|\phi\rangle|$ equals 0 or 1. In other words, either they are two orthogonal states in $\mathcal{H}$ or the same state. However, $|\chi\rangle$ and $|\phi\rangle$ are meant to be arbitrary states in $\mathcal{H}$. Thus no such unitary operator $\hat{U}$ can exist.

The next section will describe the basic process of quantum teleportation. However, it is important to also understand why teleportation is the only method by which information about an unknown quantum state can be transferred. One simple alternative could be to simply take the particle in the first location and physically move it to the second location. This could be cumbersome, and the process of moving the particle could harm the quantum state it is in. Particularly if the two locations are very far apart, this could be difficult if not impossible to do. Thus if one is stuck with having to transmit information, an alternative possibility is why not simply transmit via classical communication all the information about the particle at location A to location B? Then at location B this information could be fed into an identical particle and thus place it in the same state as the particle in location A. Underlying the inherent inability to do any options like this is the no-cloning theorem. The **no classical teleportation theorem** is a slightly stronger result than the no-cloning theorem, and states that an arbitrary quantum state cannot be converted into a sequence of classical bits. If this were possible, these classical bits could then be transmitted through a classical channel. Were this possible, it would provide a means of cloning the initial state.

Note that the no-cloning theorem is consistent with the quantum-mechanical uncertainty principle. If an unknown state could be cloned, then as many copies of that state could be made. With these copies one could measure each dynamical variable to arbitrary precision. Thus one could gain full information about the initial unknown state, which would then be in contradiction to the uncertainty principle.

15.4.2 Quantum Teleportation of a Qubit

Suppose the observer in location A, who for this problem is usually called Alice, has a quantum system, such as a spin-$\frac{1}{2}$ particle, photon or some such identical particle, in a quantum state $|\chi\rangle$. Although she is in possession of this system, the precise state of this system need not be known to her. Suppose at location B another observer, usually called Bob, is in possession of one of these identical particles. Alice wants to send the complete information about the quantum state $|\chi\rangle$ over to Bob and input it into the identical particle he has. The end result being that Bob's particle is now in exactly the same state as

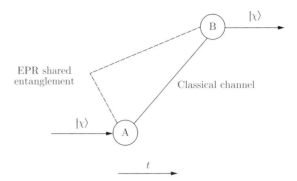

Fig. 15.9 Teleportation of state $|\chi\rangle$ from A to B requiring both a classical and a quantum channel.

the particle Alice was in possession of. Thus even though no matter is transferred over, only information, the end result is as if Bob has received the exact system Alice had in her possession. This process is called quantum teleportation[6] (secretly we know the participants are Kirk and Scotty of course). Thus, in the quantum teleportation process only information is transferred from one location to another, but no matter is being transferred.

To explain the quantum teleportation process, we will focus on the simplest case where the state $|\chi\rangle$ is from a two-state system, such as a spin-$\frac{1}{2}$ identical particle (or equally the example could apply to the polarisation of a photon). The way quantum teleportation works is first two spin-$\frac{1}{2}$ particles are prepared in the Bell state:

$$|\Psi_{AB}^{-}\rangle = \frac{1}{\sqrt{2}}\left[|\uparrow_A\downarrow_B\rangle - |\downarrow_A\uparrow_B\rangle\right]. \tag{15.65}$$

The subscripts A and B correspond to who received the respective spin-$\frac{1}{2}$ particle, Alice or Bob. This one ebit of entanglement is thus shared between Alice and Bob. Both Alice and Bob had agreed beforehand on the above entangled state. Either Alice or Bob could first prepare this entangled state and send one of the particles to the other observer and keep one for themselves. Or the source of this EPR pair could be someone else, sending to both Alice and Bob. This detail does not matter. All that is important is Alice and Bob already know beforehand that the particle they receive is part of an entangled state and they know exactly what that entangled state is. The sending of this entangled state to Alice and Bob is referred to as establishing a **shared entanglement** or sharing one ebit of entanglement between them, as shown in Fig. 15.9. Alice now has the additional spin-$\frac{1}{2}$ particle in the quantum state, $|\chi_{A'}\rangle$, and she wishes to transfer the information about this state to Bob. She is aware that it is a spin-$\frac{1}{2}$ particle, so very generally she at least knows the state of this particle has the form

$$|\chi_{A'}\rangle = c|\uparrow_{A'}\rangle + d|\downarrow_{A'}\rangle, \tag{15.66}$$

[6] C. H. Bennett *et al.*, 'Teleporting an unknown quantum state via dual classical and Einstein–Podolsky–Rosen channels', *Phys. Rev. Lett.* **70** (1993) 1895–1899, https://doi.org/10.1103/PhysRevLett.70.1895.

for amplitude coefficients c and d which may or may not be known to her, but of course normalised $|c|^2 + |d|^2 = 1$. The subscript A′ means this particle is also in Alice's possession. The complete three-particle state can be expressed in the form

$$|\Phi_{A'AB}\rangle = c|\uparrow_{A'}\rangle|\Psi_{AB}^-\rangle + d|\downarrow_{A'}\rangle|\Psi_{AB}^-\rangle$$

$$= \frac{c}{\sqrt{2}}[|\uparrow_{A'}\rangle|\uparrow_A\rangle|\downarrow_B\rangle - |\uparrow_{A'}\rangle|\downarrow_A\rangle|\uparrow_B\rangle]$$

$$+ \frac{d}{\sqrt{2}}[|\downarrow_{A'}\rangle|\uparrow_A\rangle|\downarrow_B\rangle - |\downarrow_{A'}\rangle|\downarrow_A\rangle|\uparrow_B\rangle]. \tag{15.67}$$

Thus Alice has two spin-$\frac{1}{2}$ particles A and A′ and Bob has one B. We can now express the product of the two spin-$\frac{1}{2}$ particles Alice has in terms of the Bell basis of states Eq. (15.30). This is simply utilising the property in quantum mechanics to describe a given state in a different choice of basis. So for example the state

$$|\uparrow_{A'}\rangle|\uparrow_A\rangle = \frac{1}{\sqrt{2}}\left[|\Phi_{A'A}^+\rangle + |\Phi_{A'A}^-\rangle\right], \tag{15.68}$$

and similarly for the other product states in Eq. (15.67) (see problems at the end of this chapter). Thus in this basis Eq. (15.67) can be written as

$$|\Phi_{A'AB}\rangle = \tfrac{1}{2}[|\Psi_{A'A}^-\rangle(-c|\uparrow_B\rangle - d|\downarrow_B\rangle) + |\Psi_{A'A}^+\rangle(-c|\uparrow_B\rangle + d|\downarrow_B\rangle)$$

$$+ |\Phi_{A'A}^-\rangle(c|\downarrow_B\rangle + d|\uparrow_B\rangle) + |\Phi_{A'A}^+\rangle(c|\downarrow_B\rangle - d|\uparrow_B\rangle)]. \tag{15.69}$$

Both Alice and Bob have adequate information to be able to write the above expansion. Notice that in this change of basis, the states at B now have similar form to the state $|\chi\rangle$. In fact, the first of the above terms has exactly the state $|\chi\rangle$ up to an overall phase of -1. The other three states are related to $|\chi\rangle$ in a simple way. In particular, a unitary transformation of those three states will bring it into the state $|\chi\rangle$. At this point, if Alice now makes an observation of one of the above four Bell states at her side, this will collapse the above state vector. Whichever of the four Bell states she observes, she can communicate that information to Bob via a **classical channel** which requires just two classical bits of information (since she only needs to tell Bob which of four possibilities she has measured). Since Bob was also able to write the expansion of their mutual state in the form of Eq. (15.69), once he receives that information Alice sent from the classical channel, which thus can come no faster than the speed of light, he will know exactly what state his particle is in. Since at the onset he knew the state of $|\chi\rangle$ that Alice has was of the form Eq. (15.66), he will at this point know whether the state he has is the same as that state up to an overall minus sign, or one of the other three states. In all cases he can then perform an appropriate unitary operation on the state to get it exactly in the original state $|\chi\rangle$. Thus at this point the state $|\chi\rangle$ has been teleported from Alice to Bob. Note also that in this process of teleportation of the state $|\chi\rangle$ to Bob, it has been destroyed at Alice's side, thereby not violating the no-cloning theorem.

Exercise 15.4.1   Go through the same quantum teleportation process for teleporting the state $|\chi\rangle$ in Eq. (15.66) but now for the cases where the shared entangled states are the other three Bell states $|\Phi^\pm\rangle$ and $|\Psi^+\rangle$.

A few details should be clarified. It has been mentioned above that the precise state of the system Alice is in possession of, $|\chi_{A'}\rangle$, which she wishes to teleport may or may not be known to her. This feature is important to understand the quantum teleportation process. Suppose the precise state of $|\chi_{A'}\rangle$ is unknown to Alice. If she then made any observation of that state, and thus gained information about the state, by the nature of quantum mechanics, she would have altered the state. Perhaps the state Alice has is a qubit coming out of a quantum computer and she wants to send precisely that state over to Bob, where he can further process those qubits in his quantum computer. Or Alice might have a spin-$\frac{1}{2}$ particle, photon or some atomic system in a certain quantum state, and wants to send exactly that state to Bob. Thus she does not want to observe the state, since that will destroy it, but she simply wants it sent over to Bob.

On the other hand, note that if the system Alice were given was in a known eigenstate of some observable, then there would be no need to teleport that information. All Alice would need to do is tell Bob the eigenstate the system is in and he could create that system himself. For example, if Alice somehow already knew the spin-$\frac{1}{2}$ particle she was given was in a particular eigenstate of spin, then she could tell Bob that information via a classical communication channel. He could then use an appropriate Stern–Gerlach apparatus to create such a state himself. However, it could be that for security reasons Alice does not want to tell this information via a classical communication channel. Then, even for a known state, quantum teleportation may be useful for sending the information on the state over to Bob.

Secure Communication

An application of quantum teleportation that is of particular interest is sending information securely from one location to another. There are various procedures that have been developed that utilise teleportation to achieve this goal of transferring information, and these are called **protocols**. In such protocols, quantum teleportation is utilised to send to Bob a state known to Alice (or vice versa), since it is information she knows that she wishes to send securely.

An example of such a protocol, named quantum secure direct communication,[7] works as follows. Suppose Alice wants to send some binary message ↑↑↓↑ . . . securely to Bob. If she sends this directly via a classical channel, she risks that some third party, Eve, could intercept her message. Alternatively, Alice and Bob initially could store up a bank of entangled states between them, in amount equal to the size of the binary message she wishes to send. With each of these entangled states, Alice can now successively place her state $|\chi\rangle$ into the desired binary state, either ↑ or ↓, and send to Bob. Since Alice and Bob both know in advance what basis the $|\chi\rangle$ state is expanded in, Bob can also measure the teleported state in that same basis, thus not alter the state he received by his process of measurement. Alice still needs to make a Bell measurement of her state, and send to Bob via a classical channel the outcome. This requires transmitting two bits of information.

[7] F. L. Yan and X. Q. Zhang, 'A scheme for secure direct communication using EPR pairs and teleportation', *Eur. Phys. J.* **B41** (2004) 75–78. https://doi.org/10.1140/epjb/e2004-00296-4.

However, without having possession of the shared entanglement, which only Alice and Bob have, any third party listening to the classical channel for this transmitted information will have no idea how that correlates with the information on the binary state Alice is sending to Bob via the shared entanglement. In this sense the communication is secure. Moreover, since Bob is only interested in whether Alice is sending a $\uparrow$ or $\downarrow$ classical bit of information in her message, and he is not concerned about the phase information in his teleported state, actually only one bit of information needs to be sent via the classical channel to Bob. For example, suppose Alice wishes to send a $\uparrow$ bit of data to Bob. So she prepares her $|\chi\rangle$ state in Eq. (15.66) with $c = 1$ and $d = 0$. She then does her Bell measurement of the state in the form of Eq. (15.69). If she reports to Bob that she measured either of the Bell states $\Psi^{\pm}_{AA'}$, when Bob measures his teleported state, which will lead to the result $|\uparrow\rangle$, he will know that in Alice's original state $c = 1$ and so she is trying to send a $\uparrow$ bit of data. And if Alice reports measuring either $\Phi^{\pm}_{AA'}$, then when Bob measures his teleported $|\chi\rangle$ state, which will lead to the result $|\downarrow\rangle$, he will know in Alice's original $|\chi\rangle$ state she had $c = 1$. So again he will know that even though the state he measured was $|\downarrow\rangle$, actually Alice was attempting to send a $\uparrow$ bit of data. As such, for Alice both choices $\Psi^{\pm}_{AA'}$ are equivalent for the information she needs to convey and the same is the case also between $\Phi^{\pm}_{AA'}$, so she really just needs to send one bit of information via the classical channel.

Whereas just listening to the classical channel will not permit reading the message being transmitted, there are possible problems if the third party also got access to the shared entanglement. Suppose when the entangled particles are being sent to Alice and Bob, a third party, Eve, intercepted the particles. She could secretly record their states and along with the information from the classical channel could then read the message being sent. Eve could then send particles to Alice and Bob in hopes they do not recognise she has been eavesdropping. In doing this Eve would send particles to Alice and Bob that would have had their quantum states altered, and the entanglement between them would have been lost. Thus if Alice and Bob did tests on their entangled particles, they could check if the correlation between them is lost, and thereby know that a third party is interfering with their communication. As such, this means of communication can be completely secure. There are many protocols that have been developed for secure communication using quantum teleportation. However, this example highlights the common ideas of how quantum teleportation can provide secure communication. It includes the fact that communication requires information via two very different channels, and if a third party does not have access to one of the channels, it then ensures full security. Moreover, if a third party attains access to both channels, then the two parties who are attempting the secure communication will be able to detect the eavesdropper.

Another protocol that utilises teleportation for communicating information, called remote state preparation,[8] works as follows. Suppose Alice wishes to send Bob the state

$$|\chi\rangle = c|\uparrow\rangle + d|\downarrow\rangle, \tag{15.70}$$

[8] C. H. Bennett *et al.*, 'Remote state preparation', *Phys. Rev. Lett.* **87** (2001) 077902, https://doi.org/10.1103/PhysRevLett.87.077902. Erratum, https://doi.org/10.1103/PhysRevLett.88.099902.

where she knows exactly the coefficients c and d. Once again Alice and Bob share the entangled state Eq. (15.65), which they agreed on beforehand. This state can be expanded in the basis of the state $|\chi\rangle$ and its orthogonal state

$$|\chi^{\perp}\rangle = d^*|\uparrow\rangle - c^*|\downarrow\rangle, \tag{15.71}$$

to give

$$|\Psi^-\rangle = \frac{1}{\sqrt{2}} \left[-|\chi_A \chi_B^{\perp}\rangle + |\chi_A^{\perp} \chi_B\rangle \right]. \tag{15.72}$$

Alice now makes a measurement of her state A in this basis. If she measures $\chi_A^{\perp}$ then the state Bob will get will be the desired state χ_B. On the other hand, if Alice measures χ_A then Bob will have the state $\chi_B^{\perp}$. Alice would need to send just one bit of information via a classical channel to let Bob know which of the two states she measured, at which point Bob would know which of the two states he has. As before, if a third party does not have access to both the shared entanglement and the classical channel, they are unable to know the information being transferred from Alice to Bob. Moreover, if they try to read the quantum channel, they will destroy the entanglement, which Alice and Bob can do tests to check for. Thus this means of communication is fully secure if a third party does not have access to the shared entanglement. Also, it is secure even if the third party has this access, in the sense that Alice and Bob would know the communication has been interfered with.

15.4.3 Superdense Coding

Quantum teleportation utilises a shared entanglement and a classical channel to communicate a qubit of quantum information. In contrast, superdense coding[9] uses a quantum channel to communicate classical information. Consider sending a classical bit of information by using a qubit. Alice could prepare her qubit for example in the state $|1\rangle$ (we will switch notation here to the computational basis) and send it to Bob, who then measures the qubit to find the state $|1\rangle$, which he understood to read also as one classical bit of information. What is not possible by this approach is for Alice to send two classical bits of information to Bob using just one qubit. What is called the Holevo bound establishes that for an n-qubit state, even though there is considerable information contained in that state due to quantum superposition, the maximum amount of classical information that can be obtained from that state is n classical bits.

Superdense coding is a procedure that utilises an entangled state between two participants in a way that one participant can transmit two bits of information through the act of sending over just one qubit. For this to work, the two participants, Alice and Bob, must initially share an entangled state. Thus, this does not violate the Holevo bound since in total two qubits are sent. However, what is interesting is that one of the qubits, that establishes the shared entanglement, could have been sent long before the two bits

[9] C. H. Bennett and S. J. Wiesner, 'Communication via one- and two-particle operators on Einstein–Podolsky–Rosen states', *Phys. Rev. Lett.* **69** (1992) 2881–2884, https://doi.org/10.1103/PhysRevLett.69.2881.

of classical information that one wants to send have even been decided. When one is ready to send those two bits of classical information, due to the already established quantum channel, it only requires one qubit to be sent. This will be shown here with an example.

Suppose Alice and Bob share the following entangled (Bell) state of two spins:

$$|\Phi^+\rangle = \frac{1}{\sqrt{2}} (|00\rangle + |11\rangle), \tag{15.73}$$

with the first entry of each state associated with the qubit of Alice (A) and the second with the qubit of Bob (B). Suppose Alice has two bits, m and n, that she wishes to communicate to Bob by sending just one qubit over to him. The superdense coding protocol for achieving this is as follows. If $m = 1$, Alice applies the unitary transformation $\hat{\sigma}_z$ to her A qubit and if $m = 0$, she does nothing. If $n = 1$, she applies the unitary transformation $\hat{\sigma}_x$ to her A qubit and if $n = 0$, she does nothing. Alice now sends her A qubit over to Bob. For example, if the qubit is created from a spin-$\frac{1}{2}$ particle, she physically sends over that particle to Bob, making sure the spin state of the particle remains unaffected. Bob now applies what is called a control-NOT (CNOT) operation, which in matrix form is

$$\hat{C} = \begin{pmatrix} 1 & 0 & 0 & 0 \\ 0 & 1 & 0 & 0 \\ 0 & 0 & 0 & 1 \\ 0 & 0 & 1 & 0 \end{pmatrix}, \tag{15.74}$$

to the combined two-qubit state. The CNOT matrix has matrix elements ordered in each row and column for the qubit AB states 00, 01, 10 and 11. Bob applies what is called a Hadamard operation (these operators will be discussed further in Section 15.5), defined as

$$\hat{H}_d \equiv \frac{1}{\sqrt{2}} \begin{pmatrix} 1 & 1 \\ 1 & -1 \end{pmatrix}, \tag{15.75}$$

to the A qubit now in his possession. Bob now measures the state of the two qubits and will find A is in the state m and B is in the state n.

Let us go through these steps for one of the four cases, with the other three left as exercises. Suppose $(m, n) = (1, 1)$.

Step 1: Since $m = 1$, Alice acts on qubit A with operator σ_z:

$$\begin{pmatrix} 1 & 0 \\ 0 & -1 \end{pmatrix} \otimes \begin{pmatrix} 1 & 0 \\ 0 & 1 \end{pmatrix} \frac{1}{\sqrt{2}} \left[\begin{pmatrix} 1 \\ 0 \end{pmatrix} \otimes \begin{pmatrix} 1 \\ 0 \end{pmatrix} + \begin{pmatrix} 0 \\ 1 \end{pmatrix} \otimes \begin{pmatrix} 0 \\ 1 \end{pmatrix} \right]$$

$$= \frac{1}{\sqrt{2}} \left[\begin{pmatrix} 1 \\ 0 \end{pmatrix} \otimes \begin{pmatrix} 1 \\ 0 \end{pmatrix} + \begin{pmatrix} 0 \\ -1 \end{pmatrix} \otimes \begin{pmatrix} 0 \\ 1 \end{pmatrix} \right]. \tag{15.76}$$

Recall here that the tensor product symbol means the matrix on the left only acts on the associated spinors on the left, etc.

Step 2: Since $n = 1$, this means Alice needs to take the state resulting from Step 1 and act on qubit A with operator σ_x:

$$\begin{pmatrix} 0 & 1 \\ 1 & 0 \end{pmatrix} \otimes \begin{pmatrix} 1 & 0 \\ 0 & 1 \end{pmatrix} \frac{1}{\sqrt{2}} \left[\begin{pmatrix} 1 \\ 0 \end{pmatrix} \otimes \begin{pmatrix} 1 \\ 0 \end{pmatrix} + \begin{pmatrix} 0 \\ -1 \end{pmatrix} \otimes \begin{pmatrix} 0 \\ 1 \end{pmatrix} \right]$$

$$= \frac{1}{\sqrt{2}} \left[\begin{pmatrix} 0 \\ 1 \end{pmatrix} \otimes \begin{pmatrix} 1 \\ 0 \end{pmatrix} + \begin{pmatrix} -1 \\ 0 \end{pmatrix} \otimes \begin{pmatrix} 0 \\ 1 \end{pmatrix} \right]. \tag{15.77}$$

Step 3: Send qubit A to Bob. At this stage the entire entangled state at the end of Step 2 is in Bob's possession.

Step 4: Apply the CNOT operator on the state in Bob's possession. For this, it is more convenient to write the state in terms of the four basis states formed by the tensor product of two qubits 00, 01, 10 and 11. In accordance with how the CNOT matrix is arranged, the 00 state, for example, is the first entry in the column vector, 01 the second, 10 the third and 11 the fourth. In this representation the state emerging from Step 2, which is at this stage in Bob's possession, is

$$\frac{1}{\sqrt{2}} \begin{pmatrix} 0 \\ -1 \\ 1 \\ 0 \end{pmatrix}. \tag{15.78}$$

Applying the CNOT gate, Eq. (15.74), to it gives

$$\hat{C} \frac{1}{\sqrt{2}} (|10\rangle - |01\rangle) = \begin{pmatrix} 1 & 0 & 0 & 0 \\ 0 & 1 & 0 & 0 \\ 0 & 0 & 0 & 1 \\ 0 & 0 & 1 & 0 \end{pmatrix} \frac{1}{\sqrt{2}} \begin{pmatrix} 0 \\ -1 \\ 1 \\ 0 \end{pmatrix}$$

$$= \frac{1}{\sqrt{2}} \begin{pmatrix} 0 \\ -1 \\ 0 \\ 1 \end{pmatrix} = \frac{1}{\sqrt{2}} (|11\rangle - |01\rangle)$$

$$= \frac{1}{\sqrt{2}} (|1\rangle - |0\rangle) |1\rangle. \tag{15.79}$$

Step 5: Apply the Hadamard operator Eq. (15.75) on the A qubit for the state emerging from Step 4:

$$\frac{1}{\sqrt{2}} \begin{pmatrix} 1 & 1 \\ 1 & -1 \end{pmatrix} \otimes \begin{pmatrix} 1 & 0 \\ 0 & 1 \end{pmatrix} \frac{1}{\sqrt{2}} \left[\begin{pmatrix} 0 \\ 1 \end{pmatrix} \otimes \begin{pmatrix} 0 \\ 1 \end{pmatrix} - \begin{pmatrix} 1 \\ 0 \end{pmatrix} \otimes \begin{pmatrix} 0 \\ 1 \end{pmatrix} \right]$$

$$= - \begin{pmatrix} 0 \\ 1 \end{pmatrix} \otimes \begin{pmatrix} 0 \\ 1 \end{pmatrix} = -|11\rangle. \tag{15.80}$$

If Bob now measures the state of the two qubits A and B, he will find each in the state 1, thus he obtains two bits of information 1 and 1, which is the two bits Alice wanted to communicate to Bob.

15.5 Quantum Computing

A classical computer usually manipulates electric current (although there can also be mechanical examples) in such a manner that information can be processed. The building blocks of a classical computer are logic gates. These are switches for which certain inputs of electric currents either allow or do not allow an electric current to flow out. The relation to information comes from associating the flow or lack of flow of an electric current with the binary options 1 or 0. The placement of zeros and ones and their location in the string gives a unique correspondence to a decimal number. Thus for example the decimal number 0 is the binary number 0, 1 is 1, 2 is 10, 3 is 11, 4 is 100, and so on. The culmination of all logical operations that can be done on binary numbers is called Boolean algebra. A given logical operation of Boolean algebra can be expressed through a truth table. For example, suppose one wants to add two decimal numbers 0 and 1, which correspond respectively to the binary numbers 0 and 1. Then the four possible addition operations between these numbers are expressed in Table 15.1. Here a and b represent the two input binary numbers, s is the sum and c is the carry value. So for the decimal sum $0+1$ it corresponds to summing the binary numbers 0 and 1, which leads to the binary sum 1 with zero carry value. Whereas the decimal sum $1 + 1$ leads to sum 0 with a 1 carried over to the next entry, thus the binary number 10 which corresponds to the decimal number 2.

The transformation of binary numbers as they flow through a computer circuit occurs as the outcome of Boolean operations the computer is being tasked to do. These operations are implemented within a classical computer through logic gates. Gates that are useful to classical computing are NOT, AND, OR, XOR, NAND and NOR. Figure 15.10 shows all these logic gates and the logic operations each of them do, where 1 corresponds to the presence of a current and 0 to no current, and where the input is on the left of the gate and the output on the right. Thus, for example, in an AND gate, if there is a current (1) at both inputs, then there results a current (1) at the output. However, if either input has no current (0), then the output also has no current (0). The total set of logic operations that these gates perform comprises Boolean algebra. The logic gates that represent the Boolean algebra are combined into complicated circuits that form abstract representations of operations done by a classical computer. In fact, the set of gates in Fig. 15.10 is over-complete. NAND and NOR are universal gates, which means all other gates and thus any logical function can be constructed from circuits with either of these two gates.

Table 15.1 Truth table for the binary operation of adding numbers a and b to give sum s and carry value c

a	b	s	c
0	0	0	0
0	1	1	0
1	0	1	0
1	1	0	1

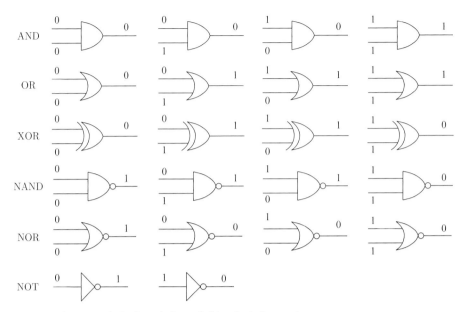

Fig. 15.10 Logic gates implementing the Boolean algebra underlying classical computing.

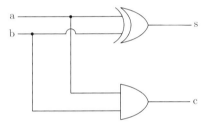

Fig. 15.11 Gate circuit called a half-adder that implements the truth table of Table 15.1.

The construction of an actual physical computer requires using physical devices that can perform the functions of these logic gates. Basically the logic gates are just switches, each functioning in a specific way. Thus any device that can serve as these switches can be used to form a computer. These could be mechanical devices, as was the case in the earliest classical computers, such as the Harvard Mark 1 built in 1944. They could be realised through vacuum tubes such as the Colossus computer built to decipher the Enigma code. In modern computing, the logic gates are realised through electronic components, in particular transistors and integrated circuits.

Returning to the example of adding two binary numbers given in Table 15.1, the corresponding gate circuit, called a half-adder, is shown in Fig. 15.11. The idea is one logic gate, XOR, adds and the other, AND, gives the carry value to the next place of the string of binary numbers. For example, the decimal addition operation 0 + 1 has binary inputs of 0 and 1 (at a and b) for which the XOR output is 1 (at s) and the AND output is 0 (at c). The meaning of this is that number 1 appears in the first place in the binary string with no carry value. All combined this corresponds to the decimal number 1. This is a consistent realisation of the corresponding entry in the truth table (Table 15.1). Similarly, all the other three addition operations from this truth table are realised by this gate circuit. For the decimal addition process $1 + 1$, the binary inputs are 1 and 1. These lead to the XOR output 0 and the AND output 1, which implies 0 for the first place on the binary string and 1 in the carry value, so the second position on the binary string, thus the binary number 10, corresponding to the decimal number 2. Similar types of circuits are used for other addition operations. For example, a full-adder differs from a half-adder in that it has three rather than two inputs, with the third having a carry-in (carry value of the input bits).

The field of quantum computing basically tries to follow the general ideas of classical computing. However, rather than electric currents moving bits of data around, in quantum computing the quantum-mechanical state vector is evolved and the goal of the subject is to somehow extract the information contained in the state vector in a form useful for computing purposes. In many ways the subject of quantum computing has evolved in the opposite way to classical computing. In the latter the basic logic structures of the subject were in place first, already in the nineteenth century in the form of the Boolean algebra. Subsequently this algebra was realised by gates and the resulting complex chain of logical operations was realised by gate circuits. In quantum computing the analogue of the bit has been identified as the aptly named qubit, and its evolution is governed by the laws of quantum mechanics. The analogue of the logic gates of classical computing is simply any linear operators that can act on the state vector. Various elementary and universal quantum gates are known. However, what is missing in quantum computing is a comprehensive quantum logic framework similar to the Boolean algebra of classical computing. This is one direction for further development of quantum computing.

The unique aspect of measurement in quantum mechanics adds additional directions for attaining computational advantages. As already discussed in the early part of the book, not all observables in a quantum system can be simultaneously measured but rather there are certain sets of compatible observables. This can be said another way; that for a given observable, what value is actually detected depends on what other observables are measured. In this way of thinking, it is referred to as **contextuality** in quantum mechanics; that the outcome of a measurement of a given observable depends on the context in which it is measured. This is easy to see. As an example, suppose there are three observables $\mathcal{A}$, $\mathcal{B}$ and $\mathcal{C}$ with corresponding quantum operators $\hat{A}$, $\hat{B}$ and $\hat{C}$. Suppose that $\hat{A}$ commutes with either $\hat{B}$ or $\hat{C}$ but that $\hat{B}$ and $\hat{C}$ do not commute with each other. Then by the compatibility theorem discussed in Section 3.7, there is a common eigenbasis for either the operators $\hat{A}$ and $\hat{B}$ or $\hat{A}$ and $\hat{C}$, but these would be two different eigenbases. Thus there is no reason to expect that the outcome of measuring $\hat{A}$ should be the same if done in conjunction with $\hat{B}$ versus $\hat{C}$. In particular, the expansion of the quantum state prior to measurement would have different probability amplitudes associated with each eigenstate of $\hat{A}$ in the two bases. This means different information can be extracted from the state vector depending on how it is measured, which thus could have computational relevance. For example, an n-qubit state occupies 2^n states in the Hilbert space. Depending on how the measurement of the state is done, that will dictate which of the 2^n states the state vector eventually collapses onto and in what basis. Thus, different information can be extracted from the state vector, which could have a computational relevance. This feature of contextuality is new to quantum computing, with no analogue in classical computing. Thus it can be used alongside quantum logic to achieve a computable outcome. How to best use these properties of quantum computing to achieve useful computational outcomes is today a developing area of research.

The origins of quantum computing arose from a more mundane but important concern about the immense heat that classical computations produce. Specifically, standard classical computing is based on irreversible logic operations. For any of the binary operations represented by logic gates such as AND, OR, XOR, NAND and NOR, two values are inputted and one is outputted. To physically realise such operations, minimally it would require one particle (electron) going into each input channel and only one coming out, leaving one particle to be dissipated away. This leads to an entropy increase of $\ln 2$. If operating at temperature T, the released particle would imply a generation of heat of amount $kT \ln 2$. In a real computer it is in fact a full electric current and not just a single electron entering and exiting logic gates. Thus the heat generated is far more. However, for any many-to-one operation, it is irreversible, and so has an associated thermodynamic cost. Classical computers are subject to this consequence. As computing technology has advanced, heat produced from binary operations has gone down substantially from millions bigger than the minimum to thousands, and this should improve even further. Nevertheless, inevitably there is still an intrinsic minimum dictated by thermodynamics. A related concern about classical computing was the inevitable breakdown of Moore's law, that continually shrinking the size of computer circuitry would eventually bring it to the atomic level, so dictated by quantum mechanics.

From these problems arose the idea of reversible computing first within the context of classical computing, such as with the universal Toffoli gate. However, it also led to Paul Benioff[10] and Richard Feynman[11] in the 1980s suggesting to use quantum mechanics to build a reversible computing machine. A related motivation offered by these authors and also Yuri Manin[12] was that a reversible quantum computing machine could be utilised to provide a realistic simulation of quantum-mechanical systems. Subsequently, the interest in quantum computing has evolved from these initial motivations. Since the mid-1980s quantum computing algorithms have been developed that could realise certain mathematical manipulations orders of magnitude faster than can be achieved from a classical computation. Thus quantum computing by now has an existence of its own, with its own special qualities. It utilises three basic ingredients inherent to quantum mechanics, superposition, interference and entanglement. Quantum computing works with a set of n-qubits. This implies a Hilbert space of 2^n basis states. These states are then acted upon by unitary operators that lead to various linear combinations of the basis states. The unitary operators within the quantum computing syntax are called gates and the combination of such operators form together to be called a circuit. Thus the state vector in general is

[10] P. Benioff, 'The computer as a physical system: A microscopic quantum mechanical Hamiltonian model of computers as represented by Turing machines', *J. Stat. Phys.* **22** (1980) 563–591 DOI: 10.1007/BF01011339.

[11] R. P. Feynman, 'Simulating physics with computers', *Int. J. Theoret. Phys.* **21** (1982) 467-488, DOI: 10.1007/BF02650179; 'Quantum mechanical computers', *Found. Phys.* **16** (1986) 507-531 (originally in *Optics News.* February 1985, 11–20, DOI: 10.1007/BF01886518).

[12] Yu. Manin, 'Computable and uncomputable' (in Russian), Sovet-skoye Radio, Moscow, 1980.

the superposition of 2^n basis states, which in principle could contain useful information for computation. Through use of interference, the state vector can enhance or destroy certain states in favour of others. This general process will also lead to entanglement of the qubits. The goal is to find suitable combinations of all these basic operations that produce something of computational utility, and these are called quantum algorithms.

In this section, we will discuss quantum gates and circuits. We will then show two examples of quantum computing in action, the Deutsch and Grover algorithms. The Deutsch algorithm is simple and has limited practical use, but is excellent for demonstrating some of the key features of quantum computing algorithms. The Grover algorithm is one of the most successful of the early algorithms to have been developed for quantum computing. The goal of this section is to give an elementary introduction to quantum computing and so it is very restricted to just a few topics which help illustrate salient features. There are many other important topics not covered here. A few are noteworthy to briefly mention here. One is the Shor algorithm, which finds the prime factors of an integer. The other topic that is important to mention is quantum error correction. Quantum systems are subject to decohering effects from interaction with the environment. This could have devastating effects on practical implementation of quantum computing. In the early days it was believed such effects would make quantum computing impossible to realise. However, what came to be called quantum threshold theorems were derived that showed one can do robust quantum computation in a noisy environment using quantum error correction. The interested reader can follow up these topics after reading our elementary introduction to quantum computing in the books of Mermin and Nielsen and Chuang referenced at the end of this chapter. Finally, another promising direction for quantum computing which was already mentioned earlier is to do simulations of quantum systems.

15.5.1 Quantum Register and Logic Gates

In classical computing the bits of data are held in a register and logic gates then access the register and act on the bits of data. A quantum computer has a similar structure of a quantum register and quantum gates. The quantum register is simply a state vector. A one-qubit register is the qubit itself, $|\psi_1\rangle = a_0|0\rangle + a_1|1\rangle$. One comment on notation. In the previous sections of this chapter we have represented the two-state system with the states $|\uparrow\rangle$ and $|\downarrow\rangle$, but in this section, to match the more standard notation of quantum computing, the two states will be expressed as $|0\rangle$ instead of $|\uparrow\rangle$ and $|1\rangle$ instead of $|\downarrow\rangle$. A two-qubit register is

$$|\psi_2\rangle = a_{00}|00\rangle + a_{10}|10\rangle + a_{01}|01\rangle + a_{11}|11\rangle, \tag{15.81}$$

and so on for higher qubit registers.

A standard approach, or to follow the quantum computing syntax, protocol, is to write the quantum register in the form

$$\sum_{c,t} a_{ct}|c\rangle_n|t\rangle_m, \tag{15.82}$$

where the first terms, labelled c, are called the control or input register and the second terms, labelled t, are called the target or output register. This is just nomenclature, and often in implementation, these names do not make literal sense. The key point is that such a product form can prove useful in quantum computing manipulations, as will be seen in some of the examples given below. The subscript above means respectively that the kets are states labelled by n and m bits. Thus, for example, $|c\rangle_4$ is a control ket specified by a binary number of length 4.

The quantum logic gates are in general linear operators that act on the quantum register. Primarily quantum computing is interested in reversible computing, thus the gates must be unitary linear operators. Recall for a unitary operator $\hat{U}$ that $\hat{U}^\dagger\hat{U} = \hat{U}\hat{U}^\dagger = I$, and so such operators preserve the norm of the quantum register. The time evolution operator of a quantum-mechanical system is a unitary operator. One can think of any unitary operation within a quantum computer as simply evolving the quantum register or state vector of the system and maintaining the overall normalisation of the state vector.

Non-unitary quantum operations have some role to play. A non-unitary gate is equivalent to a measurement on the quantum system. For example, in quantum teleportation, as discussed in Section 15.4.2, at one point Alice makes a Bell measurement on her two-photon state, which was a non-unitary operation, and is essential to the whole process. Similarly, in quantum computing, at the end of a process a measurement needs to be made to extract information. Environmental effects on a quantum system that lead to decoherence are other examples of non-unitary operations. Some research has looked at utilising non-unitary gates in a controllable way to counter effects from decoherence. Moreover, non-unitary gates have also been looked at for the development of quantum algorithms. In understanding the breadth of the subject it is useful to be aware of these points. Here the goal is to present the basic ideas of quantum computing, and so we will restrict our discussion to only quantum logic gates that are unitary operators.

Although any unitary operator could be a quantum gate, there are a few that have been found to be particularly useful. For a single qubit, perhaps the most used example is the **Hadamard operator** already introduced in Eq. (15.75):

$$-\boxed{\hat{H}_d}- = \frac{1}{\sqrt{2}}\begin{pmatrix} 1 & 1 \\ 1 & -1 \end{pmatrix} = \frac{1}{\sqrt{2}}\left[|0\rangle\langle 0| + |0\rangle\langle 1| + |1\rangle\langle 0| - |1\rangle\langle 1|\right], \qquad (15.83)$$

but here we give the name of the operator, its symbolic representation in circuits, its matrix representation and its expression in Dirac notation. The latter thus makes clear what each row and column in the matrix corresponds to. Observe that the Hadamard operator, when applied to either state $|0\rangle$ or $|1\rangle$, leads to an equal superposition of both states. A comment on notation. In quantum mechanics, the Hamiltonian operator holds a privileged role with standard notation $\hat{H}$. In quantum computing, the Hadamard operator is one of the most widely used, and is denoted the same. As these two subjects have an obvious overlap, so as to avoid ambiguity, we denote the Hadamard operator with the additional d subscript.

Another useful set of quantum logic gates are the three Pauli gates

$$-\boxed{\hat{X}}- = \begin{pmatrix} 0 & 1 \\ 1 & 0 \end{pmatrix} = |1\rangle\langle 0| + |0\rangle\langle 1|,$$

$$-\boxed{\hat{Y}}- = \begin{pmatrix} 0 & -i \\ i & 0 \end{pmatrix} = i|1\rangle\langle 0| - i|0\rangle\langle 1|,$$

$$-\boxed{\hat{Z}}- = \begin{pmatrix} 1 & 0 \\ 0 & -1 \end{pmatrix} = |0\rangle\langle 0| - |1\rangle\langle 1|.$$

These gates can be further generalised. A more general example of the Pauli-Z gate would be the R_ϕ-gate:

$$-\boxed{\hat{R}_\phi}- = \begin{pmatrix} 1 & 0 \\ 0 & e^{i\phi} \end{pmatrix} = |0\rangle\langle 0| + e^{i\phi}|1\rangle\langle 1|.$$

There are various common quantum gates acting on two-qubit or more states, called controlled operations. One of the most common, already used earlier, is the CNOT-gate:

$$-\boxed{\hat{C}}- = \begin{pmatrix} 1 & 0 & 0 & 0 \\ 0 & 1 & 0 & 0 \\ 0 & 0 & 0 & 1 \\ 0 & 0 & 1 & 0 \end{pmatrix}, \tag{15.84}$$

which in Dirac notation corresponds to

$$|00\rangle\langle 00| + |01\rangle\langle 01| + |10\rangle\langle 11| + |11\rangle\langle 10|. \tag{15.85}$$

What this gate does is, if the control state is $|0\rangle$, then the target state remains unchanged by this gate. However, if the control state is $|1\rangle$, then the target state flips so if initially $|0\rangle$, it becomes $|1\rangle$ and vice versa. This gate can also be written in the more detailed symbolic form

$$|c\rangle \quad\longrightarrow\quad |c\rangle$$
$$|t\rangle \quad\longrightarrow\quad |t \oplus c\rangle.$$

Here the $\oplus$ symbol in the ket is addition modulo 2. Somewhat confusing, but the same looking (target) symbol in the diagram is commonly used for this CNOT operation. There are only three outcomes possible, which will under this summation give $0 \oplus 0 = 0$, $1 \oplus 0 = 1$ and $1 \oplus 1 = 0$.

15.5.2 Deutsch's Algorithm

This is one of the simplest examples of a quantum computing algorithm. It is beneficial to examine it in close detail, since it illustrates many of the key features generally utilised in quantum computing.

The Deutsch algorithm answers the following question. Suppose one is given a function $f(x)$, which acts on a single bit, so x can just take on the values 0 or 1. The Deutsch algorithm answers the question whether $f(0) = f(1)$ or $f(0) \neq f(1)$. In other words, whether $f(x)$ is a constant function or not. In order to answer this question with a classical algorithm, it would require two operations, one to evaluate $f(0)$ and one $f(1)$, after which one can compare and answer the initial question. As it turns out, a quantum computer can answer the question in just one run. It will not tell us the individual values of $f(x)$ at the two points, but will only answer the question asked. This already illustrates a generic feature about quantum computing, that one needs to ask the right question in order to exploit quantum computing to produce a more efficient algorithm to classical computing. The improvement here is that quantum computing halves the number of runs compared to classical computing, but only once the right question for the function of interest, $f(x)$, is asked.

For addressing any problem in quantum computing, all manipulations are done on a quantum state. The basic steps of any quantum computing operation are to define a particular initial quantum state, act on this state with various unitary operators, and then at the end measure the resulting state and from that extract the relevant information the algorithm is meant to obtain. For the case at hand, the information that the final measurement will provide is whether the function $f(x)$ is constant or not.

In a quantum computing algorithm there is often present a unitary operator, usually called the oracle, which we will generally denote $\hat{O}$. This operator encodes the details of the mathematical quantity under study. In the example here $\hat{O}$ contains all the details about the function $f(x)$. When it was said there is an unknown function $f(x)$, in a computing context it has to be given in a form amenable to the format of the computing. In the case of quantum computing all quantities can be given either as states or as unitary operators. Thus in particular, the unknown function $f(x)$ is given in the form of this oracle operator $\hat{O}$. It is handed to the user, and the goal is for the user to use the laws of quantum mechanics to extract the information about the function $f(x)$ contained in this operator.

Turning now to the specifics of the Deutsch algorithm, the quantum state one works with is where both the input and output registers are one-qubit states, $|c\rangle|t\rangle$ and the action of the oracle on this state is

$$\hat{O}_f|c\rangle|t\rangle = |c\rangle|t \oplus f(c)\rangle. \tag{15.86}$$

Thus $|c\rangle$ can be in only one of the two-bit states, either $|0\rangle$ or $|1\rangle$, the $\oplus$ symbol means addition modulo 2 and the unknown function $f_i(x)$ is at least understood to take on only the two values 0 or 1. Thus there can be only four different types of functions here, all of which are given in Table 15.2.

The oracle as given in Eq. (15.86) can also be represented diagrammatically as

$$
\begin{array}{c}
|c\rangle \quad \underline{\hspace{0.5cm}} \begin{array}{|c|} \hline \; \\ \hat{O}_f \\ \hline \end{array} \underline{\hspace{0.5cm}} \quad |c\rangle \\
|t\rangle \quad \underline{\hspace{0.5cm}} \quad |t \oplus f(c)\rangle.
\end{array}
$$

Table 15.2 The unknown functions f_i of the Deutsch algorithm

f_i	$x = 0$	$x = 1$
f_1	0	0
f_2	0	1
f_3	1	0
f_4	1	1

For example, for the function $f_2(x)$, the action of $\hat{O}_{f_2}$ is

$$\hat{O}_{f_2}|0\rangle|0\rangle = |0\rangle|0 \oplus f_2(0)\rangle = |0\rangle|0\rangle,$$
$$\hat{O}_{f_2}|0\rangle|1\rangle = |0\rangle|1 \oplus f_2(0)\rangle = |0\rangle|1\rangle,$$
$$\hat{O}_{f_2}|1\rangle|0\rangle = |1\rangle|0 \oplus f_2(1)\rangle = |1\rangle|1\rangle,$$
$$\hat{O}_{f_2}|1\rangle|1\rangle = |1\rangle|1 \oplus f_2(1)\rangle = |1\rangle|0\rangle.$$

The next step is to set up the relevant initial quantum state and act on it with the necessary unitary operators. There is no fixed prescription for how to design quantum algorithms. This is the step of trial, error and insight, and researchers over the years have found various types of useful quantum algorithms. As such, for the Deutsch algorithm, here we simply state the necessary quantum register and operators and will then go through the steps to show it achieves the stated goal. The Deutsch algorithm is

$$\left(\hat{H}_d \otimes \hat{I}\right)\hat{O}_{f_i}\left(\hat{H}_d \otimes \hat{H}_d\right)\left(\hat{X} \otimes \hat{X}\right)|0\rangle|0\rangle. \tag{15.87}$$

The tensor product symbol $\otimes$ here means each of the two operators in the product act on the respective states in the product:

$$(\hat{A} \otimes \hat{B})|x\rangle|y\rangle \equiv \hat{A}|x\rangle\hat{B}|y\rangle. \tag{15.88}$$

Going through the operations in Eq. (15.87) in the order they act on the state from right to left:

$$\left(\hat{X} \otimes \hat{X}\right)|0\rangle|0\rangle = |1\rangle|1\rangle, \tag{15.89}$$

followed by

$$\left(\hat{H}_d \otimes \hat{H}_d\right)|1\rangle|1\rangle = \frac{1}{\sqrt{2}}\left[|0\rangle - |1\rangle\right]\frac{1}{\sqrt{2}}\left[|0\rangle - |1\rangle\right]$$
$$= \tfrac{1}{2}\left[|0\rangle|0\rangle - |0\rangle|1\rangle - |1\rangle|0\rangle + |1\rangle|1\rangle\right]. \tag{15.90}$$

This state is now acted on by the oracle:

$$\hat{O}_{f_i}\tfrac{1}{2}\left[|0\rangle|0\rangle - |0\rangle|1\rangle - |1\rangle|0\rangle + |1\rangle|1\rangle\right]$$
$$= \tfrac{1}{2}\left[|0\rangle|f_i(0)\rangle - |0\rangle|1 \oplus f_i(0)\rangle - |1\rangle|f_i(1)\rangle + |1\rangle|1 \oplus f_i(1)\rangle\right]$$
$$= \tfrac{1}{2}|0\rangle\left[|f_i(0)\rangle - |1 \oplus f_i(0)\rangle\right] - \tfrac{1}{2}|1\rangle\left[|f_i(1)\rangle - |1 \oplus f_i(1)\rangle\right]. \tag{15.91}$$

The final operation in Eq. (15.87) (leftmost) is to apply the Hadamard operator to the input state, which leads to

$$= \frac{1}{2\sqrt{2}}|0\rangle\left[|f_i(0)\rangle - |1 \oplus f_i(0)\rangle - |f_i(1)\rangle + |1 \oplus f_i(1)\rangle\right]$$
$$+ \frac{1}{2\sqrt{2}}|1\rangle\left[|f_i(0)\rangle - |1 + f_i(0)\rangle + |f_i(1)\rangle - |1 \oplus f_i(1)\rangle\right]. \tag{15.92}$$

Thus for $f_i(0) \neq f_i(1)$ this leads to

$$= \frac{1}{\sqrt{2}}|0\rangle\left[|f_i(0)\rangle - |1 \oplus f_i(0)\rangle\right], \tag{15.93}$$

and for $f_i(0) = f_i(1)$ it leads to

$$= \frac{1}{\sqrt{2}} |1\rangle \left[|f_i(0)\rangle - |1 \oplus f_i(0)\rangle \right]. \tag{15.94}$$

At this point if one measures the resulting input state, depending on whether one gets $|0\rangle$ or $|1\rangle$, one will know if $f_i(0) \neq f_i(1)$ or $f_i(0) = f_i(1)$.

Constructing the Oracle

The last remaining step in constructing this quantum algorithm is to show that all operations involved in it are done with unitary operators. Moreover, it is important to show that these operators can be written in terms of some combination of elementary unitary operators. The reason is that the ultimate goal would be to realise an actual physical quantum computer. Classical computers are made up of a network of elementary gates that are realised through electronic components. A similar idea applies in the thinking of a quantum computer, where developers seek to have elementary devices that can realise the individual quantum gates, which can then be combined to form complex quantum computing circuits. Thus it is important to understand how to write the operators of a quantum computing algorithm in terms of elementary quantum gates.

Many of those operators used in the Deutsch algorithm, such as the Hadamard and CNOT gate, are already explicitly shown above. What remains is the construction of the oracle. Equation (15.86) specifies what the oracle needs to do, but we need to show that it can be written as a unitary operator, explicitly composed of elementary quantum gates. For the four functions in Table 15.2, one can check that the unitary operators that can perform this task can be written in terms of elementary quantum gates as

$$\hat{O}_{f_1} = \hat{I},$$
$$\hat{O}_{f_2} = \hat{C},$$
$$\hat{O}_{f_3} = \hat{C}(\hat{I} \otimes \hat{X}),$$
$$\hat{O}_{f_4} = (\hat{I} \otimes \hat{X}).$$

Represented as matrices, the oracle for this problem is a 4×4 matrix. Equation (15.86) tells us that the oracle acts on both the input and output registers, and here both are two-dimensional spaces taking on the values 0 or 1. Thus we can explicitly write out the oracle in matrix form for each of the four functions f_i, where the ctth entry in the matrix corresponds to a value of the control (c) and target (t) with the two choices 0 or 1. We need to first specify our convention on what each row and column of the matrix corresponds to, thus we will arrange the matrices so the rows and columns in order correspond to the $00, 01, 10$ and 11 entries. The column corresponds to the ct values at input and the rows correspond to the $c, t \oplus f_i(c)$ values at output. Below are all four of these operators explicitly written in this matrix representation. For each of the four operators, the operator specification is given, followed by the quantum computing notation for the operator, followed by the explicit matrix representation based on the convention specified above, and then the diagrammatic representation. Thus

$$\hat{O}_{f_1} = \hat{I} = \begin{pmatrix} 1 & 0 & 0 & 0 \\ 0 & 1 & 0 & 0 \\ 0 & 0 & 1 & 0 \\ 0 & 0 & 0 & 1 \end{pmatrix}$$

$$= |c\rangle \underline{\hspace{2cm}} |c\rangle$$
$$|t\rangle \underline{\hspace{2cm}} |t \oplus f_1(c)\rangle,$$

$$\hat{O}_{f_2} = \hat{C} = \begin{pmatrix} 1 & 0 & 0 & 0 \\ 0 & 1 & 0 & 0 \\ 0 & 0 & 0 & 1 \\ 0 & 0 & 1 & 0 \end{pmatrix}$$

$$= |c\rangle \underline{\hspace{1cm}}\bullet\underline{\hspace{1cm}} |c\rangle$$
$$|t\rangle \underline{\hspace{0.3cm}}\boxed{\hat{X}}\underline{\hspace{0.3cm}} |t \oplus f_2(c)\rangle,$$

$$\hat{O}_{f_3} = \hat{C}(\hat{I} \otimes \hat{X}) = \begin{pmatrix} 0 & 1 & 0 & 0 \\ 1 & 0 & 0 & 0 \\ 0 & 0 & 1 & 0 \\ 0 & 0 & 0 & 1 \end{pmatrix}$$

$$= |c\rangle \underline{\hspace{1.5cm}}\bullet\underline{\hspace{0.8cm}} |c\rangle$$
$$|t\rangle \underline{\hspace{0.2cm}}\boxed{\hat{X}}\underline{\hspace{0.2cm}}\boxed{\hat{X}}\underline{\hspace{0.2cm}} |t \oplus f_3(c)\rangle,$$

$$\hat{O}_{f_4} = \hat{I} \otimes \hat{X} = \begin{pmatrix} 0 & 1 & 0 & 0 \\ 1 & 0 & 0 & 0 \\ 0 & 0 & 0 & 1 \\ 0 & 0 & 1 & 0 \end{pmatrix}$$

$$= |c\rangle \underline{\hspace{2cm}} |c\rangle$$
$$|t\rangle \underline{\hspace{0.4cm}}\boxed{\hat{X}}\underline{\hspace{0.4cm}} |t \oplus f_4(c)\rangle.$$

15.5.3 Grover's Algorithm

This is typically stated as a search algorithm for finding a particular item in a list. In the quantum computing context, more precisely it can find a particular state within a quantum register. In fact the Grover algorithm can be viewed as being able to address any of the following three problems. It can be used to find a state within a quantum register, it can be used to determine which particular state an unknown oracle is singling out or it can be used to do both at the same time. The way this algorithm is most typically presented is for the second purpose, to determine what state the unknown quantum oracle is picking out. The problem is set up as follows. As an example, for the case of $N = 8$ items, construct the state

$$\hat{H}_d \otimes \hat{H}_d \otimes \hat{H}_d|000\rangle \equiv |\psi\rangle = \frac{1}{2\sqrt{2}}|000\rangle + \frac{1}{2\sqrt{2}}|001\rangle + \frac{1}{2\sqrt{2}}|010\rangle + \frac{1}{2\sqrt{2}}|011\rangle$$

$$+ \frac{1}{2\sqrt{2}}|100\rangle + \frac{1}{2\sqrt{2}}|101\rangle + \frac{1}{2\sqrt{2}}|110\rangle + \frac{1}{2\sqrt{2}}|111\rangle, \quad (15.95)$$

where $\hat{H}_d$ is the Hadamard operator. By construction this state has equal probability for all the possible three-bit states. Thus there is no need for a search algorithm to tell us if any of these exists in this state, since by construction we know they do. However, suppose an oracle is handed to you and the goal is to determine what state this oracle is picking out. Let us see how Grover's algorithm solves that problem.

The algorithm involves two steps which are then iterated. The first step is to apply the oracle, $\hat{O}$. In the Grover algorithm the operation done by the oracle is to flip the sign of the state $|s\rangle$ it is singling out:

$$\hat{O} = I - 2|s\rangle\langle s|. \quad (15.96)$$

Thus in matrix representation $\hat{O}$ is a diagonal matrix that has 1 along all the diagonal entries except for the diagonal entry associated with the state $|s\rangle$, which has entry -1.

Suppose for our example the state $|s\rangle$ it singles out is $|110\rangle$. Then

$$\hat{O}|\psi\rangle \equiv |\psi_-\rangle = \frac{1}{2\sqrt{2}}|000\rangle + \frac{1}{2\sqrt{2}}|001\rangle + \frac{1}{2\sqrt{2}}|010\rangle + \frac{1}{2\sqrt{2}}|011\rangle$$

$$+ \frac{1}{2\sqrt{2}}|100\rangle + \frac{1}{2\sqrt{2}}|101\rangle - \frac{1}{2\sqrt{2}}|110\rangle + \frac{1}{2\sqrt{2}}|111\rangle, \quad (15.97)$$

so that the sign in front of the state $|110\rangle$ now is minus. The second step is to perform the Grover operation

$$\hat{G} \equiv 2|\psi\rangle\langle\psi| - I \quad (15.98)$$

on the state $|\psi_-\rangle$, which leads to

$$[2|\psi\rangle\langle\psi| - I]|\psi_-\rangle \equiv |\psi_d\rangle = \frac{1}{4\sqrt{2}}|000\rangle + \frac{1}{4\sqrt{2}}|001\rangle + \frac{1}{4\sqrt{2}}|010\rangle + \frac{1}{4\sqrt{2}}|011\rangle$$

$$+ \frac{1}{4\sqrt{2}}|100\rangle + \frac{1}{4\sqrt{2}}|101\rangle + \frac{5}{4\sqrt{2}}|110\rangle + \frac{1}{4\sqrt{2}}|111\rangle. \quad (15.99)$$

Notice the effect of these two operations $\hat{G}\hat{O}$ is to increase the amplitude of the state $|110\rangle$. By repeating these two operations $\hat{G}\hat{O}$ an order $\sqrt{N}$ times, where N is the number of basis states in the initial state $|\psi\rangle$, it leads to one state, in this example $|110\rangle$, having its maximum amplitude and much bigger than the amplitude of all the other states. If one continues to iterate beyond that point, the amplitude once again decreases. The upshot is, if after the optimal iterations one measures this state, one finds that the maximum amplitude is in $|110\rangle$ and thereby one concludes that the oracle was picking out that particular state.

The Grover iterations have a simple geometrical interpretation. Observe that the action of both operators $\hat{G}$ and $\hat{O}$ on $|\psi\rangle$ and $|s\rangle$ leads to linear combinations of these states:

$$\hat{O}|s\rangle = -|s\rangle,$$

$$\hat{O}|\psi\rangle = |\psi\rangle - \frac{2}{N^{1/2}}|s\rangle,$$

$$\hat{G}|s\rangle = \frac{2}{N^{1/2}}|\psi\rangle - |s\rangle,$$

$$\hat{G}|\psi\rangle = |\psi\rangle. \tag{15.100}$$

Thus the action of the operators $\hat{G}\hat{O}$ on $|\psi\rangle$ will be a state within the two-dimensional space spanned by $|\psi\rangle$ and $|s\rangle$.

To further understand the action of these operators, consider an arbitrary state decomposed as $|u\rangle \equiv a|s_\perp\rangle + b|s\rangle$, where $s_\perp$ is a vector orthogonal to $|s\rangle$, $\langle s_\perp|s\rangle = 0$. Then we find

$$\hat{O}|u\rangle = a|s_\perp\rangle - b|s\rangle. \tag{15.101}$$

Thus the action of $\hat{O}$ on an arbitrary state is to take the component of it along the $|s\rangle$ direction and reflect it about the remaining component of u, that by definition is orthogonal to it, $|s_\perp\rangle$. Similarly, consider an arbitrary state $|u'\rangle \equiv a|\psi\rangle + b|\psi_\perp\rangle$, where $|\psi_\perp\rangle$ is a vector orthogonal to $|\psi\rangle$, $\langle\psi|\psi_\perp\rangle = 0$. Then

$$\hat{G}|u'\rangle = a|\psi\rangle - b|\psi_\perp\rangle. \tag{15.102}$$

Thus the action of $\hat{G}$ on an arbitrary state is to take that component of it that is orthogonal to $|\psi\rangle$, $|\psi_\perp\rangle$, and reflect it about the $|\psi\rangle$ vector.

The first step of Grover's algorithm is to act on the state $|\psi\rangle$ by the oracle $\hat{O}$. For the state of interest here, the projection of $|s\rangle$ on $|\psi\rangle$ is $\langle s|\psi\rangle = 1/\sqrt{N}$, which for large N is very tiny. Decomposing $|\psi\rangle = \sqrt{\frac{N-1}{N}}|s_\perp\rangle + \frac{1}{\sqrt{N}}|s\rangle$, where in this case $|s_\perp\rangle$ is in the plane spanned by $|\psi\rangle$ and $|s\rangle$, then $\hat{O}|\psi\rangle = \sqrt{\frac{N-1}{N}}|s_\perp\rangle - \frac{1}{\sqrt{N}}|s\rangle$. In the plane defined by $|\psi\rangle$ and $|s\rangle$, let θ be the angle between $|\psi\rangle$ and $|s_\perp\rangle$ as shown in Fig. 15.12. Then the action of $\hat{O}$ on $|\psi\rangle$ is to rotate it by an angle 2θ from one side of $|s_\perp\rangle$ to the other, as shown in Fig. 15.12 by the resulting state $\hat{O}|\psi\rangle$. The action of $\hat{G}$ on this resulting state is to reflect it about the vector $|\psi\rangle$, which amounts to rotating this state by an angle 4θ from one side of $|\psi\rangle$ to the other, as shown by the resulting state $\hat{G}\hat{O}|\psi\rangle$ in Fig. 15.12. The outcome of these operations is to take the original state $|\psi\rangle$ and rotate it by an angle 2θ closer to the direction of the vector $|s\rangle$. If we now act on the resulting state $\hat{G}\hat{O}|\psi\rangle$ with the operator $\hat{O}$, it will again rotate about the vector $|s_\perp\rangle$, this time by an angle 6θ from one side of $|s_\perp\rangle$ to the other. And then the action of $\hat{G}$ will be to rotate about the direction $|\psi\rangle$ by an angle 8θ, with the resulting vector $\hat{G}\hat{O}\hat{G}\hat{O}|\psi\rangle$ now 4θ closer to the $|s\rangle$ direction, as shown in Fig. 15.12. And so on this set of operations goes.

Figure 15.12 also allows one to assess how many iterations are needed to align the original vector $|\psi\rangle$ with the direction $|s\rangle$. The angle θ is related to the probability amplitudes of $|s\rangle$ and $|s_\perp\rangle$ in the state $|\psi\rangle$, as shown in the triangle in Fig. 15.13, so that

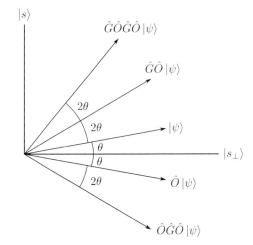

Fig. 15.12 Actions of $\hat{O}$ and $\hat{G}$ iteratively on the initial state $|\psi\rangle$ as it moves towards $|s\rangle$.

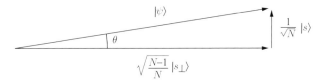

Fig. 15.13 Decomposition of the state ψ into the state that is being singled out $|s\rangle$ and the state orthogonal to it $|s_\perp\rangle$.

$$\sin\theta = \frac{1}{\sqrt{N}}.$$

For N large, so θ very small, it means $\theta \approx 1/\sqrt{N}$. Also for large N the projection of $|\psi\rangle$ is almost completely in the direction of $|s_\perp\rangle$, with only a tiny component along the $|s\rangle$ direction. Thus $|\psi\rangle$ needs to be rotated by an angle $\pi/2$ to be aligned with the $|s\rangle$ direction, with each Grover iteration rotating the vector by 2θ. Thus the number of iterations one needs is $\pi/(4\theta) = \pi\sqrt{N}/4$, which is the $\sqrt{N}$ dependence mentioned earlier.

As with any quantum computing algorithm, one would like to express all operations in terms of appropriate combinations of basic gates, as we showed for the Deutsch problem. Similar can be done for the operators $\hat{O}$ and $\hat{G}$. This will be more complicated, since these are much higher-dimensional matrices; for a list of length N it implies operators that would be represented by $N \times N$ matrices. As a matrix the oracle $\hat{O}$ is rather simple; it is a diagonal matrix with all 1s down the diagonal except one -1 entry. To create such a matrix using basic gates can be done through appropriate use of CNOT gates. Similarly the $\hat{G}$ operator can be constructed from Hadamard, $\hat{X}$ and $\hat{Z}$ gates. This is a technical detail that takes us beyond the focus of this chapter. The interested reader can find these details for example in Mermin's book referenced at the end of this chapter.

The above analysis has shown that the number of steps needed using the Grover iteration to find an item in a list goes as $O(\sqrt{N})$ rather than $O(N)$ as for a classical search algorithm. However, each iteration in the Grover algorithm requires calling many basic gate operations. Thus one may wonder if all those operations combined really make this algorithm more efficient. The real question one wants to ask is which approach is faster. Each iteration will take a certain time interval both for the classical Δt_c and quantum Δt_q case. This time interval will depend on the details of how fast the devices are for implementing that iteration. For an actual computer, classical or quantum, the practical question of interest is how much time in total does it take to complete the task. This would be the time interval for one iteration multiplied by the number of iterations. In general Δt_c and Δt_q will be different, with the former associated with the operation of comparing two quantities whereas the latter is associated with performing unitary transformations on a quantum state. The key point to note is that the time interval is just an overall multiplicative factor and it is multiplied by the number of iterations the full operation takes. Because the number of iterations for the quantum operation scales as $\sqrt{N}$ and the classical as N, it means no matter what the other multiplicative factors are, above some large enough value of N, the quantum operation will become faster than the classical, since it scales slower on the number of iterations required.

It was mentioned at the start of this section that Grover's algorithm can be used to find an item in a list and in fact this is the most common application that is discussed in most literature on this algorithm. Let us place in the proper context when this application is possible. Suppose one is given a quantum register $|\phi\rangle$ containing some linear combination of states, but one does not know any details about this state. One can use Grover's algorithm to find out if a particular state is in this register. To do that one constructs the oracle to pick out that particular state and then performs the Grover iterations. If the basis state one is searching for is in fact in the state vector $|\psi\rangle$, then after $O(\sqrt{N})$ iterations, it has a much bigger amplitude than all the other basis states. And if that basis state is not in the expansion of $|\psi\rangle$ it will have no amplitude in the iterated state. Of course, if one wishes to simply find a particular state in an unknown state vector, one could construct an operator which projects out that state and annihilates all other states. In a matrix representation it would be a diagonal matrix with 1 in the diagonal entry for the state of interest and 0 everywhere else. The difference in this approach is that it would not be a unitary operation, which means after this operation is performed, it is not reversible. All information originally about the state would be lost. On the other hand, Grover's algorithm is unitary. Thus, even though it leads to one state having a very large amplitude, one can still perform a series of inverse unitary operations and take the state back to its original form if needed. Of course, if one performed a measurement on the resulting state after the Grover iteration, that also would then be an irreversible step. However, perhaps one does not wish to perform a measurement. Perhaps one wishes to locate a particular state, enhance its amplitude and then feed that resulting state into further quantum computing manipulations, all the while not losing any information contained in the original state. If that is the goal, then Grover's algorithm could achieve it.

It is not completely clear what problems quantum computing can actually solve. The key feature of quantum computing that most of the interest centres around is that it might be able to explore many solutions at the same time. The observation being here that a quantum system in general is a superposition of many possible quantum states. For example, n spin-$\frac{1}{2}$ particles have a total of $N = 2^n$ states they can be in. Thus for 10 such spinors it corresponds to 2^{10} states, for 100, 2^{100} states and so on. As such, the number of states grows enormously, in particular exponentially, with respect to the number of spinors n. If each of these states corresponded to a potential solution to a problem, it seems possible that quantum mechanics might be able to test an enormous number of possible solutions all at once. This is vaguely the idea one is considering when looking at the potential that quantum computing might have to offer.

At its most optimistic, the hope is that quantum computing might improve the ability to solve what are called NP problems. This stands for 'nondeterministic polynomial time', which is technical terminology, but the basic meaning is problems in which once a solution is found, it can be recognised as correct in polynomial time, even though finding the solution may be very hard to do. For example, suppose one is trying to find a particular number within the set of all n-digit numbers that are randomly listed. If one simply tested each number in the list, the time it would take to find the number grows exponentially with respect to n. On the other hand, if the number one is seeking is pointed out within the list, it only takes verifying n digits to check it is indeed the correct one. Where quantum mechanics might help is if one supposed each number somehow corresponded to one of the 2^n spin states of the n spin-$\frac{1}{2}$ particles, and that some quantum algorithm could then be used to locate the desired number.

Classical computing has extreme trouble in solving NP problems. A quantum computer is intrinsically different in how it works. It manipulates qubits and as mentioned already an n-qubit state is like having 2^n possible solutions. None are immediately accessible, since an observation on the state would collapse the wave function/state vector into a single state as dictated by the rules of quantum mechanics. However, quantum mechanics can also manipulate the amplitude coefficients of all these states through the action of unitary operators. One of the key features in quantum mechanics is interference, which can allow for the probability amplitude of certain states to be enhanced and others to be diminished. If this feature could somehow be utilised to enhance the correct answer from amongst the 2^n possible solutions and suppress all others, that would be a very different way to manipulate information in contrast to classical computing. This is a key basic idea behind quantum computing.

As seen, Grover's algorithm does precisely that to manipulate the probability amplitudes of the quantum state, and does so in such a way that it is able to speed the search process so it scales as $\sqrt{N}$ rather than as N. It does not transform an exponential increase in time with respect to N to a polynomial time, but it does still give a non-trivial speeding up.

15.A Appendix: Klein's Inequality

Let $\hat{\rho} = \sum_i p_i |i\rangle\langle i|$ and $\hat{\sigma} = \sum_j q_j |j\rangle\langle j|$ be two density matrices with just orthogonal nonvanishing elements in the respective basis $\{|i\rangle\}$ and $\{|j\rangle\}$. The relative entropy of $\hat{\rho}$ with respect to $\hat{\sigma}$ is a measure of closeness between them, and is defined as

$$S_{vN}(\hat{\rho}||\hat{\sigma}) \equiv \mathrm{Tr}[\hat{\rho}\log_2\hat{\rho} - \hat{\rho}\log_2\hat{\sigma}] = \sum_i p_i \log_2 p_i - \sum_i \langle i|\hat{\rho}\log_2\hat{\sigma}|i\rangle. \qquad (15.103)$$

The Klein inequality states that the quantum relative entropy of $\hat{\rho}$ with respect to $\hat{\sigma}$ is non-negative:

$$S_{vN}(\hat{\rho}||\hat{\sigma}) \geq 0, \tag{15.104}$$

or equivalently

$$S_{vN}(\hat{\rho}) \leq -\text{Tr}[\hat{\rho} \log_2 \hat{\sigma}]. \tag{15.105}$$

Proof: From Eq. (15.103) note that

$$\langle i| \log_2 \hat{\sigma} |i\rangle = \langle i| \sum_j \log_2 q_j |j\rangle\langle j||i\rangle = \sum_j \log_2 q_j P_{ij}, \tag{15.106}$$

where $P_{ij} \equiv \langle i|j\rangle\langle j|i\rangle \geq 0$. Using completeness and orthogonality, it can easily be verified that the matrix P_{ij} satisfies $\sum_i P_{ij} = \sum_j P_{ij} = 1$. Substituting Eq. (15.106) into Eq. (15.103) gives

$$S_{vN}(\hat{\rho}||\hat{\sigma}) = \sum_i p_i \left(\log_2 p_i - \sum_j P_{ij} \log_2 q_j \right). \tag{15.107}$$

Here $\log_2 x$ is a concave function, meaning it satisfies $\log_2[(1-a)x + ay] \geq (1-a)\log_2 x + a \log_2 y$ where $a \in [0, 1]$. Thus

$$\sum_j P_{ij} \log_2 q_j \leq \log_2 r_i < 0, \tag{15.108}$$

where $r_i = \sum_j P_{ij} q_j$. Equality only occurs if for some j, $P_{ij} = 1$, which means all other entries of P_{ij} must be zero. Plugging into Eq. (15.107) gives

$$S_{vN}(\hat{\rho}||\hat{\sigma}) \geq \sum_i p_i \log_2 \frac{p_i}{r_i} = -\sum_i p_i \log_2 \left(\frac{r_i}{p_i} \right). \tag{15.109}$$

The following relation is now useful:

$$-\log_2(x) \ln 2 = -\ln(x) \geq 1 - x, \tag{15.110}$$

for $x > 0$, with equality if and only if $x = 1$. Using Eq. (15.110) in Eq. (15.109) gives

$$S(\hat{\rho}||\hat{\sigma}) \geq -\frac{1}{\ln 2} \sum_i p_i \left(1 - \frac{r_i}{p_i} \right) = \frac{1}{\ln 2} \sum_i (r_i - p_i) = 0, \tag{15.111}$$

thus proving Eq. (15.104) (from Nielsen & Chuang, 2010).

Summary

- The local hidden variables theories are an alternative proposal to quantum mechanics. (15.1)
- The Einstein, Podolsky and Rosen (EPR) paradox illustrates the property of quantum entanglement. (15.1.1)
- Bell's inequality provides an experimental test to distinguish between quantum mechanics and local hidden variables theories. (15.2)
- Experimental tests of Bell's inequality have shown that local hidden variable theories are ruled out. (15.2.2)

- The qubit is a two-component quantum state and is the basic unit of quantum information. (15.3.1)
- The Bell states are the four possible two-qubit entangled states. (15.3.1)
- The Shannon entropy is, according to classical information theory, the minimum bits per symbol needed to encode a string of symbols. (15.3.2)
- The von Neumann entropy serves the analogue role for quantum information as the Shannon entropy does for classical information. (15.3.3)
- The entanglement entropy is a measure of the degree of entanglement for a quantum state. (15.3.4)
- The measurement process can be used in a variety of ways to determine quantum behaviour with quantities such as entanglement entropy, mutual information and quantum discord. (15.3.5)
- The no-cloning theorem states that it is impossible to create an identical copy of an arbitrary unknown quantum state. (15.4.1)
- Quantum teleportation allows the complete information about an unknown quantum state possessed by one observer to be transferred into the quantum state of a similar system (such as a photon) possessed by another observer. (15.4.2)
- Superdense coding utilises an entangled state between two observers to allow one participant to transmit two bits of classical information. (15.4.3)
- Quantum computing manipulates a quantum state (register) using quantum operators (gates) in a manner to allow processing of information utilising the quantum-mechanical properties of superposition, entanglement and interference. (15.5)
- The Deutsch algorithm is the simplest example of a quantum computing algorithm for an idealised problem. It answers the question of whether a function with two values is equal or not in just one step. (15.5.2)
- The Grover algorithm searches for a particular state from a set of quantum states. (15.5.3)

Reference

The proof of Klein's inequality and our discussion of subadditivity follows from the below reference. This is also a good book to learn more about quantum information.

Nielsen, M. A. and Chuang, I. L. (2010). *Quantum Computation and Quantum Information*. Cambridge University Press, Cambridge.

Further Reading

To learn more about quantum computing and quantum information, these books are recommended.

Mermin, N. D. (2007). *Quantum Computing Science: An Introduction*. Cambridge University Press, Cambridge.

Nielsen, M. A. and Chuang, I. L. (2010). *Quantum Computation and Quantum Information*. Cambridge University Press, Cambridge.

Preskill, J. (2015). *Lecture Notes for Physics 229: Quantum Information and Computation*.

Wilde, M. M. (2017). *Quantum Information Theory*. Cambridge University Press, Cambridge.

Problems

15.1 Show that the four Bell states are orthonormal. Express the four basis elements $|\uparrow\uparrow\rangle$, $|\uparrow\downarrow\rangle$, $|\downarrow\uparrow\rangle$ and $|\downarrow\downarrow\rangle$ as linear combinations of Bell states.

Consider the two-qubit state vector $|\Psi\rangle = \frac{1}{2}(|\uparrow\uparrow\rangle + |\uparrow\downarrow\rangle + |\downarrow\uparrow\rangle + |\downarrow\downarrow\rangle)$. Show that this state is separable, so can be expressed as $|\Psi\rangle = |\Psi_A\rangle \otimes |\Psi_B\rangle$, and determine $|\Psi_A\rangle$ and $|\Psi_B\rangle$.

15.2 Suppose Alice (A) and Bob (B) share a pair of electrons prepared in the state

$$|\chi_{AB}\rangle = \frac{1}{\sqrt{30}}\left[5|\uparrow_A\downarrow_B\rangle + 2|\downarrow_A\uparrow_B\rangle - |\downarrow_A\downarrow_B\rangle\right].$$

Show that this state is entangled by showing that it cannot be written as a tensor product between the states $a|\uparrow_A\rangle + b|\downarrow_A\rangle$ and $c|\uparrow_B\rangle + d|\downarrow_B\rangle$, for any choice of coefficients a, b, c and d.

Suppose Alice measures her electron with respect to the x-axis, where $|+\rangle \equiv |x_+\rangle = (|\uparrow\rangle + |\downarrow\rangle)/\sqrt{2}$ and $|-\rangle \equiv |x_-\rangle = (|\uparrow\rangle - |\downarrow\rangle)/\sqrt{2}$. Determine the probability of Alice measuring each outcome with respect to the x-axis, spin-up $|+_A\rangle$ and spin-down $|-_A\rangle$.

For each of the two possibilities, determine the state onto which the electron at B will project.

15.3 In the quantum teleportation problem described in the main text, Alice had the unknown state $|\chi_{A'}\rangle = c|\uparrow_{A'}\rangle + d|\downarrow_{A'}\rangle$ that resulted in one of four possible states teleported to Bob $|\chi_1\rangle = -c|\uparrow\rangle - d|\downarrow\rangle$, $|\chi_2\rangle = -c|\uparrow\rangle + d|\downarrow\rangle$, $|\chi_3\rangle = c|\downarrow\rangle + d|\uparrow\rangle$ and $|\chi_4\rangle = c|\downarrow\rangle - d|\uparrow\rangle$. For each of the four cases, state the unitary transformation that takes $|\chi_i\rangle$ into $|\chi_{A'}\rangle$.

15.4 Show that, if the density matrix is in diagonal form with just orthonormal states, then the von Neumann entropy equals the Shannon entropy. Show that if the density matrix is for a single pure state $|\phi\rangle$, the von Neumann entropy is zero.

Suppose the density matrix is written in the basis of three states, $|\phi_1\rangle$, $|\phi_2\rangle$ and $|\phi_3\rangle$. In this basis, write the density matrix for the pure state $|\phi\rangle = a_1|\phi_1\rangle + a_2|\phi_2\rangle + \sqrt{1 - a_1^2 - a_2^2}|\phi_3\rangle$ and compute the resulting von Neumann entropy.

Suppose the density matrix is comprised of two spin states, one is spin-up in the z-direction with probability p_1 and the other is spin-up in a direction at angle θ with respect to the z-axis, in the z-x plane, with probability p_2. Compute the von Neumann entropy and compare to the Shannon entropy.

15.5 Consider the two-spin state, $|\Psi\rangle = \cos\theta| \uparrow_A\downarrow_B\rangle + \sin\theta| \downarrow_A\uparrow_B\rangle$. Compute the entanglement entropy S_A and S_B. Show that the state of maximal entanglement is for $\theta = \pi/4$ and that the separable tensor product state where $S_A = S_B = 0$ is for $\theta = 0, \pi/2$.

15.6 The density matrix

$$\rho_{Werner} = \kappa|\Psi^-\rangle\langle\Psi^-| + \frac{1-\kappa}{3}\left[|\Psi^+\rangle\langle\Psi^+| + |\Phi^-\rangle\langle\Phi^-| + |\Phi^+\rangle\langle\Phi^+|\right]$$

is called a **Werner state**. Here $|\Psi^-\rangle = 1/\sqrt{2}[| \uparrow_A\downarrow_B\rangle - | \downarrow_A\uparrow_B\rangle]$, and so forth for the other Bell states. Show that the above density matrix can equivalently be written as

$$\rho_{Werner} = a|\Psi^-\rangle\langle\Psi^-| + bI,$$

where I is a 4×4 identity matrix, and determine a and b.

Calculate $\text{Tr}\rho^2_{Werner}$. Explain why, if $\text{Tr}\rho^2 < 1$, then that density matrix is a mixed state. For what κ is $\text{Tr}\rho^2_{Werner}$ smallest? Write this maximally mixed state.

Compute the entanglement and von Neumann entropies for ρ_{Werner}.

15.7 Two photons are initially prepared at a source and move along the y-direction in opposite directions toward polarisers at A and B. They are initially in the state

$$|\Psi\rangle = \frac{1}{\sqrt{2}}[|V_A V_B\rangle + |H_A H_B\rangle],$$

where $|V_I\rangle$ and $|H_I\rangle$ are the two linear polarisation states, vertical and horizontal, of the photon moving to the polariser at $I = A$ or B. To be explicit, we will fix the vertical polarisation direction here as the $\hat{z}$-direction. In particular, the V state is in the direction of $\hat{z}$ and the H state is in the $\hat{x}$-direction, thus perpendicular to both the $\hat{z}$-direction and the direction of motion for the photon. Let $\underline{n}_I$ be other directions at polarisers $I = A$ or B that are perpendicular to the direction of motion of the photons, so in the $\hat{x}$–$\hat{z}$ plane, and at an angle θ_{Az} to the $\hat{z}$-direction. If photon I is measured through a polariser oriented with respect to the $\underline{n}_I$-direction:

$$|V_{I,I}\rangle = \cos\theta_{Iz}|V\rangle - \sin\theta_{Iz}|H\rangle,$$
$$|H_{I,I}\rangle = \sin\theta_{Iz}|V\rangle + \cos\theta_{Iz}|H\rangle.$$

The two photons initially in the state $|\Psi\rangle$ are then measured respectively at A and B, with polarisers oriented along the $\underline{n}_A$ and $\underline{n}_B$-directions, respectively. Let $P_{VV}(\underline{n}_A, \underline{n}_B)$ be the probability that the vertical polarisation of the photon at A is measured along direction $\underline{n}_A$ for polarisers rotated by angle θ_{Az} and for the photon at B the vertical polarisation is measured along direction $\underline{n}_B$ for polarisers rotated by angle θ_{Bz}. Show that

$$P_{VV}(\underline{n}_A, \underline{n}_B) = |\langle V_{A,A} V_{B,B}|\Psi\rangle|^2 = \tfrac{1}{2}\cos^2(\theta_{B1} - \theta_{A1}).$$

Also compute $P_{HH}(\underline{n}_A, \underline{n}_B)$, $P_{VH}(\underline{n}_A, \underline{n}_B)$ and $P_{HV}(\underline{n}_A, \underline{n}_B)$.

Show that

$$E(\underline{n}_A, \underline{n}_B) \equiv P_{VV}(\underline{n}_A, \underline{n}_B) + P_{HH}(\underline{n}_A, \underline{n}_B) - P_{VH}(\underline{n}_A, \underline{n}_B) - P_{HV}(\underline{n}_A, \underline{n}_B)$$
$$= \cos[2(\theta_{Az} - \theta_{Bz})].$$

For the expression

$$S = E(\underline{n}_A, \underline{n}_B) - E(\underline{n}_A, \underline{n}'_B) + E(\underline{n}'_A, \underline{n}_B) + E(\underline{n}'_A, \underline{n}'_B),$$

a realistic local theory gives the Bell inequality $-2 \le S \le 2$. Show that this inequality is violated for the quantum-mechanical expressions for photons derived here. For this choose the orientations

$$\theta_{Bz} - \theta_{Az} = \theta_{A'z} - \theta_{Bz} = \theta_{B'z} - \theta_{A'z} = \phi \quad \text{and} \quad \theta_{B'z} - \theta_{Az} = 3\phi,$$

$$\underline{n}_i \cdot \underline{n}_j = \underline{n}'_i \cdot \underline{n}_j = \underline{n}'_i \cdot \underline{n}'_j = \cos\phi \quad \text{and} \quad \underline{n}_i \cdot \underline{n}'_j = \cos 3\phi$$

and plot as a function of ϕ. Determine the range of angles for ϕ for which Bell's inequality is violated.

15.8 Show that the quantum mutual information between two systems A and B is zero, $I_{vN}(A, B) = 0$, when $\hat{\rho}_{AB} = \hat{\rho}_A \otimes \hat{\rho}_B$.

Show that all probabilistic mixtures of the Bell states can be written in the form

$$\hat{\rho}_{Bell} = \frac{1}{4}\left(I + \sum_{i=1}^{3} c_i \sigma_i \otimes \sigma_i\right),$$

for some real constants $\{c_i\}$. Compute the eigenvalues of $\hat{\rho}_{AB}$. Compute the quantum mutual information in $\hat{\rho}_{AB}$.

15.9 Let $\{\hat{\Pi}_j\}$ be a complete set of orthogonal projectors representing measurement and let $\hat{\rho}$ be a density operator. Define the resulting density operator after measurement $\hat{\rho}' = \sum_j \hat{\Pi}_j \hat{\rho} \hat{\Pi}_j$. Show that $S_{vN}(\hat{\rho}') \ge S_{vN}(\hat{\rho})$. In words, the von Neumann entropy of a system increases due to the act of measurement. Discuss why entropy increases from measurement as this result suggests.

15.10 Form an alternative proof of the no-cloning theorem to the one given in the text, by assuming that $\hat{U}$ is a unitary operator that copies an arbitrary quantum state $|\phi\rangle$. Perform this operation on both the state and a basis decomposition of the state and assess the consequences. For simplicity, assume the Hilbert space has only two states, $|\phi\rangle = a|\uparrow\rangle + b|\downarrow\rangle$.

15.11 The Greenberger, Horne and Zeilinger (GHZ) state vector is a three-qubit entangled state, which can be written as

$$|GHZ\rangle = \frac{1}{\sqrt{2}}(|000\rangle + |111\rangle).$$

Use unitary transformations to show that an alternative form is

$$|\Gamma\rangle = \frac{1}{2}(|000\rangle - |011\rangle - |101\rangle - |110\rangle). \tag{15.112}$$

15.12 A kindly quantum mechanics professor helps the three students in his class who are struggling with the course by offering a game, which, if they win, they will be allowed to pass the course. The names of the three students are obviously Alice (A), Bob (B) and Charlie (C). The three students will be placed in three separate rooms and not allowed to communicate with anyone but the professor (yes, *even* their smartphones are taken from them). The professor goes to every room and hands each student a slip of paper which has written on it either Heisenberg (H) or Schrödinger (S). The professor has explained to the students that either he will hand out two slips or zero slips with Heisenberg written on them. After the slips of paper are handed out, the students are instructed to shout out either 1 or −1. To win the game, if S was handed out to all three students, the product of what they shout out needs to be 1 and if two students had H handed to them, the product of what they shout out needs to be −1. To summarise, the four possible combinations of names on the slips and the associated product of what needs to be shouted out are respectively (SSS, SHH, HSH, HHS) (ordered as Alice, Bob, Charlie, ABC) and $(1, -1, -1, -1)$. The students can work together to devise any plan they like before the game starts, but after that no communication is allowed. Show that no classical deterministic or classical probabilistic strategy can guarantee winning the game. However, show that by utilising quantum mechanics, there is a foolproof solution for winning this game and it involves this entangled GHZ state, Eq. (15.112).

15.13 Apply the superdense coding protocol. Suppose Alice and Bob share the entangled two-qubit Bell state

$$|\Phi^+\rangle = \tfrac{1}{2}\left(|00\rangle + |11\rangle\right),$$

with the first entry of each state associated with the qubit of Alice (A) and the second with the qubit of Bob (B). Work through the protocol for the three examples not treated in the text, the cases where Alice wants to communicate the two bits of information $(m, n) = (0, 0), (1, 0)$ and $(0, 1)$ through sending just one qubit to Bob.

15.14 Show that the action of the Hadamard operator on a single qubit can be expressed as

$$\hat{H}_d|x\rangle_1 = \frac{1}{\sqrt{2}} \sum_{y=0}^{1} (-1)^{xy} |y\rangle,$$

where $x = 0, 1$.

Generalise this for an n-qubit register to show

$$\hat{H}_d^{\otimes n}|x_n \ldots x_1\rangle_n = \frac{1}{2^{n/2}} \sum_{y_n=0}^{1} \cdots \sum_{y_1=0}^{1} (-1)^{\sum_{i=1}^{n} x_i y_i} |y_n \ldots y_1\rangle,$$

where $x_i = 0, 1$ for $i = 1, \ldots, n$.

15.15 The Deutsch algorithm can be generalised to many input bits and is then called the Deutsch–Josza algorithm. Suppose there is an n-bit function $f_{DJ}(\{x\}) : \{0, 1\}^n \rightarrow \{0, 1\}$, where $\{0, 1\}^n$ means a string of n integers each taking the value either 0 or 1. Suppose we know that either $f_{DJ}(\{x\})$ is the same, either 0 or 1, for all n-bit inputs, called constant, or $f_{DJ}(\{x\}) = 0$ for precisely half the $\{x\} \in \{0,1\}^n$ and $f_{DJ}(\{x\}) = 1$ for the other half, called balanced. Classically one would need to sample $f_{DJ}(\{x\})$ $2^{n-1} + 1$ times to determine with certainty whether $f_{DJ}(\{x\})$ was constant or balanced. The quantum computing Deutsch–Josza algorithm, which will be developed here, can determine this with one iteration. Define the oracle

$$\hat{O}|x\rangle_n = (-1)^{f_{DJ}(\{x\})}|x\rangle_n.$$

Show that

$$|\psi_1\rangle \equiv \hat{H}_d^{\otimes n}\hat{O}\hat{H}_d^{\otimes n}|0\rangle_n = \sum_{y\in\{0,1\}^n}\sum_{x\in\{0,1\}^n}\frac{(-1)^{f_{DJ}(\{x\})+x\cdot y}}{2^n}|y\rangle_n,$$

where the sum for x or y is over a string of n integers each taking the value either 0 or 1.

Examine the amplitude $_n\langle 0|\psi_1\rangle$. Show that if $f_{DJ}(\{x\})$ is constant, then this amplitude is ± 1 depending on whether the constant is 0 or 1 and this amplitude is 0 if $f_{DJ}(\{x\})$ is balanced.

15.16 Let a be an unknown non-negative integer less than 2^n. Let $f(\{x\}) = a \cdot x = \sum_{j=1}^n a_j x_j$. Classically, if we wanted to determine what a is, we would need to evaluate $f(\{x\})$ n times. There is a quantum computing algorithm, Bernstein–Vazirani, in which a single operation is enough to determine a. For this define the oracle acting on an n-qubit register as

$$\hat{O}|x\rangle_n = (-1)^{f(\{x\})}|x\rangle_n.$$

Show that

$$\hat{H}_d^{\otimes n}\hat{O}\hat{H}_d^{\otimes n}|0\rangle_n = |a\rangle_n.$$

Note: The above operations will first lead to the same form as in the previous question. From that form, then consider the specific form of $f_{BV}(\{x\})$ to obtain the above expression.

Suppose each qubit were a spinor. Explain how one measures the final result. Why does the measurement process not alter the fact that this problem is solved in one operation?

Show that at no stage in this algorithm is the state of the system entangled.

16   Time-Independent Perturbation Theory

Very few problems in quantum mechanics can be solved exactly. For example, in the case of the helium atom, including the inter-electron electrostatic repulsion term in the Hamiltonian changes the problem into one which cannot be solved analytically. Perturbation theory provides a method for finding approximate energy eigenvalues and eigenstates for a system whose Hamiltonian is of the form

$$\hat{H} = \hat{H}_0 + \hat{H}'.$$

The system for which we want to find the eigenstates and eigenvalues is $\hat{H}$, yet the system for which we know exactly the eigenstates and eigenvalues is only $\hat{H}_0$. The addition of $\hat{H}'$ to $\hat{H}_0$ now makes the problem intractable to solve exactly. Perturbation theory provides a systematic method for obtaining approximate solutions to $\hat{H}$ as an expansion in powers of the perturbation Hamiltonian $\hat{H}'$. In this chapter this method will be explained. There is a distinction between systems with and without degeneracies and the perturbation expansion for both cases will be developed here. Applications of perturbation theory will be examined for the hydrogen and helium atoms. Their basic properties were discussed in earlier chapters, and here perturbation corrections will be examined.

16.1   Nondegenerate Time-Independent Perturbation Theory

To set up the problem, for the exactly solvable Hamiltonian, $\hat{H}_0$, denote the eigenvalues, $E_n^{(0)}$, and corresponding eigenstates, $|n^{(0)}\rangle$, all of which we know. The Hamiltonian $\hat{H}'$ is treated as a small, time-independent perturbation. $\hat{H}$, $\hat{H}_0$ and $\hat{H}'$ are all Hermitian operators.

The basic idea is to get approximate solutions of the TISE for problems which are 'close' to exactly solvable systems. Since the approximation works in powers of the perturbation Hamiltonian, it is convenient to introduce a bookkeeping parameter λ as

$$\hat{H} = \hat{H}_0 + \lambda \, \hat{H}', \tag{16.1}$$

which is a real parameter that conveniently can keep track of the number of powers of the perturbation Hamiltonian in the intermediate stages of the calculation, but then set to one at the end of the analysis. We shall assume for now that $\hat{H}$ and $\hat{H}_0$ possess discrete, nondegenerate eigenvalues only.

We write

$$\hat{H}_0 \, |n^{(0)}\rangle = E_n^{(0)} \, |n^{(0)}\rangle \qquad (16.2)$$

in Dirac notation. The states $|n^{(0)}\rangle$ are orthonormal. The effect of the perturbation will be to modify each state and its corresponding energy slightly; $|n^{(0)}\rangle$ will go over to $|n\rangle$ and $E_n^{(0)}$ will shift to E_n, where

$$\hat{H} \, |n\rangle = E_n \, |n\rangle. \qquad (16.3)$$

It is helpful to see pictorially what perturbation theory is attempting to do. For this let us imagine the Hilbert space spanned by the eigenstates $\{|n^{(0)}\rangle\}$ of $\hat{H}_0$. This is in general an infinite-dimensional space, but we can make a pictorial representation as shown in Fig. 16.1. All these states we can calculate exactly, and the set of all these eigenstates $\{|n^{(0)}\rangle\}$ spans the Hilbert space. On the other hand, the eigenstates of $\hat{H}$ are what we want to know, yet we cannot calculate them exactly. What we do know is that they also will lie in the same Hilbert space as those spanned by the eigenvectors of $\hat{H}_0$, $\{|n^{(0)}\rangle\}$. Thus the set of states $\{|n\rangle\}$ can be expressed as some linear combination of the basis states $\{|n^{(0)}\rangle\}$. The starting premise for using perturbation theory is that $\hat{H}'$ is in some respect 'small' (and we will quantify what this means a little later), which means we therefore imagine that each of the exact eigenstates $|n\rangle$ is close to one of the approximate eigenstates $|n^{(0)}\rangle$ as depicted in Fig. 16.1. In this geometric way of looking at the problem, the goal of perturbation theory is to determine the small corrections needed in each of the vectors $|n^{(0)}\rangle$ so it becomes the corresponding vector $|n\rangle$. Perturbation theory offers a systematic procedure for doing that.

Let us obtain the expressions for the case where there are no degeneracies in the eigenstate spectrum of $\hat{H}_0$. We solve the full eigenvalue problem by assuming that we can expand E_n and $|n\rangle$ in a series as follows:

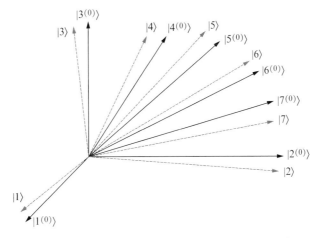

Fig. 16.1 Illustration of the Hilbert space of state vectors. The dashed vectors represent eigenstates of the exact Hamiltonian $\hat{H}$ and the solid vectors the eigenstates of the unperturbed Hamiltonian $\hat{H}_0$.

$$E_n = E_n^{(0)} + \lambda\, E_n^{(1)} + \lambda^2\, E_n^{(2)} + \cdots, \tag{16.4}$$

$$|n\rangle = |n^{(0)}\rangle + \lambda\, |n^{(1)}\rangle + \lambda^2\, |n^{(2)}\rangle + \cdots, \tag{16.5}$$

where the correction terms $E_n^{(1)}$, $E_n^{(2)}, \ldots$ and $|n^{(1)}\rangle$, $|n^{(2)}\rangle, \ldots$ are of successively higher order of smallness, with the power of the parameter λ keeping track of this for us. The correction terms $|n^{(1)}\rangle$, $|n^{(2)}\rangle, \ldots$ are *not* normalised to unity.

We can substitute these expansions into the eigenvalue equation for the full Hamiltonian:

$$(\hat{H}_0 + \lambda\,\hat{H}') \left(|n^{(0)}\rangle + \lambda\,|n^{(1)}\rangle + \lambda^2\,|n^{(2)}\rangle + \cdots \right)$$
$$= (E_n^{(0)} + \lambda\, E_n^{(1)} + \lambda^2\, E_n^{(2)} + \cdots) \left(|n^{(0)}\rangle + \lambda\,|n^{(1)}\rangle + \lambda^2\,|n^{(2)}\rangle + \cdots \right),$$

and equate terms of the same degree in λ, giving

$$
\begin{aligned}
\lambda^0 &: \hat{H}_0\,|n^{(0)}\rangle &&= E_n^{(0)}\,|n^{(0)}\rangle \\
\lambda^1 &: \left(\hat{H}_0 - E_n^{(0)}\right)|n^{(1)}\rangle &&= \left(E_n^{(1)} - \hat{H}'\right)|n^{(0)}\rangle \\
\lambda^2 &: \left(\hat{H}_0 - E_n^{(0)}\right)|n^{(2)}\rangle &&= \left(E_n^{(1)} - \hat{H}'\right)|n^{(1)}\rangle + E_n^{(2)}\,|n^{(0)}\rangle \\
\lambda^3 &: \left(\hat{H}_0 - E_n^{(0)}\right)|n^{(3)}\rangle &&= \left(E_n^{(1)} - \hat{H}'\right)|n^{(2)}\rangle + E_n^{(2)}\,|n^{(1)}\rangle + E_n^{(3)}\,|n^{(0)}\rangle.
\end{aligned}
\tag{16.6}
$$

The first of these equations tells us nothing new; it simply restates that $|n^{(0)}\rangle$ is an eigenstate of $\hat{H}_0$ belonging to the eigenvalue $E_n^{(0)}$.

16.1.1 First-Order Correction

The second equation is more interesting; taking the scalar product with, say, the kth unperturbed eigenstate yields

$$\langle k^{(0)}| \left(\hat{H}_0 - E_n^{(0)}\right)|n^{(1)}\rangle = \langle k^{(0)}| \left(E_n^{(1)} - \hat{H}'\right)|n^{(0)}\rangle. \tag{16.7}$$

The first term on the left-hand side can be rewritten

$$
\begin{aligned}
\langle k^{(0)}|\hat{H}_0\,|n^{(1)}\rangle &= \langle n^{(1)}|\hat{H}_0^\dagger|k^{(0)}\rangle^* \\
&= \langle n^{(1)}|\hat{H}_0\,|k^{(0)}\rangle^* && \text{since } \hat{H}_0 \text{ is Hermitian} \\
&= E_k^{(0)}\langle n^{(1)}|k^{(0)}\rangle^* && \text{since } E_k^{(0)} \text{ is real} \\
&= E_k^{(0)}\langle k^{(0)}|n^{(1)}\rangle.
\end{aligned}
$$

Thus the equation becomes

$$\left(E_k^{(0)} - E_n^{(0)}\right)\langle k^{(0)}|n^{(1)}\rangle = E_n^{(1)}\,\delta_{k,n} - \langle k^{(0)}|\hat{H}'\,|n^{(0)}\rangle. \tag{16.8}$$

Since we have chosen k arbitrarily, let's see what happens when we make the particular choice $k = n$. The left-hand side vanishes, and the Kronecker delta on the right-hand side is equal to one, yielding

$$E_n^{(1)} = \langle n^{(0)}|\hat{H}'\,|n^{(0)}\rangle \equiv H'_{nn}, \tag{16.9}$$

which is possibly *the most useful result in quantum mechanics*. It tells us how to compute the change in the nth energy eigenvalue, to first order:

Fig. 16.2   Effect of a perturbation on a nondegenerate level of energy $E_n^{(0)}$.

The shift in energy induced by a perturbation is given to first order by the expectation value of the perturbation with respect to the unperturbed state.

Since $E_n^{(1)}$ represents the *shift* in the energy to first order, it is sometimes also written $\Delta E_n^{(1)}$, or just ΔE_n, as shown in Fig. 16.2.

If, on the other hand, we choose $k \neq n$, we can find the first-order correction to the eigenstates. We have

$$\left(E_k^{(0)} - E_n^{(0)}\right) \langle k^{(0)}|n^{(1)}\rangle = -\langle k^{(0)}|\hat{H}'|n^{(0)}\rangle \qquad k \neq n. \tag{16.10}$$

We can expand the first-order correction term $|n^{(1)}\rangle$ in terms of the complete set of unperturbed energy eigenstates:

$$|n^{(1)}\rangle = \sum_m a_{nm}^{(1)} |m^{(0)}\rangle. \tag{16.11}$$

Taking the scalar product with $\langle k^{(0)}|$ yields

$$\langle k^{(0)}|n^{(1)}\rangle = \sum_m a_{nm}^{(1)} \langle k^{(0)}|m^{(0)}\rangle$$

$$= \sum_m a_{nm}^{(1)} \delta_{k,m}$$

$$= a_{nk}^{(1)},$$

so that the coefficients are given by

$$a_{nk}^{(1)} = \frac{\langle k^{(0)}|\hat{H}'|n^{(0)}\rangle}{(E_n^{(0)} - E_k^{(0)})} \qquad k \neq n. \tag{16.12}$$

Note that there is no need for the coefficient at $k = n$. This freedom will be used in the next section to normalise the eigenstates $|n\rangle$ computed from perturbation theory. From this it will emerge that the first-order correction term for $k = n$ will be zero. Thus we end up with the following expression for the first-order correction to the nth energy eigenstate:

$$|n^{(1)}\rangle = \sum_{k \neq n} \frac{\langle k^{(0)}|\hat{H}'|n^{(0)}\rangle}{\left(E_n^{(0)} - E_k^{(0)}\right)} |k^{(0)}\rangle \equiv \sum_{k \neq n} \frac{H'_{kn}}{\left(E_n^{(0)} - E_k^{(0)}\right)} |k^{(0)}\rangle. \tag{16.13}$$

We speak of the perturbation *mixing the unperturbed eigenstates*, since the effect is to add to the unperturbed eigenstate, $|n^{(0)}\rangle$, a small amount of each of the other unperturbed eigenstates. We see that, unlike the formula for the energy shift, we are faced in general with evaluating an infinite sum to find the correction to the eigenstates.

Notes:

- For the first-order changes to the eigenstate to be small, it had better be the case that

$$\left|H'_{kn}\right| \ll \left|E_n^{(0)} - E_k^{(0)}\right| \qquad \text{for all } k \neq n. \tag{16.14}$$

- Similarly, we require that the level shift be small compared to the level spacing in the unperturbed system:

$$\left|E_n^{(1)}\right| \ll \min \left|E_n^{(0)} - E_k^{(0)}\right|. \tag{16.15}$$

- These conditions may break down *if there are degeneracies in the unperturbed system.* However, we need only assume that the *particular energy level whose shift we are calculating is nondegenerate* for the preceding analysis to be correct.

Example 16.1 A particle moves in the one-dimensional potential

$$V(x) = \infty, \quad |x| > a, \qquad V(x) = V_0 \cos(\pi x/2a), \quad |x| \leq a.$$

Calculate the ground-state energy to first order in perturbation theory.

Here we take the unperturbed Hamiltonian, $\hat{H}_0$, to be that of the infinite square well, for which we already know the eigenvalues and eigenstates:

$$E_n^{(0)} = \frac{\pi^2 \hbar^2 n^2}{8ma^2}, \qquad u_n^{(0)} = \frac{1}{\sqrt{a}} \left\{ \begin{array}{c} \cos \\ \sin \end{array} \right\} \frac{n\pi x}{2a} \quad \text{for } n \left\{ \begin{array}{c} \text{odd} \\ \text{even} \end{array} \right\}.$$

The perturbation $\hat{H}'$ is $V_0 \cos(\pi x/2a)$, which is small provided $V_0 \ll E_2^{(0)} - E_1^{(0)}$. To first order, then:

$$\Delta E \equiv E_1^{(1)} = H'_{11} = \int_{-\infty}^{\infty} u_1^{(0)} \hat{H}' u_1^{(0)} \, dx = \frac{V_0}{a} \int_{-a}^{a} \cos^3 \frac{\pi x}{2a} \, dx.$$

Evaluating the integral is straightforward and yields the result

$$\Delta E = \frac{8V_0}{3\pi} = 0.85 \, V_0.$$

16.1.2 Second-Order Energy Correction

It may turn out that the first-order calculation yields zero shift, for example on symmetry grounds. In this case we must consider what happens at second order. Going back to our results for the λ^2 terms in the expansion, we saw that

$$\left(\hat{H}_0 - E_n^{(0)}\right) |n^{(2)}\rangle = \left(E_n^{(1)} - \hat{H}'\right) |n^{(1)}\rangle + E_n^{(2)} |n^{(0)}\rangle. \tag{16.16}$$

Proceeding as before and taking the scalar product with an unperturbed eigenstate gives

$$\langle k^{(0)}| \left(\hat{H}_0 - E_n^{(0)}\right) |n^{(2)}\rangle = \langle k^{(0)}| \left(E_n^{(1)} - \hat{H}'\right) |n^{(1)}\rangle + E_n^{(2)} \delta_{k,n}. \tag{16.17}$$

The left-hand side may be simplified as before, replacing $\hat{H}_0$ by the eigenvalue $E_k^{(0)}$ so that

$$\left(E_k^{(0)} - E_n^{(0)}\right)\langle k^{(0)}|n^{(2)}\rangle = \langle k^{(0)}|\left(E_n^{(1)} - \hat{H}'\right)|n^{(1)}\rangle + E_n^{(2)}\,\delta_{k,n}. \tag{16.18}$$

When $k = n$ this simplifies to give

$$E_n^{(2)} = \langle n^{(0)}|\hat{H}'|n^{(1)}\rangle - E_n^{(1)}\langle n^{(0)}|n^{(1)}\rangle$$

$$= \langle n^{(0)}|\hat{H}'|n^{(1)}\rangle \qquad \text{using the orthogonality of } |n^{(1)}\rangle \text{ and } |n^{(0)}\rangle$$

$$= \sum_{m\neq n} a_{nm}^{(1)}\,H'_{nm} \qquad \text{using the expansion of } |n^{(1)}\rangle \text{ in Eq. (16.11)}$$

$$= \sum_{m\neq n} \frac{H'_{mn}\,H'_{nm}}{\left(E_n^{(0)} - E_m^{(0)}\right)} \qquad \text{using Eq. (16.12)}$$

$$= \sum_{m\neq n} \frac{|H'_{mn}|^2}{\left(E_n^{(0)} - E_m^{(0)}\right)} \qquad \text{since } H'_{nm} = (H'_{mn})^*.$$

Again we are in the position in general of having to evaluate an infinite sum of terms to get the desired result, although sometimes there are reasons, such as symmetry, why only a finite number of terms is non-zero. Sometimes we may obtain a sufficiently accurate answer by including in the sum only a few terms for which $\left|E_n^{(0)} - E_m^{(0)}\right|$ is small.

16.1.3 Normalisation and Orthogonality of the Eigenstates to First Order

The perturbed eigenstates $|n\rangle$ of the full Hamiltonian $\hat{H}$ need to be orthonormalised, $\langle m|n\rangle = \delta_{mn}$. The perturbed eigenstates as computed by the perturbation theory procedure still leave the freedom to multiply the whole perturbed eigenstates by an overall factor. Moreover, there is freedom to choose the expansion coefficient $a_{nn}^{(k)}$ at every order k. This was already noted at first order $k = 1$, but in fact it holds at all orders. To normalise the perturbed state, at first order we impose the condition,

$$\lambda^1 : \langle n^{(0)}|n^{(1)}\rangle + \langle n^{(1)}|n^{(0)}\rangle = 0, \tag{16.19}$$

which implies $a_{nn}^{(1)} + a_{nn}^{(1)*} = 0$, thus $\mathrm{Re}(a_{nn}^{(1)}) = 0$. This means that the imaginary part of $a_{nn}^{(k)}$ is left indeterminate. This pattern persists to higher orders – this normalisation condition implies order-by-order conditions on the real parts of $a_{nn}^{(k)}$ but leaves the imaginary parts indeterminate, which can be set to zero.

Let us now check the orthogonality of the perturbed states up to first order in λ. For states $n \neq m$, consider to first order (here the bookkeeping parameter λ has been set to one)

$$\langle n|m\rangle = \langle n^{(0)} + n^{(1)} + \cdots|m^{(0)} + m^{(1)} + \cdots\rangle$$

$$= \langle n^{(0)}|m^{(1)}\rangle + \langle n^{(1)}|m^{(0)}\rangle$$

$$= \langle n^{(0)}|\sum_{k\neq m} a_{mk}^{(1)}|k^{(0)}\rangle + \sum_{k\neq n} a_{nk}^{(1)*}\langle k^{(0)}|m^{(0)}\rangle$$

$$= a_{mn}^{(1)} + a_{nm}^{(1)*}$$

$$= \frac{H'_{nm}}{E_m^{(0)} - E_n^{(0)}} + \frac{H'^*_{mn}}{E_n^{(0)} - E_m^{(0)}}$$

$$= H'_{nm} \left(\frac{1}{E_m^{(0)} - E_n^{(0)}} + \frac{1}{E_n^{(0)} - E_m^{(0)}} \right) = 0. \tag{16.20}$$

Thus to first order the perturbed eigenstates are orthogonal.

16.1.4 Higher-Order Corrections

Here we will construct the second-order corrections to the perturbed eigenstates and then use that to determine the third-order correction to the energy.

Second-Order Eigenstate Correction, Normalisation and Orthogonality

Similar to the first-order case in Eq. (16.11), we expand the second-order correction term $|n^{(2)}\rangle$ in terms of the complete set of unperturbed energy eigenstates

$$|n^{(2)}\rangle = \sum_m a_{nm}^{(2)} |m^{(0)}\rangle. \tag{16.21}$$

This expansion, along with the first-order expansion Eq. (16.11), is substituted into the second-order Eq. (16.18) to give

$$(E_k^{(0)} - E_n^{(0)})a_{nk}^{(2)} + \langle k^{(0)}|\hat{H}'| \sum_{m \neq n} a_{nm}^{(1)}|m^{(0)}\rangle - E_n^{(1)}a_{nk}^{(1)} - E_n^{(2)}\delta_{nk} = 0, \tag{16.22}$$

which becomes

$$(E_k^{(0)} - E_n^{(0)})a_{nk}^{(2)} + \sum_{m \neq n} H'_{km}a_{nm}^{(1)} - E_n^{(1)}a_{nk}^{(1)} - E_n^{(2)}\delta_{nk} = 0. \tag{16.23}$$

For the case when $k \neq n$, we find

$$a_{nk}^{(2)} = -\sum_{m \neq n} \frac{H'_{km}}{(E_k^{(0)} - E_n^{(0)})} a_{nm}^{(1)} + \frac{E_n^{(1)}}{(E_k^{(0)} - E_n^{(0)})} a_{nk}^{(1)}$$

$$= -\sum_{m \neq n} \frac{H'_{km}}{(E_k^{(0)} - E_n^{(0)})} \frac{H'_{mn}}{(E_n^{(0)} - E_m^{(0)})} + \frac{H'_{nn}}{(E_k^{(0)} - E_n^{(0)})} \frac{H'_{kn}}{(E_n^{(0)} - E_k^{(0)})}. \tag{16.24}$$

In order to normalise the perturbed eigenstates up to second order, the freedom of choosing the expansion coefficients $a_{nn}^{(k)}$ is used again, now at second order to give the condition

$$\lambda^2 : \langle n^{(0)}|n^{(2)}\rangle + \langle n^{(2)}|n^{(0)}\rangle + \langle n^{(1)}|n^{(1)}\rangle = 0, \tag{16.25}$$

which implies $a_{nn}^{(2)} + a_{nn}^{(2)*} + \sum_i |a_{ni}^{(1)}|^2 = 0$, or equivalently $2\mathrm{Re}(a_{nn}^{(2)}) + \sum_i |a_{ni}^{(1)}|^2 = 0$. Since all the coefficients $a_{ni}^{(1)}$ are known from the $O(\lambda)$ calculation, this equation can then be used to determine $\mathrm{Re}(a_{nn}^{(2)})$. Once again at this order $\mathrm{Im}(a_{nn}^{(2)})$ is indeterminant and we set it equal to zero. We get

$$a_{nn}^{(2)} = -\frac{1}{2} \sum_{m \neq n} \frac{H'_{nm} H'_{mn}}{(E_n^{(0)} - E_m^{(0)})^2}. \tag{16.26}$$

Let us check the orthogonality of the perturbed eigenstates at second order in λ. We have for $n \neq m$:

$$\langle n|m \rangle = \langle n^{(0)} + n^{(1)} + n^{(2)} \cdots |m^{(0)} + m^{(1)} + m^{(2)} \cdots \rangle$$
$$= \langle n^0 | m^{(2)} \rangle + \langle n^{(2)} | m^{(0)} \rangle + \langle n^{(1)} | m^{(1)} \rangle, \tag{16.27}$$

where we note that all terms to zeroth and first order in λ have already been shown to be orthogonal. Using the expansion of the perturbed states:

$$\langle n^0 | m^{(2)} \rangle + \langle n^{(2)} | m^{(0)} \rangle + \langle n^{(1)} | m^{(1)} \rangle = \sum_k \langle n^{(0)} | a_{mk}^{(2)} k^{(0)} \rangle + \sum_k \langle a_{nk}^{(2)} k^{(0)} | m^{(0)} \rangle$$
$$+ \sum_{kk'} \langle a_{nk}^{(1)} k^{(0)} | a_{mk'}^{(1)} k'^{(0)} \rangle$$
$$= a_{mn}^{(2)} + a_{nm}^{(2)*} + \sum_k a_{nk}^{(1)*} a_{mk}^{(1)}$$
$$= \sum_{k \neq m} \frac{H'_{nk} H'_{km}}{(E_m^{(0)} - E_n^{(0)})(E_m^{(0)} - E_k^{(0)})} - \frac{H'_{mm} H'_{nm}}{(E_m^{(0)} - E_n^{(0)})^2}$$
$$+ \sum_{k \neq n} \frac{H'^*_{mk} H'^*_{kn}}{(E_n^{(0)} - E_m^{(0)})(E_n^{(0)} - E_k^{(0)})} - \frac{H'^*_{nn} H'^*_{mn}}{(E_m^{(0)} - E_n^{(0)})^2}$$
$$+ \sum_k a_{nk}^{(1)*} a_{mk}^{(1)}. \tag{16.28}$$

Noting in the second sum on the last step above that $H'^*_{mk} = H'_{km}$ and $H'^*_{kn} = H'_{nk}$, the first four terms in the last step above combine to give

$$\sum_{k \neq n,m} \left(\frac{H'_{nk} H'_{km}}{(E_m^{(0)} - E_n^{(0)})(E_m^{(0)} - E_k^{(0)})} + \frac{H'_{nk} H'_{km}}{(E_n^{(0)} - E_m^{(0)})(E_n^{(0)} - E_k^{(0)})} \right) + \frac{H'_{nn} H'_{nm}}{(E_m^{(0)} - E_n^{(0)})^2}$$
$$- \frac{H'_{mm} H'_{nm}}{(E_m^{(0)} - E_n^{(0)})^2} + \frac{H'_{mm} H'_{nm}}{(E_n^{(0)} - E_m^{(0)})^2} - \frac{H'_{nn} H'_{nm}}{(E_n^{(0)} - E_m^{(0)})^2}. \tag{16.29}$$

The last four terms cancel and the first two combine to give

$$\sum_{k \neq m,n} \frac{H'_{nk} H'_{km}}{(E_m^{(0)} - E_k^{(0)})(E_n^{(0)} - E_k^{(0)})} = \sum_k a_{nk}^{(1)*} a_{mk}^{(1)}. \tag{16.30}$$

This term therefore cancels the last term in the last step of Eq. (16.28), which then shows $\langle n^0 | m^{(2)} \rangle + \langle n^{(2)} | m^{(0)} \rangle + \langle n^{(1)} | m^{(1)} \rangle = 0$. Thus we have verified that at second order in the perturbation expansion the perturbed eigenstates are orthogonal. At higher order one can continue this method, utilising the freedom both to choose the expansion coefficients $a_{nn}^{(k)}$ and to have an overall normalisation factor N to orthonormalise the perturbed eigenstates.

Third-Order Energy Correction

With the second-order corrections to the perturbed eigenstates in Eqs (16.24) and (16.26), the third-order energy correction $E_n^{(3)}$ can now be obtained from the $O(\lambda^3)$ term in Eq. (16.6) and, taking a scalar product with the state $\langle n^{(0)}|$, gives

$$
\begin{aligned}
E_n^{(3)} &= \langle n^{(0)}|\hat{H}'|n^{(2)}\rangle - E_n^{(1)}\langle n^{(0)}|n^{(2)}\rangle - E_n^{(2)}\langle n^{(0)}|n^{(1)}\rangle \\
&= \sum_m H'_{nm} a_{nm}^{(2)} - E_n^{(1)} a_{nn}^{(2)},
\end{aligned}
\tag{16.31}
$$

where we use that $\langle n^{(0)}|n^{(1)}\rangle = 0$. Using Eqs (16.24) and (16.26), this gives

$$
E_n^{(3)} = \sum_{k,m\neq n} \frac{H'_{nm} H'_{mk} H'_{kn}}{(E_n^{(0)} - E_m^{(0)})(E_n^{(0)} - E_k^{(0)})} - H'_{nn} \sum_{m\neq n} \frac{H'_{nm} H'_{mn}}{(E_m^{(0)} - E_n^{(0)})^2}.
\tag{16.32}
$$

We see that the normalisation condition does not affect the eigenstates $|n\rangle$ to order λ and the energy corrections to order λ^2.

16.1.5 Properties of the Perturbation Expansion

The perturbation series for a quantity, such as the energy, is a sum of terms order-by-order in powers of the interaction Hamiltonian matrix elements

$$
E = E^{(0)} + E^{(1)} + E^{(2)} + E^{(3)} + \cdots,
\tag{16.33}
$$

where $E^{(k)}$ generally involves a sum of terms each with $k-1$-powers of the energy ratio $\sim H'_{ab}/(E_a^{(0)} - E_b^{(0)})$ that are themselves summed over all intermediate states. Any interaction Hamiltonian generally can be characterised by a dimensionless coupling times a particular set of operators. For example, for atomic perturbations the coupling parameter is the fine-structure constant α, so the interaction Hamiltonian has the form $\hat{H}' = \alpha\hat{\mathcal{H}}'$, for which the kth-order correction to say the energy will have dependence $\alpha^k \tilde{E}^{(k)}$. The full perturbation expansion is then

$$
E(\alpha) = E^{(0)} + \sum_{k=1}^{\infty} \alpha^k \tilde{E}^{(k)}.
\tag{16.34}
$$

If $\alpha < 1$, then increasingly higher powers of this coupling lead to an ever smaller factor. Thus, if $\tilde{E}^{(k)}$ were roughly of the same magnitude for all k, the expansion Eq. (16.34) would converge. However, the structure of perturbation theory is such that this is typically not the case. The correction terms $\tilde{E}^{(k)}$ generally tend to grow extremely rapidly, in fact typically as $k!$, and this growth then renders the perturbation expansion as non-convergent.

It takes a detailed analysis of the perturbation expansion for a specified Hamiltonian to determine exactly the behaviour of the terms at each order and many studies have been done which the reader can examine (see suggested reading at the end of this chapter). However, one can get some idea of why the expansion is non-convergent simply by examining the order-by-order equations of perturbation theory Eq. (16.6). For example, computing the energy, one obtains the first-order Eq. (16.9), second-order Eq. (16.19) and third-order Eq. (16.32) expressions. First order is just a single matrix element. At second order it involves a single sum over all intermediate states of matrix elements H'_{ab} and a factor of the energy ratio $\sim H'_{ab}/(E_a^{(0)} - E_a^{(0)})$. The single factor of H' has dimensions of energy and the ratio is dimensionless. Moreover, provided that for the energy ratio $H'_{ab} \ll |E_a^{(0)} - E_b^{(0)}|$, which is one of the stated conditions for perturbation theory, then this energy ratio acts like a suppression factor to the single matrix element of the interaction Hamiltonian. There is of course then a sum, in general infinite, over all intermediate states. In practice the matrix elements are dominated by a few states with remaining terms in the sum being negligible. Accepting this, then if the corresponding coupling associated with the interaction Hamiltonian is adequately small, the second-order correction to the energy can indeed be smaller than the first-order correction. This would be the best-case scenario for the perturbation expansion to converge. If we look now at the third-order correction to the energy Eq. (16.32), it involves two factors of the energy ratio with $H'_{ab} \ll |E_a^{(0)} - E_b^{(0)}|$ and two intermediate sums over all states. If the single sum at second order controlled the growth, then it's plausible this will also be the case with this double sum. However, there is one other feature at third order in that there are two terms now present, each with double sums. It is this latter feature that is also important in understanding why the perturbation expansion is generally not convergent. As one goes to order k in the perturbation expansion, there are in general increasingly more terms involving $k - 1$ products of energy ratios and $k - 1$ sums over intermediate states. It requires a detailed analysis of a specific interaction to determine exactly how many such terms there are and how exactly quantitatively they affect the perturbation expansion. However, it is worth inspecting the general perturbation expansion at order k, since a few observations can be made that make it plausible that this expansion is gaining increasingly more terms. At order k the perturbation equation from Eq. (16.6) is

$$\lambda^k : \left(\hat{H}_0 - E_n^{(0)}\right) |n^{(k)}\rangle = \left(E_n^{(1)} - \hat{H}'\right) |n^{(k-1)}\rangle + E_n^{(2)} |n^{(k-2)}\rangle$$
$$+ E_n^{(3)} |n^{(k-3)}\rangle + \cdots + E_n^{(k)} |n^{(0)}\rangle. \tag{16.35}$$

From this equation, at kth order the expression for the energy correction is

$$E_n^{(k)} = \langle n^{(0)}|\hat{H}' - E^{(1)}|n^{(k-1)}\rangle - E_n^{(k-2)}\langle n^{(0)}|n^{(2)}\rangle$$
$$- E_n^{(k-3)}\langle n^{(0)}|n^{(3)}\rangle - \cdots - E_n^{(k-1)}\langle n^{(0)}|n^{(1)}\rangle, \tag{16.36}$$

and writing the kth-order correction to the perturbed energy eigenstate as

$$|n^{(k)}\rangle = \sum_{n'} a_{nn'}^{(k)} |n'^{(0)}\rangle \tag{16.37}$$

gives

$$a_{nm}^{(k)} = \frac{1}{E_m^{(0)} - E_n^{(0)}} \left[-\sum_{n'} H'_{mn'} a_{nn'}^{(k-1)} + E_n^{(1)} a_{nm}^{(k-1)} + E_n^{(2)} a_{nm}^{(k-2)} + \cdots + E_n^{(k-1)} a_{nm}^{(1)} \right].$$
$$\tag{16.38}$$

What can be seen from this expansion for the kth-order expression for the energy Eq. (16.36) is that there are k terms that are summed over. Each of those terms involve lower-order expressions for the energy and eigenstate. The eigenstate expansion at order k,

Eq. (16.38), also has about k terms that are summed. Thus the product of energy correction and eigenstate in each term in the sum in Eq. (16.36) has around $(k - s)s \stackrel{>}{\sim} k$ terms and there are k such terms, so at least $k(k - 1)$ terms altogether. This process of counting terms cascades down at each lower order until all terms are just energy ratios and sums over intermediate states. This type of branching structure to the perturbation expansion makes it plausible that the resulting number of such terms grows very quickly with order k in perturbation theory. The arguments given here are not sufficient to determine that the growth rate is at least of order $k!$, but it's clear the growth is very rapid.

This argument makes it plausible that there is rapid growth behaviour of the perturbation expansion. To go further requires careful study of a specific perturbation Hamiltonian to show that indeed the perturbation expansion grows approximately as $k!$. Accepting that these perturbation expansions are non-convergent does not imply the expansion is meaningless. Methods are available, notably the Borel transform, that can render meaning to the perturbation expansion. These details would take us into a mathematical discussion which goes away from the main focus of this text. Our purpose here was only to highlight to the reader the characteristics of the perturbation expansion that is generally encountered in quantum mechanics.

16.2 Degenerate Perturbation Theory

We saw in the nondegenerate case discussed in the previous section that the perturbation can be thought of as inducing a small mixing of the unperturbed states. The perturbed states $\{|n\rangle\}$ are in one-to-one correspondence with the unperturbed states $\{|n^{(0)}\rangle\}$ and go over smoothly to them in the limit $\lambda \to 0$.

Recall that degeneracy means there are several eigenstates with the same energy eigenvalue. If there are g such degenerate states, we say there is a g-fold degeneracy. In the presence of degeneracy, it is not at all clear to which of the infinitely many degenerate unperturbed states the perturbed states return as $\lambda \to 0$. Suppose there is a g-fold degeneracy. It then means any linear combination of the corresponding g unperturbed eigenstates is still an eigenstate with the same energy eigenvalue. Thus there is a g-dimensional subspace in which we can define any g orthogonal directions and they are just as good as any other g orthogonal directions in being the eigenstates of $\hat{H}_0$. The question then is for a given perturbation $\hat{H}'$, which are the correct directions in this subspace from which to perturb. In other words, out of the infinitely many directions, which are the directions the perturbed states return to as $\lambda \to 0$. Degenerate perturbation theory addresses this problem. The basic idea is within the g-dimensional degenerate subspace of states, to diagonalise the perturbation Hamiltonian $\hat{H}'$. Since all directions of this g-dimensional subspace are equally good eigenstates of $\hat{H}_0$, diagonalising $\hat{H}'$ selects particular g-directions, which now up to first order in perturbation theory are eigenstates. Furthermore, if the perturbation $\hat{H}'$ is such that at first order the shifts in energy corresponding to these g directions are all different, then the degeneracy has been lifted at first order. Let us develop this approach in full detail.

To set up the problem, for simplicity, we will assume that the unperturbed Hamiltonian, $\hat{H}_0$, has one level which is g-fold degenerate, with all other levels being nondegenerate, and if necessary, we can relabel all of the eigenstates so that the first g states are the degenerate ones. Thus

$$
\begin{aligned}
\hat{H}_0|E_n^{(0)}\rangle &= E^{(0)}|E_n^{(0)}\rangle && n = 1, \ldots, g, \\
\hat{H}_0|E_n^{(0)}\rangle &= E_n^{(0)}|E_n^{(0)}\rangle && n > g,
\end{aligned}
\tag{16.39}
$$

where $E^{(0)}$ is the degenerate eigenvalue, $E_n^{(0)} \neq E^{(0)}$ for $n > g$ and $E_n^{(0)} \neq E_r^{(0)}$ when $n \neq r$ for $n, r > g$. Note that we use $E_n^{(0)}$ (as opposed to $n^{(0)}$) to label the unperturbed energy eigenstates. As we will see, these degenerate states will require additional labels to specify them completely.

We can without loss of generality choose an initial fiducial basis of g mutually orthogonal degenerate states $\{|E_n^{(0)}\rangle\}$, if necessary by using the Gram–Schmidt procedure. Then any linear combination of the g degenerate states of the form

$$
|E_{\{B^m\}}^{(0)}\rangle \equiv \sum_{n=1}^{g} B_n^m |E_n^{(0)}\rangle
\tag{16.40}
$$

is also an eigenstate of $\hat{H}_0$ with eigenvalue $E^{(0)}$. We now attempt to find a set of degenerate states $\{|E_{\{B^m\}}^{(0)}\rangle\}$, with $m = 1, \ldots g$, which are an appropriate linear combination of the fiducial set of degenerate states as shown in Eq. (16.40), which only undergo small mixing by the perturbation and have the correct limiting behaviour as the perturbation is switched off.

Proceeding as before, we assume that

$$
\left(\hat{H}_0 + \lambda \hat{H}'\right)|E_{\{B^m\}}\rangle = E_{\{B^m\}}|E_{\{B^m\}}\rangle,
\tag{16.41}
$$

where

$$
E_{\{B^m\}} = E^{(0)} + \lambda E_{\{B^m\}}^{(1)} + \lambda^2 E_{\{B^m\}}^{(2)} + \cdots,
\tag{16.42}
$$

and

$$
|E_{\{B^m\}}\rangle = |E_{\{B^m\}}^{(0)}\rangle + \lambda |E_{\{B^m\}}^{(1)}\rangle + \lambda^2 |E_{\{B^m\}}^{(2)}\rangle + \cdots.
\tag{16.43}
$$

We pause to explain the somewhat cumbersome notation. In the subscripts, the notation $\{B^m\}$ is meant to imply a particular choice of linear combination of the fiducial zeroth-order eigenstates with m indexing from 1 to g the respective states. For each such choice, it will imply a particular perturbation expansion, leading to different first- and higher-order corrections to the energy eigenstates and values of the energy. Only the zeroth-order value for the energy eigenstates is in general always the same, as that is the defining reason that these states are degenerate in the first place.

Substituting these expansions into the eigenvalue equation and equating terms of the same degree in λ, we find:

$$\lambda^0 : \left(\hat{H}_0 - E^{(0)} \right) |E^{(0)}_{\{B^m\}}\rangle = 0, \tag{16.44}$$

$$\lambda^1 : \left(\hat{H}_0 - E^{(0)} \right) |E^{(1)}_{\{B^m\}}\rangle + \left(\hat{H}' - E^{(1)}_{\{B^m\}} \right) |E^{(0)}_{\{B^m\}}\rangle = 0, \tag{16.45}$$

$$\lambda^2 : \left(\hat{H}_0 - E^{(0)} \right) |E^{(2)}_{\{B^m\}}\rangle + \left(\hat{H}' - E^{(1)}_{\{B^m\}} \right) |E^{(1)}_{\{B^m\}}\rangle - E^{(2)}_{\{B^m\}} |E^{(0)}_{\{B^m\}}\rangle = 0. \tag{16.46}$$

As before, the first of these tells us what we already know, that the state $|E^{(0)}_{\{B^m\}}\rangle$ is an eigenstate of $\hat{H}_0$ belonging to the eigenvalue $E^{(0)}$.

16.2.1 First-Order Degenerate Perturbation Correction

Turning to the second equation, we take the scalar product with the kth unperturbed state of the fiducial basis:

$$\langle E^{(0)}_k | \left(\hat{H}_0 - E^{(0)} \right) |E^{(1)}_{\{B^m\}}\rangle + \langle E^{(0)}_k | \left(\hat{H}' - E^{(1)}_{\{B^m\}} \right) |E^{(0)}_{\{B^m\}}\rangle = 0, \tag{16.47}$$

but, as before, we note that $\langle E^{(0)}_k | \hat{H}_0 = E^{(0)}_k \langle E^{(0)}_k |$ to simplify the first term, whilst in the second term we substitute the expansion

$$|E^{(0)}_{\{B^m\}}\rangle = \sum_{n=1}^{g} B^m_n |E^{(0)}_n\rangle \tag{16.48}$$

to obtain

$$\left(E^{(0)}_k - E^{(0)} \right) \langle E^{(0)}_k | E^{(1)}_{\{B^m\}}\rangle + \sum_{n=1}^{g} B^m_n \left(H'_{kn} - E^{(1)}_{\{B^m\}} \delta_{kn} \right) = 0, \tag{16.49}$$

where we have used the orthonormality of the unperturbed basis

$$\langle E^{(0)}_k | E^{(0)}_n \rangle = \delta_{kn} \tag{16.50}$$

and

$$H'_{kn} \equiv \langle E^{(0)}_k | \hat{H}' | E^{(0)}_n \rangle \tag{16.51}$$

is the matrix element of the perturbation between unperturbed states of the fiducial basis.

Now k was chosen arbitrarily, so consider $k \leq g$. In this case, $E^{(0)}_k = E^{(0)}$ and thus the first term vanishes, leaving

$$\sum_{n=1}^{g} \left(H'_{kn} - E^{(1)}_{\{B^m\}} \delta_{kn} \right) B^m_n = 0, \qquad k = 1, \ldots, g, \tag{16.52}$$

which we can regard as a set of g linear equations for the B^m_n, with coefficients which are the elements of a $g \times g$ matrix. For each $m = 1, \ldots, g$, there will be a set of solutions $\{B^m_n\}$ and corresponding eigenvalues $E^{(1)}_{\{B^m\}}$. Recall that, for a non-trivial solution to a set of linear equations, the determinant of the $g \times g$ coefficient matrix must vanish:

$$\det \left(H'_{kn} - E^{(1)}_{\{B^m\}} \delta_{kn} \right) = 0, \tag{16.53}$$

which has the following structure:

$$
\begin{vmatrix}
H'_{11} - E^{(1)}_{\{B^m\}} & H'_{12} & \cdots & H'_{1g} \\
H'_{21} & H'_{22} - E^{(1)}_{\{B^m\}} & & \vdots \\
\vdots & & \ddots & \vdots \\
H'_{g1} & \cdots & \cdots & H'_{gg} - E^{(1)}_{\{B^m\}}
\end{vmatrix} = 0.
$$

The determinant is thus a polynomial of degree g in $E^{(1)}_{\{B^m\}}$, and therefore has g 'roots' or solutions $E^{(1)}_{\{B^m\}}$, with $m = 1, \ldots, g$, and for each solution m there corresponds a set of coefficients $\{B^m_n\}$ introduced in Eq. (16.40). Note that there is no requirement for the coefficients $\{B^m_n\}$ to be small.

Equation (16.52) may be regarded as a standard eigenvalue problem for the perturbation $\hat{H}'$ in the g-dimensional degenerate subspace. The g roots $E^{(1)}_{\{B^m\}}$, $m = 1, \ldots, g$, are real, since H' is a Hermitian matrix, and give the values of the first-order shift in energy from the degenerate value $E^{(0)}$. At this stage, a new basis has been obtained within the degenerate subspace. To summarise, the new unperturbed eigenstates basis is Eq. (16.48), with corresponding unperturbed energy E and as before for $n > g$, $|E_n\rangle$ with corresponding unperturbed energy E_n. The first-order energy correction is

$$
\langle E^{(0)}_{\{B^n\}} | \hat{H}' | E^{(0)}_{\{B^m\}} \rangle \equiv H'_{B^n B^m} = E^{(1)}_{\{B^m\}} \delta_{nm} \quad n, m \le g, \tag{16.54}
$$

$$
\langle E^{(0)}_n | \hat{H}' | E^{(0)}_n \rangle = E^{(1)}_n \quad n > g, \tag{16.55}
$$

and in general $\langle E^{(0)}_{\{B^m\}} | \hat{H}' | E^{(0)}_n \rangle \ne 0$ for $m \le g$ and $n > g$ and $\langle E^{(0)}_n | \hat{H}' | E^{(0)}_m \rangle \ne 0$ for $m, n > g$. Above, the matrix element $H'_{\{B^n\}\{B^m\}}$ is the slightly cumbersome notation we adopt to refer to matrix elements of the perturbation Hamiltonian in the rotated basis Eq. (16.48).

It is helpful to see what the diagonalisation of $\hat{H}'$ within the degenerate sector does in the wider context of the full Hamiltonian. For illustrative purposes, let us suppose the linear operator for the Hamiltonian is a matrix of size $m \times m$, which is bigger than the degenerate subspace $m > g$. As before, the degenerate subspace will be the first $g \times g$ sector of the Hamiltonian matrix. Initially, before diagonalising the degenerate sector in the perturbation Hamiltonian, the Hamiltonian has the form

$$
\hat{H}_0 = \left(
\begin{array}{ccccc|ccc}
E & 0 & 0 & \cdots & 0 & 0 & 0 & 0 \\
0 & E & \cdots & 0 & 0 & 0 & 0 & 0 \\
0 & 0 & E & \cdots & \vdots & \vdots & 0 & 0 \\
\vdots & & & \ddots & & & \vdots & \vdots \\
0 & 0 & 0 & \cdots & E & 0 & \cdots & 0 \\
\hline
0 & 0 & \cdots & 0 & 0 & E^{(0)}_{g+1} & \cdots & 0 \\
\vdots & & & & \vdots & \cdots & \ddots & \vdots \\
0 & 0 & 0 & 0 & 0 & 0 & \cdots & E^{(0)}_m
\end{array}
\right)
$$

and

$$
\hat{H}' = \left(
\begin{array}{ccccc|ccc}
H'_{11} & H'_{12} & H'_{13} & \cdots & H'_{1g} & H'_{1g+1} & \cdots & H'_{1m} \\
H'_{21} & H'_{22} & H'_{23} & \cdots & H'_{2g} & H'_{2g+1} & \cdots & H'_{2m} \\
H'_{31} & \cdots & & \cdots & & \vdots & & H'_{3m} \\
\vdots & & & \ddots & & & \ddots & \vdots \\
H'_{g1} & \cdots & & \cdots & H'_{gg} & H'_{gg+1} & \cdots & H'_{gm} \\
- & - & - & - & - & - & - & - \\
H'_{g+11} & & \cdots & & H'_{g+1g} & H'_{g+1,g+1} & \cdots & H'_{g+1,m} \\
\vdots & \ddots & & & \vdots & \vdots & \ddots & \vdots \\
H'_{m1} & \cdots & & & H'_{mg} & H'_{mg+1} & \cdots & H'_{mm}
\end{array}
\right).
$$

After the diagonalisation of $\hat{H}'$ in the degenerate sector, the $\hat{H}_0$ Hamiltonian remains the same as above, but the perturbation Hamiltonian matrix now has the form

$$
\hat{H}' = \left(
\begin{array}{ccccc|ccc}
H'_{B^1 B^1} & 0 & 0 & \cdots & 0 & H'_{B^1 g+1} & \cdots & H'_{B^1 m} \\
0 & H'_{B^2 B^2} & 0 & \cdots & 0 & H'_{B^2 g+1} & \cdots & H'_{B^2 m} \\
0 & 0 & H'_{B^3 B^3} & \cdots & \vdots & \vdots & & H'_{B^3 m} \\
\vdots & & & \ddots & & & \ddots & \vdots \\
0 & \cdots & & \cdots & H'_{B^g B^g} & H'_{B^g g+1} & \cdots & H'_{B^g m} \\
- & - & - & - & - & - & - & - \\
H'_{g+1 B^1} & & \cdots & & H'_{g+1 B^g} & H'_{g+1,g+1} & \cdots & H'_{g+1,m} \\
\vdots & \ddots & & & \vdots & \vdots & \ddots & \vdots \\
H'_{m B^1} & \cdots & & & H'_{m B^g} & H'_{mg+1} & \cdots & H'_{mm}
\end{array}
\right).
$$

In particular, the degenerate $g \times g$ sector is diagonal by design. The top right and bottom left sectors in general still have non-zero matrix elements, but because the degenerate states have been reordered into new linear combinations, these matrix elements will now differ from before. Finally, the bottom right sector, which is completely unaffected by the degenerate state, is unaltered from before. Thus in general, $\hat{H}'$ is diagonal in the $g \times g$ sector, but that does not mean the entire Hamiltonian becomes diagonalised.

The result of this diagonalisation process of the degenerate subspace leaves several cases to consider:

1. *All roots are distinct.* In this case, as shown in Fig. 16.3, the g-fold degeneracy is completely lifted at first order and the degenerate level splits into g distinct levels.
2. *Some roots are equal.* Here, as shown in Fig. 16.4, some degeneracy remains at first order.
3. *All roots are equal.* As shown in Fig. 16.5, the degeneracy remains completely at first order. If the roots are non-zero, there is a shift in energy.

For the first case above, one can use the new basis to compute higher-order corrections using nondegenerate perturbation theory. For the latter two cases, some states are still

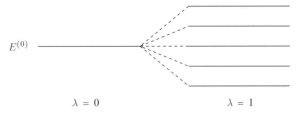

Fig. 16.3 Degenerate energy levels split at first order in perturbation theory so that all g initially degenerate levels are now distinct.

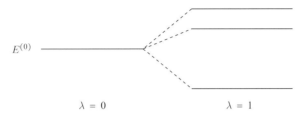

Fig. 16.4 Degenerate energy levels split at first order in perturbation theory so that some of the g initially degenerate levels still remain so.

Fig. 16.5 Degenerate energy levels all remain also at first order in perturbation theory.

degenerate, and one needs to go to second or even higher order still to break this degeneracy. This will be discussed further below. In all cases it will be useful to obtain the first-order correction to the energy eigenstates. For states with $n > g$, this is the same as before. For $m \leq g$ from Eq. (16.49) we expand the first-order correction as

$$|E^{(1)}_{\{B^m\}}\rangle = \sum_{n>g} a^{(1)}_{\{B^m\}n}|E^{(0)}_n\rangle + \sum_{n\leq g} a^{(1)}_{B^m B^n}|E^{(0)}_{B^n}\rangle. \tag{16.56}$$

In general, at first order there can be mixing with states within the initial degenerate subspace, and the second term above accounts for those. Substituting the above into Eq. (16.47) and using $k > g$ gives

$$a^{(1)}_{\{B^m\}k} = \frac{-H'_{k\{B^m\}}}{E^{(0)}_k - E}, \tag{16.57}$$

so that

$$|E^{(1)}_{\{B^m\}}\rangle = \sum_{n>g} \frac{-H'_{n\{B^m\}}}{E_n - E}|E^{(0)}_n\rangle + \sum_{n\leq g} a^{(1)}_{B^m B^n}|E^{(0)}_{B^n}\rangle. \tag{16.58}$$

To obtain the second term above, the second-order equation in Eq. (16.46) needs to be examined. Considering the scalar product of this second-order equation with the state $\langle E^{(0)}_{B^n}|$ gives

$$\langle E^{(0)}_{B^n}| \left(\hat{H}' - E^{(1)}_{\{B^m\}}\right) |E^{(1)}_{\{B^m\}}\rangle - E^{(2)}_{\{B^m\}}\delta_{B^n B^m} = 0. \tag{16.59}$$

Using the expression Eq. (16.58) for $|E^{(1)}_{\{B^m\}}\rangle$ gives

$$(E^{(1)}_{B^n} - E^{(1)}_{B^m})a^{(1)}_{B^m B^n} - \sum_{k>g} \frac{H'_{B^n k} H'_{k\{B^m\}}}{E_k - E} - E^{(2)}_{\{B^m\}}\delta_{B^n B^m} = 0, \tag{16.60}$$

so for $n \neq m$ and $n, m \leq g$ this gives

$$a^{(1)}_{B^m B^n} = \sum_{k>g} \frac{H'_{B^n k} H'_{k\{B^m\}}}{(E_k - E)(E^{(1)}_{B^n} - E^{(1)}_{B^m})}. \tag{16.61}$$

Although there are two powers of the perturbation Hamiltonian matrix element, the expression is also divided by the first-order energy corrections, thus making the overall expression still first order.

There are a couple of points to note here. First, for any state $\langle k^{(0)}|$ that is not degenerate, when following the above procedure of taking a scalar product of it with the second-order expression in Eq. (16.46), unlike above, there will also be a contribution from the first term involving the difference in zeroth-order energies. This will lead to an expression for the second-order expansion coefficients, $a^{(2)}_{B^m k}$, in exactly the same way as done in Section 16.1.4 for the nondegenerate case. Second, the mixing of the degenerate states as found from using the second-order equation to obtain the first-order correction to those states in Eq. (16.58) will similarly happen for higher-order corrections to these states with a second-order contribution from the degenerate states obtained from the third-order equation, and so on.

Special Case

If it should so happen that the matrix H' is diagonal, that is, $H'_{kn} = 0$ for $k \neq n$, then the determinant reduces to

$$\left(H'_{11} - E^{(1)}\right) \times \left(H'_{22} - E^{(1)}\right) \times \cdots \times \left(H'_{gg} - E^{(1)}\right) = 0. \tag{16.62}$$

In this case the initial, or fiducial, basis for the degenerate states already diagonalises the perturbation Hamiltonian. Therefore the g first-order energy shifts are just given by essentially the same formula as derived earlier for the nondegenerate case:

$$E^{(1)}_n \equiv \Delta E^{(1)}_n = \langle E^{(0)}_n|\hat{H}'|E^{(0)}_n\rangle \qquad n = 1, \ldots, g.$$

Can we arrange for this to be the case? Suppose that we can find an observable represented by an operator $\hat{A}$ such that

$$[\hat{H}_0, \hat{A}] = [\hat{H}', \hat{A}] = 0, \tag{16.63}$$

then there are simultaneous eigenstates of $\hat{H}_0$ and $\hat{A}$:

$$\hat{H}_0|E^{(0)}, A_i\rangle = E^{(0)}|E^{(0)}, A_i\rangle \quad \text{and} \tag{16.64}$$

$$\hat{A}|E^{(0)}, A_i\rangle = A_i|E^{(0)}, A_i\rangle . \tag{16.65}$$

If and only if the eigenvalues A_i are distinct for $i = 1, \ldots, g$, we choose these states as our basis states, that is, we take

$$|E_n^{(0)}\rangle = |E^{(0)}, A_n\rangle \qquad n = i, \ldots, g. \tag{16.66}$$

The vanishing of the commutator of $\hat{H}'$ and $\hat{A}$ then means that

$$\langle E_k^{(0)}|[\hat{H}', \hat{A}]|E_n^{(0)}\rangle = 0. \tag{16.67}$$

Writing out the commutator gives

$$\langle E_k^{(0)}|\hat{H}'\hat{A}|E_n^{(0)}\rangle - \langle E_k^{(0)}|\hat{A}\hat{H}'|E_n^{(0)}\rangle = (A_n - A_k)\langle E_k^{(0)}|\hat{H}'|E_n^{(0)}\rangle$$
$$= (A_n - A_k)H_{kn}' = 0, \tag{16.68}$$

where we have used the eigenvalue equations $\hat{A}|E_n^{(0)}\rangle = A_n|E_n^{(0)}\rangle$ and $\langle E_k^{(0)}|\hat{A} = A_k\langle E_k^{(0)}|$. For $k \neq n$, $A_k \neq A_n$ so we must have

$$H_{kn}' = 0. \tag{16.69}$$

Note: we do *not* require that $[\hat{H}_0, \hat{H}'] = 0$, since that would imply that $\hat{H}_0$ and $\hat{H}'$ have simultaneous eigenstates, which would be *exact solutions of the full, perturbed problem.*

Of course, if the A_i are not distinct, we search for a further operator, $\hat{B}$, such that

$$[\hat{A}, \hat{B}] = [\hat{H}_0, \hat{A}] = [\hat{H}_0, \hat{B}] = [\hat{H}', \hat{A}] = [\hat{H}', \hat{B}] = 0, \tag{16.70}$$

i.e. $\{\hat{H}_0, \hat{A}, \hat{B}\}$ are a commuting set of operators, and so on.

Example 16.2 Central potential, spin-$\frac{1}{2}$, spin–orbit interaction:

$$\hat{H} = \hat{p}^2/2m + \hat{V}(r) + f(r)\underline{\hat{L}} \cdot \underline{\hat{S}} = \hat{H}_0 + \hat{H}'.$$

We know that

$$[\hat{H}', \hat{L}_z] \neq 0 \quad \text{and} \quad [\hat{H}', \hat{S}_z] \neq 0 \quad \text{but that} \quad [\hat{H}', \hat{J}_z] = 0,$$

and so if we use the *coupled basis*, $\{|n, \ell, s, j, m_j\rangle\}$, we can use the nondegenerate theory to compute the first-order energy shifts. Note that $[\hat{H}_0, \underline{\hat{L}} \cdot \underline{\hat{S}}] = 0$, but that $[\hat{H}_0, \hat{H}'] \neq 0$ because of the $f(r)$ term in $\hat{H}'$.

16.2.2 Second-Order Degenerate Perturbation Energy Correction

It is possible that some of the degeneracy is not broken at first order. In that case one would need to go to higher orders to try and break the remaining degeneracy. From the initial degenerate states that diagonalised the perturbation Hamiltonian, $|E_{\{B^m\}}^{(0)}\rangle$ for $m = 1, \ldots g$, suppose $g' \leq g$ of these states still remain degenerate at first order. In other words, for all these g' states, the energy corrected up to first order $E^{(0)} + E^{(1)}$ is the same. Without loss of generality, relabel so that the states $|E^{(0)}_{\{B^m\}}\rangle$, for $m = 1, \ldots g'$, are the states still degenerate at first order and for any remaining states that were degenerate at zeroth order, they no longer are with the first-order energy corrections added. Any linear combination of these $m \leq g'$ states degenerate at first order will still be eigenstates of $\hat{H}'$ within the subspace restricted to these g' states. Thus, the task now is similar to the one at first order, to find a suitable linear combination of these degenerate states,

$$|E^{(0)}_{\{BC^m\}}\rangle = \sum_{n=1}^{g'} C_n^m |E^{(0)}_{\{B^n\}}\rangle, \qquad (16.71)$$

that can break the degeneracy at second order. Note that for any set of g' orthogonal states it still holds that the perturbation Hamiltonian is diagonal in both the g'- and g-dimensional subspace, $\langle E^{(0)}_{\{B^m\}}|\hat{H}'|E^{(0)}_{\{BC^n\}}\rangle = C_m^n H'_{\{B^m\}\{B^m\}}$ for $m, n \leq g'$, $\langle E^{(0)}_{\{B^m\}}|\hat{H}'|E^{(0)}_{\{BC^n\}}\rangle = 0$ for $m > g'$ and $n \leq g'$, and $\langle E^{(0)}_{\{B^m\}}|\hat{H}'|E^{(0)}_{\{B^n\}}\rangle = H'_{\{B^m\}\{B^m\}}\delta_{mn}$ for $m, n > g'$. As the perturbation Hamiltonian matrix elements between states within the degenerate subspace are still zero, it means the issue of terms blowing up due to energy denominators will still not occur. However, within the $g' \times g'$ subspace we still need to find the right directions that the true eigenvalues will reduce to when the interaction term goes to zero. As before, this is a nonperturbative step requiring diagonalising the relevant second-order perturbation matrix.

The expansion of the zeroth-order states in the new basis is given in Eq. (16.71) and similarly the first-order corrections would then be

$$|E^{(1)}_{\{BC^n\}}\rangle = \sum_{m=1}^{g'} C_m^n |E^{(1)}_{\{B^m\}}\rangle, \qquad (16.72)$$

where the expansion for $|E^{(1)}_{\{B^m\}}\rangle$ is given above in Eq. (16.56). As we will see below, there will be no contribution in this expression from the zeroth-order degenerate states. Using the λ^2 equation in Eq. (16.46) and taking a scalar product of it with one of the states $\langle E^{(0)}_{\{B^m\}}|$ for $m, n \leq g'$, so that the first term in this equation vanishes, leads to

$$\langle E^{(0)}_{\{B^m\}}|\hat{H}' - E^{(1)}_{\{B^n\}}|E^{(1)}_{\{BC^n\}}\rangle - E^{(2)}_{\{BC^n\}}\langle E^{(0)}_{\{B^m\}}|E^{(0)}_{\{BC^n\}}\rangle = 0. \qquad (16.73)$$

Using the expansion for $|E^{(1)}_{\{BC^n\}}\rangle$ from Eqs (16.72), (16.56) and (16.57), the first thing to note is that in the expansion of the state $|E^{(1)}_{\{BC^n\}}\rangle$, the terms involving the degenerate states $|E^{(0)}_{B^m}\rangle$ for $m \leq g'$ will not contribute in the first term above since the first-order energy corrections are also degenerate. Thus the above expression becomes

$$\langle E^{(0)}_{\{B^m\}}|\hat{H}' \sum_{m'=1}^{g'} C_{m'}^n \sum_{k>g} \frac{-H'_{k\{B^{m'}\}}}{E_k - E}|E^{(0)}_k\rangle - E^{(2)}_{\{BC^n\}} C_m^n = 0, \qquad (16.74)$$

which can be written in matrix form as

$$\sum_{m'=1}^{g'} C_{m'}^n \left(\sum_{k>g} H'_{\{B^m\}k} \frac{H'_{k\{B^{m'}\}}}{E_k - E} + E^{(2)}_{\{BC^n\}}\delta_{m'm} \right) = 0, \qquad (16.75)$$

which will lead to $n = 1, \ldots g'$ eigenvalues $E^{(2)}_{\{BC^n\}}$ and corresponding sets of expansion coefficients $\{C^n_m\}$. If all the $E^{(2)}_{\{BC^n\}}$ are different, then the remaining g' degeneracy at first order is now broken at second order.

16.3 Applications

Time-independent perturbation will be applied to two important applications in atomic systems. The first is the fine-structure corrections to the hydrogen atom and the second is to the ground state of the helium atom.

16.3.1 Hydrogen Fine Structure

As our first major example of perturbation theory in action, we consider the fine structure of hydrogen, taking account of relativistic effects which give rise to corrections to the results previously obtained by solving the Schrödinger equation for the Coulomb potential.

The unperturbed Hamiltonian[1] for a hydrogen-like atom with nuclear charge Ze is

$$\hat{H}_0 = \frac{\hat{p}^2}{2m} + \hat{V}(r), \qquad \text{where} \quad \hat{V}(r) = -\frac{Ze^2}{r}, \tag{16.76}$$

with eigenvalues given by the Bohr formula, which can be written in a number of ways:

$$E^{(0)}_n = -\frac{m}{2\hbar^2} \frac{(Ze^2)^2}{n^2} = -\frac{e^2}{2a_0} \frac{Z^2}{n^2} = -mc^2 \frac{(Z\alpha)^2}{2n^2}. \tag{16.77}$$

Here, a_0 is the Bohr radius and α is the fine-structure constant

$$a_0 = \frac{\hbar^2}{me^2} \qquad \text{and} \qquad \alpha \equiv \frac{e^2}{\hbar c} \simeq \frac{1}{137}. \tag{16.78}$$

Units: We remark that in addition to the Gaussian cgs units used above, the Rydberg unit of energy is often used in discussing atomic physics,

$$1 \text{ Ry} \equiv \frac{e^2}{2a_0} = 13.6 \text{ eV}. \tag{16.79}$$

The ground-state energy of hydrogen is thus -1 Ry $= -13.6$ eV.

The perturbation consists of three contributions:

$$\hat{H}' = \hat{H}'_{\text{KE}} + \hat{H}'_{\text{S--O}} + \hat{H}'_{\text{Darwin}}. \tag{16.80}$$

Here the first term is a relativistic correction to the kinetic energy, obtained by expanding the relativistic expression for the kinetic energy in powers of $(p/mc)^2$,

$$\hat{H}'_{\text{KE}} = -\frac{\hat{p}^4}{8m^3c^2}. \tag{16.81}$$

---

[1] Assuming an infinitely heavy nucleus. In reality we replace the electron mass, m, by the reduced mass, μ, of the electron–nucleus system, and a_0 by a_μ, the Bohr radius for reduced mass μ (corrections $< 0.05\%$).

The second term is the **spin–orbit term**, whose physical origin is the interaction between the intrinsic magnetic dipole moment of the electron and the magnetic field due to the electron's orbital motion in the electric field of the nucleus (apparent magnetic field the electron sees due to the nucleus),

$$\hat{H}'_{S-O} = f(r)\,\underline{\hat{L}} \cdot \underline{\hat{S}}, \tag{16.82}$$

where

$$f(r) = \frac{1}{2m^2c^2r}\frac{dV(r)}{dr}. \tag{16.83}$$

Finally the third term, known as the **Darwin term**, is a relativistic correction to the potential energy,

$$\hat{H}'_{\text{Darwin}} = \frac{\hbar^2}{8m^2c^2}\,\nabla^2 V(\underline{r}) = Ze^2\frac{\pi\hbar^2}{2m^2c^2}\delta(\underline{r}). \tag{16.84}$$

The origin of these correction terms from relativistic theory is derived in Appendix 16.A at the end of this chapter.

Kinetic Energy Correction

The unperturbed level with energy $E_n^{(0)}$ is $2n^2$-fold degenerate so on the face of it we should use degenerate perturbation theory. However, the perturbation contains no spin operators and so trivially commutes with $\hat{S}^2$ and $\hat{S}_z$. Furthermore, it commutes with the components of orbital angular momentum and so it is diagonal in the uncoupled basis and we can use the nondegenerate formula to find the energy shift:

$$\Delta E_{\text{KE}} = \langle n, \ell, m_\ell, s, m_s|\left(-\frac{\hat{p}^4}{8m^3c^2}\right)|n, \ell, m_\ell, s, m_s\rangle$$

$$= -\frac{1}{2mc^2}\langle n, \ell, m_\ell, s, m_s|\hat{T}^2|n, \ell, m_\ell, s, m_s\rangle. \tag{16.85}$$

Because $\hat{T} = \hat{p}^2/(2m)$ contains no spin operators, there is no dependence of ΔE_{KE} on the spin quantum numbers and we can simply write

$$\Delta E_{\text{KE}} = -\frac{1}{2mc^2}\langle n, \ell, m_\ell|\hat{T}^2|n, \ell, m_\ell\rangle. \tag{16.86}$$

Now $\hat{T} = \hat{H}_0 - \hat{V}(r)$ and so

$$\Delta E_{\text{KE}} = -\frac{1}{2mc^2}\left(E_n^{(0)2} - 2E_n^{(0)}\langle\hat{V}\rangle_{n\ell m_\ell} + \langle\hat{V}^2\rangle_{n\ell m_\ell}\right). \tag{16.87}$$

To evaluate the right-hand side we need the expectation values of $1/r$ and $1/r^2$, which don't actually depend on m_ℓ since they have no ϕ dependence. A detailed calculation gives

$$\left\langle\frac{1}{r}\right\rangle_{n\ell} = \frac{Z}{a_0}\frac{1}{n^2} \quad \text{and} \quad \left\langle\frac{1}{r^2}\right\rangle_{n\ell} = \frac{Z^2}{a_0^2}\frac{1}{n^3\left(\ell + \frac{1}{2}\right)}. \tag{16.88}$$

Substituting these in gives

$$\Delta E_{\text{KE}} = -E_n^{(0)} \frac{(Z\alpha)^2}{n^2} \left[\frac{3}{4} - \frac{n}{\ell + \frac{1}{2}} \right]. \tag{16.89}$$

Spin–Orbit Correction

In the case of the spin–orbit interaction in the hydrogenic atom, we know that the degeneracy of a level with given n and ℓ is $(2\ell + 1) \times 2$, since, for a given ℓ, there are $(2\ell + 1)$ possible values of m_ℓ and two possible values of m_s. However, if we choose to work with states of the coupled basis $|n, \ell, s, j, m_j\rangle$, rather than with states of the uncoupled basis $|n, \ell, m_\ell, s, m_s\rangle$, we can use nondegenerate theory. Firstly, we note that we can rewrite the spin–orbit term as

$$\hat{H}'_{S-O} = f(r) \underline{\hat{L}} \cdot \underline{\hat{S}} = \tfrac{1}{2} f(r) \{ \hat{J}^2 - \hat{L}^2 - \hat{S}^2 \}, \tag{16.90}$$

using the fact that $\hat{J}^2 \equiv (\underline{\hat{L}} + \underline{\hat{S}})^2 = \hat{L}^2 + \hat{S}^2 + 2\underline{\hat{L}} \cdot \underline{\hat{S}}$. Noting that

$$\{\hat{J}^2 - \hat{L}^2 - \hat{S}^2\}|n, \ell, s, j, m_j\rangle = \{j(j+1) - \ell(\ell+1) - s(s+1)\}\hbar^2|n, \ell, s, j, m_j\rangle, \tag{16.91}$$

we see that $\hat{H}'$ is diagonal in the uncoupled basis and has expectation value

$$\Delta E_{S-O} = \langle n, \ell, s, j, m_j | \hat{H}_{S-O} | n, \ell, s, j, m_j \rangle = \tfrac{1}{2}\{j(j+1) - \ell(\ell+1) - s(s+1)\}\hbar^2 \langle f(r) \rangle. \tag{16.92}$$

Now $s = \frac{1}{2}$ for an electron, so that j can have two values for a given ℓ, namely, $j = (\ell + \frac{1}{2})$ and $j = (\ell - \frac{1}{2})$, which means that a state of given n and ℓ separates into a doublet when the spin–orbit interaction is present, except in the case $\ell = 0$, where $j = s = \frac{1}{2}$ so there is no level shift.

Since $f(r)$ is independent of the angular variables θ, ϕ and of the spin, the expectation value of $f(r)$ may be written

$$\langle f(r) \rangle = \frac{Ze^2}{2m^2c^2} \int_0^\infty \frac{1}{r^3} |R_{n\ell}(r)|^2 r^2 \, dr. \tag{16.93}$$

The integral can be evaluated exactly using the hydrogenic radial functions and gives

$$\left\langle \frac{1}{r^3} \right\rangle_{n\ell} = \frac{Z^3}{a_0^3} \frac{1}{n^3 \ell \, (\ell + \frac{1}{2})(\ell + 1)}. \tag{16.94}$$

Thus for $\ell \neq 0$ we obtain

$$\Delta E_{S-O} = -E_n^{(0)} \frac{(Z\alpha)^2}{2n} \frac{1}{\ell(\ell + \frac{1}{2})(\ell + 1)} \times \begin{cases} \ell & \text{for } j = \ell + \frac{1}{2}, \\ -\ell - 1 & \text{for } j = \ell - \frac{1}{2}. \end{cases} \tag{16.95}$$

Darwin Correction

The Darwin term contains no spin dependence and commutes with $\hat{L}^2$ and $\hat{L}_z$. By virtue of the three-dimensional Dirac delta function, $\delta(\underline{r})$, it is non-zero only at the origin $\underline{r} = 0$.

Since the radial functions $R_{n\ell}(r)$ vanish at the origin when $\ell \neq 0$, we only need to consider the case $\ell = 0$:

$$
\begin{aligned}
\Delta E_{\text{Darwin}} &= Ze^2 \frac{\pi \hbar^2}{2m^2c^2} \langle n, 0, 0 | \delta(\underline{r}) | n, 0, 0 \rangle \\
&= Ze^2 \frac{\pi \hbar^2}{2m^2c^2} \int u_{n00}^*(\underline{r}) \delta(\underline{r}) u_{n00}(\underline{r}) d^3r = Ze^2 \frac{\pi \hbar^2}{2m^2c^2} |u_{n00}(0)|^2.
\end{aligned}
$$

Now

$$
|u_{n00}(0)|^2 = \frac{1}{4\pi} |R_{n0}(0)|^2 = \frac{Z^3}{\pi a_0^3 n^3} \tag{16.96}
$$

and so

$$
\Delta E_{\text{Darwin}} = -E_n^{(0)} \frac{(Z\alpha)^2}{n}. \tag{16.97}
$$

We can combine the contributions of the three terms to write for all ℓ (problem):

$$
\Delta E_{nj} = E_n^{(0)} \frac{(Z\alpha)^2}{n^2} \left(\frac{n}{j + \frac{1}{2}} - \frac{3}{4} \right), \tag{16.98}
$$

where the subscripts nj indicate that the correction depends on n and on j but not on ℓ.

Russell–Saunders Notation

There is yet another piece of notation used widely in the literature, the so-called **Russell–Saunders** or **term notation**. The states that arise in coupling orbital angular momentum ℓ and spin s to give total angular momentum j are denoted

$$
{}^{(2s+1)}l_j, \tag{16.99}
$$

where l denotes the letter (which can be either lower or upper case) corresponding to the ℓ value in the usual way, $l = s, p, d, f, \ldots$, and the factor $(2s + 1)$ is the spin multiplicity, i.e. the number of allowed values of m_s. Sometimes, the principal quantum number n is prepended as well, giving the notation

$$
n \, {}^{(2s+1)}l_j.
$$

For one-electron atoms like hydrogen, $s = \frac{1}{2}$ and the spin multiplicity is therefore always 2, so is usually omitted.

Example 16.3 The ground state of hydrogen has $n = 1$, $\ell = 0$, $s = \frac{1}{2}$ and $j = \frac{1}{2}$ and therefore corresponds to the single term $1s_{\frac{1}{2}}$.

Figure 16.6 shows the fine-structure splittings of the $n = 2$ and $n = 3$ levels of hydrogen (expressed in Russell–Saunders notation). The relative corrections to the unperturbed energy levels are proportional to α^2, where α is the fine-structure constant, $\alpha \simeq 1/137$ and therefore $< O(10^{-4})$.

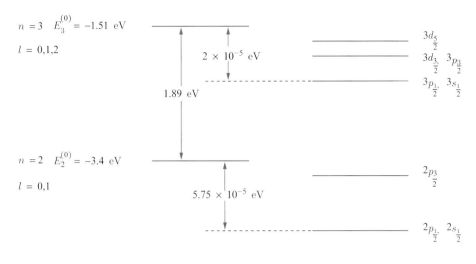

$n = 3$ $E_3^{(0)} = -1.51$ eV

$l = 0,1,2$

2×10^{-5} eV

$3d_{\frac{5}{2}}$

$3d_{\frac{3}{2}}, \ 3p_{\frac{3}{2}}$

$3p_{\frac{1}{2}}, \ 3s_{\frac{1}{2}}$

1.89 eV

$n = 2$ $E_2^{(0)} = -3.4$ eV

$l = 0,1$

5.75×10^{-5} eV

$2p_{\frac{3}{2}}$

$2p_{\frac{1}{2}}, \ 2s_{\frac{1}{2}}$

(a) The unperturbed levels $E_n^{(0)}$ (b) The fine structure

Fig. 16.6 The fine structure of the $n = 2$ and $n = 3$ levels of hydrogen (not to scale). The Lamb shift between the $2s_{\frac{1}{2}}$ and $2p_{\frac{1}{2}}$ levels is not shown.

The degeneracy between levels with $\ell = j \pm \frac{1}{2}$ for given n and j is in fact removed by taking account of quantum electrodynamic effects. This small additional shift, known as the *Lamb shift*, was discovered experimentally in 1947 by Lamb and Retherford. In atomic hydrogen the $2p_{\frac{1}{2}}$ level lies roughly 4×10^{-6} eV below the $2s_{\frac{1}{2}}$ level.

16.3.2 Helium Atom Perturbation Treatment of Electron Repulsion Term

The helium atom has already been discussed in Chapter 13, where the ground-state two-electron state vector was constructed and an estimate of the ground-state energy was obtained. We can now attempt to incorporate the effect of the inter-electron Coulomb repulsion by treating it as a perturbation. The Hamiltonian for helium once again is

$$\hat{H} = \hat{H}_0 + \hat{H}', \tag{16.100}$$

where

$$\hat{H}_0 = \hat{H}_1 + \hat{H}_2 \tag{16.101}$$

with

$$\hat{H}_i = \frac{\hat{p}_i^2}{2m} - \frac{2e^2}{r_i} \tag{16.102}$$

and

$$\hat{H}' = \frac{e^2}{|\underline{r}_1 - \underline{r}_2|}. \tag{16.103}$$

The ground-state wave function that was written in Section 13.4 is an eigenfunction of the unperturbed Hamiltonian, $\hat{H}_0$

$$\Psi(\text{ground state}) = u_{100}(\underline{r}_1)\, u_{100}(\underline{r}_2)\, \chi_{0,0}. \tag{16.104}$$

To compute the first-order correction to the ground-state energy, we have to evaluate the expectation value of the perturbation, $\hat{H}'$, with respect to this wave function:

$$\Delta E_1 = e^2 \int u_{100}^*(\underline{r}_1)\, u_{100}^*(\underline{r}_2)\, \chi_{0,0}^* \frac{1}{r_{12}} u_{100}(\underline{r}_1)\, u_{100}(\underline{r}_2)\, \chi_{0,0}\; d\tau_1 d\tau_2. \tag{16.105}$$

The scalar product of $\chi_{0,0}$ with its conjugate is one, since it is normalised. Putting in the explicit form of the hydrogenic $1s$ wave function

$$u_{100}(\underline{r}) = \frac{1}{\sqrt{\pi}} (Z/a_0)^{3/2}\, \exp(-Zr/a_0), \tag{16.106}$$

thus yields the expression

$$\Delta E_1 = \left(\frac{Z^3 e}{\pi a_0^3}\right)^2 \int \frac{1}{r_{12}} \exp\{-2Z(r_1 + r_2)/a_0\} d^3 r_1 d^3 r_2. \tag{16.107}$$

This integral can be evaluated analytically with the result

$$\Delta E_1 = \frac{5}{4} Z \,\text{Ry} = \frac{5}{2} \,\text{Ry} = 34 \text{ eV}, \tag{16.108}$$

giving for the first-order estimate of the ground-state energy

$$E_1 = -108.8 + 34 \text{ eV} = -74.8 \text{ eV} = -5.5 \text{ Ry}, \tag{16.109}$$

to be compared with the experimentally measured value of -79.005 eV.

16.A Appendix: Derivation of the Fine-Structure Terms from Relativistic Theory

The fine-structure corrections of hydrogen are relativistic effects. The Schrödinger equation is only applicable for non-relativistic velocities. The equation that describes the relativistic motion of electrons is called the **Dirac equation**. That equation is beyond the scope of this book for a thorough discussion. However, we will present a brief introduction to the Dirac equation. The Dirac equation for a free particle is

$$i\hbar \frac{\partial}{\partial t} \psi(\underline{r}, t) = \hat{H}\psi(\underline{r}, t), \tag{16.110}$$

with

$$\hat{H} = c\underline{\alpha} \cdot \hat{\underline{p}} + \beta\, mc^2, \tag{16.111}$$

leading to the Dirac differential equation for a free particle

$$i\hbar \frac{\partial}{\partial t} \psi(\underline{r}, t) = \left(-i\hbar c\underline{\alpha} \cdot \underline{\nabla} + \beta mc^2\right) \psi(\underline{r}, t), \tag{16.112}$$

where

$$\beta = \begin{pmatrix} 1 & 0 \\ 0 & -1 \end{pmatrix} \qquad \underline{\alpha} = \begin{pmatrix} 0 & \sigma \\ \sigma & 0 \end{pmatrix}, \tag{16.113}$$

where $\underline{\sigma}$ are the Pauli matrices.

Exercise 16.A.1   Verify the anticommutation relations

$$\{\alpha_i, \alpha_j\} = 2\delta_{ij},$$
$$\{\alpha_i, \beta_j\} = 0.$$

Also verify $\beta^2 = 1$.

The Dirac equation emerges from requiring relativistic invariance to be maintained, i.e. $E^2 = |p|^2 c^2 + m^2 c^4$. Dirac therefore demanded that

$$\hat{H}^2 \psi(\underline{r}, t) = \left(c^2 |\hat{p}|^2 + m^2 c^4\right) \psi(\underline{r}, t). \tag{16.114}$$

From Eq. (16.111) we have

$$\hat{H}^2 \psi(\underline{r}, t) = \left\{c\underline{\alpha} \cdot \hat{p} + \beta mc^2\right\} \left\{c\underline{\alpha} \cdot \hat{p} + \beta mc^2\right\} \psi(\underline{r}, t). \tag{16.115}$$

Expanding the right-hand side of this equation, being very careful about the ordering of the quantities α^i and β, gives

$$\hat{H}^2 \Psi(\underline{r}, t) = \left\{c^2 \left[(\alpha^1)^2 (\hat{p}^1)^2 + (\alpha^2)^2 (\hat{p}^2)^2 + (\alpha^3)^2 (\hat{p}^3)^2\right] + m^2 c^4 \beta^2\right\} \psi(\underline{r}, t)$$

$$+ \left\{\left(\alpha^1 \alpha^2 + \alpha^2 \alpha^1\right) \hat{p}^1 \hat{p}^2 + \left(\alpha^2 \alpha^3 + \alpha^3 \alpha^2\right) \hat{p}^2 \hat{p}^3 + \left(\alpha^3 \alpha^1 + \alpha^1 \alpha^3\right) \hat{p}^1 \hat{p}^3\right\} \psi(\underline{r}, t)$$

$$+ mc^3 \left\{\left(\alpha^1 \beta + \beta \alpha^1\right) \hat{p}^1 + \left(\alpha^2 \beta + \beta \alpha^2\right) \hat{p}^2 + \left(\alpha^3 \beta + \beta \alpha^3\right) \hat{p}^3\right\} \psi(\underline{r}, t).$$

Condition Eq. (16.114) is satisfied if

$$(\alpha^1)^2 = (\alpha^2)^2 = (\alpha^3)^2 = \beta^2 = 1$$
$$\alpha^i \alpha^j + \alpha^j \alpha^i = 0 \qquad (i \neq j), \tag{16.116}$$
$$\alpha^i \beta + \beta \alpha^i = 0, \tag{16.117}$$

or, more compactly, these are the anticommutation relations above Eq. (16.114).

Returning to the Dirac Eq. (16.112) and looking for plane-wave solutions of the form

$$\psi(\underline{r}, t) = \exp(-ik_\mu x^\mu) u(\underline{p})$$

$$\equiv \exp\left\{-\frac{i}{\hbar} \left(cp_0 t - \underline{p} \cdot \underline{r}\right)\right\} u(\underline{p}), \tag{16.118}$$

with $u(\underline{p})$ a four-component column vector, leads, upon substitution into Eq. (16.112), to the matrix equation for $u(\underline{p})$:

$$p^0 u = \left(\underline{\alpha} \cdot \underline{p} + \beta mc\right) u. \tag{16.119}$$

This is an eigenvalue problem with four solutions. Two of these solutions have the same eigenvalue $p^0 = (m^2c^2 + |\underline{p}|^2)^{1/2} > 0$, which is positive and

$$u = \begin{pmatrix} \phi \\ \chi \end{pmatrix} \quad \text{with} \quad \chi = \frac{c\,\underline{\sigma}\cdot\underline{p}}{E + mc^2}\,\phi, \tag{16.120}$$

with two possible choices for ϕ:

$$\phi^1 = \begin{pmatrix} 1 \\ 0 \end{pmatrix} \quad \text{and} \quad \phi^2 = \begin{pmatrix} 0 \\ 1 \end{pmatrix}. \tag{16.121}$$

The two other solutions have negative eigenvalue $p^0 = -(m^2c^2 + |\underline{p}|^2)^{1/2} > 0$ and eigenvectors

$$u(\underline{p}) = \begin{pmatrix} \dfrac{c\,\underline{\sigma}\cdot\underline{p}}{E + mc^2}\,\phi^{1,2} \\ \phi^{1,2} \end{pmatrix}. \tag{16.122}$$

Exercise 16.A.2   Substitute the solutions of Eqs (16.120) and (16.122) in the Dirac Eq. (16.119) and verify the eigenvalues are as expected.

The Dirac equation can be extended for a particle interacting with an electromagnetic field A^μ through the minimal coupling prescription (for an electron $q = -e$)

$$p^\mu \to p^\mu - \frac{q}{c}A^\mu \quad \text{where} \quad A^\mu = (A^0, \underline{A}). \tag{16.123}$$

This leads to the Dirac equation for a particle of charge q interacting with an EM field A^μ:

$$i\hbar\frac{\partial}{\partial t}\psi(\underline{r},t) = \left\{ c\underline{\alpha}\cdot\left(\hat{\underline{p}} - \frac{q}{c}\underline{A}\right) + \beta mc^2 + qA^0 \right\}\psi(\underline{r},t) = \hat{H}_D\psi(\underline{r},t). \tag{16.124}$$

From this equation we will now obtain the Schrödinger equation for an electron interacting with this field in the non-relativistic limit. We will then consider this electromagnetic field as that created from a nucleus in the hydrogen atom. This will then allow us to derive the leading relativistic corrections to the Schrödinger equation, which, as will be shown, will be precisely the fine-structure corrections. As with the free particle Dirac equation, $\psi(\underline{r},t)$ has four components, and as shown above it can be written in terms of two two-component spinors. Anticipating that, similar to the free particle case, there will be both positive and negative energy solutions, as we are interested in the electron, we will focus only on the positive energy solutions. We write the trial solution in the form

$$\psi = \begin{pmatrix} \Psi \\ \Phi \end{pmatrix}, \tag{16.125}$$

where in analogy to the free particle case we will regard the lower components being 'small' in the non-relativistic limit. This can be seen from the free particle case Eq. (16.120). Since the speed of a relativistic particle is $v = |\underline{p}|/\gamma m = c^2|\underline{p}|/E$, it follows that $\chi \sim (v/c)\phi$. This will now be shown in the full derivation that follows.

In anticipation of the NR limit, subtract $mc^2\psi$ from both sides of the Dirac equation:

$$i\hbar\frac{\partial}{\partial t}\psi(\underline{r},t) - mc^2\psi(\underline{r},t) = \hat{H}_1\psi(\underline{r},t)$$

$$= \left\{c\underline{\alpha}\cdot\left(-i\hbar\underline{\nabla} - \frac{q}{c}\underline{A}\right) + \beta mc^2 + qA^0 - mc^2\right\}\psi(\underline{r},t).$$

In two-component notation:

$$\hat{H}_1\begin{pmatrix}\Psi \\ \Phi\end{pmatrix} = \begin{pmatrix} 0 & c\,\underline{\sigma}\cdot\left(-i\hbar\underline{\nabla} - \frac{q}{c}\underline{A}\right) \\ c\,\underline{\sigma}\cdot\left(-i\hbar\underline{\nabla} - \frac{q}{c}\underline{A}\right) & 0 \end{pmatrix}\begin{pmatrix}\Psi \\ \Phi\end{pmatrix}$$

$$- 2mc^2\begin{pmatrix}0 \\ \Phi\end{pmatrix} + q\,A^0\begin{pmatrix}\Psi \\ \Phi\end{pmatrix}, \tag{16.126}$$

which, on multiplying out, becomes

$$i\hbar\frac{\partial}{\partial t}\Psi - mc^2\Psi = \hat{H}_1\Psi = c\underline{\sigma}\cdot\left(-i\hbar\underline{\nabla} - \frac{q}{c}\underline{A}\right)\Phi + qA^0\Psi, \tag{16.127}$$

$$i\hbar\frac{\partial}{\partial t}\Phi - mc^2\Phi = \hat{H}_1\Phi = c\underline{\sigma}\cdot\left(-i\hbar\underline{\nabla} - \frac{q}{c}\underline{A}\right)\Psi + qA^0\Phi - 2mc^2\Phi. \tag{16.128}$$

We can rewrite Eq. (16.128) as

$$\left(\hat{H}_1 - qA^0 + 2mc^2\right)\Phi = c\underline{\sigma}\cdot\left(-i\hbar\underline{\nabla} - \frac{q}{c}\underline{A}\right)\Psi.$$

Thus, provided all matrix elements of $\hat{H}_1 \ll mc^2$ and $|qA^0| \ll mc^2$, we can drop them on the left-hand side as a first approximation, and so

$$\Phi \simeq \frac{c\underline{\sigma}\cdot\left(-i\hbar\underline{\nabla} - \frac{q}{c}\underline{A}\right)}{2mc^2}\Psi, \tag{16.129}$$

and we can rewrite Eq. (16.127) as

$$\hat{H}_1\Psi = \frac{1}{2m}\left\{\underline{\sigma}\cdot\left(-i\hbar\underline{\nabla} - \frac{q}{c}\underline{A}\right)\right\}^2\Psi + qA^0\Psi. \tag{16.130}$$

The operator on the right-hand side is called the **Pauli Hamiltonian**. Multiplying out the first term on the right-hand side, we obtain

$$\frac{1}{2m}\left\{\underline{\sigma}\cdot\left(-i\hbar\underline{\nabla} - \frac{q}{c}\underline{A}\right)\right\}^2\Psi$$

$$= \left\{\frac{1}{2m}(\underline{\sigma}\cdot\underline{\hat{p}})^2 - \frac{q}{2mc}(\underline{\sigma}\cdot\underline{\hat{p}})(\underline{\sigma}\cdot\underline{A}) - \frac{q}{2mc}(\underline{\sigma}\cdot\underline{A})(\underline{\sigma}\cdot\underline{\hat{p}}) + \frac{q^2}{2mc^2}(\underline{\sigma}\cdot\underline{A})^2\right\}\Psi.$$

Using the (anti-)commutation relations for the Pauli matrices, $\left[\sigma_i,\sigma_j\right] = 2i\epsilon_{ijk}\sigma_k$ and $\left\{\sigma_i,\sigma_j\right\} = 2\delta_{ij}$, we can derive the following identity for any two vectors $\underline{V}$ and $\underline{W}$:

$$(\underline{\sigma}\cdot\underline{V})(\underline{\sigma}\cdot\underline{W}) = \underline{V}\cdot\underline{W} + i\underline{\sigma}\cdot(\underline{V}\times\underline{W}). \tag{16.131}$$

Remembering that $\underline{A}$ and $\hat{p}$ do not in general commute, we can write

$$(\underline{\sigma} \cdot \hat{p})(\underline{\sigma} \cdot \underline{A}) = \hat{p} \cdot \underline{A} + i\underline{\sigma} \cdot \hat{p} \times \underline{A}$$

$$= \underline{A} \cdot \hat{p} - i\hbar(\underline{\nabla} \cdot \underline{A}) + i\underline{\sigma} \cdot (-i\hbar\underline{\nabla} \times \underline{A})$$

$$= \underline{A} \cdot \hat{p} - i\hbar(\underline{\nabla} \cdot \underline{A}) + \hbar\underline{\sigma} \cdot \underline{B} - i\underline{\sigma} \cdot (\underline{A} \times \hat{p}).$$

Similarly

$$(\underline{\sigma} \cdot \underline{A})(\underline{\sigma} \cdot \hat{p}) = \underline{A} \cdot \hat{p} + i\underline{\sigma} \cdot (\underline{A} \times \hat{p}).$$

Exercise 16.A.3 Verify the identity Eq. (16.131) and the above relations that follow it. Take careful note of the ordering for $\hat{p}$ as it can act on both $\underline{A}$ and the wave function Ψ.

One can also work this out with explicit indices for the first term in Eq. (16.130), which gives

$$\left\{\underline{\sigma} \cdot \left(\hat{p} - \frac{q}{c}\underline{A}\right)\right\}^2 = (\delta^{ij} + i\epsilon^{ijk}\sigma^k)(\hat{p}^i - \frac{q}{c}A^i)(\hat{p}^j - \frac{q}{c}A^j)$$

$$= (\hat{p} - \frac{q}{c}\underline{A})^2 + i\underbrace{\epsilon^{ijk}}_{\text{antisym}}\sigma^k(\underbrace{\hat{p}^i\hat{p}^j}_{\text{sym}} + (\frac{q}{c})^2\underbrace{A^iA^j}_{\text{sym}} - \frac{q}{c}(\hat{p}^iA^j + A^i\hat{p}^j))$$

$$= (\hat{p} - \frac{q}{c}\underline{A})^2 + i\epsilon^{ijk}\sigma^k(-\frac{q}{c})(-i\hbar)(\partial^iA^j + \underbrace{A^j\partial^i + A^i\partial^j}_{\text{sym}})$$

$$= (\hat{p} - \frac{q}{c}\underline{A})^2 - \frac{q\hbar}{c}\underline{\sigma} \cdot \underline{B}. \tag{16.132}$$

Here the identity $\sigma^i\sigma^j = \delta^{ij} + i\epsilon^{ijk}\sigma^k$ for the Pauli matrices was used. Also remembering that $\hat{p}$ can act on $\underline{A}$, the symmetric terms vanish due to the ϵ^{ijk} tensor, and

$$\underline{B} = \underline{\nabla} \times \underline{A} \tag{16.133}$$

is the magnetic field.

Substituting back into the Pauli Hamiltonian, we obtain

$$\hat{H}_1\Psi = \left\{\frac{\hat{p}^2}{2m} - \frac{q}{mc}(\underline{A} \cdot \hat{p}) + \frac{iq\hbar}{2mc}(\underline{\nabla} \cdot \underline{A}) - \frac{q\hbar}{2mc}(\underline{\sigma} \cdot \underline{B}) + \frac{q^2A^2}{2mc^2} + qA^0\right\}\Psi, \tag{16.134}$$

where $\underline{B} = \underline{\nabla} \times \underline{A}$ is the magnetic field. Note for the $(\underline{\nabla} \cdot \underline{A})$ term on the right-hand side above, the ∇ operator only acts on the vector potential and not anything to its right, which is what the parentheses around this expression are meant to imply. The 'spin term' is

$$-\frac{q}{mc}(\underline{\hat{S}} \cdot \underline{B}) = +g_s\frac{e}{2mc}(\underline{\hat{S}} \cdot \underline{B}), \tag{16.135}$$

for an electron of charge $q = -e$. Thus the Dirac equation predicts that $g_s = 2$, in good agreement with experiment. Recall that $g = 1$ in classical physics and therefore $g_s = 2$ has to be put in 'by hand' when doing non-relativistic atomic physics.

It is often claimed that to get $g_s = 2$ requires relativity. This is of course true in the sense that it falls out *naturally* from the Dirac equation. However, it is worth pointing out that one can start out (somewhat artificially) with the two-component wave equation

$$\frac{(\underline{\sigma} \cdot \hat{p})^2}{2m}\Psi = i\hbar\frac{\partial}{\partial t}\Psi, \tag{16.136}$$

which is equivalent to Schrödinger's equation for a free particle (check it!). The minimal coupling prescription then yields the Pauli Hamiltonian, Eq. (16.130), and thence $g_s = 2$. Note that we can write the Pauli equation in the compact form

$$i\hbar\frac{\partial}{\partial t}\Psi = \left\{ \frac{(\hat{p} - q\underline{A}/c)^2}{2m} - \frac{q\hbar}{2mc}(\underline{\sigma} \cdot \underline{B}) + qA^0 + mc^2 \right\}\Psi. \tag{16.137}$$

Finally we now factor out the energy associated with the mass, so define the non-relativistic wave function as $\Psi \equiv \Psi_{NR}\exp(-imc^2t/\hbar)$ which, when substituted above, gives the Pauli equation

$$i\hbar\frac{\partial}{\partial t}\Psi_{NR} = \left\{ \frac{(\hat{p} - q\underline{A}/c)^2}{2m} - \frac{q\hbar}{2mc}(\underline{\sigma} \cdot \underline{B}) + qA^0 \right\}\Psi_{NR}. \tag{16.138}$$

The Fine Structure of Hydrogen

Let us apply the Dirac equation to the case of the hydrogen atom, namely:

$$qA^0 = V = -\frac{e^2}{r}. \tag{16.139}$$

There are two approaches to the solution of this problem:

1. Solve the Dirac equation exactly.
2. Perform the non-relativistic reduction in the hope of 'predicting' the fine structure obtained in non-relativistic atomic physics.

Approach 1 involves a long technical calculation similar to the non-relativistic solution. Let's follow approach 2 as it's simpler and gives us some physical insight. If we look for energy eigenstates of the Dirac equation

$$\psi(\underline{r}, t) = \psi(\underline{r})\exp(-iEt/\hbar), \tag{16.140}$$

and substitute into Eqs (16.127) and (16.128) using Eq. (16.139) for the vector potential (so with $\underline{A} = 0$), we find, in two-component notation

$$(E - V - mc^2)\Psi(\underline{r}) - c(\underline{\sigma} \cdot \hat{p})\Phi(\underline{r}) = 0, \tag{16.141}$$

$$(E - V + mc^2)\Phi(\underline{r}) - c(\underline{\sigma} \cdot \hat{p})\Psi(\underline{r}) = 0. \tag{16.142}$$

We can solve the second equation for Φ,

$$\Phi = (E - V + mc^2)^{-1}c(\underline{\sigma} \cdot \hat{p})\Psi(\underline{r}). \tag{16.143}$$

Substitute back into the first equation, being careful about ordering (since $[\hat{p}, V] \neq 0$):

$$(E - V - mc^2)\Psi = c(\underline{\sigma} \cdot \hat{p})\left[\frac{1}{E - V + mc^2}\right]c(\underline{\sigma} \cdot \hat{p})\Psi. \tag{16.144}$$

If we approximate $E - V + mc^2$ by $2mc^2$, we simply get Schrödinger's equation. Defining the quantity $E' = E - mc^2$, we expand the term in square brackets as follows:

$$\frac{1}{2mc^2 + E' - V} = \frac{1}{2mc^2}\left(1 - \frac{E' - V}{2mc^2}\right)^{-1}, \tag{16.145}$$

$$\simeq \frac{1}{2mc^2} - \frac{E' - V}{4m^2c^4}. \tag{16.146}$$

Substituting back into Eq. (16.144) we get

$$E'\Psi = \left[\frac{|\hat{p}|^2}{2m} + V - \frac{(\underline{\sigma}\cdot\hat{\underline{p}})\,(E' - V)(\underline{\sigma}\cdot\hat{\underline{p}})}{4m^2c^2}\right]\Psi. \tag{16.147}$$

This is not yet in Schrödinger form, since E' appears on both sides. Note that, since the 'non-relativistic' energy E' is $O(v^2)$, the third term is expected to be $O(v^4)$ (strictly $O(v^2(v^2/c^2))$). The two $(\underline{\sigma}\cdot\hat{p})$ terms use up a factor of v^2, so we only need $E' - V$ through $O(v^2)$. We can get this from the same equation truncated to lowest order:

$$(E' - V)\Psi = \frac{|\hat{p}|^2}{2m}\Psi. \tag{16.148}$$

Now

$$(E' - V)(\underline{\sigma}\cdot\hat{\underline{p}})\Psi = (\underline{\sigma}\cdot\hat{\underline{p}})(E' - V)\Psi + \underline{\sigma}\cdot\left[(E' - V), \hat{\underline{p}}\right]\Psi$$

$$= (\underline{\sigma}\cdot\hat{\underline{p}})\frac{|\hat{p}|^2}{2m}\Psi + \underline{\sigma}\cdot\left[\hat{\underline{p}}, V\right]\Psi. \tag{16.149}$$

Substituting back into Eq. (16.147), we get

$$E'\Psi = \left\{\frac{|\hat{p}|^2}{2m} + V - \frac{|\hat{p}|^4}{8m^3c^2} - \frac{(\underline{\sigma}\cdot\hat{\underline{p}})\left(\underline{\sigma}\cdot\left[\hat{\underline{p}}, V\right]\right)}{4m^2c^2}\right\}\Psi$$

$$= \left\{\frac{|\hat{p}|^2}{2m} + V - \frac{|\hat{p}|^4}{8m^3c^2} - \frac{i\underline{\sigma}\cdot\hat{\underline{p}}\times\left[\hat{\underline{p}}, V\right]}{4m^2c^2} - \frac{\hat{\underline{p}}\cdot\left[\hat{\underline{p}}, V\right]}{4m^2c^2}\right\}\Psi$$

$$= \hat{H}'\Psi, \tag{16.150}$$

where we used the identity Eq. (16.131) to get the second line. We recognise the third term on the right-hand side to be just the relativistic correction to the kinetic energy. The fourth term is the spin–orbit interaction,

$$\frac{-i\underline{\sigma}\cdot\hat{\underline{p}}\times\left[\hat{\underline{p}}, V\right]}{4m^2c^2} = \frac{-i\underline{\sigma}\cdot\hat{\underline{p}}\times\left[-i\hbar\underline{\nabla}(-e^2/r)\right]}{4m^2c^2}$$

$$= \frac{-\hbar e^2\underline{\sigma}\cdot\hat{\underline{p}}\times\underline{r}}{4m^2c^2\,r^3} = \frac{\hbar e^2}{4m^2c^2\,r^3}\,\underline{\sigma}\cdot\underline{r}\times\hat{\underline{p}}$$

$$= \frac{e^2}{2m^2c^2r^3}\,\hat{\underline{S}}\cdot\hat{\underline{L}} = \hat{H}_{S-O}. \tag{16.151}$$

Note that in the first line, we used the general result $[\hat{p}, f(x)] = -i\hbar\frac{df}{dx}$.

The expression for $\hat{H}_{S-O}$ is familiar to atomic physics. In the non-relativistic Schrödinger equation it has to be put in 'by hand'. In a more careful classical calculation, it turns out that there is an extra 'Thomas precession factor' of $\frac{1}{2}$ due to the electron being in a rotating frame with respect to the stationary nucleus. This is cancelled by $g_s = 2$ to give the naive classical result! The important point is that the Dirac equation gets it right because it is relativistically covariant.

The fifth and final term in $\hat{H}$ is a problem because it isn't Hermitian. A non-Hermitian Hamiltonian means non-unitary evolution. Thus the integral of the would-be probability density

$$\int \Psi^\dagger \Psi d^3 r \neq \text{constant in time.} \qquad (16.152)$$

However, the full probability density from the Dirac equation is $\psi^\dagger \psi$, where ψ is the four-component wave function, therefore the correct conservation law is

$$\int \psi^\dagger \psi d^3 r = \int \left(\Psi^\dagger \Psi + \Phi^\dagger \Phi \right) d^3 r = \text{constant in time.} \qquad (16.153)$$

So Ψ is not a good candidate for the Schrödinger wave function to this order (it was fine through $O(v^2)$). We have

$$\Phi = (E - V + mc^2)^{-1} c(\underline{\sigma} \cdot \hat{\underline{p}})\Psi \simeq \frac{\underline{\sigma} \cdot \hat{\underline{p}}}{2mc}\Psi, \qquad (16.154)$$

therefore

$$\Phi^\dagger \Phi \simeq \frac{\Psi^\dagger (\underline{\sigma} \cdot \hat{\underline{p}})(\underline{\sigma} \cdot \hat{\underline{p}})\Psi}{(2mc)^2} = \Psi^\dagger \frac{|\hat{\underline{p}}|^2}{4m^2 c^2}\Psi. \qquad (16.155)$$

Substituting into Eq. (16.153), we can write

$$\int \Psi^\dagger \left(1 + \frac{|\hat{\underline{p}}|^2}{4m^2 c^2} \right) \Psi d^3 r = \int \left[\left(1 + \frac{|\hat{\underline{p}}|^2}{8m^2 c^2} \right) \Psi \right]^\dagger \cdot \left(1 + \frac{|\hat{\underline{p}}|^2}{8m^2 c^2} \right) \Psi d^3 r,$$

where we used $(1 + x) = (1 + x/2)^2 + O(x^2)$ and the hermiticity of $\hat{\underline{p}}$. A good candidate for the Schrödinger wave function to this order is then

$$\Psi_S = \left(1 + \frac{|\hat{\underline{p}}|^2}{8m^2 c^2} \right)\Psi. \qquad (16.156)$$

Now eliminate Ψ from Eq. (16.150)

$$E' \left(1 + \frac{|\hat{\underline{p}}|^2}{8m^2 c^2} \right)^{-1} \Psi_S = \hat{H}' \left(1 + \frac{|\hat{\underline{p}}|^2}{8m^2 c^2} \right)^{-1} \Psi_S. \qquad (16.157)$$

Rearrange into 'Schrödinger' form

$$
E' \Psi_S \simeq \left(1 + \frac{|\hat{p}|^2}{8m^2c^2}\right) \hat{H}' \left(1 - \frac{|\hat{p}|^2}{8m^2c^2}\right) \Psi_S
$$

$$
= \left(\hat{H}' + \left[\frac{|\hat{p}|^2}{8m^2c^2}, H'\right]\right) \Psi_S \qquad \text{(to this order in } v/c\text{)}
$$

$$
\equiv \hat{H}_S \Psi_S. \tag{16.158}
$$

To evaluate the commutator, we need only the $O(v^2)$ part of $\hat{H}'$, since $|\hat{p}|^2/8m^2c^2$ is $O(v^2/c^2)$ and we are working to $O(v^2 \, (v^2/c^2))$, so

$$
\hat{H}_S = \hat{H}' + \left[\frac{|\hat{p}|^2}{8m^2c^2}, V(\underline{r})\right] \tag{16.159}
$$

is the desired Schrödinger Hamiltonian. This extra term combines with the last term in Eq. (16.150) to form the Darwin term

$$
\hat{H}_{\text{Darwin}} = \frac{1}{8m^2c^2}\left(-2\underline{\hat{p}} \cdot \left[\underline{\hat{p}}, V\right] + \left[\underline{\hat{p}} \cdot \underline{\hat{p}}, V(\underline{r})\right]\right)
$$

$$
= -\frac{1}{8m^2c^2}\left[\underline{\hat{p}}, \cdot \left[\underline{\hat{p}}, V\right]\right]
$$

$$
= \frac{\hbar^2}{8m^2c^2} \nabla^2 V(\underline{r}) \qquad \left(\text{using } [\hat{p}, f(x)] = -i\hbar \frac{df}{dx} \text{ twice}\right)
$$

$$
= \frac{e^2\hbar^2\pi}{2m^2c^2} \delta^{(3)}(\underline{r}) \qquad \left(\text{since } V(\underline{r}) = -\frac{e^2}{r} \text{ and } \nabla^2(1/r) = -4\pi\delta^{(3)}(\underline{r})\right),
$$

$$
\tag{16.160}
$$

which is precisely the result stated in Eq. (16.84) in Section 16.3.1. Note that the Darwin term affects only S-states, because all other states have zero wave function at the origin. Its physical interpretation is that a quantum particle cannot be localised to within a distance less than its Compton wavelength $\hbar/mc$ (uncertainty principle), so it sees some smeared average of the potential around the point $\underline{r}$.

Notes:

- Our solution is correct through $O(v^4/c^4)$. The Dirac equation can be solved exactly for the Coulomb case (see e.g. Merzbacher for more details). The resulting energy spectrum is

$$
E_{nj} = mc^2 \left[1 + \left(\frac{\alpha}{n - (j + \frac{1}{2}) + \left[(j + \frac{1}{2})^2 - \alpha^2\right]^{1/2}}\right)^2\right]^{-1/2}. \tag{16.161}
$$

Expansion in powers of α gives the $O(v^4/c^4)$ result above – as it must! Note that all states of a given n and j are degenerate to all orders in α.

- This result is not quite the last word in the structure of atomic hydrogen:

  – The *Lamb shift*, which is due to the interaction of the electron with the vacuum fluctuations of the electromagnetic field, removes the remaining degeneracy between the (nL_j) $2S_{1/2}$ and $2P_{1/2}$ states.

– The *hyperfine splitting*, due to the magnetic spin–spin interaction between the spin-$\frac{1}{2}$ electron and the spin-$\frac{1}{2}$ proton, removes the degeneracy between the spin-singlet (total $s = 0$) and spin-triplet (total $s = 1$) states.

The agreement with experiment is then near perfect.

• The systematic reduction of the Dirac equation to a non-relativistic Schrödinger equation is called the *Foldy–Wouthuysen* transformation.

Summary

• Nondegenerate time-independent perturbation theory is a systematic expansion in powers of the perturbation Hamiltonian to compute the energy eigenvalues and eigenstates. (16.1)
• The expression for the nondegenerate perturbation expansion is done to third order in the energy eigenvalue corrections and second order in the eigenstates. (16.1)
• Normalisation and orthogonality of the perturbed eigenstates are examined. (16.1.3)
• Degeneracy means there are many eigenstates with the same eigenvalue. Degenerate perturbation theory systematically treats energy and eigenstate corrections amongst degenerate states. (16.2)
• The expression for the degenerate perturbation theory expansion is given up to second order for the energy eigenvalues and first order for the eigenstates. (16.2)
• Application of perturbation theory to hydrogen fine structure and the helium atom ground-state energy are presented. (16.3.1, 16.3.2)
• Derivation of the fine-structure corrections from the relativistic Dirac equation is presented. (16.A)

References

The derivation in Appendix 16.A of fine-structure terms from relativistic theory followed these sources. Further treatment with regard to the interaction of the electromagnetic field with a quantum system can be found here also.

Aitchison, I. J. R. (1972). *Relativistic Quantum Mechanics*. Macmillan, New York.

Gottfried, K. and Yan, T. (2003). *Quantum Mechanics: Fundamentals*, 2nd ed. Springer-Verlag, New York.

Shankar, R. (1994). *Principles of Quantum Mechanics*, 2nd ed. Plenum Press, New York.

Further Reading

Further treatment on convergence issues related to perturbation theory can be found in the following textbooks:

Fernandez, F. M. (2001). *Perturbation Theory in Quantum Mechanics*. CRC Press, Boca Raton, FL.

Konishi, K. and Paffuti, G. (2009). *Quantum Mechanics: A New Introduction*. Oxford University Press, Oxford.

For the exact solution of the Dirac equation for the Coulomb potential, please see:

Merzbacher, E. (1968). *Quantum Mechanics*, 3rd ed. Wiley, Chichester.

Problems

16.1 A particle moves in one dimension in the potential

$$V(x) = \infty, \quad |x| > a, \quad V(x) = V_0 \cos(\pi x/2a), \quad |x| \leq a.$$

Calculate the energy of the first excited state to first order in perturbation theory.

16.2 A particle moves in one dimension in the potential

$$V(x) = \infty, \quad |x| > a, \quad V(x) = V_0 \sin(\pi x/a), \quad |x| \leq a.$$

- Show that the first-order energy shift is zero.
- Obtain an expression for the second-order correction to the energy of the ground state.

16.3 The one-dimensional anharmonic oscillator: a particle of mass m is described by the Hamiltonian

$$\hat{H} = \frac{\hat{p}^2}{2m} + \tfrac{1}{2}m\omega^2\,\hat{x}^2 + \gamma\hat{x}^4.$$

- Assuming that γ is small, use first-order perturbation theory to calculate the ground-state energy.
- Show more generally that the energy eigenvalues are approximately

$$E_n \simeq (n + \tfrac{1}{2})\hbar\omega + 3\gamma\left(\frac{\hbar}{2m\omega}\right)^2(2n^2 + 2n + 1).$$

Hint: To evaluate matrix elements of powers of $\hat{x}$, write $\hat{x}$ in terms of the harmonic oscillator raising and lowering operators $\hat{a}$ and $\hat{a}^\dagger$. Recall that the raising and lowering operators are defined by

$$\hat{a} \equiv \sqrt{\frac{m\omega}{2\hbar}}\,\hat{x} + \frac{i}{\sqrt{2m\omega\hbar}}\,\hat{p} \quad \text{and} \quad \hat{a}^\dagger \equiv \sqrt{\frac{m\omega}{2\hbar}}\,\hat{x} - \frac{i}{\sqrt{2m\omega\hbar}}\,\hat{p},$$

with the properties that

$$\hat{a}|n\rangle = \sqrt{n}\,|n-1\rangle \quad \text{and} \quad \hat{a}^\dagger|n\rangle = \sqrt{n+1}\,|n+1\rangle.$$

16.4 Starting from the relativistic expression for the total energy of a single particle, $E = (m^2 c^4 + p^2 c^2)^{1/2}$, and expanding in powers of p^2, obtain the leading relativistic correction to the kinetic energy.

16.5 Consider the proton in a hydrogen atom to be a uniform sphere of charge, of radius $R \ll a_0$, rather than a point charge. What is the classical electrostatic potential energy function, $V(r)$, of the electron?

Hint: Use Gauss's theorem to find the internal and external electric fields and hence the electrostatic potential, $\Phi(r)$. Remember that the corresponding potential energy is $-e\Phi(r)$.

By treating the difference between the electrostatic potential energy of the electron in the field of such a finite-size proton and in the field of a point proton by perturbation theory, find the shift in the ground-state energy of the hydrogen atom arising from the finite size of the proton. Estimate the order of magnitude of this shift, given that $R \simeq 10^{-15}$ m.

16.6 A one-dimensional harmonic oscillator of mass m carries an electric charge, q. A weak, uniform, static electric field of magnitude $\mathcal{E}$ is applied in this x-direction. Write down an expression for the classical electrostatic potential energy. Show that, to first order in perturbation theory, the oscillator energy levels are unchanged, and calculate the second-order shift. Can you show that the second-order result is in fact exact? *Hint:* To evaluate matrix elements of $\hat{x}$, write $\hat{x}$ in terms of the harmonic oscillator raising and lowering operators $\hat{a}$ and $\hat{a}^{\dagger}$.

16.7 Quarkonium is a system consisting of a heavy quark of mass m_Q bound to its antiquark, also of mass m_Q. The inter quark potential is of the form

$$V(r) = -\frac{a}{r} + br,$$

where a, b are constants and r is the quark–antiquark separation. Given the Bohr formula for the energy levels of hydrogen

$$E_n^{(0)} = -\frac{me^4}{2\hbar^2} \frac{1}{n^2},$$

where m is the reduced mass of the electron–proton system, deduce an expression for the energy levels of quarkonium in the approximation which neglects the second term in $V(r)$.

What are the corresponding degeneracies of the lowest two energy levels?

Use first-order perturbation theory to calculate the corrections to the lowest two energy levels. Why it is not necessary to use degenerate perturbation theory for this problem?

You may assume that the radial functions for hydrogen are

$$R_{10} = 2(a_0)^{-3/2} \exp(-r/a_0),$$

$$R_{20} = 2(2a_0)^{-3/2} \left(1 - \frac{r}{2a_0}\right) \exp(-r/2a_0),$$

$$R_{21} = \frac{1}{\sqrt{3}} (2a_0)^{-3/2} \frac{r}{a_0} \exp(-r/2a_0),$$

where the Bohr radius $a_0 = \hbar^2/me^2$, and that

$$\int_0^\infty \exp(-kr)\, r^n\, dr = n!\,/k^{n+1}, \quad n > -1.$$

16.8 The change in energy levels in an atom due to the application of an external electric field is known as the **Stark effect**. The perturbation corresponding to a uniform static electric field of magnitude $\mathcal{E}$, applied in the z-direction to a hydrogen atom, is

$$\hat{H}' = e\mathcal{E}z = e\mathcal{E}r \cos\theta.$$

Use degenerate perturbation theory to calculate the effect on the four-fold degenerate $n = 2$ level of atomic hydrogen.

The relevant unperturbed eigenfunctions are

$$u_{200} = \left(8\pi a_0^3\right)^{-1/2} \left(1 - \frac{r}{2a_0}\right) \exp\left(-r/2a_0\right),$$

$$u_{211} = -\left(\pi a_0^3\right)^{-1/2} \frac{r}{8a_0} \sin\theta \exp\left(i\phi\right) \exp\left(-r/2a_0\right),$$

$$u_{210} = \left(8\pi a_0^3\right)^{-1/2} \frac{r}{2a_0} \cos\theta \exp\left(-r/2a_0\right),$$

$$u_{21-1} = \left(\pi a_0^3\right)^{-1/2} \frac{r}{8a_0} \sin\theta \exp\left(-i\phi\right) \exp\left(-r/2a_0\right).$$

16.9 The so-called hyperfine interaction in the hydrogen atom between an $\ell = 0$ electron and the spin of the proton which constitutes the nucleus can be treated by perturbation theory, with

$$\hat{H}' = \frac{4}{3}\, g_p\, \frac{m}{M}\, \alpha^4 mc^2 \frac{1}{n^3\hbar^2}\, \underline{\hat{S}}_e \cdot \underline{\hat{S}}_p \qquad (g_p = 5.56),$$

where $\underline{\hat{S}}_e$ and $\underline{\hat{S}}_p$ are the operators representing the electron and proton spins, respectively, m and M are the electron and proton masses and α is the fine-structure constant.

What are the allowed values of the total spin of the electron–proton system? Use first-order perturbation theory in the coupled basis to calculate the hyperfine splitting of the hydrogen ground state ($n = 1$) and hence show that the wavelength of radiation emitted in transitions between the triplet and singlet states is approximately 21 cm (famous in radio astronomy as the signature of hydrogen!).

16.10 Verify that the kinetic energy, spin–orbit and Darwin corrections to the energy levels of a one-electron atom may be combined for any allowed value of ℓ to give the fine-structure formula quoted in the text:

$$\Delta E_{nj} = E_n^{(0)} \frac{(Z\alpha)^2}{n^2} \left(\frac{n}{j + \frac{1}{2}} - \frac{3}{4}\right).$$

Hint: Consider the cases $\ell = 0$ and $\ell \neq 0$ separately.

16.11 The isotropic harmonic oscillator in two dimensions is described by the Hamiltonian

$$\hat{H}_0 = \sum_i \left\{ \frac{\hat{p}_i^2}{2m} + \tfrac{1}{2}m\omega^2 \hat{x}_i^2 \right\} \qquad\qquad i = 1, 2$$

and has energy eigenvalues

$$E_n = (n+1)\hbar\omega \equiv (n_1 + n_2 + 1)\hbar\omega \qquad\qquad n = 0, 1, 2, \ldots.$$

What is the degeneracy of the first excited level? Use degenerate perturbation theory to determine the splitting induced by the perturbation

$$\hat{H}' = Cx_1 x_2,$$

where C is a constant.

Hint: The matrix elements of the perturbation may be computed by using the lowering and raising operators.

16.12 The so-called 'spin Hamiltonian'

$$\hat{H} = D\hat{S}_z^2 + F\left(\hat{S}_x^2 - \hat{S}_y^2\right),$$

where D and F are positive constants, is known to describe the energy levels of an ion with $s = 1$ in many crystals.

- Given that $F \ll D$, so that the second term may be treated as a small perturbation, use the 3×3 matrix representation of the spin operators to show that the unperturbed energy eigenvalues are $D\hbar^2$ (twice) and zero.
- What are the corresponding eigenvectors and what quantum numbers may be used to label them?
- Use first-order perturbation theory to find the perturbed eigenvalues and the corresponding eigenvectors.
- Show that the perturbation theory result is an exact solution of the energy eigenvalue problem.
- Sketch an energy-level diagram showing the pattern of splitting in the two cases (i) $D > F$ and (ii) $D < F$, assuming that $D, F > 0$.

16.13 Show explicitly that the Darwin term

$$\hat{H}_{\text{Darwin}} = \frac{1}{8m^2 c^2}\left(-2\underline{\hat{p}} \cdot \left[\underline{\hat{p}}, V\right] + \left[\underline{\hat{p}} \cdot \underline{\hat{p}}, V(\underline{r})\right] \right),$$

where $V(\underline{r}) = -\dfrac{e^2}{r}$, may be written in the form

$$\hat{H}_{\text{Darwin}} = \frac{e^2 \hbar^2 \pi}{2m^2 c^2} \delta^{(3)}(\underline{r}).$$

16.14 The origin of the Darwin term can be understood from uncertainty principle arguments. In a relativistic theory, a particle cannot be localised to better than its

Compton wavelength $\delta r \approx \hbar/mc$. (Why not?) So the potential it sees is not just $V(\underline{r})$ but some smeared average about the point $\underline{r}$:

$$\overline{V(\underline{r})} \;=\; V(\underline{r}) \;+\; \sum_i \overline{\frac{\partial V}{\partial x_i}}\,\delta x_i \;+\; \tfrac{1}{2} \sum_{ij} \overline{\frac{\partial^2 V}{\partial x_i \partial x_j}}\,\delta x_i\,\delta x_j \;+\; O(\delta r^3).$$

Assuming that the fluctuations in the various directions are uncorrelated and spherically symmetric, and that $\delta r \approx \hbar/mc$, show that this extremely naive estimate gives a result which is two-thirds of the exact one.

17

Calculation Methods Beyond Perturbation Theory

There are various calculational methods beyond the perturbation theory of the previous chapter that can be applied in specific circumstances to give either exact or approximate results. In this chapter some of the most common methods are explained. We start with the Rayleigh–Ritz variational method that can be used to obtain an upper-bound estimate of the ground-state energy of a quantum-mechanical system. Next we examine multi-electron atoms. In such a case simple application of perturbation theory becomes difficult and more needs to be done. We will look at the Hartree–Fock or central field method, where the field that a given electron evolves in is created by all the other electrons and protons in the system. The idea of the approximation is to approximate this field, and then self-consistently compute the energy levels. Then the Born–Oppenheimer method will be discussed, which can be applied in situations where there is a heavy and a light mass particle that are interacting. The motion of the heavy particle in a sense decouples from the system dynamics and one can exploit that to simplify calculations. Then the Hellmann–Feynman method is explained. This method can be used to calculate the derivative of an eigenvalue of a Hamiltonian $\hat{H}$ in terms of the expectation value of the derivative of the Hamiltonian operator with respect to the same parameter. This can be a convenient way to compute some expectation values in the system. Finally the Wenzel–Kramers–Brillouin–Jeffreys (WKBJ) approximation is explained. This method is applicable when the potential varies slowly relative to the de Broglie wavelength of the particle. Since the de Broglie wavelength goes as the inverse of the momentum of the particle, the method can be applied for a given potential for particles with sufficiently high kinetic energy.

17.1 Rayleigh–Ritz Variational Method

An important alternative approximation scheme to perturbation theory, which is very useful in estimating the ground-state energy of a system, is the variational method, also known as the **Rayleigh–Ritz variational method**.

Consider a system described by a Hamiltonian, $\hat{H}$, with a complete orthonormal set of eigenstates $\{|n\rangle\}$ and corresponding energy eigenvalues $\{E_n\}$, ordered in increasing value:

$$E_1 \leq E_2 \leq E_3 \leq \cdots .$$

At say $t = 0$, an arbitrary state $|\psi\rangle$ can be expanded as a linear superposition:[1]

$$|\psi\rangle = \sum_n c_n |n\rangle \qquad \text{with} \qquad c_n = \langle n|\psi\rangle. \tag{17.1}$$

The expectation value of the energy in the state $|\psi\rangle$ is just given by

$$\langle E\rangle = \frac{\langle\psi|\hat{H}|\psi\rangle}{\langle\psi|\psi\rangle} = \frac{\sum_n |c_n|^2 E_n}{\sum_n |c_n|^2} \geq \frac{\sum_n |c_n|^2 E_1}{\sum_n |c_n|^2} \equiv E_1, \tag{17.2}$$

since $E_n \geq E_1$ for all $n > 1$. Thus we have the inequality

$$\langle E\rangle \geq E_1. \tag{17.3}$$

This is the key result which underlies the variational method. It says that *the expectation value of $\hat{H}$ for any state $|\psi\rangle$ is an upper bound on the ground-state energy.*

The variational principle is as follows: we choose a trial state, $|\psi_T\rangle$, or wave function, ψ_T, which depends on one or more parameters $\alpha_1, \alpha_2, \ldots, \alpha_r$. We then calculate

$$E(\alpha_1, \alpha_2, \ldots, \alpha_r) = \frac{\langle\psi_T|\hat{H}|\psi_T\rangle}{\langle\psi_T|\psi_T\rangle}, \tag{17.4}$$

and minimise $E(\alpha_1, \alpha_2, \ldots, \alpha_r)$ with respect to the variational parameters $\alpha_1, \alpha_2, \ldots, \alpha_r$ by solving the equations

$$\frac{\partial E(\alpha_1, \alpha_2, \ldots, \alpha_r)}{\partial \alpha_i} = 0, \qquad i = 1, 2, \ldots, r. \tag{17.5}$$

The resulting minimum value of $E(\alpha_1, \alpha_2, \ldots, \alpha_r)$ is then our best estimate of the ground-state energy for the given trial function. It is *an upper bound for the true ground-state energy.*

How should we choose the trial function, ψ_T? A good trial function is one that incorporates as many of the features of the exact wave function as possible. Some of these features can often be deduced by symmetry arguments. Note that the variational method is aimed at optimising the estimate of the energy. It may not necessarily give a good approximation to the true wave function, and could give poor results if used to calculate quantities other than the energy.

17.1.1 Ground State of Hydrogen

As an example, we consider the ground state of the hydrogen atom, for which we already know the exact solution. We know from general considerations that the ground state will have $\ell = 0$, since there is no centrifugal barrier term in the TISE for $\ell = 0$. Thus the trial

---

[1] Recall that the alternative way of expressing the completeness of the basis is by writing the operator identity $\hat{I} \equiv \sum_n |n\rangle\langle n|$. The unit or identity operator $\hat{I}$ simply reproduces any given state: $\hat{I}|\psi\rangle = |\psi\rangle$, so that

$$\hat{I}|\psi\rangle = |\psi\rangle = \sum_n |n\rangle\langle n|\psi\rangle \equiv \sum_n c_n|n\rangle \qquad \text{with} \qquad c_n = \langle n|\psi\rangle.$$

function should depend only on r, should be non-zero at the origin and should vanish for large r. A simple trial function with these properties is

$$\psi_T(r) = C \exp\left(-\beta r\right),\qquad(17.6)$$

where β is the (single) variational parameter. The normalisation constant is determined in the usual way:

$$\langle\psi_T|\psi_T\rangle = |C|^2 \int_{\phi=0}^{2\pi}\int_{\theta=0}^{\pi}\int_{r=0}^{\infty} \exp\left(-2\beta r\right) r^2 \sin\theta\, dr\, d\theta\, d\phi$$

$$= 4\pi|C|^2 \int_0^{\infty} \exp\left(-2\beta r\right) r^2 dr = 1,$$

which gives $C = \left(\beta^3/\pi\right)^{1/2}$. A similar calculation also gives the expectation value of the potential energy:

$$\langle\hat{V}\rangle = \langle V(r)\rangle = \langle\psi_T|\left(-\frac{e^2}{r}\right)|\psi_T\rangle = -e^2\beta.\qquad(17.7)$$

The expectation value of the kinetic energy can be evaluated directly or by using a trick which is useful for more complicated problems. Note the identity for any function $\psi(\underline{r})$:

$$\int_V \underline{\nabla}\cdot(\psi^*\underline{\nabla}\psi)d^3r = \int_V \psi^*\nabla^2\psi d^3r + \int_V (\underline{\nabla}\psi^*)\cdot(\underline{\nabla}\psi)d^3r,\qquad(17.8)$$

where $\int_V$ denotes the integral over all space. The left-hand side can be transformed into a surface integral at infinity by applying the divergence theorem to the vector field $\psi^*\underline{\nabla}\psi$:

$$\int_V \underline{\nabla}\cdot(\psi^*\underline{\nabla}\psi)d^3r = \int_S (\psi^*\underline{\nabla}\psi)\cdot\underline{dS}.\qquad(17.9)$$

If ψ tends to zero sufficiently rapidly as $r\to\infty$, the surface integral $\int_S$ vanishes so that

$$\int_V \psi^*\nabla^2\psi d^3r = -\int_V \left|\underline{\nabla}\psi\right|^2 d^3r.\qquad(17.10)$$

In the present case therefore

$$\langle\hat{T}\rangle = \langle\psi_T|\frac{\hat{p}^2}{2m}|\psi_T\rangle = -\frac{\hbar^2}{2m}\int_V \psi_T^*\nabla^2\psi_T d^3r = \frac{\hbar^2}{2m}\int_V \left|\underline{\nabla}\psi_T\right|^2 d^3r.\qquad(17.11)$$

Now our trial function depends only on the radial coordinate r, so $\underline{\nabla}\psi_T = \dfrac{\partial\psi_T}{\partial r}\underline{e}_r$. Hence

$$\left|\underline{\nabla}\psi_T\right|^2 = \left|\frac{\partial\psi_T}{\partial r}\right|^2 = \beta^2\left|\psi_T\right|^2,\qquad(17.12)$$

so that

$$\langle\hat{T}\rangle = \langle\psi_T|\frac{\hat{p}^2}{2m}|\psi_T\rangle = \frac{\hbar^2\beta^2}{2m}.\qquad(17.13)$$

Thus

$$E(\beta) = \langle \psi_T | \hat{H} | \psi_T \rangle = \frac{\hbar^2 \beta^2}{2m} - e^2 \beta \qquad \Rightarrow \qquad \frac{dE(\beta)}{d\beta} = \frac{\hbar^2 \beta}{m} - e^2. \tag{17.14}$$

Setting $dE(\beta)/d\beta = 0$ tells us that $E(\beta)$ has a minimum for

$$\beta = \frac{me^2}{\hbar^2} = \frac{1}{a_0}, \tag{17.15}$$

where a_0 is the Bohr radius, giving an upper bound of

$$E_{\min} = -\frac{e^2}{2a_0} = -1 \, \mathrm{Ry} = -13.6 \mathrm{eV} \tag{17.16}$$

for the ground-state energy. Of course, this is the *exact* result, because our trial function with $\beta = 1/a_0$ is the exact ground-state eigenfunction for the hydrogen atom!

17.1.2 Ground State of Helium

A more realistic example is the helium ground state. We take as trial wave function a product of $(1s)$ orbital wave functions for each electron, but with an effective nuclear charge, $\tilde{Z}$, which takes account of screening of the nuclear charge by the other electron:

$$\psi_T(r_1, r_2) = \frac{\tilde{Z}^3}{\pi a_0^3} \exp\left(-\tilde{Z}(r_1 + r_2)/a_0\right). \tag{17.17}$$

Setting $\tilde{Z} = Z$ would correspond to no screening. This wave function is already correctly normalised. $\tilde{Z}$ is our variational parameter.

For the kinetic energy expectation value we can use the previous result for hydrogen:

$$\langle \psi_T | -\frac{\hbar^2}{2m}(\nabla_1^2 + \nabla_2^2)|\psi_T\rangle = \frac{\hbar^2 \tilde{Z}^2}{ma_0^2} = 2\tilde{Z}^2 \, \mathrm{Ry} \tag{17.18}$$

[recall that $1 \, \mathrm{Ry} = e^2/2a_0 = \hbar^2/(2ma_0^2)$] and for the potential energy

$$\langle \psi_T | -Ze^2\left(\frac{1}{r_1} + \frac{1}{r_2}\right)|\psi_T\rangle = -\frac{2Ze^2}{a_0}\tilde{Z} = -4Z\tilde{Z} \, \mathrm{Ry}. \tag{17.19}$$

The electron–electron interaction term gives

$$\langle \psi_T | \frac{e^2}{r_{12}}|\psi_T\rangle = \frac{5}{4}\tilde{Z} \, \mathrm{Ry}. \tag{17.20}$$

Thus in Rydbergs, we have

$$E(\tilde{Z}) = 2\left[\tilde{Z}^2 - \left(2Z - \frac{5}{8}\right)\tilde{Z}\right] \, \mathrm{Ry}. \tag{17.21}$$

Differentiating with respect to $\tilde{Z}$ and setting the derivative to zero gives

$$\tilde{Z} = Z_{\mathrm{eff}} = \left(Z - \frac{5}{16}\right) \tag{17.22}$$

for the value of $\tilde{Z}$ which minimises the energy, so that

$$E(Z_{\text{eff}}) = -2\left(Z - \frac{5}{16}\right)^2 = -2\left(\frac{27}{16}\right)^2 = -5.695 \text{ Ry} = -77.46\text{eV}, \tag{17.23}$$

which is within 2% of the measured value -79.005 eV. This is significantly better than the first-order perturbation theory estimate of -74.8 eV.

17.1.3 Excited States

The variational method can be adapted to give bounds on the energies of excited states, under certain conditions. Suppose we choose a trial function which is orthogonal to the ground state by imposing the condition $\langle n = 1|\psi_T\rangle = 0$. The expansion of $|\psi_T\rangle$ in energy eigenstates does not then contain a term with $n = 1$ and we find that

$$\frac{\langle \psi|\hat{H}|\psi\rangle}{\langle \psi|\psi\rangle} = \frac{\sum\limits_{n=2} |c_n|^2 E_n}{\sum\limits_{n=2} |c_n|^2} \geq E_2. \tag{17.24}$$

We can apply the variational method as before to obtain an upper bound on the first excited-state energy, E_2, and so on. The drawback is that we do not in general know the exact eigenstates with respect to which our trial wave function is to be orthogonal. Using approximate eigenfunctions does not guarantee that we get an upper bound. These problems are avoided if the excited state has symmetries which differ from those of the lower-lying levels. In this case, if we choose a trial function with the correct symmetries, orthogonality is guaranteed and we will get an upper bound to the energy of the lowest-lying level with those symmetries.

Example 17.1 The ground state of helium is a ^1S state. Its first four excited states are ^3S, ^1S, ^3P and ^1P states. With the exception of the ^1S state, these all have symmetries different from those of the ground state. For example, the P states have $L = 1$ and must be orthogonal to the ground state which has $L = 0$, so simple trial functions suggested by the configurational description $(1s)(2p)$ are

$$\psi_T^{\pm}(\underline{r}_1,\underline{r}_2) = C_{\pm}[f(r_1)g(r_2)\cos\theta_1 \pm f(r_2)g(r_1)\cos\theta_2],$$

where

$$f(r) = \exp(-\beta r/a_0) \qquad \text{and} \qquad g(r) = r\exp(-\gamma r/2a_0),$$

with β and γ variational parameters.

17.2 Atomic Systems

For multi-particle systems, the simpler methods of quantum mechanics are insufficient for calculating their properties. Multi-atomic systems are a good example of such systems, and here some approaches for handling such systems will be discussed.

17.2.1 First Excited States of Helium

The helium Hamiltonian once again is

$$\hat{H} = \hat{H}_1 + \hat{H}_2 + \frac{e^2}{|\underline{r}_1 - \underline{r}_2|}, \tag{17.25}$$

where

$$\hat{H}_i = \frac{\hat{p}_i^2}{2m} - \frac{2e^2}{r_i}. \tag{17.26}$$

The first excited state of helium was briefly discussed in Chapter 13, but here we will address it in greater detail. Again ignoring the electronic repulsion for the time being, the first excited state corresponds to exciting one electron to the first hydrogen-like excited state, which has n_1 or $n_2 = 2$ and, correspondingly, ℓ_1 or $\ell_2 = 1$ or 0; the electronic configuration is denoted $(1s)(2p)$ or $(1s)(2s)$.

In this case we can construct both symmetric and antisymmetric spatial wave functions:

$$\frac{1}{\sqrt{2}} \left\{ u_{2\ell m_\ell}(\underline{r}_1) \cdot u_{100}(\underline{r}_2) \pm u_{100}(\underline{r}_1) \cdot u_{2\ell m_\ell}(\underline{r}_2) \right\} \qquad \ell = 0, 1, \tag{17.27}$$

which may be combined with singlet or triplet spin functions to give the overall antisymmetric two-electron wave functions already shown in Section 13.4:

$$\Psi_{L,M_L,S,M_S}^{\text{singlet}}(1,2) = \frac{1}{\sqrt{2}} \left\{ u_{2\ell m_\ell}(\underline{r}_1) \cdot u_{100}(\underline{r}_2) + u_{100}(\underline{r}_1) \cdot u_{2\ell m_\ell}(\underline{r}_2) \right\} \chi^{S=0},$$

$$\Psi_{L,M_L,S,M_S}^{\text{triplet}}(1,2) = \frac{1}{\sqrt{2}} \left\{ u_{2\ell m_\ell}(\underline{r}_1) \cdot u_{100}(\underline{r}_2) - u_{100}(\underline{r}_1) \cdot u_{2\ell m_\ell}(\underline{r}_2) \right\} \chi^{S=1}.$$

Here, L denotes the total orbital angular momentum quantum number, which is zero for $(1s)(2s)$ and 1 for $(1s)(2p)$. There are four degenerate singlet states, corresponding to the Russell–Saunders terms $^1S_0$ and $^1P_1$. For the triplet states, the degeneracy is 12-fold, as shown in Table 17.1, making a total of 16 degenerate states.

We can again use perturbation theory to evaluate the shift induced by the Coulomb repulsion between the two electrons. Since the perturbation commutes with $\hat{L}_z$, the shift must be independent of the m_ℓ value, and so it is sufficient to calculate for $m_\ell = 0$ states:

$$\Delta E_{\text{triplet}}^{\text{singlet}} = e^2 \int \left\{ u_{2\ell 0}(\underline{r}_1) \cdot u_{100}(\underline{r}_2) \pm u_{100}(\underline{r}_1) \cdot u_{2\ell 0}(\underline{r}_2) \right\}^*$$

$$\times \frac{1}{r_{12}} \times \left\{ u_{2\ell 0}(\underline{r}_1) \cdot u_{100}(\underline{r}_2) \pm u_{100}(\underline{r}_1) \cdot u_{2\ell 0}(\underline{r}_2) \right\} d^3 r_1 d^3 r_2,$$

Table 17.1 Degeneracies of the helium atom first excited singlet and triplet states				
S	L	J	g_J	Terms
0	0	0	1	$^1S_0$
	1	1	3	$^1P_1$
1	0	1	3	$^3S_1$
	1	0	1	$^3P_0$
	1	1	3	$^3P_1$
	1	2	5	$^3P_2$

which can be rewritten, noting that the perturbation is symmetric under $1 \leftrightarrow 2$, as

$$\Delta E_{\text{triplet}}^{\text{singlet}} = e^2 \int |u_{2\ell0}(\underline{r}_1)|^2 \cdot \frac{1}{r_{12}} \cdot |u_{100}(\underline{r}_2)|^2 d^3r_1 d^3r_2$$
$$\pm e^2 \int u_{2\ell0}^*(\underline{r}_1)u_{100}^*(\underline{r}_2) \cdot \frac{1}{r_{12}} \cdot u_{100}(\underline{r}_1)u_{2\ell0}(\underline{r}_2) d^3r_1 d^3r_2.$$

Notes:

- The first term looks similar to the ground-state shift and has the form of the electrostatic interaction between two electron clouds.
- The second term has no obvious classical interpretation; it is called the *exchange contribution* and its sign depends on whether the total spin is 0 or 1. So although the perturbation does not depend on the electron spins explicitly, the symmetry of the wave function makes the potential behave as if it were spin-dependent! Because of this, the singlet and triplet terms are no longer degenerate.

Quite generally, the splittings have the form

$$\Delta E_{n\ell}^{\text{singlet}} = A_{n\ell} + B_{n\ell}, \tag{17.28}$$

$$\Delta E_{n\ell}^{\text{triplet}} = A_{n\ell} - B_{n\ell}. \tag{17.29}$$

Clearly $A_{n\ell}$ is positive, and it turns out that $B_{n\ell}$ is too, but smaller in magnitude so that

$$\Delta E_{n\ell}^{\text{triplet}} < \Delta E_{n\ell}^{\text{singlet}}. \tag{17.30}$$

This is intuitively plausible: in triplet states, the spatial wave function vanishes when the electrons are close together, so the Coulomb repulsion has less effect. The effect of the Coulomb repulsion term on the helium spectrum is illustrated in Fig. 17.1.

17.2.2 Multi-electron Atoms

We now consider briefly the problem of an atom or ion containing a nucleus with charge Ze and N electrons. If we assume that the nucleus is infinitely heavy and we neglect all

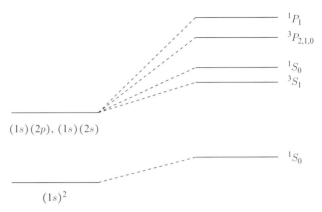

Fig. 17.1 Effect of the Coulomb repulsion on the spectrum of helium.

interactions except the Coulomb interaction between each electron and the nucleus, and the mutual electronic repulsion, the Hamiltonian is

$$\hat{H} = \sum_{i=1}^{N} \left\{ \frac{\hat{p}_i^2}{2m} - \frac{Ze^2}{r_i} \right\} + \sum_{i>j=1}^{N} \frac{e^2}{r_{ij}}. \tag{17.31}$$

We would like to solve the TISE for the energy eigenvalues and the corresponding totally antisymmetric eigenfunctions. However, the presence of the electron–electron interaction terms means that the Schrödinger equation is not separable. We are forced to use approximation methods.

Central Field Approximation

This is an *independent particle model* so the wave function will take a product form of single-particle orthonormal wave functions:

$$\psi(\underline{r}_1, \underline{r}_2, \dots, \underline{r}_N) = u_1(\underline{r}_1) u_2(\underline{r}_2) \dots u_N(\underline{r}_N). \tag{17.32}$$

As these are electrons, this function will need to be overall antisymmetrised, and that detail will be discussed later. The assumption is that each electron moves in an effective central potential, $V(r)$, which incorporates the nuclear attraction and the average effect of the repulsive interaction with the other $(N-1)$ electrons. The full Hamiltonian for this system is

$$\hat{H} = \hat{H}_c + \hat{H}', \tag{17.33}$$

where $\hat{H}_c$ corresponds to the central field approximation

$$\hat{H}_c = \sum_{i=1}^{N} \left(\frac{\hat{p}_i^2}{2m} + V(r_i) \right) = \sum_{i=1}^{N} \hat{H}_i \qquad \text{with} \qquad \hat{H}_i = \frac{\hat{p}_i^2}{2m} + V(r_i), \tag{17.34}$$

and $\hat{H}'$ is a remaining part of the Hamiltonian that contains residual interactions that cannot be treated in the central field approximation.

The effective potential in Eq. (17.34) has the form

$$V(r_i) = V_n(r_i) + V_e(r_i), \tag{17.35}$$

where

$$V_n(r_i) = -\frac{Ze^2}{r_i} \tag{17.36}$$

is the attractive potential between the electron and the nuclear core of Z protons, and

$$V_e(r_i) = e^2 \int \sum_{j \neq i} \frac{|u_j(\underline{r}_j)|^2}{r_{ji}} d^3 r_j, \tag{17.37}$$

with $r_{ji} = |\underline{r}_j - \underline{r}_i|$ the repulsive potential between electron i and all the other electrons, each distributed in accordance with its single-particle wave function. The effective potential should also have certain limiting behaviour:

$$\lim_{r \to 0} V(r) = -\frac{Ze^2}{r} \quad \text{and} \quad \lim_{r \to \infty} V(r) = -\frac{[Z - (N-1)]e^2}{r}. \tag{17.38}$$

Finding the effective potential at intermediate distances is a much more difficult problem, since it depends on the distribution of the electrons.

At this point neither the electron wave functions nor the effective potential is known. The **Hartree–Fock method** or **self-consistent field method** is used to determine these quantities iteratively. The procedure is as follows. Start with some initial approximate central field potential $V^{(1)}(r)$, where the superscript indexes each stage of the iteration.

- Compute the electron single-particle wave functions with this approximate potential $V^{(1)}(r)$:

$$\left(\frac{\hat{p}_i^2}{2m} + V^{(1)}(r_i)\right) u^{(1)}_{n_i, \ell_i, m_{\ell_i}}(\underline{r}_i) = E_{n_i, \ell_i} u^{(1)}_{n_i, \ell_i, m_{\ell_i}}(\underline{r}_i). \tag{17.39}$$

- With these single-particle wave functions $\{u^{(1)}_{n_i, \ell_i, m_{\ell_i}}(\underline{r}_i)\}$, the charge densities of the electrons can be computed, i.e. $e^2 |u^{(1)}_{n_i, \ell_i, m_{\ell_i}}(\underline{r}_i)|^2$, and used in Eq. (17.37) to compute the modified central potential $V^{(2)}(r)$.

Now repeat the above two steps until the central potential does not change appreciably with another iteration:

$$V^{(k+1)}(r) \approx V^{(k)}(r). \tag{17.40}$$

At this point this is the self-consistent effective atomic potential along with the single-particle eigenstates and eigenvalues.

A few details are important to mention. The central field Hamiltonian is the sum of N identical single-particle Hamiltonians, $\hat{H}_i$, and will thus have energy eigenfunctions which are products of single-particle eigenfunctions, so the solutions to the central field equation are

$$\left(\frac{\hat{p}_i^2}{2m} + V(r_i)\right) u_{n_i, \ell_i, m_{\ell_i}}(\underline{r}_i) = E_{n_i, \ell_i} u_{n_i, \ell_i, m_{\ell_i}}(\underline{r}_i). \tag{17.41}$$

The energy eigenvalues will be independent of the magnetic quantum number, m_{ℓ_i}, because of the spherical symmetry of the Hamiltonian – there is no preferred direction – and the so-called one-electron central-field orbitals, $u_{n_i,\ell_i,m_{\ell_i}}(\underline{r}_i)$ are products of a radial function and a spherical harmonic:

$$u_{n_i,\ell_i,m_{\ell_i}}(\underline{r}_i) = R_{n_i,\ell_i}(r_i) \cdot Y_{\ell_i,m_{\ell_i}}(\theta_i,\phi_i). \tag{17.42}$$

We can take account of spin by multiplying these spatial orbitals by a spin-$\frac{1}{2}$ function to form normalised spin orbitals:

$$u_{n_i,\ell_i,m_{\ell_i},s_i,m_{s_i}}(i) = u_{n_i,\ell_i,m_{\ell_i}}(\underline{r}_i) \cdot \chi_{s_i,m_{s_i}}, \qquad m_{s_i} = \pm\tfrac{1}{2}. \tag{17.43}$$

The label s_i is really redundant, since it is always $\frac{1}{2}$. The spin does not affect the calculation of the central field potential, although any effects of it could then be treated with the perturbation Hamiltonian $\hat{H}'$.

Note that the energy eigenvalues depend in general on both n_i and ℓ_i but not on m_{ℓ_i} or m_{s_i}; it is a peculiarity of the pure Coulomb potential that the hydrogenic energy eigenvalues depend only on the principal quantum number n, but for the self-consistent central potential dependence in general will also depend on ℓ_i. Note though that each individual electron energy level is $2(2\ell_i + 1)$-fold degenerate due to the spin and magnetic quantum number degeneracies.

The total energy of the atom in the central field approximation is thus just the sum of the individual electron energies:

$$E_c = \sum_{i=1}^{N} E_{n_i\ell_i}. \tag{17.44}$$

The corresponding total central-field wave function is a totally antisymmetric wave function built from products of single-electron spin orbitals and is called a **Slater determinant**:

$$\Psi(1,2,\ldots,N) = \frac{1}{\sqrt{N!}} \begin{vmatrix} u_\alpha(1) & u_\beta(1) & \ldots & u_\nu(1) \\ u_\alpha(2) & u_\beta(2) & \ldots & u_\nu(2) \\ \vdots & \vdots & & \vdots \\ u_\alpha(N) & u_\beta(N) & \ldots & u_\nu(N) \end{vmatrix}.$$

Here, the indices $\alpha, \beta, \ldots, \nu$ designate different sets of the four quantum numbers (n, ℓ, m_ℓ, m_s) so as to satisfy the Pauli exclusion principle.

It is generally found by this method that the order of the individual energy levels does not depend strongly on the potential, $V(r)$. It is given by

$$1s, 2s, 2p, 3s, 3p, [4s, 3d], 4p, [5s, 4d], 5p, [6s, 4f, 5d], \ldots$$

Square brackets denote levels which have nearly the same energy, so that the precise order can vary from one atom to another. Electrons in orbitals having the same n value are said to belong to the same *shell*. The shells are labelled $K, L, M, N, \ldots$ according to whether $n = 1, 2, 3, 4, \ldots$. Electrons with the same values of n and ℓ are said to belong to the same *subshell*. The maximum number of electrons in a subshell is $2(2\ell + 1)$. This is the basis for finding the ground-state configurations of neutral atoms.

17.3 Born–Oppenheimer Approximation

In systems where some masses are much bigger than others, this method can be applied, which makes the underlying assumption that the motion of the more massive and less massive particles can be separated.

17.3.1 The H_2^+ Ion and Bonding

The hydrogen molecular ion is the simplest example of a molecular system in which there are disparate mass scales. Moreover, this system displays many of the key features of more complex molecules. The hydrogen molecular ion H_2^+, shown in Fig. 17.2, is a system composed of two protons and a single electron. It is useful to use centre of mass (CM) coordinates by defining the relative position vector $\underline{R}$ of proton 2 with respect to proton 1, and the position vector $\underline{r}$ of the electron relative to the centre of mass of the two protons.

The Schrödinger equation is

$$\left[-\frac{\hbar^2}{2\mu_{12}}\nabla_R^2 - \frac{\hbar^2}{2\mu_e}\nabla_r^2 - \frac{e^2}{r_1} - \frac{e^2}{r_2} + \frac{e^2}{R} \right] \psi(\underline{r}, \underline{R}) = E\psi(\underline{r}, \underline{R}), \tag{17.45}$$

where the reduced mass of the two-proton system is $\mu_{12} = M/2$, with M the proton mass, and μ_e is the reduced mass of the electron/two-proton system:

$$\mu_e = \frac{m(2M)}{m + 2M} \simeq m, \tag{17.46}$$

where m is the electron mass.

Because nuclei are a great deal more massive than electrons, the motion of the nuclei is much slower than that of the electrons. Thus the nuclear and electronic motions can be treated more or less independently and it is a good approximation to determine the electronic states at each value of R by treating the nuclei as fixed. This is the basis of the **Born–Oppenheimer approximation**.

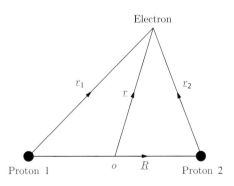

Fig. 17.2   The hydrogen molecular ion H_2^+ is a system composed of two protons and a single electron.

In this approximation, the electron is described by an eigenfunction $U_j(\underline{r}, \underline{R})$ satisfying the Schrödinger equation:

$$\hat{H}_r U_j(\underline{r}, \underline{R}) = \left[-\frac{\hbar^2}{2\mu_e} \nabla_r^2 - \frac{e^2}{r_1} - \frac{e^2}{r_2} + \frac{e^2}{R} \right] U_j(\underline{r}, \underline{R})$$
$$= E_j(\underline{R}) \, U_j(\underline{r}, \underline{R}). \tag{17.47}$$

This is solved by keeping $\underline{R}$ constant. For each $\underline{R}$, a set of energy eigenvalues $E_j(\underline{R})$ and eigenfunctions $U_j(\underline{r}, \underline{R})$ is found. The functions $U_j(\underline{r}, \underline{R})$ are known as **molecular orbitals**. The full wave function for the jth electronic level is taken to be the simple product

$$\psi(\underline{r}, \underline{R}) = F_j(\underline{R}) \, U_j(\underline{r}, \underline{R}), \tag{17.48}$$

where $F_j(\underline{R})$ is a wave function describing the nuclear motion. Substituting this form into the full Schrödinger equation and using the electronic equation yields

$$\left[-\frac{\hbar^2}{2\mu_{12}} \nabla_R^2 + E_j(\underline{R}) - E \right] F_j(\underline{R}) \, U_j(\underline{r}, \underline{R}) = 0. \tag{17.49}$$

A little vector calculus gives

$$\begin{aligned}
\nabla_R^2 \left\{ F_j(\underline{R}) \, U_j(\underline{r}, \underline{R}) \right\} &= \underline{\nabla}_R \cdot \left\{ \underline{\nabla}_R \left[F_j(\underline{R}) \, U_j(\underline{r}, \underline{R}) \right] \right\} \\
&= \underline{\nabla}_R \cdot \left\{ U_j(\underline{r}, \underline{R}) \, \underline{\nabla}_R F_j(\underline{R}) + F_j(\underline{R}) \, \underline{\nabla}_R U_j(\underline{r}, \underline{R}) \right\} \\
&= U_j(\underline{r}, \underline{R}) \nabla_R^2 F_j(\underline{R}) + F_j(\underline{R}) \nabla_R^2 U_j(\underline{r}, \underline{R}) \\
&\quad + 2 \left(\underline{\nabla}_R U_j(\underline{r}, \underline{R}) \right) \cdot \left(\underline{\nabla}_R F_j(\underline{R}) \right).
\end{aligned} \tag{17.50}$$

Assuming that the variation of the molecular orbitals with inter-proton separation $\underline{R}$ is weak, we can neglect the terms involving $\underline{\nabla}_R U_j(\underline{r}, \underline{R})$ and $\nabla_R^2 U_j(\underline{r}, \underline{R})$, leaving a single-particle-type Schrödinger equation for the nuclear motion:

$$\left[-\frac{\hbar^2}{2\mu_{12}} \nabla_R^2 + E_j(\underline{R}) - E \right] F_j(\underline{R}) = 0, \tag{17.51}$$

in which $E_j(\underline{R})$ plays the role of a potential. We will return to this later in this section.

Electronic Ground State

We now try to investigate the lowest electronic levels of H_2^+ using the Rayleigh–Ritz variational method. First we note that, since $\underline{r}_1 = \underline{r} + \underline{R}/2$ and $\underline{r}_2 = \underline{r} - \underline{R}/2$, the electronic Hamiltonian is invariant under the parity operation $\underline{r} \to -\underline{r}$, which results in $\underline{r}_1 \to -\underline{r}_2$ and $\underline{r}_2 \to -\underline{r}_1$. If $\hat{\mathcal{P}}$ denotes the parity operator, then

$$[\hat{\mathcal{P}}, \hat{H}] = 0. \tag{17.52}$$

The compatibility theorem tells us that $\hat{\mathcal{P}}$ and $\hat{H}$ have simultaneous eigenfunctions. These eigenfunctions are called **gerade** if the parity is even and **ungerade** if the parity is odd:

$$\hat{\mathcal{P}} U_j^g(\underline{r}, \underline{R}) = U_j^g(\underline{r}, \underline{R}), \qquad \hat{\mathcal{P}} U_j^u(\underline{r}, \underline{R}) = -U_j^u(\underline{r}, \underline{R}). \tag{17.53}$$

Now think about trial functions. If R is large, the system separates into a hydrogen atom and a proton. This suggests that for the ground state we try functions which go to a hydrogen ground-state function for large separation. So we take linear combinations of gerade or ungerade symmetry of $1s$ orbitals:

$$\psi^g = u_{1s}(r_1) + u_{1s}(r_2) = \frac{1}{\sqrt{\pi a_0^3}} \left[e^{-|\underline{r} + \frac{R}{2}|/a_0} + e^{-|\underline{r} - \frac{R}{2}|/a_0} \right] \qquad (17.54)$$

and

$$\psi^u = u_{1s}(r_1) - u_{1s}(r_2). = \frac{1}{\sqrt{\pi a_0^3}} \left[e^{-|\underline{r} + \frac{R}{2}|/a_0} - e^{-|\underline{r} - \frac{R}{2}|/a_0} \right], \qquad (17.55)$$

with the full normalisation taken care of below when implementing the variational method. Quite generally, this procedure of taking linear combinations of atomic orbitals is known as the LCAO method.

We calculate the expectation value of the electronic Hamiltonian using these trial LCAO molecular wave functions:

$$E^{g,u}(\underline{R}) = \frac{\langle \psi^{g,u} | \hat{H}_r | \psi^{g,u} \rangle}{\langle \psi^{g,u} | \psi^{g,u} \rangle} = \frac{\int \psi^{g,u*}(\underline{r}, \underline{R}) \, \hat{H}_r \, \psi^{g,u}(\underline{r}, \underline{R}) d^3r}{\int |\psi^{g,u}(\underline{r}, \underline{R})|^2 \, d^3r}, \qquad (17.56)$$

which gives an upper bound on the energies $E^g(\underline{R})$ and $E^u(\underline{R})$ for each value of $\underline{R}$; the latter therefore plays the role of the variational parameter. The evaluation of the integrals can be somewhat simplified. For the numerator:

$$\begin{aligned}
\langle \psi^{g,u} | \hat{H}_r | \psi^{g,u} \rangle &= \langle u_{1s}(r_1) | \hat{H}_r | u_{1s}(r_1) \rangle + \langle u_{1s}(r_2) | \hat{H}_r | u_{1s}(r_2) \rangle \\
&\quad \pm \left(\langle u_{1s}(r_1) | \hat{H}_r | u_{1s}(r_2) \rangle + \langle u_{1s}(r_2) | \hat{H}_r | u_{1s}(r_1) \rangle \right) \\
&= 2\langle u_{1s}(r_1) | \hat{H}_r | u_{1s}(r_1) \rangle \pm 2\langle u_{1s}(r_1) | \hat{H}_r | u_{1s}(r_2) \rangle.
\end{aligned} \qquad (17.57)$$

Moreover

$$\langle u_{1s}(r_1) | \hat{H}_r | u_{1s}(r_1) \rangle = E_{1s} + \frac{e^2}{R} - \langle u_{1s}(r_1) | \frac{e^2}{r_2} | u_{1s}(r_1) \rangle \qquad (17.58)$$

and

$$\langle u_{1s}(r_1) | \hat{H}_r | u_{1s}(r_2) \rangle = \left[E_{1s} + \frac{e^2}{R} \right] \langle u_{1s}(r_1) | u_{1s}(r_2) \rangle - \langle u_{1s}(r_1) | \frac{e^2}{r_1} | u_{1s}(r_2) \rangle. \qquad (17.59)$$

For the denominator

$$\langle \psi^{g,u} | \psi^{g,u} \rangle = 2 \pm 2\langle u_{1s}(r_1) | u_{1s}(r_2) \rangle. \qquad (17.60)$$

Thus the problem now reduces to three integrals, which are left as a problem at the end of the chapter. The result is

$$E^g(\underline{R}) = E_{1s} + \frac{e^2}{R} \times \frac{(1 + R/a_0) \exp(-2R/a_0) + [1 - (2/3)(R/a_0)^2] \exp(-R/a_0)}{1 + [1 + (R/a_0) + (1/3)(R/a_0)^2] \exp(-R/a_0)}$$

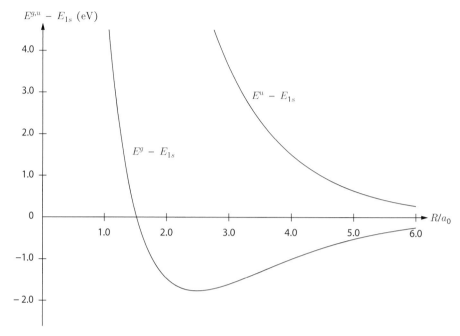

Fig. 17.3   Covalent bonding energy for gerade (g) and ungerade (u) bonds.

and

$$E^u(\underline{R}) = E_{1s} + \frac{e^2}{R} \times \frac{(1 + R/a_0)\exp(-2R/a_0) - [1 - (2/3)(R/a_0)^2]\exp(-R/a_0)}{1 - [1 + (R/a_0) + (1/3)(R/a_0)^2]\exp(-R/a_0)},$$

where a_0 is the Bohr radius and E_{1s} is the ground-state energy of atomic hydrogen.

The two curves $E^g - E_{1s}$ and $E^u - E_{1s}$ are plotted as a function of R in Fig. 17.3. Note that the curve which corresponds to the symmetric (gerade) orbital exhibits a minimum at $R = R_0$, where $R_0/a_0 \simeq 2.5$, corresponding to $E^g - E_{1s} = -1.77$ eV. Since this is an upper bound on the ground-state energy, this implies that there is a stable bound state, a molecular ion. The curve represents an effective attraction between the two protons. By contrast, the curve corresponding to the ungerade orbital has no minimum, so that a H_2^+ ion in this state will dissociate into a proton and a hydrogen atom. If we think of the protons being attracted by the electron and repelled by each other, the symmetrical state should be the more tightly bound because the electron spends more of its time between the protons, where it attracts both of them. This is an example of **covalent bonding**.

Rotational and Vibrational Modes

We can now study the effective one-body Schrödinger equation for the nuclear motion by setting $E_j(\underline{R}) = E^g(R)$ for the ground state. Because $E^g(R)$ only depends on the magnitude of R it represents an effective central potential, so the solutions are of the form

$$F^g(\underline{R}) = \frac{1}{R}\mathcal{R}_{NL}(R)Y_{LM_L}(\theta,\phi), \tag{17.61}$$

where the function $\mathcal{R}_{NL}(R)$ satisfies the radial equation

$$\left[-\frac{\hbar^2}{2\mu_{12}}\left(\frac{d^2}{dR^2} - \frac{L(L+1)}{R^2}\right) + E^g(R) - E\right]\mathcal{R}_{NL} = 0. \tag{17.62}$$

We can approximate the centrifugal barrier term by setting it equal to its value at $R = R_0$, writing

$$E_r = \frac{\hbar^2}{2\mu_{12}R_0^2} L(L+1). \tag{17.63}$$

In this approximation we are treating the molecule as a rigid rotator. We can also approximate $E^g(R)$ by Taylor expanding about $R = R_0$. Because this point is a minimum, the first derivative is zero:

$$E^g(R) \simeq E^g(R_0) + \tfrac{1}{2}k(R - R_0)^2 + \cdots, \tag{17.64}$$

where k is the value of the second derivative of E^g at $R = R_0$.

With these two approximations, the radial equation becomes

$$\left[-\frac{\hbar^2}{2\mu_{12}}\frac{d^2}{dR^2} + \tfrac{1}{2}k(R - R_0)^2 - E_N\right]\mathcal{R}_{NL} = 0, \tag{17.65}$$

where

$$E_N = E - E^g(R_0) - E_r. \tag{17.66}$$

This is the equation for a simple harmonic oscillator with energies

$$E_N = \hbar\omega_0\left(N + \tfrac{1}{2}\right), \qquad N = 0, 1, 2, \ldots, \tag{17.67}$$

where $\omega_0 = \sqrt{k/\mu_{12}}$. The vibrational energies are of the order of a few tenths of an electronvolt, whereas the rotational energies are of the order of 10^{-3} eV. Both are much smaller than the spacing of the electronic levels. Transitions between these various levels give rise to **molecular spectra**. The pure rotational spectrum consists of closely spaced lines in the infrared or microwave range. Transitions which also involve changes to the vibrational state give rise to **vibrational–rotational band spectra**.

17.4 Hellmann–Feynman Method

This method can be used to calculate the derivative of an eigenvalue of a Hamiltonian $\hat{H}$ in terms of the expectation value of the derivative of the Hamiltonian operator with respect to the same parameter. Suppose $|\psi_n\rangle$ is an eigenstate of $\hat{H}$, $\hat{H}|\psi_n\rangle = E_n|\psi_n\rangle$. This can be written as an expectation value, $E_n = \langle\psi_n|\hat{H}|\psi_n\rangle$. Let $\hat{H}$ depend on some parameter

λ, which means in turn the eigenvalues of the Hamiltonian will also be functions of this parameter. Taking a derivative of this expectation value gives

$$
\begin{aligned}
\frac{dE_n(\lambda)}{d\lambda} &= \frac{d}{d\lambda}\langle\psi_n|\hat{H}|\psi_n\rangle \\
&= \langle\frac{d\psi_n}{d\lambda}|\hat{H}|\psi_n\rangle + \langle\psi_n|\frac{d\hat{H}}{d\lambda}|\psi_n\rangle + \langle\psi_n|\hat{H}|\frac{d\psi_n}{d\lambda}\rangle \\
&= E_n(\lambda)\langle\frac{d\psi_n}{d\lambda}|\psi_n\rangle + \langle\psi_n|\frac{d\hat{H}}{d\lambda}|\psi_n\rangle + E_n(\lambda)\langle\psi_n|\frac{d\psi_n}{d\lambda}\rangle \\
&= E_n(\lambda)\frac{d}{d\lambda}\langle\psi_n|\psi_n\rangle + \langle\psi_n|\frac{d\hat{H}}{d\lambda}|\psi_n\rangle \\
&= \langle\psi_n|\frac{d\hat{H}}{d\lambda}|\psi_n\rangle.
\end{aligned}
\tag{17.68}
$$

The third line above required that $|\psi_n\rangle$ be an eigenstate of $\hat{H}$ in order to replace the action of the Hamiltonian operator with the eigenvalue, which can then be taken outside the scalar product. The last line follows from the one before by recognising that since the wave function $|\psi_n\rangle$ is normalised, the derivative of its scalar product will be zero. Thus the above expression gives the Hellmann–Feynman relation

$$
\frac{dE_n(\lambda)}{d\lambda} = \langle\psi_n|\frac{d\hat{H}}{d\lambda}|\psi_n\rangle.
\tag{17.69}
$$

The Hellmann–Feynman method can be applied to obtain expectation values of the inverse radial distance in hydrogenic atoms. The effective radial Hamiltonian for such a system is

$$
\hat{H} = -\frac{\hbar^2}{2\mu}\frac{d^2}{dr^2} - \frac{Ze^2}{r} + \frac{\ell(\ell+1)\hbar^2}{2\mu r^2},
\tag{17.70}
$$

leading to energy eigenvalue solutions

$$
E_{n_r,\ell} = -(Ze^2)^2\frac{\mu}{2\hbar^2}\frac{1}{(n_r+\ell)^2}.
\tag{17.71}
$$

Here we have kept the radial and angular quantum numbers explicit rather than replacing by the principal quantum number $n = n_r + \ell$, since we want to keep the ℓ dependence explicit.

If we select the parameter to take the derivative with respect to in Eq. (17.69) as $\lambda = e^2$, then the resulting expression that is obtained is

$$
-Z^2\frac{\mu e^2}{\hbar^2}\frac{1}{n^2} = -Z\langle\frac{1}{r}\rangle,
\tag{17.72}
$$

which rearranging leads to

$$
\langle\frac{1}{r}\rangle = Ze^2\frac{\mu}{\hbar^2 n^2} = \frac{Z}{a_0 n^2},
\tag{17.73}
$$

where $a_0 = \hbar^2/(\mu e^2)$ is the Bohr radius.

If instead the parameter selected to take the derivative with respect to in Eq. (17.69) is ℓ, then the resulting expression that is obtained is

$$(Ze^2)^2 \frac{\mu}{2\hbar^2} \frac{2}{(n_r + \ell)^3} = \frac{(2\ell + 1)\hbar^2}{2\mu} \langle \frac{1}{r^2} \rangle, \tag{17.74}$$

which after rearranging leads to

$$\langle \frac{1}{r^2} \rangle = \frac{Z^2}{\ell + \frac{1}{2}} \frac{1}{a_0^2 n^3}. \tag{17.75}$$

One can verify the correctness of both expectation values through directly calculating them using the appropriate hydrogenic wave functions. The Hellmann–Feynman method in this case requires considerably less calculation.

17.5 Wenzel–Kramers–Brillouin–Jeffreys Approximation

If the potential in the TISE is constant:

$$-\frac{\hbar^2}{2m} \frac{d^2 \psi}{dx^2} + V\psi = E\psi, \tag{17.76}$$

then this equation can be solved exactly and has eigenfunctions that are similar to those of a free particle:

$$\psi(x) = B \exp\left(\pm \frac{i}{\hbar} px \right), \tag{17.77}$$

with

$$p(x) = [2m(E - V)]^{1/2}. \tag{17.78}$$

The WKBJ method works from this limit, with the potential, rather than being constant, varying slowly. In the limit of a constant potential, the eigenfunction solutions Eq. (17.77) show that they vary substantially over a distance of the de Broglie wavelength, $\Delta x \gtrsim \hbar/p$. If now the potential were considered to vary very little over the scale of Δx, then Eq. (17.77) would still be a reasonably good approximate eigenfunction. In particular, the slowly varying requirement is $\Delta x |dp/dx| \ll p$. Equivalently, if

$$\frac{\hbar}{p^2} \left| \frac{dp}{dx} \right| = \frac{\hbar m |dV/dx|}{[2m(E - V(x))]^{3/2}} \ll 1 \tag{17.79}$$

holds throughout the spatial region of interest, then one can develop a systematic approximation to the eigenfunctions based on the limiting form Eq. (17.77). In particular, one assumes a form for the wave function

$$\psi(x) = B \exp\left(\frac{iS(x)}{\hbar} \right). \tag{17.80}$$

Substituting this into the TISE leads to the equation for $S(x)$:

$$-i\frac{\hbar}{2m}\frac{d^2 S(x)}{dx^2} + \frac{1}{2m}\left(\frac{dS}{dx}\right)^2 + V - E = 0. \tag{17.81}$$

At this point no approximation has been made. If one could solve this equation for a given specified potential $V(x)$, it would be an exact solution. It is noteworthy that this equation is non-linear, so in that respect following this approach generally seems harder. The idea is to develop a systematic approximation in powers of $\hbar$, thus expanding $S(x)$ as

$$S(x) = S_0(x) + \hbar S_1(x) + \hbar^2 S_2(x) + \cdots . \tag{17.82}$$

Substituting this into Eq. (17.81) and collecting all terms with the same number of powers in $\hbar$ leads to the following series of equations:

$$
\begin{aligned}
\hbar^0 : \quad & \frac{1}{2m}\left(\frac{dS_0}{dx}\right)^2 + V - E = 0, \\
\hbar^1 : \quad & -\frac{i}{2}\frac{d^2 S_0}{dx^2} + \frac{dS_0}{dx}\frac{dS_1}{dx} = 0, \\
\hbar^2 : \quad & -\frac{i}{2}\frac{d^2 S_1}{dx^2} + \frac{dS_0}{dx}\frac{dS_2}{dx} + \frac{1}{2}\left(\frac{dS_1}{dx}\right)^2 = 0, \\
\hbar^3 : \quad & -\frac{i}{2}\frac{d^2 S_2}{dx^2} + \frac{dS_0}{dx}\frac{dS_3}{dx} + \frac{dS_1}{dx}\frac{dS_2}{dx} = 0,
\end{aligned}
\tag{17.83}
$$

etc.

The procedure is now to solve each equation successively, starting first with the equation at $O(\hbar^0)$ to obtain $S_0(x)$ and then substituting that into the equation at $O(\hbar)$ and solving that equation for $S_1(x)$, and so forth. The WKBJ approximation is to only retain the first two equations, up to $O(\hbar)$, and so solve just for $S_0(x)$ and $S_1(x)$.

The explicit differential equation for $S_0(x)$ from Eq. (17.83) is

$$\frac{dS_0(x)}{dx} = \pm\sqrt{2m(E - V(x))} = \pm p(x), \tag{17.84}$$

for which the solution is

$$S_0(x) = \pm \int^x dx' p(x'). \tag{17.85}$$

There is an arbitrary integration constant above, which can be absorbed in the amplitude B in Eq. (17.80). The expression for $S_0(x)$ is now used in Eq. (17.83) to give the equation for $S_1(x)$ as

$$\frac{dS_1(x)}{dx} = \frac{i}{2}\frac{d\ln p}{dx}, \tag{17.86}$$

which is integrated to give

$$S_1(x) = \frac{i}{2}\ln p(x), \tag{17.87}$$

where again there is an arbitrary integration constant which as before can be absorbed into B, and so is omitted above.

The results for $S_0(x)$ and $S_1(x)$ can now be combined and form what is called the WKBJ approximation. Before looking in more detail at this result, $S_2(x)$ will be computed and it will be checked that provided the condition Eq. (17.79) holds, indeed $S_2(x)$ is more suppressed, thus justifying neglecting it and higher terms. The equation for $S_2(x)$ based on Eq. (17.83) is

$$\frac{dS_2(x)}{dx} = \pm \left[\frac{3}{8p^3} \left(\frac{dp}{dx} \right)^2 - \frac{1}{4p^2} \frac{d^2p}{dx^2} \right]. \tag{17.88}$$

The second term on the right-hand side is now integrated by parts, then combining with the first term on the right-hand side it leads to

$$S_2(x) = \mp \left[\frac{1}{4p^2} \frac{dp}{dx} + \frac{1}{8} \int^x \frac{1}{p^3} \left(\frac{dp}{dx} \right)^2 dx \right]. \tag{17.89}$$

Including the one factor of $\hbar$ based on Eqs (17.80) and (17.82) for the first term above, it satisfies the criteria Eq. (17.79) and so is negligible. For the second term above, provided the integral is significant up to a region of order $\sim \hbar/p(x)$ and $p(x)$ is slowly varying, as expected for the WKBJ approximation, then the second term above is also negligible. Thus the $S_2(x)$ term is shown to be subleading to the first two terms and so can be ignored.

Having established that only the first two terms need to be retained in Eq. (17.82), the general WKBJ solution for $E > V(x)$ is

$$\psi(x) = \frac{1}{\sqrt{p(x)}} \left[B_1 \exp\left(\frac{i}{\hbar} \int^x p(x')dx' \right) + B_2 \exp\left(-\frac{i}{\hbar} \int^x p(x')dx' \right) \right]. \tag{17.90}$$

The above analysis was done for a particle in the classically allowed region, $E > V(x)$, but a similar analysis can be done in the classically forbidden region, $E < V(x)$, and would lead to the general WKBJ solution

$$\psi(x) = \frac{1}{\sqrt{|p(x)|}} \left[B_1 \exp\left(\frac{1}{\hbar} \int^x |p(x')|dx' \right) + B_2 \exp\left(-\frac{1}{\hbar} \int^x |p(x')|dx' \right) \right], \tag{17.91}$$

where in this regime from Eq. (17.78) $p(x)$ becomes imaginary, and so the absolute value is needed for the above expression.

In the classical limit, the de Broglie wavelength vanishes. As discussed above, the WKBJ approximation is valid when the characteristic scale on which the potential varies is much bigger than the de Broglie wavelength, which is another reason why sometimes this method is also called a semiclassical or quasiclassical approximation. A related way to understand this terminology is that based on Eq. (17.82), this approximation is an expansion in $\hbar$, which is the parameter that measures the extent of quantum behaviour in a system. The limit $\hbar \to 0$ is the classical limit. Thus the WKBJ method is an expansion around the classical limit.

In the following sections, we will examine the WKBJ treatment of the potential barrier, potential well and symmetric double well.

17.5.1 Potential Barrier

Consider the case of a potential barrier as shown in Fig. 17.4. Thus $x < x_L$ and $x > x_R$ are the classically allowed regions, where the wave function is oscillatory, whereas the middle

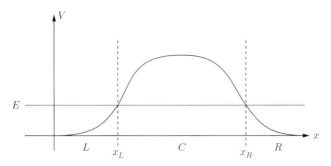

Fig. 17.4 Potential barrier with particle having energy E. The region between $x_L < x < x_R$ is classically forbidden.

is the classically forbidden region where the wave function is exponentially damped. The potential barrier WKBJ wave function in the three regions is

$$\psi(x) = \begin{cases} \frac{1}{\sqrt{p(x)}} \left(L_+ \exp\left[\frac{i}{\hbar} \int_x^{x_L} p(x')dx'\right] + L_- \exp\left[-\frac{i}{\hbar} \int_x^{x_L} p(x')dx'\right] \right) & \text{if } x < x_L, \\ \frac{1}{\sqrt{|p(x)|}} \left(C_+ \exp\left[\frac{1}{\hbar} \int_{x_L}^x |p(x')|dx'\right] + C_- \exp\left[-\frac{1}{\hbar} \int_{x_L}^x |p(x')|dx'\right] \right) & \text{if } x_L < x < x_R, \\ \frac{R_+}{\sqrt{p(x)}} \exp\left[\frac{i}{\hbar} \int_{x_R}^x p(x')dx'\right] & \text{if } x_R < x. \end{cases}$$

The amplitude coefficients of the wave functions in the three regions now need to be related amongst themselves. The points x_L and x_R are called the classical turning points. As they are approached, in fact the condition Eq. (17.79) necessary for the validity of the WKBJ approximation fails. Near these regions the potential can be well approximated as linear. Thus one can solve exactly the Schrödinger equation near these two points. These 'patching' solutions, $\psi_p(x)$, can then be used to join the WKBJ wave functions on both sides slightly away from these points. This procedure will then allow relating the amplitude coefficients of the above potential barrier WKBJ wave functions.

In general, near a turning point, which to keep notation simple we will shift to the origin, the potential can be approximated as $V(x) \approx E + V'(0)x$. The Schrödinger equation in this region under this approximation is

$$-\frac{\hbar^2}{2m} \frac{d^2\psi_p}{dx^2} + (E + V'(0)x)\psi_p = E\psi_p, \tag{17.92}$$

which simplifies to

$$\frac{d^2\psi_p}{dy^2} = y\psi_p, \tag{17.93}$$

where $y = \rho x$ and $\rho = (2mV'(0)/\hbar^2)^{1/3}$. This is the Airy equation with solutions

$$\text{Ai}[(3a)^{-1/3}x] = \frac{(3a)^{1/3}}{\pi} \int_0^\infty \cos(at^3 + xt)dt,$$

$$\text{Bi}[(3a)^{-1/3}x] = \frac{(3a)^{1/3}}{\pi} \int_0^\infty \left[\exp(-at^3 + xt) + \sin(at^3 + xt) \right] dt. \tag{17.94}$$

The asymptotic behaviour of these Airy functions is what will be useful for the analysis to follow,

$$y \gg 1: \mathrm{Ai}(y) \sim \frac{1}{2\sqrt{\pi}\,y^{1/4}} \exp\left(-\frac{2}{3}y^{3/2}\right), \ \mathrm{Bi}(y) \sim \frac{1}{\sqrt{\pi}\,y^{1/4}} \exp\left(\frac{2}{3}y^{3/2}\right),$$

$$y \ll -1: \mathrm{Ai}(y) \sim \frac{1}{\sqrt{\pi}(-y)^{1/4}} \sin\left[\frac{2}{3}(-y)^{3/2} + \frac{\pi}{4}\right],$$

$$\mathrm{Bi}(y) \sim \frac{1}{\sqrt{\pi}(-y)^{1/4}} \cos\left[\frac{2}{3}(-y)^{3/2} + \frac{\pi}{4}\right]. \tag{17.95}$$

The general solution of the second-order differential Eq. (17.93) will be a linear combination of both Airy functions:

$$\psi_P(y) = d_-^L \mathrm{Ai}(y) + d_+^L \mathrm{Bi}(y). \tag{17.96}$$

Consider now the turning point x_L, which without loss of generality will be shifted to the origin to simplify notation. We wish to examine this solution in the regime where $|y|$ is large but at the same time $|x|$ is small enough that the potential can be approximated as linear. At this turning point, the potential is upward sloping, $V' > 0$. In the C (central) region for $y \gg 1$ using the asymptotic behaviour of the Airy functions given above, the solution becomes

$$\psi_P(y) = \frac{d_+^L}{\sqrt{\pi}\,y^{1/4}} \exp\left(\frac{2}{3}y^{3/2}\right) + \frac{d_-^L}{2\sqrt{\pi}\,y^{1/4}} \exp\left(-\frac{2}{3}y^{3/2}\right). \tag{17.97}$$

The WKBJ wave function in this C-region near x_L is

$$\psi(y)\frac{1}{(\hbar^2 \rho^2 y)^{1/4}} \left[C_+ \exp\left(\frac{2}{3}y^{3/2}\right) + C_- \exp\left(-\frac{2}{3}y^{3/2}\right)\right], \tag{17.98}$$

where near this turning point $|p(x)| \approx \sqrt{2mV'(0)x}$.

On the other side of this point, which is the L (left)-region, when $x < 0$ but close enough for the potential to be well approximated as linear, but where $y \ll -1$, the patching solution Eq. (17.96) can be approximated as

$$\psi_P(y) = \frac{1}{\sqrt{\pi}(-y)^{1/4}} \left[d_-^L \sin\left(\frac{2}{3}(-y)^{3/2} + \frac{\pi}{4}\right) + d_+^L \cos\left(\frac{2}{3}(-y)^{3/2} + \frac{\pi}{4}\right)\right]$$

$$= \frac{1}{\sqrt{\pi}(-y)^{1/4}} \left[\tfrac{1}{2}(d_+^L - id_-^L)\exp\left[i\left(\frac{2}{3}(-y)^{3/2} + \frac{\pi}{4}\right)\right]\right.$$

$$\left. + \tfrac{1}{2}(d_+^L + id_-^L)\exp\left[-i\left(\frac{2}{3}(-y)^{3/2} + \frac{\pi}{4}\right)\right]\right]. \tag{17.99}$$

For the potential barrier WKBJ wave function in this L-region near x_L (shifted now to the origin), $p(x) = \sqrt{-2mV'(0)x}$ and

$$\psi(y) \approx \frac{1}{(-\hbar^2 \rho^2 y)^{1/4}} \left[L_+ \exp\left(i\frac{2}{3}(-y)^{3/2}\right) + L_- \exp\left(-i\frac{2}{3}(-y)^{3/2}\right)\right]. \tag{17.100}$$

These expressions now allow a relation between the WKBJ amplitude coefficients in the L- and C-regions:

$$L_+ = \tfrac{1}{2} \exp\left(i\frac{\pi}{4}\right)(C_+ - i2C_-),$$

$$L_- = \tfrac{1}{2} \exp\left(-i\frac{\pi}{4}\right)(C_+ + i2C_-). \tag{17.101}$$

Examining next at x_R and once again shifting the origin to it so as to simplify the notation, the potential near this point is approximately $V(x) \approx E - |V'(0)|x$. As the potential is downward sloping at this point, $V'(0) < 0$, we have written it as $-|V'(0)|$. Once again, due to this linear behaviour, near this point the wave function can be approximated as

$$\psi_p(y) = d_-^R Ai(-y) + d_+^R Bi(-y), \tag{17.102}$$

where the minus signs in the arguments of the Airy functions are due to the downward-sloping potential. In the C-region near x_R (shifted to the origin) with $x < 0$ and $y \ll -1$ (so $-y \gg 1$), the patching wave function becomes

$$\psi_p(y) = \frac{d_-^R}{2\sqrt{\pi}(-y)^{1/4}} \exp\left(-\frac{2}{3}(-y)^{3/2}\right) + \frac{d_+^R}{\sqrt{\pi}(-y)^{1/4}} \exp\left(\frac{2}{3}(-y)^{3/2}\right) \tag{17.103}$$

and the WKBJ wave function is

$$\psi(y) \approx \frac{1}{(-\hbar^2 \rho^2 y)^{1/4}} \left[C_- \exp\left(-\frac{2}{3}(-y)^{3/2}\right) + C_+ \exp\left(\frac{2}{3}(-y)^{3/2}\right)\right], \tag{17.104}$$

where $C_- = C_+ e^\beta$ and $C_+ = C_- e^{-\beta}$ with

$$\beta \equiv \frac{1}{\hbar} \int_{x_L}^{x_R} |p(x')|dx'. \tag{17.105}$$

In the R-region, $x > 0$ for $y \gg 1$ (so $-y \ll -1$), the patching wave function Eq. (17.102) is

$$\psi_p(y) = \frac{1}{\sqrt{\pi} y^{1/4}} \left[d_-^R \sin\left(\frac{2}{3}y^{3/2} + \frac{\pi}{4}\right) + d_+^R \cos\left(\frac{2}{3}y^{3/2} + \frac{\pi}{4}\right)\right]$$

$$= \frac{1}{\sqrt{\pi} y^{1/4}} \left[\tfrac{1}{2}(d_+^R - id_-^R) \exp\left[i\left(\frac{2}{3}y^{3/2} + \frac{\pi}{4}\right)\right] + \tfrac{1}{2}(d_+^R + id_-^R) \exp\left[-i\left(\frac{2}{3}y^{3/2} + \frac{\pi}{4}\right)\right]\right]$$

and the WKBJ wave function near x_R is

$$\psi(y) \approx \frac{R_+}{(\hbar^2 \rho^2 y)^{1/4}} \exp\left(i\frac{2}{3}(y)^{3/2}\right). \tag{17.106}$$

These expressions allow us to relate the WKBJ amplitude coefficients as

$$C_+ = \frac{i}{2} \exp\left(-\beta - i\frac{\pi}{4}\right)R_+ \qquad C_- = \exp\left(\beta - i\frac{\pi}{4}\right)R_+. \tag{17.107}$$

Using the above, along with Eq. (17.101), then gives

$$L_+ = \frac{i}{2}\left(\tfrac{1}{2}e^{-\beta} - 2e^\beta\right)R_+. \tag{17.108}$$

These results can be used to examine tunnelling, which was discussed in Section 5.2. The WKBJ expression for the transmission coefficient to tunnel through (tunnelling probability) is then

$$T \equiv \left| \frac{R_+}{L_+} \right|^2 = \frac{e^{-2\beta}}{\left(1 - \frac{e^{-2\beta}}{4}\right)^2}. \tag{17.109}$$

For large enough separation, $\beta \gg 1$, $T \approx e^{-2\beta}$, which agrees with the result for the potential step in Section 5.2.

17.5.2 Potential Well

The WKBJ treatment of the potential well in Fig. 17.5 will now be considered. Retaining similar notation to the previous example, the points x_L and x_R are the turning points, only now they are such for the particle inside the C-region. The L-region and R-region are for this case classically forbidden. The potential well WKBJ wave function for the three regions is

$$\psi(x) = \begin{cases} \frac{L_-}{\sqrt{|p(x)|}} \exp\left[-\frac{1}{\hbar} \int_x^{x_L} |p(x')|dx'\right] & \text{if } x < x_L, \\ \frac{1}{\sqrt{p(x)}} \left(C_+ \exp\left[\frac{i}{\hbar} \int_{x_L}^x p(x')dx'\right] + C_- \exp\left[-\frac{i}{\hbar} \int_{x_L}^x p(x')dx'\right]\right) & \text{if } x_L < x < x_R, \\ \frac{R_-}{\sqrt{|p(x)|}} \exp\left[-\frac{1}{\hbar} \int_{x_R}^x |p(x')|dx'\right] & \text{if } x_R < x. \end{cases}$$

As with the previous case, the amplitude coefficients for these three regions need to be related amongst each other. The patching solutions will once again be used near the turning points to accomplish this task.

Starting with the turning point x_L, which again will be shifted to the origin to simplify notation, for the patching solution in the C-region, the potential is approximated near the turning point as $V(x) \approx E - |V'(0)|x$, where note that $V' < 0$ at x_L. We wish to examine this solution in the regime where $|y|$ is large but at the same time $|x|$ is small enough that the potential can be approximated as linear. The patching solution in the L-region, $x < 0$,

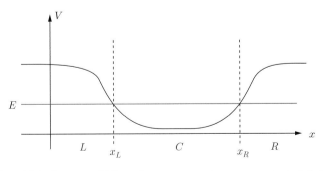

Potential well with particle having energy E. The regions $x > x_R$ and $x < x_L$ are classically forbidden.

is of the general form Eq. (17.96), but now the asymptotic behaviour is evaluated in the non-classical regime, for which

$$\psi_P(y) = \frac{d_-^L}{2\sqrt{\pi}(-y)^{1/4}} \exp\left(-\frac{2}{3}(-y)^{3/2}\right) + \frac{d_+^L}{\sqrt{\pi}(-y)^{1/4}} \exp\left(\frac{2}{3}(-y)^{3/2}\right). \qquad (17.110)$$

The WKBJ wave function near this turning point, near x_L, has $|p(x)| \approx \sqrt{\hbar^2 \rho^2(-y)}$ and

$$\psi(y) \approx \frac{L_-}{\sqrt{\hbar\rho}(-y)^{1/4}} \exp\left(-\frac{2}{3}(-y)^{3/2}\right). \qquad (17.111)$$

Comparing the above two equations, one deduces that $d_+^L = 0$ and $d_-^L = 2L_-\sqrt{\pi}/\sqrt{\hbar\rho}$. On the other side of this turning point, x_L, which goes into the C-region, when $x > 0$ but close enough for the potential to be well approximated as linear, yet $y \gg 1$ (the Airy equation (17.93) but now with a minus sign on the right-hand side), the patching solution Eq. (17.96) can be approximated as

$$\begin{aligned}
\psi_P(y) &= \frac{d_-^L}{\sqrt{\pi}y^{1/4}} \sin\left(\frac{2}{3}y^{3/2} + \frac{\pi}{4}\right) \\
&= \frac{-id_-^L}{2\sqrt{\pi}y^{1/4}}\left[\exp\left[i\left(\frac{2}{3}y^{3/2} + \frac{\pi}{4}\right)\right] - \exp\left[-i\left(\frac{2}{3}y^{3/2} + \frac{\pi}{4}\right)\right]\right]. \qquad (17.112)
\end{aligned}$$

For the potential well WKBJ wave function in this C-region, $p(x) = \sqrt{2m|V'(0)|x}$ and

$$\psi(y) \approx \frac{1}{\sqrt{\hbar\rho}y^{1/4}}\left[C_+ \exp\left(i\frac{2}{3}(y)^{3/2}\right) + C_- \exp\left(-i\frac{2}{3}(y)^{3/2}\right)\right]. \qquad (17.113)$$

These expressions now allow a relation between the WKBJ amplitude coefficients in the L- and C-regions:

$$C_+ = -i\exp\left(i\frac{\pi}{4}\right)L_-, \qquad C_- = i\exp\left(-i\frac{\pi}{4}\right)L_-. \qquad (17.114)$$

Examining next at x_R and once again shifting this point to the origin so as to simplify the notation, the potential near this point is approximately $V(x) \approx E + V'(0)x$, with $V' > 0$. Once again, due to this linear behaviour, near this point the wave function can be approximated based on this linear potential to be

$$\psi_P(y) = d_-^R Ai(y) + d_+^R Bi(y). \qquad (17.115)$$

In the R-region, $x > 0$ for $y \gg 1$, the patching wave function is

$$\psi_P(y) = \frac{d_-^R}{2\sqrt{\pi}y^{1/4}} \exp\left(-\frac{2}{3}y^{3/2}\right) + \frac{d_+^R}{\sqrt{\pi}y^{1/4}} \exp\left(\frac{2}{3}y^{3/2}\right) \qquad (17.116)$$

and the WKBJ wave function is

$$\psi(y) \approx \frac{R_-}{\sqrt{\hbar\rho}y^{1/4}} \exp\left(-\frac{2}{3}(y)^{3/2}\right). \qquad (17.117)$$

Comparing the above two expressions, it becomes clear that $d_+^R = 0$ and $d_-^R = 2\sqrt{\pi}R_-/\sqrt{\rho\hbar}$. On the other side of this turning point, x_R, which goes into the C-region, when $x < 0$ but close enough for the potential to be well approximated as linear, but where $y \ll -1$, the patching solution Eq. (17.96) can be approximated as

$$\psi_p(y) = \frac{-id_-^R}{2\sqrt{\pi}(-y)^{1/4}} \left[\exp\left[i\left(\frac{2}{3}(-y)^{3/2} + \frac{\pi}{4}\right)\right] - \exp\left[-i\left(\frac{2}{3}(-y)^{3/2} + \frac{\pi}{4}\right)\right]\right], \quad (17.118)$$

where we already used that $d_+^R = 0$. The WKBJ wave function is

$$\psi(y) \approx \frac{1}{\sqrt{\hbar\rho}(-y)^{1/4}} \left[C_- \exp\left(-i\frac{2}{3}(-y)^{3/2}\right) + C_+ \exp\left(i\frac{2}{3}(-y)^{3/2}\right)\right], \quad (17.119)$$

where $C_- = C_+ e^{i\beta}$ and $C_+ = C_- e^{-i\beta}$, with

$$\beta \equiv \frac{1}{\hbar} \int_{x_L}^{x_R} p(x')dx'. \quad (17.120)$$

Comparing amplitude coefficients gives that

$$d_-^R = \frac{-i2\sqrt{\pi}C_+ e^{i(\beta+\pi/4)}}{\sqrt{\hbar\rho}} = \frac{i2\sqrt{\pi}C_- e^{-i(\beta+\pi/4)}}{\sqrt{\hbar\rho}}. \quad (17.121)$$

Using Eq. (17.114) to substitute C_+ and C_- with L_+, the above two expressions lead to the condition

$$e^{i(\beta+\pi/2)} = e^{-i(\beta+\pi/2)}, \quad (17.122)$$

so $e^{2i\beta} = -1$, thus the quantisation condition

$$\frac{1}{\hbar} \int_{x_L}^{x_R} p(x')dx' = \left(n + \frac{1}{2}\right)\pi, \quad (17.123)$$

with $n = 0, 1, 2, \ldots$.

17.5.3 Symmetric Double Well

The potential for the symmetric double well is shown in Fig. 17.6. Retaining similarity in notation to the previous examples, the points x_L and x_R are the turning points, at the outer sides of the double well, with inner turning points at $x_{L'}$ and $x_{R'}$. Thus regions to the right of x_R, left of x_L and the C-regions are classically forbidden. The symmetric double-well WKBJ wave function for the five regions is

$$\psi(x) = \begin{cases} \frac{L_-}{\sqrt{|p(x)|}} \exp\left[-\frac{1}{\hbar}\int_x^{x_L} |p(x')|dx'\right] & \text{if } x < x_L, \\[2mm] \frac{1}{\sqrt{p(x)}} \left(L'_+ \exp\left[\frac{i}{\hbar}\int_{x_L}^x p(x')dx'\right] + L'_- \exp\left[-\frac{i}{\hbar}\int_{x_L}^x p(x')dx'\right]\right) & \text{if } x_L < x < x_{L'}, \\[2mm] \frac{1}{\sqrt{|p(x)|}} \left(C_+ \exp\left[\frac{1}{\hbar}\int_x^{x_{R'}} |p(x')|dx'\right] + C_- \exp\left[-\frac{1}{\hbar}\int_x^{x_{R'}} |p(x')|dx'\right]\right) & \text{if } x_{L'} < x < x_{R'}, \\[2mm] \frac{1}{\sqrt{p(x)}} \left(R'_+ \exp\left[\frac{i}{\hbar}\int_x^{x_R} p(x')dx'\right] + R'_- \exp\left[-\frac{i}{\hbar}\int_x^{x_R} p(x')dx'\right]\right) & \text{if } x_{R'} < x < x_R, \\[2mm] \frac{R_-}{\sqrt{|p(x)|}} \exp\left[-\frac{1}{\hbar}\int_{x_R}^x |p(x')|dx'\right] & \text{if } x_R < x. \end{cases}$$

The amplitude coefficients for these five regions are now related using patching wave functions.

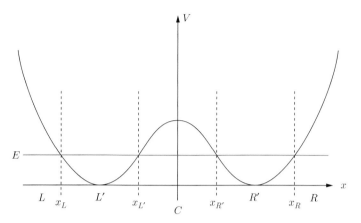

Symmetric double-well potential with particle having energy E.

The patching procedure at x_L and x_R is the same as for the potential well case. Thus, similar to the potential well Eq. (17.101), in the L'-region at x_L, we find $L'_+ = -i\exp(i\pi/4)L_-$ and $L'_- = i\exp(-i\pi/4)L_-$. Also the WKBJ wave function in the L'-region is

$$\psi(x) = -\frac{2L_-}{\sqrt{p(x)}}\sin\left(\frac{1}{\hbar}\int_x^{x_{L'}}p(x')dx' - \beta' - \frac{\pi}{4}\right), \tag{17.124}$$

and similarly in the R'-region the WKBJ wave function is

$$\psi(x) = -\frac{2R_-}{\sqrt{p(x)}}\sin\left(\frac{1}{\hbar}\int_{x_{R'}}^x p(x')dx' - \beta' - \frac{\pi}{4}\right), \tag{17.125}$$

where, due to the symmetry of the double well:

$$\beta' = \frac{1}{\hbar}\int_{x_L}^{x_{L'}}p(x')dx' = \frac{1}{\hbar}\int_{x_{R'}}^{x_R}p(x')dx'. \tag{17.126}$$

The matching at $x_{L'}$ and $x_{R'}$ is different from the cases done already, so they will be worked out in more detail. The above WKBJ wave function Eq. (17.124) near $x_{L'}$ for $x < x_{L'}$ is

$$\psi(y) \approx \frac{iL_-}{\sqrt{\hbar\rho}(-y)^{1/4}}\left[\exp\left(i\frac{2}{3}(-y)^{3/2} - i\frac{\pi}{4} - i\beta'\right) - \exp\left(-i\frac{2}{3}(-y)^{3/2} + i\frac{\pi}{4} + i\beta'\right)\right]. \tag{17.127}$$

The patching wave function near $x_{L'}$ in the region $x < x_{L'}$ is

$$\psi_p(y) = \frac{d_+^{L'}}{\sqrt{\pi}(-y)^{1/4}}\cos\left(\frac{2}{3}(-y)^{3/2} + \frac{\pi}{4}\right) + \frac{d_-^{L'}}{\sqrt{\pi}(-y)^{1/4}}\sin\left(\frac{2}{3}(-y)^{3/2} + \frac{\pi}{4}\right). \tag{17.128}$$

These equations can be compared to give the relation

$$d_+^{L'} - id_-^{L'} = 2\sqrt{\frac{\pi}{\hbar\rho}} \exp(-i\beta')L_-. \tag{17.129}$$

The patching wave function near the region $x > x_{L'}$ is

$$\psi_P(y) = \frac{d_+^{L'}}{\sqrt{\pi}(y)^{1/4}} \exp\left(\frac{2}{3}y^{3/2}\right) + \frac{d_-^{L'}}{2\sqrt{\pi}(y)^{1/4}} \exp\left(-\frac{2}{3}y^{3/2}\right), \tag{17.130}$$

whereas in the region near $x > x_{L'}$, the WKBJ wave function approximates to

$$\psi(y) \approx \frac{1}{\sqrt{\hbar\rho}y^{1/4}} \left[C_+ \exp\left(\frac{2}{3}y^{3/2}\right) + C_- \exp\left(-\frac{2}{3}y^{3/2}\right) \right]. \tag{17.131}$$

Comparing these two equations gives $C_+ = d_+^{L'}\sqrt{\hbar\rho/\pi}$ and $2C_- = d_-^{L'}\sqrt{\hbar\rho/\pi}$.

Going now to R', the WKBJ wave function in the region near $x < x_{R'}$ is

$$\psi(y) \approx \frac{1}{\sqrt{\hbar\rho}(-y)^{1/4}} \left[C_+ \exp\left(-\frac{2}{3}(-y)^{3/2} + \beta\right) + C_- \exp\left(\frac{2}{3}(-y)^{3/2} - \beta\right) \right], \tag{17.132}$$

where

$$\beta = \frac{1}{\hbar} \int_{x'_L}^{x_{R'}} p(x')dx', \tag{17.133}$$

and the patching wave function in this region is

$$\psi_P(y) = \frac{d_+^{R'}}{\sqrt{\pi}(-y)^{1/4}} \exp\left(\frac{2}{3}(-y)^{3/2}\right) + \frac{d_-^{R'}}{2\sqrt{\pi}(-y)^{1/4}} \exp\left(-\frac{2}{3}(-y)^{3/2}\right). \tag{17.134}$$

Comparing this gives $d_+^{R'} = C_- \exp(-\beta)\sqrt{\pi/(\hbar\rho)}$ and $d_-^{R'} = 2C_+ \exp(\beta)\sqrt{\pi/(\hbar\rho)}$.

In the region $x > x_{R'}$, the WKBJ wave function approximates to

$$\psi(y) \approx \frac{iR_-}{\sqrt{\hbar\rho}y^{1/4}} \left[\exp\left(i\frac{2}{3}y^{3/2} - i\frac{\pi}{4} - i\beta'\right) - \exp\left(-i\frac{2}{3}y^{3/2} + i\frac{\pi}{4} + i\beta'\right) \right] \tag{17.135}$$

and the patching wave function is

$$\psi_P(y) = \frac{d_+^{R'}}{\sqrt{\pi}y^{1/4}} \cos\left(\frac{2}{3}y^{3/2} + \frac{\pi}{4}\right) + \frac{d_-^{L'}}{\sqrt{\pi}y^{1/4}} \sin\left(\frac{2}{3}y^{3/2} + \frac{\pi}{4}\right). \tag{17.136}$$

Comparing these equations gives

$$d_+^{R'} - id_-^{R'} = 2\sqrt{\frac{\pi}{\hbar\rho}} \exp(-i\beta')R_-. \tag{17.137}$$

Taking all the above amplitude relations amongst the wave functions in the different regions, this now leads to a relation between L_- and R_-:

$$R_- = \tfrac{1}{2}L_-[(\sin(\beta')\cos(\beta')\exp(-\beta) + 4\sin(\beta')\cos(\beta')\exp(\beta))$$
$$+ i(\sin^2(\beta')\exp(-\beta) - 4\cos^2(\beta')\exp(\beta))]. \tag{17.138}$$

For the symmetric and antisymmetric wave functions, it will require $R_- = \pm L_-$. This means that the imaginary term must be zero, so that $\exp(-2\beta)\tan^2(\beta') = 4$. For the real part, as we want R_- and L_- to be equal or opposite, respectively, for the symmetric or antisymmetric solutions, it implies the respective relations $\exp(-\beta)\tan(\beta') = 2$ and $\exp(-\beta)\tan(\beta') = -2$.

Summary

- The Rayleigh–Ritz variational method uses a trial wave function with adjustable parameters which are then fixed through minimisation of the energy expectation value to give an upper-bound estimate of the ground-state energy. (17.1)
- Variational method is applied for the ground-state energy of the hydrogen and helium atoms. (17.1.1, 17.1.2)
- The first excited-state energy of the helium atom is calculated to first order in perturbation theory as an introduction to discuss multi-electron atoms. (17.2.1)
- The Hartree–Fock or self-consistent field method can be applied to a multi-particle system such as the electrons in an atom. It treats each particle independently and computes an effective potential created by all the other particles, which is then used for calculating energy levels and wave functions. (17.2.2)
- The Born–Oppenheimer approximation can be used to calculate energy levels in systems where there are some particles with masses that are heavy and others that are light. (17.3)
- The ground-state energy of the hydrogen molecular ion H_2^+ is calculated using the Born–Oppenheimer approximation. (17.3.1)
- The Hellmann–Feynman method computes a relation between the derivative of an energy eigenvalue with respect to some parameter in terms of an expectation value of the derivative of the associated Hamiltonian with respect to the same parameter. (17.4)
- The Wenzel–Kramers–Brillouin–Jeffreys (WKBJ) method is applicable for calculating the approximate energy levels and wave functions for a system where the potential varies slowly relative to the particle's de Broglie wavelength. (17.5)
- Using the WKBJ approximation, The quantisation conditions are computed for a potential well and symmetric double well and the transition probability is computed for a potential barrier. (17.5)

Further Reading

Bransden, B. H. and Joachain, C. J. (1989). *Quantum Mechanics*. Longman Scientific and Technical, London.

Griffiths, D. J. (1995). *Introduction to Quantum Mechanics*. Prentice Hall, Englewood Cliffs, NJ.

Problems

17.1 Estimate the ground-state energy of a one-dimensional simple harmonic oscillator using as trial functions

- $\psi_T(x) = C \exp(-\kappa x^2)$,
- $\psi_T(x) = \kappa^2 - x^2$ for $|x| < \kappa$, zero elsewhere,
- $\psi_T(x) = \cos \kappa x$ for $|\kappa x| < \pi/2$, zero elsewhere.

In each case, κ is the variational parameter. Don't forget the normalisation.

17.2 A particle moves in one dimension in the potential

$$V(x) = \infty, \quad |x| > a, \quad V(x) = 0, \quad |x| \le a.$$

Use a trial function of the form

$$\psi_T(x) = \begin{cases} (a^2 - x^2)(1 + cx^2), & |x| \le a, \\ 0, & |x| > a, \end{cases}$$

where c is a variational parameter, to obtain an upper bound on the ground-state energy. This can be done numerically using, for example, Maple or Python packages, to solve the minimisation problem. How does your bound compare with the exact ground-state energy?

17.3 Repeat the previous problem, taking

$$\psi_T(x) = \begin{cases} (a^2 - x^2)(x + cx^3), & |x| \le a, \\ 0 & |x| > a, \end{cases}$$

as the trial function. Why does this give an upper bound for the first excited energy level? Compare your variational result with the exact eigenvalue of the $n = 2$ level.

17.4 Obtain a variational estimate of the ground-state energy of the hydrogen atom by taking as trial function

$$\psi_T(r) = \exp(-\alpha r^2).$$

How does your result compare with the exact result?

17.5 A particle of mass m is bound in a central potential

$$V(r) = -A \exp(-r/a).$$

Use a simple trial function to obtain an upper bound for the ground-state energy.

    The exponential potential provides a simple model for the binding energy of a deuteron due to the strong nuclear force if we take $A = 32$ MeV and $a = 2.2$ fm. How does your upper bound compare with the exact solution for the ground-state energy, which is -2.245 MeV?

17.6 The variational method can be used to show that in one dimension an attractive potential always has a bound state. For this, consider the simple potential

$$V(x) = \begin{cases} -V_0, & |x| \leq a, \\ 0, & |x| > a, \end{cases}$$

where $V_0 > 0$. Consider the trial wave function

$$\psi(x) = N \exp(-\kappa x^2).$$

Normalise $\psi(x)$ and determine N.

Compute the energy E for this trial wave function. Here assume $a \ll 1/\sqrt{\kappa}$, so that

$$\int_{-a}^{a} \exp(-2\kappa x^2)dx \approx 2a.$$

Determine the condition on κ such that $E < 0$, thus showing the presence of a bound state. Determine the energy at the minimum with κ the variational parameter.

17.7 Calculate the matrix elements from Section 17.3.1, which applies the Born–Oppenheimer approximation to the H_2^+ ion. Show that (these can be done analytically and it's good to try and do that without use of any computational aides)

$$\langle u_{1s}(r_1)|\frac{e^2}{r_2}|u_{1s}(r_1)\rangle = \frac{e^2}{R}\left[1 - \left(1 + \frac{R}{a_0}\right)\exp(-2R/a_0)\right],$$

$$\langle u_{1s}(r_1)|\frac{e^2}{r_1}|u_{1s}(r_2)\rangle = \frac{e^2}{a_0}\left(1 + \frac{R}{a_0}\right)\exp(-R/a_0)$$

and

$$\langle u_{1s}(r_1)|u_{1s}(r_2)\rangle = \left[1 + \frac{R}{a_0} + \frac{1}{3}\left(\frac{R}{a_0}\right)^2\right]\exp(-R/a_0).$$

Combine these expressions and show that they lead to the results for $E^{g,u}(\underline{R})$ as given in that section.

17.8 The rigid rotator: a simple model for the rotational motion of a diatomic molecule is to treat it as a rigid dumb-bell with the nuclei of the two atoms at a fixed separation R. For rotations about any axis through the centre of mass and perpendicular to the symmetry axis of the molecule, the classical kinetic energy of rotation is

$$T = \tfrac{1}{2}I\omega^2,$$

where I is the moment of inertia and ω is the angular frequency of rotation. Express T in terms of the magnitude of the angular momentum and hence write down the quantum Hamiltonian for the system (there is no potential energy). Show that the energy eigenvalues of the rotator are given by

$$E_\ell = \frac{\hbar^2}{2I}\ell(\ell + 1), \qquad \ell = 0, 1, 2, \ldots.$$

17.9 Estimate the magnitude of the rotational energies of the hydrogen molecular ion in electronvolts. [$M_p = 1.67 \times 10^{-27}$ kg, $a_0 = 0.53 \times 10^{-10}$ m, $\hbar = 1.05 \times 10^{-34}$ J s, 1 eV $= 1.6 \times 10^{-19}$ J.]

17.10 Show that if in the hydrogen molecular ion H_2^+ the electron is replaced by a muon, which has the same electric charge as the electron but is a factor 207 times more massive, the equilibrium separation of the protons in the ground state is 6.4×10^{-13} m.

17.11 Apply the Hellmann–Feynman method to compute the expectation value $\langle x^2 \rangle_n$ for any energy eigenstate n, for the one-dimensional simple harmonic oscillator

$$V(x) = \tfrac{1}{2}m\omega^2 \hat{x}^2.$$

Compare your result to those found in Chapter 6.

17.12 Apply the Hellmann–Feynman method to compute the expectation values $\langle r^2 \rangle$ and $\langle 1/r^2 \rangle$ for any energy eigenstate for the three-dimensional isotropic harmonic oscillator. Use the results from Chapter 11 where applicable.

17.13 Determine the allowed energy levels of the simple harmonic oscillator $V(x) = m\omega^2 x^2/2$ using the WKBJ approximation.

17.14 Determine the classical turning point x_t for a simple harmonic oscillator in energy state $E_n = (n + \tfrac{1}{2})\hbar\omega$.

Let $\Delta x = x - x_t$ be the displacement from the classical turning point x_t. Determine how large Δx can be so that the error between the linear approximation to the simple harmonic oscillator potential and its exact value remains less than f.

Determine for what n, $y = \rho\Delta x \gg 1$, where $\rho = (2mV'(x_t)/\hbar^2)^{1/3}$ is the parameter introduced in treating the patching solution to the WKBJ approximation near the turning points. Thus show that there is a regime at the turning points where the linear approximation to the simple harmonic approximation is valid, but at the same time the linearised solution can be evaluated in the asymptotic regime of the Airy function.

17.15 The WKBJ expression for the tunnelling probability (17.109) can be applied to α decay. A simple model for nuclear binding is given in Fig. 17.7. It consists of a repulsive Coulomb potential (solid line) from the charge of the α particle of $2e$ and the remaining nuclear core of charge $(Z - 2)e$. Coming in from large r, this potential rises to some distance r_L from the centre of the nucleus. Then for $r < r_L$ the potential is just a flat well, where the protons and neutrons comprising the nucleus are trapped. For protons with energy $E > 0$, there is a finite probability they can tunnel through and become free at distance $r > r_R$ from the centre of the nucleus. The potential (solid line in Fig. 17.7) through which the α particle tunnels specifically is

$$V(r) = \begin{cases} -V_0 & \text{if } r < r_L, \\ \dfrac{2(Z-2)e^2}{r} & \text{if } r_L < r. \end{cases}$$

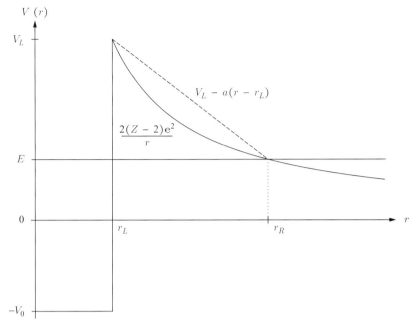

Fig. 17.7 The tunnelling of an α particle out of the nucleus.

Show that the WKBJ tunnelling probability for the above potential $T = \exp(-2\beta)$ has

$$\beta = \frac{\sqrt{2m_\alpha E}}{\hbar}\left[r_R \arctan(\sqrt{r_R/r_L}) - \sqrt{r_L(r_R - r_L)}\right]. \tag{17.139}$$

Note that although strictly this is a three-dimensional problem, for zero angular momentum, the one-dimensional expression can be applied as is.

Suppose the α particle, while confined to the potential well, has speed v. Obtain an expression for the lifetime of the parent nucleus (of charge Z) before it decays into an α particle and remaining nucleus of charge $(Z - 2)e$.

Repeat the above calculations for the tunnelling probability but now for the simpler potential, where the repulsive part is the sloped line (dashed line) in Fig. 17.7:

$$V(r) = \begin{cases} -V_0 & \text{if } r < r_L, \\ V_L - a(r - r_L) & \text{if } r_L < r, \end{cases}$$

where $V_L = 2(Z-2)e^2/r_L$ is the maximum value the Coulomb potential rises to at r_L and $a = E/r_L$ is such that the potential equals E at the same r_R as for the Coulomb potential. Using r_L and r_R the same as for the Coulomb potential then means this width of the non-classical region is the same in the two cases and the potential at r_L and r_R also matches the Coulomb potential.

17.16 Compute the lifetime for the α-decay processes U(238) $\rightarrow$ Th(234) + α and Ra(226) $\rightarrow$ Rn(222) + α from the expressions obtained in the previous question both where the repulsive part of the potential is Coulomb and the sloped line. The experimental observed values are respectively around 4×10^9 years and 1600 years.

Note that the energy of the α particle that emerges from the decay is $E = m_p c^2 - m_d c^2 - m_\alpha c^2$, where m_p is the mass of the initial (parent) nuclei, m_d is the mass of the final (daughter) nuclei that remains after the decay and m_α is the mass of the α particle. You should look up the masses of the nuclei needed for this question. You will find that the results are very sensitive to the masses used, so different sources for these masses may lead to significant differences in the lifetime calculations.

In estimating the velocity of the α particle while in the nuclear potential, equate E with the kinetic energy of the α particle in the nuclear well, $E = m_\alpha v^2/2$. This ignores potential energy effects but provides a reasonable estimate. For the size of the nuclear potential well, you can use the empirical expression $r_L = 1.4 A^{1/3} \times 10^{-13}$ cm, where A is the atomic number of the parent nuclei.

18

Time-Dependent Perturbation Theory

We now examine problems where the Hamiltonian depends on time. For such cases, energy is not conserved and so there are no stationary states. In general, for such problems it is very difficult to find exact solutions to the Schrödinger equation. Where progress can be made is in situations where time dependence is only in a small part of the Hamiltonian. Time-dependent perturbation theory is the formal approach to address such problems. There are many examples of physical relevance where this can be applied, such as to weak time-dependent fields applied to a quantum system. An important application examined here will be the interaction of a quantum system, such as an atom, with electromagnetic radiation. We will also see in Chapter 19 that this formalism is useful for treating the quantum-mechanical scattering of particles.

The objective of time-dependent perturbation theory is to compute how the system transitions in time between states in an initially prescribed basis. For perturbation theory to be valid, this basis of states needs to be chosen carefully so that the transition between the different states is small. The application of the formalism requires that the Hamiltonian be expressed in the very general form $\hat{H} = \hat{H}_0 + \hat{H}'(t)$, where $\hat{H}_0$ has no time dependence. The eigenstates of $\hat{H}_0$ are then the basis and the goal is to determine, for the system in some initial state with respect to this basis, how the effect of the perturbation Hamiltonian $\hat{H}'(t)$ changes that state as a function of time. Throughout this approach the basis states always remain the same. The state vector is expanded in this basis, and one is interested in how the amplitude coefficients of these basis states change as a function of time.

18.1 Time-Dependent Hamiltonians

Recall that for a system described by a Hamiltonian, $\hat{H}_0$, which is time-*independent*, the most general state of the system can be described by a state vector $|\Psi, t\rangle$ which satisfies the TDSE

$$i\hbar \frac{\partial}{\partial t} |\Psi, t\rangle = \hat{H}_0 |\Psi, t\rangle. \tag{18.1}$$

The Hamiltonian has a complete orthonormal set of eigenstates $\{|n^{(0)}\rangle\}$ and a corresponding set of energy eigenvalues, $\{E_n^{(0)}\}$:

$$\hat{H}_0 |n^{(0)}\rangle = E_n^{(0)} |n^{(0)}\rangle. \tag{18.2}$$

The solution to the TDSE can be written formally as an expansion in the energy eigenbasis $\{|n^{(0)}\rangle\}$ as

$$|\Psi, t\rangle = \sum_n c_n^{(0)} \exp(-iE_n^{(0)}t/\hbar) |n^{(0)}\rangle \equiv \sum_n c_n^{(0)} \exp(-i\omega_n t) |n^{(0)}\rangle, \qquad (18.3)$$

where the coefficients, $c_n^{(0)}$, are time-independent. We see that each eigenstate $|n^{(0)}\rangle$ evolves in time with its characteristic angular frequency, ω_n, defined by $E_n^{(0)} = \hbar\omega_n$. You should verify this by substituting the expansion back into the TDSE. This means, for example, that we can calculate the state vector at time t from a knowledge of the state vector at $t = 0$.

When we generalise to the case where the Hamiltonian is of the form

$$\hat{H} = \hat{H}_0 + \hat{H}'(t), \qquad (18.4)$$

we can again expand in the eigenbasis of $\hat{H}_0$:

$$|\Psi, t\rangle = \sum_n c_n(t) \exp(-iE_n^{(0)}t/\hbar) |n^{(0)}\rangle, \qquad (18.5)$$

but the coefficients c_n will now in general be time-dependent. We have chosen for later convenience to make explicit the time-dependent phase factors associated with the energy eigenstates $|n^{(0)}\rangle$. We note that the probability of finding the system in the state $|m^{(0)}\rangle$ at time t is

$$|\langle m^{(0)}|\Psi, t\rangle|^2 = |c_m(t) \exp(-iE_m^{(0)}t/\hbar)|^2 = |c_m(t)|^2, \qquad (18.6)$$

where we have used the orthonormality of the eigenstates of $\hat{H}_0$.

As always, the state vector satisfies the TDSE:

$$i\hbar \frac{\partial}{\partial t}|\Psi, t\rangle = \hat{H}|\Psi, t\rangle, \qquad (18.7)$$

so we can substitute the expansion of $|\Psi, t\rangle$ to obtain the equations satisfied by the coefficients $c_n(t)$. Denoting the time derivative of c_n by $\dot{c}_n$, we obtain

$$i\hbar \sum_n (\dot{c}_n - i\omega_n c_n) \exp(-i\omega_n t)|n^{(0)}\rangle = \sum_n (c_n \hbar\omega_n + c_n\hat{H}') \exp(-i\omega_n t)|n^{(0)}\rangle, \qquad (18.8)$$

which simplifies immediately to give

$$\sum_n (i\hbar\dot{c}_n - c_n\hat{H}') \exp(-i\omega_n t)|n^{(0)}\rangle = 0. \qquad (18.9)$$

We now take the scalar product of this equation with $\langle m^{(0)}|$ to give

$$i\hbar\dot{c}_m \exp(-i\omega_m t) - \sum_n c_n H'_{mn} \exp(-i\omega_n t) = 0. \qquad (18.10)$$

This leads to the following set of coupled, first-order differential equations for the coefficients:

$$\dot{c}_m = (i\hbar)^{-1} \sum_n c_n H'_{mn} \exp(i\omega_{mn}t), \qquad (18.11)$$

where $\omega_{mn} = \omega_m - \omega_n$ and we defined $H'_{mn} \equiv \langle m^{(0)}|\hat{H}'(t)|n^{(0)}\rangle$.

So far, everything is exact, but not terribly useful because we must, in general, solve an infinite set of coupled equations.

18.1.1 Time-Dependent Perturbation Theory

Consider the related Hamiltonian

$$\hat{H} = \hat{H}_0 + \lambda \hat{H}'(t), \tag{18.12}$$

and assume that we can expand the coefficients c_n in a power series of terms of increasing degree of smallness. Here λ is our bookkeeping parameter, as was also used for time-independent perturbation theory:

$$c_n = c_n^{(0)} + \lambda c_n^{(1)} + \lambda^2 c_n^{(2)} + \cdots . \tag{18.13}$$

We substitute in the equation for $\dot{c}_m$ derived above, remembering to replace $\hat{H}'$ by $\lambda \hat{H}'$, to give

$$\dot{c}_m^{(0)} + \lambda \dot{c}_m^{(1)} + \cdots = (i\hbar)^{-1} \lambda \sum_n c_n^{(0)} H'_{mn} \exp(i\omega_{mn}t) + \cdots . \tag{18.14}$$

We can now equate terms of the same degree in λ. The zeroth and first-order terms give

$$
\begin{aligned}
\lambda^0 : & \qquad \dot{c}_m^{(0)} & = & \quad 0, \\
\lambda^1 : & \qquad \dot{c}_m^{(1)} & = & \quad (i\hbar)^{-1} \sum_n c_n^{(0)} H'_{mn} \exp(i\omega_{mn}t).
\end{aligned}
$$

The first of these simply states the obvious, that to zeroth order the coefficients are time-independent, since to this order the Hamiltonian is time-independent, and we recover the unperturbed result.

The second equation allows us to obtain the first-order contribution by integrating the first-order differential equation to give

$$c_m^{(1)}(t) = c_m^{(1)}(t_0) + (i\hbar)^{-1} \sum_n c_n^{(0)} \int_{t_0}^t H'_{mn} \exp(i\omega_{mn}t') dt' . \tag{18.15}$$

Now consider the special case where initially the system is known to be in an eigenstate of $\hat{H}_0$, say $|k^{(0)}\rangle$, for all $t \le t_0$, then $c_k^{(0)} = 1$ and all other $c_n^{(0)} = 0$, $n \ne k$ so that the sum on the right-hand side reduces to a single term. Furthermore, if we are interested in computing the probability of finding the system in a different state $|m^{(0)}\rangle$ at time t, we have $c_m^{(1)}(t_0) = 0$, giving

$$c_m^{(1)}(t) = (i\hbar)^{-1} \int_{t_0}^t H'_{mk} \exp(i\omega_{mk}t') dt', \qquad m \ne k. \tag{18.16}$$

Thus the so-called **transition probability** of finding the system at a later time, t, in the state $|m^{(0)}\rangle$, where $m \ne k$, is given by

$$p_{mk}(t) \simeq |c_m^{(1)}(t)|^2 = \frac{1}{\hbar^2} \left| \int_{t_0}^t H'_{mk} \exp(i\omega_{mk}t') dt' \right|^2 . \tag{18.17}$$

18.2 Time-Dependent Perturbations

Here we look at two important types of perturbations that arise in time-dependent problems.

18.2.1 Time-Independent Perturbation Switched on at a Given Time

The results obtained in the last section are particularly simple when applied to the case where the perturbation, $\hat{H}'$, is actually independent of time. For example, we might wish to consider the effect of a constant external field which is switched on at $t = 0$. In this case we obtain

$$
\begin{aligned}
c_m^{(1)}(t) &= (i\hbar)^{-1} H'_{mk} \int_0^t \exp(i\omega_{mk}t')dt' \\
&= \frac{H'_{mk}}{\hbar\omega_{mk}} [1 - \exp(i\omega_{mk}t)],
\end{aligned} \tag{18.18}
$$

giving for the transition probability

$$
p_{mk}(t) = |c_m^{(1)}(t)|^2 = \frac{2}{\hbar^2} |H'_{mk}|^2 \frac{(1 - \cos\omega_{mk}t)}{\omega_{mk}^2} \equiv \frac{2|H'_{mk}|^2}{\hbar^2} f(t, \omega_{mk}), \tag{18.19}
$$

where the function $f(t, \omega_{mk})$ is defined by

$$
f(t, \omega_{mk}) \equiv \frac{(1 - \cos\omega_{mk}t)}{\omega_{mk}^2} = \frac{2\sin^2(\omega_{mk}t/2)}{\omega_{mk}^2}. \tag{18.20}
$$

The function $f(t, \omega_{mk})$ consists essentially of a large peak, centred on $\omega_{mk} = 0$, with a height proportional to t^2 and width $\simeq 2\pi/t$, as indicated in Fig. 18.1. Thus there is only a significant transition probability for those states whose energy lies in a band of width

$$
\delta E \simeq \frac{2\pi\hbar}{t} \tag{18.21}
$$

about the initial energy, $E_k^{(0)}$.

Notes: Using the standard integral $\displaystyle\int_{-\infty}^{+\infty} \frac{\sin^2 x}{x^2} dx = \pi$, it's easy to show that the function $f(t, \omega)$ has the following useful properties:

$$
\int_{-\infty}^{\infty} f(t, \omega)d\omega = \pi t \tag{18.22}
$$

and

$$
\lim_{t \to \infty} f(t, \omega) \sim \pi t \delta(\omega), \tag{18.23}
$$

where $\delta(\omega)$ is a Dirac delta function.

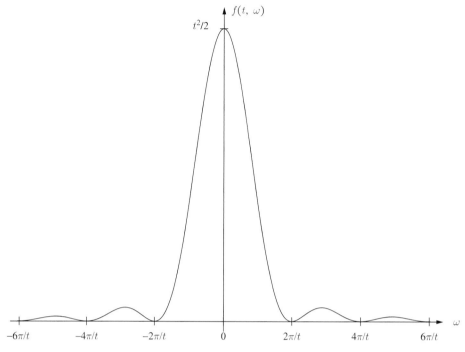

Fig. 18.1 Plot of $f(t, \omega)$ as a function of ω. Note the peak of height $t^2/2$, centred on $\omega = 0$, and the positions of the first two zeros at $\omega = \pm 2\pi/t$.

Application

Nondegenerate Case

If $\omega_{mk} \neq 0$, the total first-order probability that the system has made a transition from the initial state is

$$P^{(1)}(t) = \sum_{m \neq k} p_{mk}^{(1)}(t) = \sum_{m \neq k} \frac{4|H'_{mk}|^2}{\hbar^2 \omega_{mk}^2} \sin^2(\omega_{mk}t/2). \qquad (18.24)$$

For our perturbation theory treatment to be valid, $P^{(1)}(t) \ll 1$, and a sufficient condition for this is that

$$\sum_{m \neq k} \frac{4|H'_{mk}|^2}{\hbar^2 \omega_{mk}^2} \ll 1. \qquad (18.25)$$

This can always be satisfied for a sufficiently weak perturbation.

Degenerate Case

If the states $|k^{(0)}\rangle$ and $|m^{(0)}\rangle$ are degenerate, $\omega_{mk} = 0$ and we have

$$c_m^{(1)}(t) = -\frac{i}{\hbar} H'_{mk} t \quad \Rightarrow \quad p_{mk}^{(1)}(t) = \frac{|H'_{mk}|^2}{\hbar^2} t^2. \qquad (18.26)$$

Clearly after a sufficient length of time has elapsed this will violate the requirement that it is $\ll 1$ and so our perturbative treatment will break down.

18.2.2 Fermi's Golden Rule

We are often interested in the situation where transitions take place not to a single final state but to a group, G, of final states with energy in some range about the initial state energy:

$$E_k^{(0)} - \Delta E \le E_m^{(0)} \le E_k^{(0)} + \Delta E, \tag{18.27}$$

for example, when considering transitions to the continuous part of the energy eigenvalue spectrum. Then the total transition probability is obtained by summing the contributions of all the final states.

The number of final states in the interval $E_m \to E_m + dE_m$ is $\rho(E_m)dE_m$, where the function $\rho(E_m)$ is known as the *density of final states*. For the special case of the constant perturbation which we considered in the previous section, the total transition probability for transitions to G is then given by

$$p_G(t) = \frac{2}{\hbar^2} \int_{E_k^{(0)}-\Delta E}^{E_k^{(0)}+\Delta E} |H'_{mk}|^2 f(t, \omega_{mk}) \rho(E_m)dE_m. \tag{18.28}$$

If we assume that ΔE is small enough that we can neglect the variation of $\rho(E_m)$ and H'_{mk} with E_m, we can take them out of the integral, and evaluate the density of final states at $E_m^{(0)} = E_k^{(0)}$ to give

$$p_G(t) = \frac{2|H'_{mk}|^2}{\hbar^2} \rho(E_k^{(0)}) \int_{E_k^{(0)}-\Delta E}^{E_k^{(0)}+\Delta E} f(t, \omega_{mk})dE_m. \tag{18.29}$$

If t is large enough that $\Delta E \gg 2\pi\hbar/t$, we observe that the only significant contributions to the integral come from the energy range corresponding to the narrow central peak of the function $f(t, \omega_{mk})$. It is then a good approximation to extend the limits on the integration to $\pm\infty$, because the extra range gives almost zero contribution. Noting that $dE_m = \hbar d\omega_{mk}$ and using the result noted already, that

$$\int_{-\infty}^{\infty} f(t, \omega)d\omega = \pi t, \tag{18.30}$$

we obtain for the first-order transition probability for energy-conserving transitions

$$p_G(t) = \frac{2\pi t}{\hbar} |H'_{mk}|^2 \rho(E), \tag{18.31}$$

where $E = E_k^{(0)} = E_m^{(0)}$.

The numbers of transitions per unit time, the transition rate, R, is just the derivative of this with respect to t and is thus given by the so-called Fermi golden rule:

$$R = \frac{2\pi}{\hbar} |H'_{mk}|^2 \rho(E).$$

Later in the book we shall see how this result can be used to calculate scattering cross-sections.

18.2.3 Harmonic Perturbations

A second special case for which the transition probability amplitude takes a simple form is when the perturbation is a sinusoidal function of time which is turned on at $t = 0$. An obvious application is to the interaction of a quantum system with electromagnetic radiation. Suppose that

$$\hat{H}'(t) = \hat{\mathcal{H}}' \sin \omega t, \tag{18.32}$$

where $\hat{\mathcal{H}}'$ is a time-independent Hermitian operator. It is convenient to write this as

$$\hat{H}'(t) = \hat{A} \exp(i\omega t) + \hat{A}^{\dagger} \exp(-i\omega t), \tag{18.33}$$

where $\hat{A} = \hat{\mathcal{H}}'/2i$.

The initial condition is that, for $t \leq 0$, the system is in the unperturbed state $|k^{(0)}\rangle$, with energy $E_k^{(0)}$, so that $c_k(0) = 1$ and $c_m(0) = 0$, $m \neq k$. Then from (18.16):

$$c_m^{(1)}(t) = (i\hbar)^{-1} \left\{ A_{mk} \int_0^t \exp[i(\omega_{mk} + \omega)t']dt' + A_{mk}^{\dagger} \int_0^t \exp[i(\omega_{mk} - \omega)t']dt' \right\}, \tag{18.34}$$

where $A_{mk} = \mathcal{H}'_{mk}/2i$ and $A_{mk}^{\dagger} = A_{km}^*$. The integrals are straightforward and yield

$$c_m^{(1)}(t) = A_{mk} \left(\frac{1 - \exp[i(\omega_{mk} + \omega)t]}{\hbar(\omega_{mk} + \omega)} \right) + A_{mk}^{\dagger} \left(\frac{1 - \exp[i(\omega_{mk} - \omega)t]}{\hbar(\omega_{mk} - \omega)} \right). \tag{18.35}$$

The corresponding transition probability is given by taking the square modulus of the amplitude:

$$p_{mk}^{(1)}(t) = \left| A_{mk} \left(\frac{1 - \exp[i(\omega_{mk} + \omega)t]}{\hbar(\omega_{mk} + \omega)} \right) + A_{mk}^{\dagger} \left(\frac{1 - \exp[i(\omega_{mk} - \omega)t]}{\hbar(\omega_{mk} - \omega)} \right) \right|^2. \tag{18.36}$$

We now observe that there are two situations in which this transition probability can become large. Assuming that $E_k^{(0)} \neq E_m^{(0)}$, one or other denominator can be close to zero but not both simultaneously:

1. If $E_m^{(0)} \simeq E_k^{(0)} + \hbar\omega$, the second term will dominate and the corresponding transition probability is given by

$$p_{mk}^{(1)}(t) \simeq \frac{2}{\hbar^2} |A_{mk}^{\dagger}|^2 \, f(t, \omega_{mk} - \omega), \tag{18.37}$$

where the function $f(t, \omega_{mk} - \omega)$ is sharply peaked about $\omega_{mk} = \omega$. From our previous discussion of the constant perturbation we conclude that the transition probability is only significant if $E_m^{(0)}$ lies in an interval of width $2\pi\hbar/t$ about the value $E_k^{(0)} + \hbar\omega$. In other words, the system has *absorbed* an amount of energy given, to within $2\pi\hbar/t$, by

$$\hbar\omega = E_m^{(0)} - E_k^{(0)}. \tag{18.38}$$

When this condition is exactly satisfied, *resonance* is said to occur, and the first-order formula for the transition probability reduces to

$$p_{mk}^{(1)}(t) = \frac{|A_{mk}^\dagger|^2}{\hbar^2} t^2 = \frac{|\mathcal{H}_{mk}'|^2}{4\hbar^2} t^2. \tag{18.39}$$

2. If $E_m^{(0)} \simeq E_k^{(0)} - \hbar\omega$, the first term will dominate and the corresponding transition probability is given by

$$p_{mk}^{(1)}(t) \simeq \frac{2}{\hbar^2} |A_{mk}|^2 f(t, \omega_{mk} + \omega), \tag{18.40}$$

where the function $f(t, \omega_{mk}+\omega)$ is sharply peaked about $\omega_{mk} = -\omega$. From our previous discussion of the constant perturbation we conclude that the transition probability is only significant if $E_m^{(0)}$ lies in an interval of width $2\pi\hbar/t$ about the value $E_k^{(0)} - \hbar\omega$. In other words, the system has *emitted* an amount of energy given, to within $2\pi\hbar/t$, by

$$\hbar\omega = E_k^{(0)} - E_m^{(0)}. \tag{18.41}$$

Again, resonance occurs when this condition is exactly satisfied.

Transitions to a Group of States

As in the case of the constant perturbation, we can consider transitions to a group of states whose energy lies in

$$(E_k^{(0)} \pm \hbar\omega) - \Delta E \leq E_m^{(0)} \leq (E_k^{(0)} \pm \hbar\omega) + \Delta E, \tag{18.42}$$

with density $\rho(E_m)$. It is straightforward to show that, for transitions where the system *absorbs* energy $\hbar\omega \simeq E_m^{(0)} - E_k^{(0)}$, the transition rate, R_{mk}, is given by

$$R_{mk} = \frac{2\pi}{\hbar} |A_{mk}^\dagger|^2 \rho(E), \tag{18.43}$$

where $E = E_k^{(0)} + \hbar\omega$. For transitions where the system *emits* energy $\hbar\omega \simeq E_k^{(0)} - E_m^{(0)}$, the corresponding expression is

$$R_{mk} = \frac{2\pi}{\hbar} |A_{mk}|^2 \rho(E), \tag{18.44}$$

with $E = E_k^{(0)} - \hbar\omega$. These expressions can be regarded as a generalisation of the golden rule.

The results above can be generalised to the case of a perturbation $\hat{H}'(t)$ which is a general periodic function by expanding the perturbation in a Fourier series:

$$\hat{H}'(t) = \sum_{n=1}^{\infty} \left[\hat{A}_n \exp(i\omega_n t) + \hat{A}_n^\dagger \exp(-i\omega_n t) \right]. \tag{18.45}$$

18.3 Interaction of Radiation with Quantum Systems

A quantum system, such as an atom or a nucleus, can interact with electromagnetic radiation in a number of ways. It can emit radiation under the influence of an external radiation field in a process called **stimulated emission**. It can absorb radiation, making an internal transition to a state of higher energy, a process known as **absorption**. Finally, it can make spontaneous transitions from an excited state to a state of lower energy, with emission of photons. This last process is called **spontaneous emission** and takes place in the absence of any external field. A full treatment of all these processes requires us to quantise the electromagnetic field using quantum electrodynamics (QED), which is beyond the scope of this book. However, the so-called **semiclassical treatment**, in which the radiation field is treated classically and only the atomic system is quantised, is sufficient for treating the stimulated emission and absorption of radiation.

18.3.1 Semiclassical Treatment of Electromagnetic Radiation

A transverse, monochromatic plane wave propagating *in vacuo* corresponds to solutions of Maxwell's equations in which the electric and magnetic fields in Gaussian cgs units are

$$\underline{\mathcal{E}}(\underline{r},t) = \mathcal{E}_0 \, \underline{\varepsilon} \, \sin(\underline{k} \cdot \underline{r} - \omega t + \delta_\omega),$$

$$\underline{\mathcal{B}}(\underline{r},t) = \mathcal{E}_0 \, \frac{c}{\omega}(\underline{k} \times \underline{\varepsilon}) \, \sin(\underline{k} \cdot \underline{r} - \omega t + \delta_\omega).$$

The electric field has an amplitude $\mathcal{E}_0$ and ω is the angular frequency of the radiation. The wave vector, $\underline{k}$, defines the direction of propagation, and its magnitude, k, is the wave number, $\omega = ck$. The direction of the electric field is defined by the **polarisation vector**, $\underline{\varepsilon}$, which is a unit vector perpendicular to $\underline{k}$. Consequently, we see that $\underline{\mathcal{E}}$, $\underline{\mathcal{B}}$ and $\underline{k}$ are mutually perpendicular. δ_ω is a real constant phase.

The energy density of the field is given by

$$\frac{1}{8\pi}\left(|\underline{\mathcal{E}}|^2 + |\underline{\mathcal{B}}|^2\right) = \frac{1}{4\pi}\mathcal{E}_0^2 \sin^2(\underline{k} \cdot \underline{r} - \omega t + \delta_\omega). \tag{18.46}$$

The average of $\sin^2(\underline{k} \cdot \underline{r} - \omega t + \delta_\omega)$ over a period $T = 2\pi/\omega$ is

$$\frac{1}{T}\int_0^T \sin^2(\underline{k} \cdot \underline{r} - \omega t + \delta_\omega)dt = \tfrac{1}{2}, \tag{18.47}$$

so that the average energy density is

$$\rho = \frac{1}{8\pi}\mathcal{E}_0^2. \tag{18.48}$$

The average rate of flow of energy per unit area normal to the direction of propagation as shown in Fig. 18.2 defines the **intensity** of the radiation, I. Since the velocity of the radiation in free space is just c, we have $I = \rho c$.

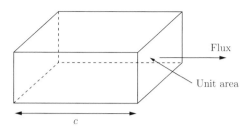

Flux

Unit area

c

   Flux of energy per unit area.

Incoherent Radiation

In practice, no radiation is perfectly monochromatic but can be represented as a super-position of monochromatic plane waves. If each frequency component has the same propagation direction and polarisation, the most general electric field is

$$\underline{\mathcal{E}}(\underline{r},t) = \underline{\varepsilon} \int_0^\infty \mathcal{E}_0(\omega)\sin(\underline{k}\cdot\underline{r} - \omega t + \delta_\omega)d\omega, \tag{18.49}$$

with a corresponding expression for $\underline{\mathcal{B}}$. Note that $\mathcal{E}_0(\omega)$ is now an amplitude *per unit angular frequency*. Typically, a source of radiation will comprise many atoms each emitting photons independently. In this case, the phases δ_ω are randomly distributed and we say that the radiation is **incoherent**. When we calculate the average energy density and intensity from these expressions for $\underline{\mathcal{E}}$ and $\underline{\mathcal{B}}$, it can be shown that the cross-terms average to zero if the phases are randomly distributed. The corresponding average energy density, $\bar{\rho}$, and intensity, $\bar{I}$, are thus given by

$$\bar{\rho} = \int_0^\infty \rho(\omega)d\omega \qquad \text{and} \qquad \bar{I} = \int_0^\infty I(\omega)d\omega, \tag{18.50}$$

where $\rho(\omega)$ and $I(\omega)$ are the energy density and intensity *per unit angular frequency*, with

$$\rho(\omega) = \frac{1}{8\pi}\mathcal{E}_0^2(\omega) \qquad \text{and} \qquad I(\omega) = c\rho(\omega). \tag{18.51}$$

18.3.2 Interaction with a One-Electron Atom

The interaction of an electron of charge $q = -e$ with the electromagnetic field created by the nucleus with vector potential $\underline{A}$ is contained in the Pauli Hamiltonian Eq. (16.130) as the term

$$\hat{H}_I = \frac{e}{2mc}[\hat{\underline{p}}\cdot\underline{A} + \underline{A}\cdot\hat{\underline{p}}]. \tag{18.52}$$

In the second term the action of the operator $\hat{\underline{p}}$ leads to two terms, with the term $\hat{\underline{p}}\cdot\underline{A} = -i\hbar\nabla\cdot\underline{A} = 0$, where the electromagnetic field $\underline{A}$ is in the Coulomb gauge. Thus the interaction Hamiltonian takes on a somewhat simpler form:

$$\hat{H}_I = \frac{e}{mc}\underline{A}\cdot\hat{\underline{p}}. \tag{18.53}$$

Considering the electromagnetic field at just one frequency and wave number, the vector potential has the expansion

$$\underline{A} = A_0 \underline{\epsilon} \cos(\underline{k} \cdot \underline{r} - \omega t)$$

$$= \underline{\epsilon} \frac{A_0}{2} (e^{i(\underline{k} \cdot \underline{r} - \omega t)} + e^{-i(\underline{k} \cdot \underline{r} - \omega t)}), \tag{18.54}$$

which gives

$$\hat{H}_I = \frac{e}{2mc} A_0 (e^{i(\underline{k} \cdot \underline{r} - \omega t)} + e^{-i(\underline{k} \cdot \underline{r} - \omega t)}) \underline{\epsilon} \cdot \hat{\underline{p}}.$$

Computing the matrix element for this interaction Hamiltonian:

$$\langle f | \hat{H}_I | i \rangle = \frac{e}{mc} \langle f | \underline{A} \cdot \hat{\underline{p}} | i \rangle$$

$$= \frac{e}{2mc} A_0 \underline{\epsilon} \cdot \langle f | \hat{\underline{p}} | i \rangle (e^{i(\underline{k} \cdot \underline{r} - \omega t)} + e^{-i(\underline{k} \cdot \underline{r} - \omega t)}).$$

Recall that

$$\hat{\underline{p}} = \frac{im}{\hbar} [\hat{H}_0, \hat{\underline{r}}]$$

gives

$$\langle f | \hat{\underline{p}} | i \rangle = \frac{im}{\hbar} \langle f | \hat{H}_0 \hat{\underline{r}} - \hat{\underline{r}} \hat{H}_0 | i \rangle$$

$$= \frac{im}{\hbar} (E_f^{(0)} - E_i^{(0)}) \langle f | \hat{\underline{r}} | i \rangle$$

$$= im\omega_{fi} \langle f | \hat{\underline{r}} | i \rangle,$$

with $\omega_{fi} \equiv (E_f^{(0)} - E_i^{(0)}/\hbar$. Thus

$$\langle f | \hat{H}_I | i \rangle = \frac{e}{mc} \langle f | \underline{A} \cdot \hat{\underline{p}} | i \rangle$$

$$= \frac{ie\omega_{fi}}{2c} A_0 \underline{\epsilon} \cdot \langle f | \hat{\underline{r}} | i \rangle (e^{i(\underline{k} \cdot \underline{r} - \omega t)} + e^{-i(\underline{k} \cdot \underline{r} - \omega t)}). \tag{18.55}$$

The time-dependent perturbation calculation examines how the electron either absorbs or emits energy to this electromagnetic field at the frequency ω of the field, so that $\omega_{fi} = \pm\omega$ are the energy differences where transitions are possible. Thus we can equally replace ω_{fi} by ω in the above expression with a minus sign between the two terms. In particular, based on the expression for the first-order transition for the harmonic perturbation in (18.34), the first term above will lead to an oscillatory exponential $\exp[i(\omega_{fi} - \omega)]$ and the second term to $\exp[i(\omega_{fi} + \omega)]$. For fixed external driving frequency ω, this means that the first expression has transitions with energy difference such that $\omega_{fi} = \omega$ and the second with $\omega_{fi} = -\omega$. Thus with the time-dependent calculation in mind, we can write (18.55) equivalently as

$$\langle f | \hat{H}_I | i \rangle = \frac{e}{mc} \langle f | \underline{A} \cdot \hat{\underline{p}} | i \rangle$$

$$= \frac{ie\omega}{2c} A_0 \underline{\epsilon} \cdot \langle f | \hat{\underline{r}} | i \rangle (e^{i(\underline{k} \cdot \underline{r} - \omega t)} - e^{-i(\underline{k} \cdot \underline{r} - \omega t)}). \tag{18.56}$$

Recalling that the electric field is

$$\underline{\mathcal{E}} = -\frac{1}{c}\frac{\partial \underline{A}}{\partial t} \tag{18.57}$$

and using (18.54) gives

$$\underline{\mathcal{E}} = \frac{i\omega}{2c}A_0\underline{\epsilon}(e^{i(\underline{k}\cdot\underline{r}-\omega t)} - e^{-i(\underline{k}\cdot\underline{r}-\omega t)}). \tag{18.58}$$

Comparing with (18.55) implies that we can also write the interaction Hamiltonian in terms of the electric field as

$$\hat{H}' = \underline{\mathcal{E}} \cdot e\hat{\underline{r}}. \tag{18.59}$$

There is a simple intuitive argument that can lead to this interaction term. The force on an electron (charge $-e$) with velocity $\underline{v}$ in electric fields $\underline{\mathcal{E}}$ and $\underline{\mathcal{B}}$ is given by

$$\underline{F} = -e[\underline{\mathcal{E}} + \frac{\underline{v}}{c} \times \underline{\mathcal{B}}]. \tag{18.60}$$

We shall ignore the force on the nucleus, which is much more massive than the electron. Since for light atoms at least v/c is small, that means the magnetic term is subdominant and can be ignored in the Lorentz force.

Dipole Approximation

We begin by studying the interaction of a monochromatic plane wave. The wavelength of the radiation is typically much greater than the size of the atom and so we can to a good approximation take the spatial distribution of the electric field to be uniform across the atom and equal to its value at the nucleus, which we take to be at $\underline{r} = 0$. In this approximation:

$$\underline{F}(t) = -e\underline{\mathcal{E}} = e\mathcal{E}_0\,\underline{\epsilon}\,\sin(\omega t - \delta_\omega), \tag{18.61}$$

corresponding to an interaction energy

$$\hat{H}' = e\underline{\mathcal{E}} \cdot \underline{r} \equiv -\underline{\mathcal{E}} \cdot \underline{D}, \tag{18.62}$$

where $\underline{D} \equiv -e\underline{r}$ is the **electric dipole operator** for a one-electron atom. This is the **dipole approximation**. It is equivalent to the approximation,

$$\exp(i\underline{k} \cdot \underline{r}) \simeq 1. \tag{18.63}$$

Absorption

According to the earlier analysis, the probability of finding the atom, initially in the state $|k^{(0)}\rangle$, in the state $|m^{(0)}\rangle$ when irradiated for a time t with radiation of angular frequency ω, where $E_m^{(0)} > E_k^{(0)}$, is

$$p_{mk}^{(1)}(t) = \frac{2}{\hbar^2}|A_{mk}^\dagger|^2 f(t, \omega_{mk} - \omega) = \frac{1}{2\hbar^2}|\mathcal{H}_{mk}'|^2 f(t, \omega_{mk} - \omega)$$

$$= \frac{1}{2}\left(\frac{\mathcal{E}_0}{\hbar}\right)^2 |\langle m^{(0)}|\underline{\varepsilon}\cdot\underline{D}|k^{(0)}\rangle|^2 f(t, \omega_{mk} - \omega)$$

$$= \frac{1}{2}\left(\frac{\mathcal{E}_0}{\hbar}\right)^2 |\underline{\varepsilon}\cdot\underline{D}_{mk}|^2 f(t, \omega_{mk} - \omega), \qquad (18.64)$$

where $\underline{D}_{mk} \equiv \langle m^{(0)}|\underline{D}|k^{(0)}\rangle$ is known as a **dipole matrix element**. In terms of the intensity, we can rewrite the transition probability as

$$p_{mk}^{(1)}(t) = \frac{4\pi I}{c\hbar^2}|\underline{\varepsilon}\cdot\underline{D}_{mk}|^2 f(t, \omega_{mk} - \omega) = \frac{4\pi I}{c\hbar^2}\cos^2\theta\,|D_{mk}|^2 f(t, \omega_{mk} - \omega), \quad (18.65)$$

where θ is the angle between the polarisation vector, $\underline{\varepsilon}$, and the electric dipole vector, $\underline{D}$. In terms of Cartesian components of the dipole moment:

$$|D_{mk}|^2 = e^2\left\{|\langle m^{(0)}|x|k^{(0)}\rangle|^2 + |\langle m^{(0)}|y|k^{(0)}\rangle|^2 + |\langle m^{(0)}|z|k^{(0)}\rangle|^2\right\}. \qquad (18.66)$$

For radiation which is an incoherent superposition of frequencies, we can add the probabilities for different frequencies, that is, integrate over the distribution of intensities, $I(\omega)$:

$$p_{mk}^{(1)}(t) = \frac{4\pi}{c\hbar^2}\cos^2\theta\,|D_{mk}|^2\int_0^\infty I(\omega)\,f(t, \omega_{mk} - \omega)d\omega. \qquad (18.67)$$

If the intensity distribution is slowly varying over the peak of $f(t, \omega_{mk} - \omega)$, we can replace it by its value at $\omega = \omega_{mk}$. We can extend the limits on the integral to $\pm\infty$ because the integrand is small except in the region $\omega \simeq \omega_{mk}$. Thus we obtain

$$p_{mk}^{(1)}(t) = \frac{4\pi^2 I(\omega_{mk})}{c\hbar^2}\cos^2\theta\,|D_{mk}|^2 t. \qquad (18.68)$$

As usual, this will only be valid for times sufficiently short that $p_{mk}^{(1)}(t) \ll 1$. In this regime, we can define a *transition rate* for absorption in the dipole approximation:

$$R_{mk} = \frac{4\pi^2 I(\omega_{mk})}{c\hbar^2}\cos^2\theta\,|D_{mk}|^2.$$

If the radiation is isotropic and unpolarised, we can replace the factor of $\cos^2\theta$ by its average value, namely

$$\overline{\cos^2\theta} = \frac{1}{2}\int_{-1}^{+1}\cos^2\theta\,(d\cos\theta) = \frac{1}{3}, \qquad (18.69)$$

giving

$$R_{mk} = \frac{4\pi^2 I(\omega_{mk})}{3c\hbar^2}|D_{mk}|^2.$$

Stimulated Emission

We can repeat the analysis of the previous section for the case where the first term in the probability amplitude dominates and

$$p_{mk}^{(1)}(t) = \frac{2}{\hbar^2}|A_{mk}|^2 f(t, \omega_{mk} + \omega) = \frac{1}{2\hbar^2}|\mathcal{H}'_{mk}|^2 f(t, \omega_{mk} + \omega)$$

$$= \tfrac{1}{2}\left(\frac{\mathcal{E}_0}{\hbar}\right)^2 |\langle m^{(0)}|\underline{\varepsilon} \cdot \underline{D}|k^{(0)}\rangle|^2 \, f(t, \omega_{mk} + \omega).$$

Thus for radiation which is an incoherent superposition of frequencies with intensity per unit frequency $I(\omega)$, we obtain

$$p_{mk}^{(1)}(t) = \frac{4\pi^2 I(-\omega_{mk})}{c\hbar^2} \cos^2\theta \, |D_{mk}|^2 t \tag{18.70}$$

and

$$R_{mk} = \frac{4\pi^2 I(-\omega_{mk})}{c\hbar^2} \cos^2\theta \, |D_{mk}|^2. \tag{18.71}$$

Detailed Balance

Note that $\underline{D}_{mk} = \underline{D}^*_{km}$ so that $|D_{mk}|^2 = |D_{km}|^2$. Hence, for a *given pair of states*, labelled by k, m, we see that,

$$R_{mk}^{\text{abs}} = R_{km}^{\text{em}}, \tag{18.72}$$

so that for a given radiation field, the number of transitions per second exciting an atom from state k to state m is the same as the number de-exciting the atom from state m to state k. This is essentially the thermodynamical principle of *detailed balance*.

We can attempt to relate the energy density to the number of photons per unit volume. If we suppose that the number of photons of angular frequency ω in a volume V is $N(\omega)$, then the energy density will be $\hbar\omega N(\omega)/V$, which enables us to relate the amplitude of the electric field to the photon density:

$$\mathcal{E}_0 = \left(8\pi\rho(\omega)\right)^{1/2} = [8\pi\hbar\omega N(\omega)/V]^{1/2}. \tag{18.73}$$

Spontaneous Emission

Although we said that a full quantum treatment of spontaneous emission really requires quantum electrodynamics, we can calculate the rate of spontaneous transitions indirectly by using a thermodynamic argument proposed by Einstein.

Consider an enclosure containing atoms of a single species together with radiation in equilibrium at temperature T. Suppose that k and m label two atomic states with energies E_k and E_m, respectively, and assume that $E_k < E_m$.

Absorption

The rate of transition from k to m through absorption of radiation of angular frequency $\omega = \omega_{mk}$ will be proportional to the number of atoms in the state k and to the energy density per unit angular frequency, $\rho(\omega_{mk})$. We can write

$$\dot{N}_{mk} = B_{mk} N_k \rho(\omega_{mk}), \tag{18.74}$$

where the coefficient B_{mk} is called the **Einstein coefficient for absorption**. In terms of the transition rate per atom, R_{mk}, which we calculated earlier:

$$\dot{N}_{mk} = R_{mk} N_k, \tag{18.75}$$

so that

$$B_{mk} = R_{mk}/\rho(\omega_{mk}) = c R_{mk}/I(\omega_{mk}). \tag{18.76}$$

In the dipole approximation, we have for unpolarised radiation

$$B_{mk} = \frac{4\pi^2}{3\hbar^2} |D_{mk}|^2. \tag{18.77}$$

Emission

The rate of transition from m to k, $\dot{N}_{km}$, is the sum of the rate of spontaneous transition, which is independent of ρ, and the rate of stimulated transition, which is proportional to ρ:

$$\dot{N}_{km} = A_{km} N_m + B_{km} N_m \rho(\omega_{mk}). \tag{18.78}$$

Here A_{km} is the **Einstein coefficient for spontaneous emission** and B_{km} is the **Einstein coefficient for stimulated emission**.

In thermal equilibrium, $\dot{N}_{mk} = \dot{N}_{km}$, so that

$$\frac{N_k}{N_m} = \frac{A_{km} + B_{km}\rho(\omega_{mk})}{B_{mk}\rho(\omega_{mk})}. \tag{18.79}$$

But we know from thermodynamics that

$$\frac{N_k}{N_m} = \exp[-(E_k - E_m)/k_B T] = \exp(\hbar\omega_{mk}/k_B T), \tag{18.80}$$

where k_B is Boltzmann's constant. Thus, eliminating the ratio N_k/N_m gives

$$\rho(\omega_{mk}) = \frac{A_{km}}{B_{mk} \exp(\hbar\omega_{mk}/k_B T) - B_{km}}. \tag{18.81}$$

We now appeal to Planck's law, which says that, at temperature T, the energy density per unit wavelength is given by

$$n(\lambda) = \frac{8\pi hc}{\lambda^5} \frac{1}{\exp(hc/\lambda k_B T) - 1}. \tag{18.82}$$

To express this in terms of angular frequency, we note that $\lambda = 2\pi c/\omega$ and $|n(\lambda)d\lambda| = |\rho(\omega)d\omega|$, so that

$$\rho(\omega) = n(\lambda)\left|\frac{d\lambda}{d\omega}\right| = n(\lambda)\frac{2\pi c}{\omega^2}, \tag{18.83}$$

and hence, from Eq. (18.82):

$$\rho(\omega_{mk}) = \frac{\hbar\omega_{mk}^3}{\pi^2 c^3}\frac{1}{\exp(\hbar\omega_{mk}/k_B T) - 1}. \tag{18.84}$$

Comparing this with the previous expression for $\rho(\omega_{mk})$ gives

$$B_{km} = B_{mk} \quad \text{and} \quad A_{km} = \frac{\hbar\omega_{mk}^3}{\pi^2 c^3} B_{km}. \tag{18.85}$$

Thus, in the dipole approximation, the transition rate for spontaneous emission is

$$R_{mk}^{\text{spon}} = \frac{4\omega_{mk}^3}{3c^3\hbar}|D_{mk}|^2.$$

One obtains the same result in QED.

18.3.3 Selection Rules

We have seen that in the dipole approximation, transition rates between atomic states depend on the dipole matrix elements $\underline{D}_{mk}$. In certain cases, we find that the dipole matrix element vanishes independently of the details of the atomic wave functions; a **selection rule** is said to apply in such a case.

As a first example, we consider a hydrogenic atom. As the electric dipole operator is spin-independent, we can work in the uncoupled basis and ignore the spin quantum numbers. Consider the commutator

$$[\hat{L}_z, \hat{z}] = [\hat{x}\hat{p}_y - \hat{y}\hat{p}_x, \hat{z}] = 0. \tag{18.86}$$

If we take matrix elements of this between hydrogenic states, we have

$$\langle n'\ell'm'|\,[\hat{L}_z, \hat{z}]\,|n\ell m\rangle = 0, \tag{18.87}$$

but the left-hand side of this can be written

$$\langle n'\ell'm'|\left(\hat{L}_z\hat{z} - \hat{z}\hat{L}_z\right)|n\ell m\rangle = (m' - m)\,\hbar\langle n'\ell'm'|\hat{z}|n\ell m\rangle = 0. \tag{18.88}$$

From this we can conclude that

$$\langle n'\ell'm'|\hat{z}|n\ell m\rangle = 0 \quad \textbf{unless} \quad m' - m \equiv \Delta m = 0. \tag{18.89}$$

We can derive similar selection rules for the x and y components of the dipole operator. For reasons that will become apparent, it is convenient to consider the complex linear combinations $\hat{x} \pm i\hat{y}$:

$$[\hat{L}_z, \hat{x} \pm i\hat{y}] = [\hat{L}_z, \hat{x}] \pm i[\hat{L}_z, \hat{y}]$$

$$= [\hat{x}\hat{p}_y - \hat{y}\hat{p}_x, \hat{x}] \pm i[\hat{x}\hat{p}_y - \hat{y}\hat{p}_x, \hat{y}]$$

$$= -\hat{y}[\hat{p}_x, \hat{x}] \pm i\hat{x}[\hat{p}_y, \hat{y}]$$

$$= i\hbar\hat{y} \pm \hbar\hat{x}$$

$$= \pm\hbar(\hat{x} \pm i\hat{y}), \tag{18.90}$$

where we have made use of the basic commutation relations $[\hat{p}_x, \hat{x}] = -i\hbar$ and $[\hat{p}_y, \hat{y}] = -i\hbar$. We now take matrix elements of this operator equality:

$$\langle n'\ell'm'|[\hat{L}_z, \hat{x} \pm i\hat{y}]|n\ell m\rangle = \pm\hbar\langle n'\ell'm'|(\hat{x} \pm i\hat{y})|n\ell m\rangle. \tag{18.91}$$

The left-hand side reduces to $(m' - m)\,\hbar\,\langle n'\ell'm'|(\hat{x} \pm i\hat{y})|n\ell m\rangle$ and so we have that

$$(m' - m \mp 1)\,\hbar\,\langle n'\ell'm'|(\hat{x} \pm i\hat{y})|n\ell m\rangle = 0, \tag{18.92}$$

so we deduce that

$$\langle n'\ell'm'|(\hat{x} \pm i\hat{y})|n\ell m\rangle = 0 \quad \textbf{unless} \quad m' = m \pm 1, \tag{18.93}$$

giving the selection rule $\Delta m = \pm 1$.

Thus it can be concluded that electric dipole transitions are forbidden unless

$$\Delta m = 0, \pm 1. \tag{18.94}$$

Parity Selection Rule

Under the parity operation $\underline{r} \to -\underline{r}$, the electric dipole operator is odd. Thus, for the matrix element to be non-zero, the initial-state parity and final-state parity must be opposite:

$$(-1)^{\ell'} = -(-1)^{\ell}, \tag{18.95}$$

which implies that $\Delta\ell \equiv \ell' - \ell$ must be odd. This argument restricts the possible values of $\Delta\ell$ to $\pm 1, \pm 3, \ldots$ but doesn't tell us the complete story.

Selection Rules for the Orbital Angular Momentum

It is harder to derive selection rules for $\Delta\ell$. One approach is to use properties of the spherical harmonics, but we will give an algebraic derivation, similar to the arguments used for Δm. We start from the identity, which involves a rather lengthy proof, left as a problem at the end of this chapter:

$$[\hat{L}^2, [\hat{L}^2, \hat{z}]] = 2\hbar^2\left\{\hat{L}^2\hat{z} + \hat{z}\hat{L}^2\right\}, \tag{18.96}$$

and take matrix elements between hydrogenic states:

$$\langle n'\ell'm'|[\hat{L}^2, [\hat{L}^2, \hat{z}]]|n\ell m\rangle = 2\hbar^2\langle n'\ell'm'|\left(\hat{L}^2\hat{z} + \hat{z}\hat{L}^2\right)|n\ell m\rangle. \tag{18.97}$$

The left-hand side reduces to

$$\{\ell'(\ell'+1) - \ell(\ell+1)\}\, \hbar^2 \langle n'\ell'm'|\left(\hat{L}^2\hat{z} - \hat{z}\hat{L}^2\right)|n\ell m\rangle$$
$$= \{\ell'(\ell'+1) - \ell(\ell+1)\}^2\, \hbar^4\, \langle n'\ell'm'|\hat{z}|n\ell m\rangle, \tag{18.98}$$

whilst the right-hand side is

$$2\hbar^4\, \{\ell'(\ell'+1) + \ell(\ell+1)\}\, \langle n'\ell'm'|\hat{z}|n\ell m\rangle. \tag{18.99}$$

Thus, provided that $\langle n'\ell'm'|\hat{z}|n\ell m\rangle \neq 0$, we have

$$\ell'^2(\ell'+1)^2 - 2\ell'\ell(\ell'+1)(\ell+1) + \ell^2(\ell+1)^2 - 2\ell'(\ell'+1) - 2\ell(\ell+1) = 0, \tag{18.100}$$

which reduces after a lot of algebra to

$$(\ell'+\ell+2)(\ell'+\ell)(\ell'-\ell+1)(\ell'-\ell-1) = 0. \tag{18.101}$$

Now, since $\ell', \ell \geq 0$, the first factor is always non-zero. The second factor can only vanish if $\ell' = \ell = 0$ but in that case, we must have $m' = m = 0$ and we know from the Δm selection rule that $\langle n'00|\hat{x}|n00\rangle = \langle n'00|\hat{y}|n00\rangle = 0$. Spherical symmetry tells us therefore that $\langle n'00|\hat{z}|n00\rangle = 0$ also in this case. Alternatively, the parity argument guarantees that the dipole matrix elements vanish if $\ell' = \ell$. So we conclude that

$$(\ell'-\ell+1)(\ell'-\ell-1) = 0, \tag{18.102}$$

which implies the selection rule

$$\Delta\ell \equiv \ell' - \ell = \pm 1 \quad \text{only.} \tag{18.103}$$

18.4 Time-Dependent Perturbation Theory at Higher Order

In the previous sections a method was developed from which the first-order perturbation expression was obtained. One can iterate the perturbation expressions obtained there and systematically determine higher-order terms. However, there is a more elegant formulation that will be developed here.

18.4.1 Time Dependence in Quantum Mechanics

There are three common descriptions of time evolution utilised in quantum mechanics.

Schrödinger Picture

According to Postulate 6 in Chapter 4, the time development of a quantum system is determined by Schrödinger's equation:

$$\hat{H}|\Psi,t\rangle = i\hbar\frac{d}{dt}|\Psi,t\rangle. \tag{18.104}$$

Since $\hat{H}$ is a *linear* operator in Hilbert space, and $i\hbar\dfrac{d}{dt}$ is a linear differential operator, the time evolution of the state vector $|\Psi, t\rangle$ should also be governed by a linear operator:

$$|\Psi, t\rangle = \hat{U}(t, t_0)\,|\Psi, t_0\rangle. \tag{18.105}$$

The probability interpretation requires that the state vector be normalised at all times:

$$\langle \Psi, t_0|\Psi, t_0\rangle = 1$$
$$\text{and} \quad \langle \Psi, t|\Psi, t\rangle = \langle \Psi, t_0|\hat{U}^\dagger(t, t_0)\,\hat{U}(t, t_0)|\Psi, t_0\rangle = 1. \tag{18.106}$$

This must be true for all state vectors, therefore

$$\hat{U}^\dagger(t, t_0)\,\hat{U}(t, t_0) = \hat{1}$$
$$\text{i.e.} \qquad\qquad \hat{U}^\dagger(t, t_0) = \hat{U}^{-1}(t, t_0) = \hat{U}(t_0, t).$$

$\hat{U}$ is called the **time evolution operator in the Schrödinger picture** and is a **unitary operator**. [Recall that a *matrix* M is unitary if $M^{-1} = M^\dagger = (M^*)^T =$ the transpose of the complex conjugate of M.]

Formal Relation between $\hat{U}$ and $\hat{H}$

From the definition of $\hat{U}$ above, it follows that

$$\frac{|\Psi, t + \Delta t\rangle - |\Psi, t\rangle}{\Delta t} = \frac{\left[\hat{U}(t + \Delta t, t) - \hat{1}\right]|\Psi, t\rangle}{\Delta t}. \tag{18.107}$$

In the limit $\Delta t \to 0$:

$$\frac{d}{dt}|\Psi, t\rangle = \lim_{\Delta t \to 0} \frac{\left[\hat{U}(t + \Delta t, t) - \hat{1}\right]|\Psi, t\rangle}{\Delta t}, \tag{18.108}$$

which may be written as

$$\hat{H}(t)\,|\Psi, t\rangle = i\hbar \frac{d}{dt}|\Psi, t\rangle, \tag{18.109}$$

with

$$\hat{H}(t) \equiv i\hbar \lim_{\Delta t \to 0} \left\{\frac{\hat{U}(t + \Delta t, t) - \hat{1}}{\Delta t}\right\}. \tag{18.110}$$

Formal Solution for Conservative Systems

A **conservative system** is one for which $\hat{H}$ is independent of time t. In this case

$$\hat{H}\,|\Psi, t\rangle = i\hbar \frac{d}{dt}|\Psi, t\rangle \tag{18.111}$$

can be solved formally to give

$$|\Psi, t\rangle = \exp\left\{-i\hat{H}\,(t - t_0)/\hbar\right\}|\Psi, t_0\rangle$$
$$\text{or} \qquad \hat{U}(t, t_0) = \exp\left\{-i\hat{H}\,(t - t_0)/\hbar\right\}, \tag{18.112}$$

where the exponential of an operator, e.g. $\hat{A}$, is *defined* by its power series expansion, i.e. $\exp \hat{A} = \hat{1} + \hat{A} + \frac{1}{2} \hat{A}^2 + \cdots$.

Exercise 18.4.1   Check the solution above by explicit substitution into Eq. (18.111).

The (time-independent) Hamiltonian $\hat{H}$ possesses a complete, orthonormal set of eigenstates, $\{|n\rangle\}$, and a corresponding set of real eigenvalues, $\{E_n\}$:

$$\hat{H}|n\rangle = E_n|n\rangle,$$

so we may write $|\Psi, t_0\rangle = \sum_n c_n|n\rangle$ (using completeness),

then

$$|\Psi, t\rangle = \sum_n c_n \exp\left\{-i\hat{H}\ (t - t_0)/\hbar\right\} |n\rangle$$

$$= \sum_n c_n \exp\{-iE_n\ (t - t_0)/\hbar\} |n\rangle. \tag{18.113}$$

Special Case: $c_n = 0\ \forall n$ except $n = r$, so $c_r = 1$ (normalisation). Thus

$$|r, t\rangle = \exp\{-iE_r(t - t_0)/\hbar\} |r\rangle$$

and $\langle \hat{O} \rangle = \langle r, t|\hat{O}|r, t\rangle = \langle r|\hat{O}|r\rangle$ *independent* of t (the phases cancel).

This is the special case of a *stationary state*.

General Non-conservative Problem

If the Hamiltonian $\hat{H}$ depends on time, we may write the Schrödinger equation as

$$i\hbar \frac{d}{dt} \hat{U}(t, t_0) |\Psi, t_0\rangle = \hat{H}(t)\, \hat{U}(t, t_0) |\Psi, t_0\rangle. \tag{18.114}$$

This equation holds for all state vectors $|\Psi, t_0\rangle$, therefore

$$i\hbar \frac{d}{dt} \hat{U}(t, t_0) = \hat{H}(t)\, \hat{U}(t, t_0). \tag{18.115}$$

Integrating with respect to t from t_0 to t gives

$$i\hbar \hat{U}(t, t_0)\ -\ i\hbar \hat{U}(t_0, t_0) = \int_{t_0}^{t} \hat{H}(t')\, \hat{U}(t', t_0)\, dt', \tag{18.116}$$

where t' is a dummy integration variable. However, the *initial condition* is $\hat{U}(t_0, t_0) = \hat{1}$, therefore

$$\hat{U}(t, t_0) = \hat{1} - \frac{i}{\hbar} \int_{t_0}^{t} \hat{H}(t')\, \hat{U}(t', t_0)\, dt', \tag{18.117}$$

which is an *integral equation* for $\hat{U}$ equivalent to Schrödinger's equation.

Heisenberg Picture

In the **Schrödinger picture** we have studied thus far, the state vector $|\Psi, t\rangle$ depends on time, whilst the Hermitian operators corresponding to physical observables, e.g. $\hat{x}$ and $\hat{p}$, are independent of time. The **Heisenberg picture** is obtained by performing a time-dependent **unitary transformation** on all kets (and on bras and operators) such that state vectors become independent of time t. Recall that

$$|\Psi, t\rangle_S = \hat{U}(t, t_0) \, |\Psi, t_0\rangle_S, \tag{18.118}$$

where the subscript S means states in the Schrödinger picture. To remove the time dependence, define

$$|\Psi\rangle_H \equiv \hat{U}^\dagger(t, t_0)|\Psi, t\rangle_S = \hat{U}^\dagger(t, t_0) \, \hat{U}(t, t_0)|\Psi, t_0\rangle_S = |\Psi, t_0\rangle_S. \tag{18.119}$$

What happens to *operators*?

- The physics must be unchanged, so expectation values must remain *invariant*:

$$_H\langle\Psi|\hat{O}_H(t)|\Psi\rangle_H \equiv {}_S\langle\Psi, t|\hat{O}_S|\Psi, t\rangle_S$$

$$= {}_H\langle\Psi|\hat{U}^\dagger(t, t_0) \, \hat{O}_S \, \hat{U}(t, t_0)|\Psi\rangle_H \quad \text{(which must hold } \forall \, |\Psi\rangle_H)$$

$$\Rightarrow \qquad \hat{O}_H(t) = \hat{U}^\dagger(t, t_0) \, \hat{O}_S \, \hat{U}(t, t_0),$$

  i.e. operators are *time-dependent* in the Heisenberg picture.
- Eigenvalues of observables are preserved:

$$\hat{O}_S|\alpha\rangle_S = \alpha \, |\alpha\rangle_S \qquad \text{(multiply this on the left by } \hat{U}^\dagger(t, t_0))$$

$$\Rightarrow \qquad \hat{U}^\dagger \, \hat{O}_S \, \hat{U} \, \hat{U}^\dagger \, |\alpha\rangle_S = \alpha \, \hat{U}^\dagger \, |\alpha\rangle_S$$

$$\text{i.e.} \qquad \hat{O}_H(t)|\alpha, t\rangle_H = \alpha|\alpha, t\rangle_H,$$

$$\text{where} \qquad |\alpha, t\rangle_H \equiv \hat{U}^\dagger|\alpha\rangle_S.$$

Note: $|\alpha, t\rangle_H$ is the eigenstate of $\hat{O}_H(t)$ with eigenvalue α, and *not* the result of letting the state $|\alpha\rangle_S$ evolve for a time t. Note that its time dependence is determined by $\hat{U}^\dagger$ rather than $\hat{U}$.

The Heisenberg Equations of Motion

The dynamical equation in the Heisenberg picture governs the time dependence of *operators* (cf. the time dependence of *state vectors* in the Schrödinger picture is governed by the Schrödinger equation). Start with the definition and differentiate:

$$\hat{O}_H(t) = \hat{U}^\dagger(t, t_0) \, \hat{O}_S \, \hat{U}(t, t_0)$$

$$i\hbar \frac{d}{dt} \hat{O}_H(t) = \left\{ i\hbar \frac{d}{dt} \hat{U}^\dagger(t, t_0) \right\} \hat{O}_S \, \hat{U}(t, t_0) \; + \; \hat{U}^\dagger(t, t_0) \, \hat{O}_S \left\{ i\hbar \frac{d}{dt} \hat{U}(t, t_0) \right\}$$

$$+ \; i\hbar \, \hat{U}^\dagger(t, t_0) \left\{ \frac{\partial \hat{O}_S}{\partial t} \right\} \hat{U}(t, t_0). \tag{18.120}$$

However, noting that for any two operators $(\hat{A}\hat{B})^\dagger = \hat{B}^\dagger \hat{A}^\dagger$, and recalling

$$i\hbar \frac{d}{dt} \hat{U}(t,t_0) = \hat{H}_S \, \hat{U}(t,t_0),$$

hence $i\hbar \frac{d}{dt} \hat{U}^\dagger(t,t_0) = -\hat{U}^\dagger(t,t_0) \, \hat{H}_S,$

therefore $i\hbar \frac{d}{dt} \hat{O}_H = \hat{U}^\dagger(t,t_0)\big(\hat{O}_S \hat{H}_S - \hat{H}_S \hat{O}_S\big)\hat{U}(t,t_0) + i\hbar \, \hat{U}^\dagger(t,t_0) \left(\frac{\partial \hat{O}_S}{\partial t}\right) \hat{U}(t,t_0).$

However (dropping the labels t and t_0 for brevity):

$$\hat{U}^\dagger \big(\hat{O}_S \hat{H}_S - \hat{H}_S \hat{O}_S\big) \hat{U} = \hat{U}^\dagger \hat{O}_S \hat{U} \hat{U}^\dagger \hat{H}_S \hat{U} - \hat{U}^\dagger \hat{H}_S \hat{U} \hat{U}^\dagger \hat{O}_S \hat{U}$$

$$= \left[\hat{O}_H, \hat{H}_H\right], \tag{18.121}$$

so that, *defining* the partial derivative with respect to t of $\hat{O}_H$ by

$$\hat{U}^\dagger \frac{\partial \hat{O}_S}{\partial t} \hat{U} \equiv \frac{\partial \hat{O}_H}{\partial t}, \tag{18.122}$$

we obtain the final result

$$i\hbar \frac{d}{dt} \hat{O}_H = \left[\hat{O}_H, \hat{H}_H\right] + i\hbar \frac{\partial \hat{O}_H}{\partial t}, \tag{18.123}$$

which is the **Heisenberg equation of motion** for the operator $\hat{O}_H$. It is directly comparable with the *classical* equation of motion for the *observable* O_{cl} in Hamiltonian dynamics:

$$\frac{d}{dt} O_{cl} = \{O_{cl}, H_{cl}\}_{PB} + \frac{\partial O_{cl}}{\partial t}, \tag{18.124}$$

where $\{\}_{PB}$ is the Poisson bracket.

Constants of the Motion: If the operator is not time-dependent, then it is a *constant of the motion* if

$$i\hbar \frac{d}{dt} \hat{O}_H = 0, \tag{18.125}$$

i.e. if it commutes with the Hamiltonian:

$$\left[\hat{O}_H, \hat{H}_H\right] = 0. \tag{18.126}$$

Since the transformation between the two pictures preserves the commutation relations, it is equally true that if

$$\left[\hat{O}_S, \hat{H}_S\right] = 0, \tag{18.127}$$

then $\hat{O}_S$ represents a constant of the motion. Note that $\hat{H}_H = \hat{H}_S$. An example of an operator that is *explicitly* time-dependent is the periodic disturbance

$$\hat{V}(t) = 2\hat{H}' \cos(\omega t), \tag{18.128}$$

where $\hat{H}'$ is Hermitian and time-independent.

Interaction Picture

In the **interaction** or **Dirac picture**, the state vector is defined by the unitary transformation

$$|\Psi, t\rangle_I \equiv \hat{U}_0^\dagger(t, t_0)|\Psi, t\rangle_S = \exp(i\hat{H}_0\,(t - t_0)/\hbar)\,|\Psi, t\rangle_S, \qquad (18.129)$$

i.e. we attempt to remove the time dependence due to $\hat{H}_0$ (not $\hat{H}$) from the state vector. The requirement that

$$_I\langle \Psi, t|\hat{O}_I(t)|\Psi, t\rangle_I = {}_S\langle \Psi, t|\hat{O}_S|\Psi, t\rangle_S \qquad (18.130)$$

yields

$$\hat{O}_I(t) = \hat{U}_0^\dagger(t, t_0)\hat{O}_S\hat{U}_0(t, t_0)$$

$$= \exp\left(i\hat{H}_0\,(t - t_0)/\hbar\right)\,\hat{O}_S\,\exp\left(-i\hat{H}_0\,(t - t_0)/\hbar\right).$$

Equations of Motion in the Interaction Picture

State Vectors: From the definition Eq. (18.129) of $|\Psi, t\rangle_I$, we have

$$i\hbar\frac{d}{dt}\,|\Psi, t\rangle_I = -\hat{H}_0\,|\Psi, t\rangle_I + \exp\left(i\hat{H}_0\,(t - t_0)/\hbar\right)\,i\hbar\frac{d}{dt}\,|\Psi, t\rangle_S$$

$$= -\hat{H}_0\,|\Psi, t\rangle_I + \exp\left(i\hat{H}_0\,(t - t_0)/\hbar\right)\hat{H}_S|\Psi, t\rangle_S \quad \text{(by the Schrödinger equation)}$$

$$= -\hat{H}_0\,|\Psi, t\rangle_I + \hat{H}_I\,|\Psi, t\rangle_I \qquad \text{(by inserting } \hat{1} = \hat{U}_0\,\hat{U}_0^\dagger \text{ after the } \hat{H}_S).$$

Now trivially $\hat{H}_0 \equiv \hat{H}_{0S} = \hat{H}_{0I}$ (convince yourself), therefore

$$i\hbar\frac{d}{dt}\,|\Psi, t\rangle_I = \hat{V}_I(t)\,|\Psi, t\rangle_I. \qquad (18.131)$$

Operators:

$$\hat{O}_I(t) = \exp\left(i\hat{H}_0\,(t - t_0)/\hbar\right)\,\hat{O}_S\,\exp\left(-i\hat{H}_0\,(t - t_0)/\hbar\right).$$

Therefore $\quad i\hbar\,\dfrac{d}{dt}\,\hat{O}_I(t) = \left[\hat{O}_I(t),\,\hat{H}_0\right] + i\hbar\,\dfrac{\partial\hat{O}_I(t)}{\partial t}.$

(Derivation of this result is almost identical to that for operators in the Heisenberg picture.)

To summarise, in the interaction picture:

- Evolution of state vectors $|\Psi, t\rangle_I$ is governed by $\hat{V}_I(t)$ only.
- Evolution of operators $\hat{O}_I(t)$ is governed by $\hat{H}_0$ only.

Thus, the interaction picture is intermediate between the Schrödinger and Heisenberg pictures.

18.4.2 Time Evolution Operator in the Interaction Picture

Relation between $\hat{U}(t, t_0)$ and $\hat{U}_I(t, t_0)$

Define $\hat{U}_I(t, t_0)$ by

$$|\Psi, t\rangle_I = \hat{U}_I(t, t_0)|\Psi, t_0\rangle_I. \qquad (18.132)$$

From the definition of $|\Psi, t\rangle_I$:

$$|\Psi, t\rangle_I = \hat{U}_0^\dagger(t, t_0) |\Psi, t\rangle_S$$
$$= \hat{U}_0^\dagger(t, t_0) \hat{U}(t, t_0) |\Psi, t_0\rangle_S. \tag{18.133}$$

But, from the definition of $|\Psi, t\rangle_I$, the two pictures coincide at $t = t_0$, so that

$$|\Psi, t_0\rangle_I \equiv |\Psi, t_0\rangle_S, \tag{18.134}$$

therefore the evolution operator in the interaction picture is simply

$$\hat{U}_I(t, t_0) = \hat{U}_0^\dagger(t, t_0) \hat{U}(t, t_0)$$
$$\Leftrightarrow \qquad \hat{U}(t, t_0) = \hat{U}_0(t, t_0) \hat{U}_I(t, t_0). \tag{18.135}$$

Integral Equation for $\hat{U}_I(t, t_0)$

The equation of motion for state vectors

$$i\hbar \frac{d}{dt} |\Psi, t\rangle_I = \hat{V}_I(t) |\Psi, t\rangle_I$$

$$\Rightarrow \qquad i\hbar \frac{d}{dt} \hat{U}_I(t, t_0) |\Psi, t_0\rangle_I = \hat{V}_I(t) \hat{U}_I(t, t_0) |\Psi, t_0\rangle_I$$

$$\text{so} \qquad i\hbar \frac{d}{dt} \hat{U}_I(t, t_0) = \hat{V}_I(t) \hat{U}_I(t, t_0),$$

subject to the boundary condition $\qquad \hat{U}_I(t_0, t_0) = \hat{1}.$

We can recast this into an *integral equation* for $\hat{U}_I(t, t_0)$ by integrating (with respect to t) from t_0 to t:

$$\hat{U}_I(t, t_0) = \hat{1} - \frac{i}{\hbar} \int_{t_0}^{t} dt_1 \, \hat{V}_I(t_1) \hat{U}_I(t_1, t_0), \tag{18.136}$$

where t_1 is a dummy integration variable.

Iterative Solution: We can solve Eq. (18.136) iteratively for $\hat{U}_I(t, t_0)$ by 'substituting the equation into itself'. First, replace $\hat{U}_I(t_1, t_0)$ on the right-hand side by

$$\hat{U}_I(t_1, t_0) = \hat{1} - \frac{i}{\hbar} \int_{t_0}^{t_1} dt_2 \, \hat{V}_I(t_2) \hat{U}_I(t_2, t_0), \tag{18.137}$$

where t_2 is another dummy integration variable, to give

$$\hat{U}_I(t, t_0) = \hat{1} - \frac{i}{\hbar} \int_{t_0}^{t} dt_1 \, \hat{V}_I(t_1) + \left(-\frac{i}{\hbar}\right)^2 \int_{t_0}^{t} dt_1 \int_{t_0}^{t_1} dt_2 \, \hat{V}_I(t_1) \hat{V}_I(t_2) \hat{U}_I(t_2, t_0).$$

$$\tag{18.138}$$

By repeating this *ad infinitum* we obtain a perturbation series in $\hat{V}_I$:

Zeroth order: $\hat{U}_I(t, t_0) \approx \hat{1}$.

First order: $\hat{U}_I(t, t_0) \approx \hat{1} - \frac{i}{\hbar} \int_{t_0}^{t} \hat{V}_I(t_1) \, dt_1$.

Second order: $\hat{U}_I(t, t_0) \approx \hat{1} - \frac{i}{\hbar} \int_{t_0}^{t} \hat{V}_I(t_1) \, dt_1 + \left(-\frac{i}{\hbar}\right)^2 \int_{t_0}^{t} dt_1 \int_{t_0}^{t_1} dt_2 \, \hat{V}_I(t_1)\hat{V}_I(t_2)$.

$$(18.139)$$

By induction, the evolution operator in the interaction picture to all orders in perturbation theory is

$$\hat{U}_I(t, t_0) = \hat{1} + \sum_{n=1}^{\infty} \hat{U}_I^{(n)}(t, t_0)$$

where $\hat{U}_I^{(n)}(t, t_0) \equiv \left(-\frac{i}{\hbar}\right)^n \int_{t_0}^{t} dt_1 \int_{t_0}^{t_1} dt_2 \cdots \int_{t_0}^{t_{n-1}} dt_n \, \hat{V}_I(t_1) \cdots \hat{V}_I(t_n),$

where $t > t_1 > t_2 > \cdots > t_n > t_0$. From this we can obtain the evolution operator in the Schrödinger picture:

$$\hat{U}(t, t_0) = \hat{U}_0(t, t_0) \, \hat{U}_I(t, t_0),$$

i.e. $\hat{U}(t, t_0) = \hat{U}_0(t, t_0) + \sum_{n=1}^{\infty} \hat{U}^{(n)}(t, t_0),$

where $\hat{U}^{(n)}(t, t_0) = \left(-\frac{i}{\hbar}\right)^n \int_{t_0}^{t} dt_1 \cdots \int_{t_0}^{t_{n-1}} dt_n \, \hat{U}_0(t, t_0) \left(\hat{U}_0^{\dagger}(t_1, t_0) \, \hat{V}(t_1) \, \hat{U}_0(t_1, t_0)\right)$

$$\times \left(\hat{U}_0^{\dagger}(t_2, t_0) \, \hat{V}(t_2) \, \hat{U}_0(t_2, t_0)\right) \cdots \left(\hat{U}_0^{\dagger}(t_n, t_0) \, \hat{V}(t_n) \, \hat{U}_0(t_n, t_0)\right).$$

$$(18.140)$$

We can simplify this expression using the 'group properties'

$$\hat{U}_0(t, t_0) \, \hat{U}_0^{\dagger}(t_1, t_0) = \hat{U}_0(t, t_0) \, \hat{U}_0(t_0, t_1) \quad = \hat{U}_0(t, t_1)$$

and $\hat{U}_0(t_{i-1}, t_0) \, \hat{U}_0^{\dagger}(t_i, t_0) = \hat{U}_0(t_{i-1}, t_0) \, \hat{U}_0(t_0, t_i) = \hat{U}_0(t_{i-1}, t_i),$

$$(18.141)$$

to obtain the elegant result

$$\hat{U}^{(n)}(t, t_0) = \left(-\frac{i}{\hbar}\right)^n \int_{t_0}^{t} dt_1 \int_{t_0}^{t_1} dt_2 \cdots \int_{t_0}^{t_{n-1}} dt_n \, \hat{U}_0(t, t_1) \, \hat{V}(t_1) \, \hat{U}_0(t_1, t_2) \, \hat{V}(t_2) \cdots$$

$$\cdots \hat{U}_0(t_{n-1}, t_n) \, \hat{V}(t_n) \, \hat{U}_0(t_n, t_0), \qquad (18.142)$$

again with $t > t_1 > t_2 > \cdots > t_n > t_0$. This remarkable result allows us to write down an expression for transition probabilities to any desired order in perturbation theory.

Transition Probabilities

We wish to calculate the *transition probabilities*

$$w(r \to s) = \left|\langle s|\hat{U}(t, t_0)|r\rangle\right|^2. \qquad (18.143)$$

From the expression for $\hat{U}(t, t_0)$ above, we have

$$\langle s|\hat{U}(t, t_0)|r\rangle = \langle s|\hat{U}_0(t, t_0)|r\rangle + \sum_{n=1}^{\infty}\langle s|\hat{U}^{(n)}(t, t_0)|r\rangle, \qquad (18.144)$$

where

$$\hat{U}_0(t, t_0) = \exp\left(-i\hat{H}_0(t - t_0)/\hbar\right). \qquad (18.145)$$

Thus

$$\langle s|\hat{U}(t, t_0)|r\rangle = \exp\left(-iE_r^{(0)}(t - t_0)/\hbar\right)\delta_{sr} + \sum_{n=1}^{\infty}\langle s|\hat{U}^{(n)}(t, t_0)|r\rangle. \qquad (18.146)$$

Clearly, if $r \neq s$, then

$$\langle s|\hat{U}(t, t_0)|r\rangle = \sum_{n=1}^{\infty}\langle s|\hat{U}^{(n)}(t, t_0)|r\rangle. \qquad (18.147)$$

The *first* term in the expansion is

$$\langle s|\hat{U}^{(1)}(t, t_0)|r\rangle = -\frac{i}{\hbar}\int_{t_0}^{t} dt_1 \langle s|\hat{U}_0(t, t_1)\,\hat{V}(t_1)\,\hat{U}_0(t_1, t_0)|r\rangle$$

$$= -\frac{i}{\hbar}\int_{t_0}^{t} dt_1 \langle s|\exp\left(-iE_s^{(0)}(t - t_1)/\hbar\right)\hat{V}(t_1)\exp\left(-iE_r^{(0)}(t_1 - t_0)/\hbar\right)|r\rangle.$$

If we define $V_{sr}(t) \equiv \langle s|\hat{V}(t)|r\rangle$, we obtain the **first-order result**

$$\langle s|\hat{U}^{(1)}(t, t_0)|r\rangle = -\frac{i}{\hbar}\int_{t_0}^{t} dt_1 \exp\left(-iE_s^{(0)}(t - t_1)/\hbar\right) V_{sr}(t_1)\exp\left(-iE_r^{(0)}(t_1 - t_0)/\hbar\right).$$

$$(18.148)$$

The advantage of the new formalism is that we can immediately write down the **second-order term** in the expansion:

$$\langle s|\hat{U}^{(2)}(t, t_0)|r\rangle = \left(-\frac{i}{\hbar}\right)^2 \int_{t_0}^{t} dt_1 \int_{t_0}^{t_1} dt_2 \langle s|\hat{U}_0(t, t_1)\,\hat{V}(t_1)\,\hat{U}_0(t_1, t_2)\,\hat{V}(t_2)\,\hat{U}_0(t_2, t_0)|r\rangle$$

$$= \left(-\frac{i}{\hbar}\right)^2 \int_{t_0}^{t} dt_1 \int_{t_0}^{t_1} dt_2 \langle s|\exp\left(-iE_s^{(0)}(t - t_1)/\hbar\right)\hat{V}(t_1)$$

$$\times \exp\left(-i\hat{H}_0(t_1 - t_2)/\hbar\right)\sum_k |k\rangle\langle k|\,\hat{V}(t_2)\exp\left(-iE_r^{(0)}(t_2 - t_0)/\hbar\right)|r\rangle$$

$$= \left(-\frac{i}{\hbar}\right)^2 \sum_k \int_{t_0}^{t} dt_1 \int_{t_0}^{t_1} dt_2 \exp\left(-iE_s^{(0)}(t - t_1)/\hbar\right) V_{sk}(t_1)$$

$$\times \exp\left(-iE_k^{(0)}(t_1 - t_2)/\hbar\right) V_{kr}(t_2)\exp\left(-iE_r^{(0)}(t_2 - t_0)/\hbar\right),$$

where we inserted a complete set of energy eigenstates $\sum_k |k\rangle\langle k| = \hat{1}$ and used $\exp\left(-i\hat{H}_0(t_1 - t_2)/\hbar\right)|k\rangle = \exp\left(-iE_k^{(0)}(t_1 - t_2)/\hbar\right)|k\rangle$.

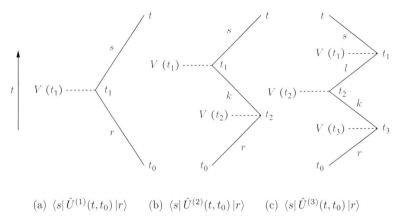

(a) $\langle s|\hat{U}^{(1)}(t,t_0)|r\rangle$ (b) $\langle s|\hat{U}^{(2)}(t,t_0)|r\rangle$ (c) $\langle s|\hat{U}^{(3)}(t,t_0)|r\rangle$

Fig. 18.3   Diagrammatic representation of the perturbation series.

Diagrammatic Representation: We can represent each term in the *perturbation series* by a diagram as shown in Fig. 18.3. From this figure we can read off a set of 'rules' to find $\langle s|\hat{U}^{(n)}|r\rangle$:

1. Draw diagram with n 'vertices' labelled $t_n, \ldots, t_1$.
2. Label 'intermediate states' $k, l, m, \ldots$.
3. For every line, associate a phase factor $\exp\left(-iE_s^{(0)}(t-t_1)/\hbar\right)$, etc.
4. For every 'vertex' a matrix element $V_{sk}(t_1)$, etc.
5. Integrate over all $t_1, \ldots, t_n$.
6. Sum over all intermediate states $k, l, m, \ldots$.
7. Multiply by $(-i/\hbar)^n$.

Thus

$$w(r \to s) = \left|\langle s|\hat{U}^{(1)}(t,t_0)|r\rangle + \langle s|\hat{U}^{(2)}(t,t_0)|r\rangle + \cdots\right|^2 \qquad (r \neq s). \qquad (18.149)$$

The above series could be described by the following words (we stop a little short of calling it a physical picture). Consider the second-order term:

1. Between t_0 and t_2, the system is in the state $|r\rangle$ and evolves according to $\hat{H}_0$.
2. At t_2 the system interacts with the perturbation, and there is an amplitude V_{kr} of converting it to state $|k\rangle$. We have to sum over all possible *intermediate* or *virtual* states $|k\rangle$ and integrate over all possible times the transition can take place.
3. Between t_2 and t_1, the system is in the state $|k\rangle$ and evolves according to $\hat{H}_0$.
4. etc.

Notes:

• The higher-order amplitudes have a similar interpretation. Evidently, the nth-order term in the perturbation series is an n-stage process of 'virtual transitions'.
• In quantum electrodynamics, electrons interact with each other via the exchange of virtual photons in a perturbation series similar to the one derived here.

First-Order Transition Probability

$$w(r \to s) \approx \left| \langle s | \hat{U}^{(1)}(t, t_0) | r \rangle \right|^2$$

$$= \frac{1}{\hbar^2} \left| \int_{t_0}^{t} dt_1 \, \exp\left(-iE_s^{(0)}(t - t_1)/\hbar \right) \, V_{sr}(t_1) \, \exp\left(-iE_r^{(0)}(t_1 - t_0)/\hbar \right) \right|^2$$

$$= \frac{1}{\hbar^2} \left| \int_{t_0}^{t} dt_1 \, \exp\left(i(E_s^{(0)} - E_r^{(0)})t_1)/\hbar \right) \, V_{sr}(t_1) \right|^2$$

i.e. $$w(r \to s) = \frac{1}{\hbar^2} \left| \int_{t_0}^{t} dt_1 \, \exp\left(i\omega_{sr}t_1 \right) \, V_{sr}(t_1) \right|^2,$$

where

$$\omega_{sr} \equiv (E_s^{(0)} - E_r^{(0)})/\hbar \qquad (18.150)$$

is the 'transition frequency' or Bohr frequency associated with the transition. This is in agreement with the result obtained in the previous sections.

Note: To first order

$$w(r \to s) = w(s \to r). \qquad (18.151)$$

For a constant perturbation that is switched on at a given time, the above expression for $w(r \to s)$ gives the same result as found using the direct perturbation expansion in Section 18.2.1.

Summary

- Time-dependent perturbation theory is a systematic treatment of transitions between quantum states when the Hamiltonian depends on time. (18.1)
- Time-dependent perturbation theory to first order is computed for the case of a perturbation that is constant in time but switched on at some particular time. (18.2.1)
- Fermi's golden rule gives the transition rate to a group of final states at first order. (18.2.2)
- Time-dependent perturbation theory to first order is computed for the case of a harmonic perturbation. (18.2.3)
- The interaction of a quantum system with an electromagnetic radiation field that is treated classically is examined and the dipole approximation is derived. (18.3)
- The Einstein A and B coefficients are derived treating absorption, stimulated emission and spontaneous emission. (18.3)
- Selection rules are based on symmetry considerations and give cases where the dipole matrix element vanishes independently of the details of the atomic wave function. (18.3.3)

- The time evolution operator evolves a quantum state over a finite interval of time. A systematic method for computing it is derived which gives time-dependent perturbation theory to all orders. (18.4)
- In the Schrödinger picture the time evolution is of the quantum state whereas in the Heisenberg picture the time evolution is in all the operators. The interaction picture is somewhere in between, with the quantum operator only evolving with respect to the unperturbed Hamiltonian. These pictures are used in deriving an all-order time-dependent perturbation theory formalism. (18.4.1)
- The transition probability is computed from the time evolution operator and is shown at first order to agree with the expression derived earlier in the chapter. (18.4.2)

Problems

18.1 A particle moving in the infinite one-dimensional square well potential

$$V(x) = 0 \quad \text{for } |x| < a, \quad V(x) = \infty \quad \text{for } |x| > a$$

is in the state described by the wave function

$$\Psi(x, 0) \equiv \psi(x) = [u_1(x) + u_2(x)]/\sqrt{2},$$

where $u_1(x)$, $u_2(x)$ are the energy eigenfunctions corresponding to the energy eigenvalues E_1 and E_2, respectively. What is the wave function at time t?

Calculate the probabilities P_1 and P_2 that at $t = 0$ a measurement of the total energy yields the results E_1 and E_2, respectively. Do P_1 and P_2 change with time?

Calculate the probabilities $P_+(t)$ and $P_-(t)$ that at time t the particle is in the intervals $0 < x < a$ and $-a < x < 0$, respectively, and try to interpret your results.

18.2 A uniform (time-dependent) electric field is suddenly applied at time $t = 0$ to a charged one-dimensional harmonic oscillator in the ground state. By writing the potential energy of the oscillator in the field in the form

$$V(x) = \tfrac{1}{2}m\omega^2(x - x_0)^2 + \text{constant},$$

show that the wave functions of the stationary states of the perturbed oscillator are $\psi_n(x - x_0)$, where $\psi_n(x)$ are the unperturbed eigenfunctions and hence determine the probabilities of transition to excited states at time $t = 0^+$, i.e. immediately after the application of the perturbation.

Hint: The interaction Hamiltonian of a particle of charge q, in a uniform electric field in one dimension, is $-qEx$. You do not need time-dependent perturbation theory for this problem. Think about the postulates of quantum mechanics.

18.3 A system has just two independent states, $|1\rangle$ and $|2\rangle$, represented by the column matrices

$$|1\rangle \longrightarrow \begin{pmatrix} 1 \\ 0 \end{pmatrix} \quad \text{and} \quad |2\rangle \longrightarrow \begin{pmatrix} 0 \\ 1 \end{pmatrix}.$$

With respect to these two states, the Hamiltonian has a time-independent matrix representation

$$\hat{H} = \begin{pmatrix} E & U \\ U & E \end{pmatrix} = EI + U\sigma_1,$$

where E and U are both real constants, I is a 2×2 unit matrix and σ_1 is a Pauli matrix.

- Show that the probability of a transition from the state $|1\rangle$ to the state $|2\rangle$ in the time interval t is given *without approximation* by

$$p(t) = \sin^2\left(\frac{Ut}{\hbar}\right).$$

  Hint: Expand the general state $|\Psi, t\rangle$ in terms of $|1\rangle$ and $|2\rangle$ and substitute in the TDSE. Note that $|1\rangle$ and $|2\rangle$ are not energy eigenstates!
- Taking the unperturbed Hamiltonian to be $\hat{H}_0 = EI$, compute the transition probability using first-order time-dependent perturbation theory and, by comparing with the exact result, deduce the conditions under which you expect the approximation to be good.
- Use the expansion of the time evolution operator in the Schrödinger picture, Eqs (18.140) and (18.142), to show that to second order in $\hat{U}$:

$$\hat{U}(t,0) = \left[1 + \frac{1}{2}\left(-\frac{i}{\hbar}Ut\right)^2 + \left(-\frac{i}{\hbar}Ut\right)\sigma_1 + O(U^2)\right]\exp\left(\frac{i}{\hbar}Et\right).$$

- Find the nth-order contribution to the time evolution operator and show that the perturbation theory can be summed to all orders to give

$$\hat{U}(t,0) = \left[\cos\left(\frac{Ut}{\hbar}\right)I - i\sin\left(\frac{Ut}{\hbar}\right)\sigma_1\right]\exp\left(\frac{i}{\hbar}Et\right).$$

- Substitute the full Hamiltonian $\hat{H}$ into the time evolution operator in the Schrödinger picture, (18.112):

$$\hat{U}(t,0) = \exp\left(-\frac{i}{\hbar}\hat{H}t\right),$$

  and sum the series that arises to show this agrees with the results from above.

18.4 The Hamiltonian which describes the interaction of a static spin-$\frac{1}{2}$ particle with an external magnetic field, $\underline{B}$, is

$$\hat{H} = -\underline{\hat{\mu}} \cdot \underline{B}.$$

When $\underline{B}$ is a static uniform magnetic field in the z-direction, $\underline{B}_0 = (0, 0, B_0)$, the matrix representation of $\hat{H}_0$ is simply

$$-\tfrac{1}{2}\gamma B_0\hbar \begin{pmatrix} 1 & 0 \\ 0 & -1 \end{pmatrix},$$

with eigenvalues $\mp\frac{1}{2}\gamma B_0\hbar$ and for this time-independent Hamiltonian, the energy eigenstates are represented by the two-component column matrices

$$|\uparrow\rangle \rightarrow \begin{pmatrix} 1 \\ 0 \end{pmatrix} \quad \text{and} \quad |\downarrow\rangle \rightarrow \begin{pmatrix} 0 \\ 1 \end{pmatrix}.$$

Now consider superimposing on the static field $\underline{B}_0$ a time-dependent magnetic field of constant magnitude B_1, rotating in the x–y plane with constant angular frequency ω:

$$\underline{B}_1(t) = (B_1\cos\omega t, B_1\sin\omega t, 0).$$

If the Hamiltonian is now written as $\hat{H}(t) = \hat{H}_0 + \hat{H}'(t)$, write down a matrix representation of $\hat{H}'(t)$.

Any spin state can be written

$$|\Psi, t\rangle = c_1(t)\exp(-iE_\uparrow t/\hbar)|\uparrow\rangle + c_2(t)\exp(-iE_\downarrow t/\hbar)|\downarrow\rangle.$$

Obtain, without approximation, the coupled equations for the amplitudes $c_1(t), c_2(t)$.

If initially at $t = 0$ the system is in the spin-down state, show that the probability that at time t the system is in the spin-up state is given without approximation by

$$p_1(t) = |c_1(t)|^2 = A\sin^2\left\{\frac{1}{2}\left[(\gamma B_1)^2 + (\omega + \gamma B_0)^2\right]^{1/2} t\right\},$$

where

$$A = \frac{(\gamma B_1)^2}{\{[(\gamma B_1)^2 + (\omega + \gamma B_0)^2]\}}.$$

What is the corresponding probability, $p_2(t)$, that the system is in the spin-down state? Sketch $p_1(t)$ and $p_2(t)$ as functions of time.

18.5 A hydrogen atom is placed in a uniform but time-dependent electric field of magnitude

$$\mathcal{E} = 0 \quad \text{for } t < 0, \qquad \mathcal{E} = \mathcal{E}_0\exp(-t/\tau) \quad \text{for } t \geq 0 \quad (\tau > 0),$$

where $\mathcal{E}_0$ is a constant. At time $t = 0$, the atom is in the ground ($1s$) state. Show to lowest order in perturbation theory that, as $t \to \infty$, the probability the atom is in the $2p$ state in which the component of the orbital angular momentum in the direction of the field is zero, is given by

$$p_{1s \to 2p} = |c(\infty)|^2 = \frac{2^{15}}{3^{10}} \frac{(e\mathcal{E}_0 a_0)^2}{(E_{2p} - E_{1s})^2 + (\hbar/\tau)^2}.$$

Hint: Take the field direction to be the z-direction. Write down the potential energy of the electron in the given field and treat as a time-dependent perturbation.

The hydrogen eigenfunctions are

$$u_{100} = (\pi a_0^3)^{-1/2} \exp(-r/a_0),$$

$$u_{211} = -\left(\pi a_0^3\right)^{-1/2} \frac{r}{8a_0} \sin\theta \exp\left(i\phi\right) \exp\left(-r/2a_0\right),$$

$$u_{210} = \left(8\pi a_0^3\right)^{-1/2} \frac{r}{2a_0} \cos\theta \exp\left(-r/2a_0\right),$$

$$u_{21-1} = \left(\pi a_0^3\right)^{-1/2} \frac{r}{8a_0} \sin\theta \exp\left(-i\phi\right) \exp\left(-r/2a_0\right),$$

and

$$\int_0^\infty \exp(-br)\, r^n\, dr = n!\,/b^{n+1}, \quad n > -1.$$

18.6 A one-dimensional harmonic oscillator of charge q is acted upon by a uniform electric field which may be considered to be a perturbation and which has time dependence of the form

$$\mathcal{E}(t) = \frac{C}{\sqrt{\pi}\,\tau} \exp\left\{-(t/\tau)^2\right\}.$$

Assuming that when $t = -\infty$, the oscillator is in its ground state, evaluate the probability that it is in its first excited state at $t = +\infty$ using time-dependent perturbation theory. You may assume that

$$\int_{-\infty}^\infty \exp(-y^2)\, dy = \sqrt{\pi},$$

$$\langle n+1|\hat{x}|n\rangle = \sqrt{\frac{(n+1)\hbar}{2m\omega}}.$$

Discuss the behaviour of the transition probability and the applicability of the perturbation theory result when (a) $\tau \ll \frac{1}{\omega}$ and (b) $\tau \gg \frac{1}{\omega}$.

18.7 The electric dipole moment operator is $\underline{D} \equiv -e\underline{r}$. The position vector can be written

$$\underline{r} = r\left\{\underline{e}_1 \sin\theta \cos\phi + \underline{e}_2 \sin\theta \sin\phi + \underline{e}_3 \cos\theta\right\},$$

where $\underline{e}_i$, $i = 1, 2, 3$, are the usual Cartesian unit vectors in the x, y, z-directions and θ, ϕ are the polar and azimuthal angles in spherical polar coordinates.

Calculate the dipole matrix elements for the radiative transition from the $2p$ states to the $1s$ state of atomic hydrogen.

Hence show that the spontaneous transition rate for the $2p \to 1s$ transition is given by

$$R_{2p\to 1s} = \left(\frac{2}{3}\right)^8 \frac{mc^2}{\hbar}\, \alpha^5,$$

where $\alpha \equiv e^2/(\hbar c)$ is the fine-structure constant.

You may assume that the initial $2p$ state is unpolarised; that is, each of the three possible values of m_ℓ is equally likely.

18.8 Show that the transition rate for spontaneous emission of a photon of energy $\hbar\omega$, wave vector $\underline{k}$ and polarisation $\underline{n}$ in the transition $i \to f$ may be written as

$$R(i \to f) = \frac{2\pi}{\hbar^2} \left| H'_{if} \right|^2 \delta(\omega + \omega_{fi}),$$

where

$$\hat{H}' = -\frac{q}{m} \left(\frac{2\pi\hbar}{\omega L^3} \right)^{1/2} \exp\left(-i\underline{k} \cdot \underline{r} \right) \underline{n} \cdot \underline{\hat{p}}.$$

Given that the number of modes of the radiation field with wave vectors within the volume element d^3k in $\underline{k}$-space is $k^2 \, dk \, d\Omega \, (L/2\pi)^3$ and that the photon has only two linearly independent polarisation states which may be represented by orthogonal unit vectors $\underline{e}^{(1)}$ and $\underline{e}^{(2)}$ perpendicular to $\underline{k}$, show that the *total* transition rate of spontaneous emission, irrespective of the polarisation of the emitted photon, is given in the dipole approximation by

$$R^{\text{tot}} = \frac{4q^2\omega^3}{3\hbar c^3} \left| \underline{r}_{if} \right|^2.$$

18.9 The probability amplitude for absorption of radiation is proportional to the matrix element

$$\langle f | \exp\left(i\underline{k} \cdot \underline{r} \right) \underline{n} \cdot \underline{\hat{p}} | i \rangle = \langle f | \underline{n} \cdot \underline{\hat{p}} | i \rangle + \langle f | \left(i\underline{k} \cdot \underline{r} \right) \underline{n} \cdot \underline{\hat{p}} | i \rangle + \cdots.$$

As we have seen, the first term on the right-hand side gives the dipole approximation.

Show that the operator in the second term can be written as

$$\left(i\underline{k} \cdot \underline{r} \right) \underline{n} \cdot \underline{\hat{p}} = i \left(\tfrac{1}{2} k_i n_j \epsilon_{ijk} \hat{L}_k + \tfrac{1}{2} k_i n_j m \frac{d}{dt} \left(x_i x_j \right) \right),$$

where $\hat{L}_k$ is the kth component of the orbital angular momentum operator. Transitions induced by these terms are called *magnetic dipole* and *electric quadrupole* transitions, respectively.

Hint: Split $\left(i\underline{k} \cdot \underline{r} \right) \underline{n} \cdot \underline{\hat{p}}$ into symmetric and antisymmetric parts, and use the Heisenberg equation of motion to obtain the second term.

18.10 Verify that

$$[\hat{L}^2, \hat{z}] = \hat{L}_x[\hat{L}_x, \hat{z}] + [\hat{L}_x, \hat{z}]\hat{L}_x + \hat{L}_y[\hat{L}_y, \hat{z}] + [\hat{L}_y, \hat{z}]\hat{L}_y.$$

Hint: Expand the commutators on the right-hand side and note that $[\hat{L}_z^2, \hat{z}] = 0$.

Show that

$$[\hat{L}_x, \hat{z}] = -i\hbar\hat{y}, \qquad [\hat{L}_y, \hat{z}] = i\hbar\hat{x}, \qquad [\hat{L}_y, \hat{x}] = -i\hbar\hat{z} \quad \text{and} \quad [\hat{L}_x, \hat{y}] = i\hbar\hat{z},$$

and hence prove that

$$[\hat{L}^2, \hat{z}] = 2i\hbar \left(\hat{L}_y\hat{x} - \hat{L}_x\hat{y} + i\hbar\hat{z} \right) = 2i\hbar(\hat{x}\hat{L}_y - \hat{L}_x\hat{y}),$$

and so on, cyclically.

Hint: Recall that $\hat{L}_x = \hat{y}\hat{p}_z - \hat{z}\hat{p}_y$ and so on, and that $[\hat{p}_z, \hat{z}] = -i\hbar$.

Hence, establish the result stated in the chapter:

$$[\hat{L}^2, [\hat{L}^2, \hat{z}]] = 2\hbar^2 \left\{ \hat{L}^2 \hat{z} + \hat{z}\hat{L}^2 \right\}.$$

18.11 Suppose that the Schrödinger picture state vector $|\Psi, t\rangle_S$ of a certain quantum system coincides with the eigenstate $|\alpha_i\rangle_S$, with eigenvalue α_i, of the Schrödinger-picture operator $\hat{A}_S$ at time $t = t_i$. Deduce that the transition amplitude, that at some later time t_f the state vector will coincide with some other eigenstate $|\alpha_f\rangle_S$ of $\hat{A}_S$, is given by the matrix element

$$_S\langle \alpha_f | \Psi, t_f \rangle_S = {}_S\langle \alpha_f | \hat{U}(t_f, t_i) | \alpha_i \rangle_S.$$

Show that this transition amplitude can be written in the Heisenberg picture as

$$_H\langle \alpha_f, t_f | \alpha_i, t_i \rangle_H,$$

where $|\alpha_i, t_i\rangle$ is the eigenvector of the Heisenberg operator $\hat{A}_H(t_i)$ at time t_i, etc.

Hence deduce that the probability amplitude for a particle, initially at position x_i at time t_i, to be found at position x_f at time t_f, is given by

$$G(x_f, t_f; x_i, t_i) \equiv {}_S\langle x_f | \hat{U}(t_f, t_i) | x_i \rangle_S = {}_H\langle x_f, t_f | x_i, t_i \rangle_H.$$

By inserting a complete set of momentum eigenstates (in either picture), show that the transition amplitude for a *free* particle is given by

$$G_0(x_f, t_f; x_i, t_i) = \left(\frac{m}{2\pi i\hbar(t_f - t_i)} \right)^{1/2} \exp\left(\frac{im(x_f - x_i)^2}{2\hbar(t_f - t_i)} \right).$$

19

Quantum Scattering Theory

One of the most common methods for studying the structure of matter and fundamental particles is to scatter particles and measure the outcome of the collision. For the collision between two particles, the problem is posed such that initially both particles are widely separated and moving towards each other but are assumed to evolve independently of any influence from each other. Then the two particles come within close proximity and interact with each other, resulting in their initial state being altered. Subsequently, the two particles in their altered form separate and some distance away this altered state is measured.

In scattering problems one is not interested in the detailed time history of the evolving system during the collision process. Rather, one measures only the initial and final states. The theory of scattering thus only needs to answer for a given initial state what are the probabilities for the various possible final states. Results already developed for time-dependent perturbation theory in Chapter 18 will be utilised for this problem.

There are two types of scattering processes, elastic and inelastic scattering. In elastic scattering the initial and final state particles are the same and only their momenta change. In inelastic scattering one or both of the initial particles participating in the collision will change in property, such as one or both particles break apart into new particles or their internal quantum state changes. Here we will focus only on elastic scattering.

19.1 Scattering Kinematics

The simplest kinematics in scattering processes is when one particle has a mass much bigger than the other, such that the heavy particle effectively is treated at rest both before and after the collision. This is called the target particle. In this case the target particle is simply regarded as a scattering centre with some potential $V(\underline{r})$ associated with it. The other, light particle, has a change in the direction of its momentum, but the magnitude of momentum, and thus its energy, remains the same. In a real scattering experiment, there is a beam containing many light particles, all of which strike some material composed of many target particles. The assumption is made that any particle in the beam will only interact with at most one target particle. In such a case, there is no interference in the scattering events between different beam particles and targets. As we will show below, the individual scattering process between a beam particle and a target can be expressed in terms of a cross-section, which is a measure of area around a target in which a beam particle will get scattered. Thus the cross-section contains all the information about the underlying dynamics which induces the scattering. Provided the density of particles in the beam and

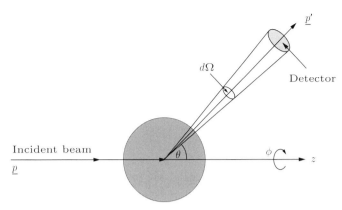

Fig. 19.1 A beam of particles of momentum $\underline{p} = \hbar \underline{k}$ is incident upon a scattering centre at $\underline{r} = 0$. The shading denotes the region where there is an interaction between the particles and the scattering centre.

the density of target particles in the material is small compared to the cross-section for interaction, this approximation for single scattering will be good.

For this case of a beam of particles, each of momentum $\underline{p} \equiv \hbar \underline{k}$, incident upon a scattering centre at the origin $\underline{r} = 0$ of polar coordinates (r, θ, ϕ), the kinematics of the scattering process is shown in Fig. 19.1. The direction of a scattered particle of momentum $\underline{p}' \equiv \hbar \underline{k}'$ is specified by angles θ and ϕ.

The scattering process is characterised by two quantities, the *incident flux* and the *scattered flux*, defined as follows:

- The **incident flux** is the number of incident particles crossing a unit area perpendicular to the beam direction per unit time.
- The **scattered flux** is the number of scattered particles scattered into the element of solid angle $d\Omega$ about the direction θ, ϕ per unit time per unit solid angle.

Note: The incident flux has dimensions $[L]^{-2}[T]^{-1}$, the scattered flux has dimensions $[T]^{-1}$.

19.1.1 The Differential Cross-Section

The differential cross-section is usually denoted $\dfrac{d\sigma}{d\Omega}$ and is defined to be the ratio of the scattered flux to the incident flux:

$$\frac{d\sigma}{d\Omega} \equiv \frac{\text{scattered flux}}{\text{incident flux}}. \tag{19.1}$$

The differential cross-section thus has dimensions of *area*: $[L]^2$.

Total Cross-Section

The total cross-section is denoted σ_T and is defined by

$$\sigma_T \equiv \int_{4\pi} \frac{d\sigma}{d\Omega}\, d\Omega = \iint \frac{d\sigma}{d\Omega} \sin\theta\, d\theta\, d\phi. \tag{19.2}$$

It is possible to address this problem by both the time-dependent and time-independent approaches and we will examine both in order.

19.2 Evolution of a Wave Packet

In quantum mechanics a particle is described through a wave packet. As such it is interesting to understand how a localised particle is described, how such a particle evolves and in what way the uncertainty principle appears. This picks up on the discussion in Section 3.10.3 and develops it further in the context of scattering. We have seen that the solution of the free-particle Schrödinger equation leads to plane waves

$$\psi(\underline{x}, t) = N \exp[\underline{k} \cdot \underline{x} - \omega(k)t], \tag{19.3}$$

with $k = |\underline{k}|$ and $\omega(k) = \hbar k^2/(2m)$. However, plane wave states are not normalisable, and in fact have equal probability to be at all points in space. The superposition property of quantum mechanics allows us to take a linear combination of plane wave states to build a spatially localised wave packet. Suppose at $t = 0$ the particle is distributed in a linear combination of plane wave states with momentum centred around $\underline{k}_c$ in the z-direction with this Gaussian amplitude:

$$g_{\underline{k}_c}(\underline{k}) = \left(\frac{2^{3/2}\pi^{1/2}}{\Delta k}\right)^{3/2} \exp\left(-\frac{k_x^2 + k_y^2 + (k_z - k_c)^2}{\Delta k^2}\right), \tag{19.4}$$

thus the k-space wave function is

$$\psi(k_x, k_y, k_z, t) = \left(\frac{2^{3/2}\pi^{1/2}}{\Delta k}\right)^{3/2} \exp\left(-\frac{k_x^2 + k_y^2 + (k_z - k_c)^2}{\Delta k^2}\right)$$
$$\times \exp\left[-i\hbar\left(\frac{k_x^2 + k_y^2 + k_z^2}{2m}t\right)\right]. \tag{19.5}$$

The corresponding position space function, where in the z-direction its origin is shifted by z_0 to account for a general initial position of the wave function, is obtained by Fourier transforming:

$$\psi(x, y, z, t) = \int \frac{dk_x dk_y dk_z}{(2\pi)^3} \exp\left[i(k_x x + k_y y + k_z(z - z_0))\right] \psi(k_x, k_y, k_z, t), \tag{19.6}$$

which gives

$$\psi(x, y, z - z_0, t) = \frac{1}{\pi^{3/4} 2^{3/4} \Delta k^{3/2}} \left(\frac{1}{\frac{1}{\Delta k^2} + \frac{i\hbar t}{2m}}\right)^{3/2} \exp\left[ik_c(z - z_0) - \frac{i\hbar k_c^2}{2m}t\right]$$
$$\times \exp\left[-\frac{x^2 + y^2 + (z - z_0 - \frac{\hbar k_c t}{m})^2}{4\left(\frac{1}{\Delta k^2} + \frac{i\hbar t}{2m}\right)}\right]. \tag{19.7}$$

We see that this wave function propagates in time with velocity $\hbar k_c/m$ in the z-direction but the width also spreads with time.

We can compute the uncertainty in position and momentum for this state. For example, in the z-direction $\Delta z = \sqrt{\langle z^2 \rangle - \langle z \rangle^2}$ with

$$\langle z \rangle = \int_{-\infty}^{\infty} dx\, dy\, dz\ \psi^*(x, y, z - z_0, t) z \psi(x, y, z - z_0, t)$$

$$= z_0 + \frac{\hbar k_c t}{m} \tag{19.8}$$

and

$$\langle z^2 \rangle = \int_{-\infty}^{\infty} dx\, dy\, dz\ \psi^*(x, y, z - z_0, t) z^2 \psi(x, y, z - z_0, t)$$

$$= \left(z_0 + \frac{\hbar k_c t}{m}\right)^2 + \Delta k^2 \left(\frac{1}{\Delta k^4} + \frac{\hbar^2 t^2}{4m^2}\right), \tag{19.9}$$

giving

$$\Delta z = \Delta k \left(\frac{1}{\Delta k^4} + \frac{\hbar^2 t^2}{4m^2}\right)^{1/2}. \tag{19.10}$$

For the momentum, we get

$$\langle \hat{p}_z \rangle = \int dx\, dy\, dz\ \psi^*(x, y, z, t) \left(-i\hbar \frac{\partial}{\partial z}\right) \psi(x, y, z, t) = \hbar k_c \tag{19.11}$$

and

$$\langle \hat{p}_z^2 \rangle = \int dx\, dy\, dz\ \psi^*(x, y, z, t) \left(-\hbar \frac{\partial^2}{\partial^2 z}\right) \psi(x, y, z, t) = \hbar^2 k_0^2 + \frac{\hbar^2}{4} \Delta k^2, \tag{19.12}$$

so that

$$\Delta p_z = \frac{\hbar}{2} \Delta k. \tag{19.13}$$

Thus

$$\Delta z \Delta p_z = \frac{\hbar \Delta k^2}{2} \left(\frac{1}{\Delta k^4} + \frac{\hbar^2 t^2}{4m^2}\right)^{1/2}, \tag{19.14}$$

which at $t = 0$ is $\hbar/2$ and at $t \to \infty$ is $\hbar^2 \Delta k^2 t/(4m)$. So at $t = 0$ the wave packet satisfies the minimal uncertainty relation and in the long time limit the uncertainty grows linearly with time. The uncertainties can also be computed in the other two directions transverse to the direction of motion of the wave packet:

$$\langle \hat{x} \rangle = \langle \hat{y} \rangle = 0, \tag{19.15}$$

$$\langle \hat{x}^2 \rangle = \langle \hat{y}^2 \rangle = \Delta k^2 \left(\frac{1}{\Delta k^4} + \frac{\hbar^2 t^2}{4m^2}\right), \tag{19.16}$$

$$\langle \hat{p}_x \rangle = \langle \hat{p}_y \rangle = 0 \tag{19.17}$$

and

$$\langle \hat{p}_x^2 \rangle = \langle \hat{p}_y^2 \rangle = \frac{\hbar^2}{4} \Delta k^2. \tag{19.18}$$

What we learn is that the uncertainty in the momentum remains the same over time, which should be the case by construction. However, the time evolution of this wave packet state indicates that the spatial width of the particle increases over time. Thus a particle prepared initially to be localised within some spatial range will, over time, become more delocalised. Based on (19.10), the wave packet has spread significantly from its initial shape once $\hbar t/2m \gg 1/\Delta k^2$.

19.3 Time-Dependent Approach to Scattering

We will first use the time-dependent approach to obtain the simple Born approximation expression for the cross-sections for elastic scattering. We will then develop the full method via the time-independent approach.

19.3.1 The Born Approximation

We can use the results of time-dependent perturbation theory to do an approximate calculation of the cross-section. Provided the interaction between particle and scattering centre is *localised* to the region around $r = 0$, we can regard the incident and scattered particles as free when they are far from the scattering centre. We just need the result that we obtained for a *constant* perturbation, Fermi's golden rule from Chapter 18, to compute the rate of transitions between the initial state (free particle of momentum $\underline{p}$) and the final state (free particle of momentum $\underline{p}'$). We write the Hamiltonian as

$$\hat{H} = \hat{H}_0 + \hat{V}(\underline{r}), \tag{19.19}$$

where $\hat{H}_0 = \hat{p}^2/2m$ is the kinetic energy operator, and treat the potential energy operator, $\hat{V}(\underline{r})$, as the perturbation which induces transitions between the eigenstates of $\hat{H}_0$, which are plane waves.

In a realistic scattering experiment the incoming particle will be represented as a wave packet. If we suppose the scattering centre is located at the origin $\underline{r} = 0$ and the particle crosses this scattering centre at $t = 0$, then the state of the particle at time t is the wave packet

$$\psi_{\underline{k}_c}(\underline{r}, t) = \frac{1}{(2\pi)^3} \int g_{\underline{k}_c}(\underline{k}) \exp[i\underline{k} \cdot (\underline{r} - \underline{r}_i) - i\omega(\underline{k})(t - t_i)] d^3k, \tag{19.20}$$

where $\omega(\underline{k}) = \hbar k^2/(2m)$ and $g_{\underline{k}_c}(\underline{k})$ is a smooth function, peaked around some particular momentum $\underline{p}_c = \hbar \underline{k}_c$, with some narrow width Δk. In position space therefore $\psi_{\underline{k}_c}(\underline{r}, t)$ is a wave packet with a width around $\underline{r} = \underline{r}_i$ at time $t = t_i < 0$. Both $\underline{r}_i$ and $\underline{k}_c$ are assumed to be along a line but in opposite directions. To fix a definite picture, suppose along the line defined by both $\underline{k}_c$ and $\underline{r}_i$, the scattering centre is at the origin $\underline{r} = \underline{0}$ and $\underline{r}_i$ is to the left of the scattering centre. If we consider computing a matrix element between an initial state $\psi_{\underline{k}_c}(\underline{r}, t_i)$ to some final state of momentum $\underline{k}'$, the initial wave packet state is dominated by the plane wave states around wave number $\underline{k}_c$. Thus, to a good approximation, the scattering of the incoming wave packet state to an outgoing state can be computed just by looking at the transition between the respective incoming and outgoing plane wave states.

If we label the initial and final plane wave states by their respective wave vectors we have, for the rate of transitions (hereafter we will call $\underline{k}_c$ simply $\underline{k}$):

$$R = \frac{2\pi}{\hbar} |\langle \underline{k}'|\hat{V}|\underline{k}\rangle|^2 \, \rho(E_{k'}), \qquad (19.21)$$

where $\rho(E_{k'})$ is the density of final states. $\rho(E_{k'})dE_{k'}$ is the number of final states with energy in the range $E_{k'} \to E_{k'} + dE_{k'}$. The quantity $\langle \underline{k}'|\hat{V}|\underline{k}\rangle$ is just the matrix element of the perturbation and is usually abbreviated thus:

$$V_{\underline{k}'\underline{k}} \equiv \langle \underline{k}'|\hat{V}|\underline{k}\rangle = \int u_{\underline{k}'}^*(\underline{r}) \, V(\underline{r}) \, u_{\underline{k}}(\underline{r}) d^3 r. \qquad (19.22)$$

Technical Aside

The plane wave states have wave functions of the form

$$u_{\underline{k}}(\underline{r}) = C \exp(i\underline{k} \cdot \underline{r}) = C \exp(i\underline{p} \cdot \underline{r}/\hbar), \qquad (19.23)$$

with C a normalisation constant. Because plane wave states are not properly normalisable, we employ the trick of normalising them in a large but finite box, for example a cube of side L with periodic boundary conditions, and then taking the limit $L \to \infty$ at the end of the calculation. Thus we require that

$$\int_{\text{box}} u_{\underline{k}}^*(\underline{r}) u_{\underline{k}}(\underline{r}) d^3 r = |C|^2 \int_{\text{box}} d^3 r = |C|^2 L^3 = 1, \qquad (19.24)$$

giving for the normalised eigenfunctions

$$u_{\underline{k}}(\underline{r}) = L^{-3/2} \exp(i\underline{k} \cdot \underline{r}). \qquad (19.25)$$

Of course, enclosing the system in a finite box has the consequence that the allowed momentum eigenvalues are no longer continuous but *discrete*. With periodic boundary conditions

$$u(-L/2, \, y, \, z) = u(+L/2, \, y, \, z), \quad \text{etc.} \qquad (19.26)$$

the momentum eigenvalues are forced to be of the form

$$\underline{p} \equiv \hbar\underline{k} = \frac{2\pi\hbar}{L}(n_x, n_y, n_z), \quad \text{with } n_x, \, n_y, \, n_z = 0, \pm 1, \pm 2, \cdots. \qquad (19.27)$$

For sufficiently large L, we can approximate the continuous spectrum arbitrarily close.

Density of Final States

Any possible final-state wave vector, $\underline{k}'$, corresponds to a point in k-space or *wave vector space* with coordinates (k'_x, k'_y, k'_z). The points form a cubic lattice with lattice spacing $2\pi/L$. Thus the volume of k'-space per lattice point is $(2\pi/L)^3$, and the number of states in a volume element d^3k' is given by

$$\left(\frac{L}{2\pi}\right)^3 d^3k' \;=\; \left(\frac{L}{2\pi}\right)^3 k'^2 dk'\, d\Omega, \tag{19.28}$$

where we have expressed the volume element in spherical polar coordinates.

We require $\rho(E_{k'})$, the density of states per unit energy, where

$$E_{k'} = \frac{\hbar^2 k'^2}{2m} \tag{19.29}$$

is the energy corresponding to wave vector $\underline{k}'$. Now, the wave vectors in the range $\underline{k}' \to \underline{k}' + d\underline{k}'$ correspond to the energy range $E_{k'} \to E_{k'} + dE_{k'}$, so that

$$\rho(E_{k'})dE_{k'} \;=\; \left(\frac{L}{2\pi}\right)^3 k'^2 dk'\, d\Omega \tag{19.30}$$

is the number of states with energy in the desired interval and with wave vector, $\underline{k}'$, pointing into the solid angle $d\Omega$ about the direction (θ, ϕ). Noting that

$$dE_{k'} \;=\; \frac{\hbar^2 k'}{m} dk' \tag{19.31}$$

yields the final result for the density of states:

$$\rho(E_{k'}) \;=\; \frac{L^3\, mk'}{8\pi^3 \hbar^2} d\Omega. \tag{19.32}$$

Incident Flux

The box normalisation corresponds to one particle per volume L^3, so that the number of particles at velocity $\underline{v}$ crossing a unit area perpendicular to the beam per unit time as shown in Fig. 19.2 is just given by the magnitude of the incident velocity divided by L^3:

$$\text{incident flux} \;=\; \frac{|\underline{v}|}{L^3} \;=\; \frac{\hbar k}{mL^3}. \tag{19.33}$$

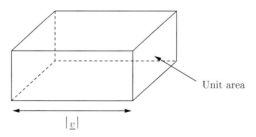

Unit area

$|\underline{v}|$

Fig. 19.2   Flux of particles at velocity $\underline{v}$ crossing a unit area perpendicular to the beam.

Scattered Flux

Using the golden rule, we have that the rate of transition between the initial state of wave vector $\underline{k}$ and the final states whose wave vectors $\underline{k}'$ lie in the element of solid angle $d\Omega$ about the direction (θ, ϕ) of the wave vector $\underline{k}'$ is given by

$$R = \frac{2\pi}{\hbar} |V_{\underline{k}'\underline{k}}|^2 \frac{L^3}{8\pi^3} \frac{mk'}{\hbar^2} \, d\Omega, \tag{19.34}$$

but this is just the number of particles scattered into $d\Omega$ per unit time. To get the scattered flux we simply divide by $d\Omega$, which then gives the number of particles per unit time per unit solid angle.

The Differential Cross-Section for Elastic Scattering

We now have all the ingredients, the scattered flux and the incident flux, to compute the cross-section:

$$\frac{d\sigma}{d\Omega} \equiv \frac{\text{scattered flux}}{\text{incident flux}} = \frac{mL^3}{\hbar k} \frac{2\pi}{\hbar} |V_{\underline{k}'\underline{k}}|^2 \frac{L^3}{8\pi^3} \frac{mk'}{\hbar^2}. \tag{19.35}$$

For a real potential, energy conservation tells us that we have **elastic scattering**, i.e. $k = k'$, and we obtain finally the **Born approximation** for the differential cross-section:

$$\frac{d\sigma}{d\Omega} = \frac{m^2}{4\pi^2\hbar^4} L^6 \left| \langle \underline{k}'|\hat{V}|\underline{k} \rangle \right|^2, \tag{19.36}$$

where the matrix element $V_{\underline{k}'\underline{k}} \equiv \langle \underline{k}'|\hat{V}|\underline{k} \rangle$ is given by

$$\langle \underline{k}'|\hat{V}|\underline{k} \rangle = \frac{1}{L^3} \int V(\underline{r}) \exp\left(-i\underline{q} \cdot \underline{r}\right) d^3r, \tag{19.37}$$

with $\underline{q} \equiv \underline{k}' - \underline{k}$ the so-called **wave vector transfer**. Thus the required matrix element in the Born approximation is just the three-dimensional Fourier transform of the potential energy function. Observe that the final result for the differential cross-section is independent of the box size, L, which we used to normalise the plane wave states.

Scattering by Central Potentials

For a central potential, $V(\underline{r}) \equiv V(r)$ is spherically symmetric, and some simplifications can be made. We can partially evaluate the scattering matrix element irrespective of the particular functional form of $V(r)$ by working in polar coordinates (Θ, Φ), *which refer to the wave vector transfer $\underline{q}$ as polar axis*, so that the scalar product $\underline{q} \cdot \underline{r}$ becomes $qr \cos\Theta$:

$$\int V(r) \exp\left(-i\underline{q} \cdot \underline{r}\right) d^3r = \int_0^{2\pi} \int_0^{\pi} \int_0^{\infty} V(r) \exp\left(-iqr \cos\Theta\right) r^2 \, dr \, \sin\Theta \, d\Theta \, d\Phi$$

$$= 2\pi \int_{-1}^{+1} \int_0^{\infty} V(r) \exp\left(-iqr \cos\Theta\right) r^2 \, dr \, d(\cos\Theta)$$

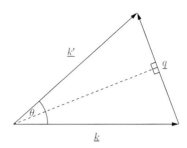

Fig. 19.3 The geometry of elastic scattering: $|\underline{k}'| = |\underline{k}|$.

$$= 2\pi \int_0^\infty V(r) \left[\frac{\exp(-iqr\cos\Theta)}{-iqr} \right]_{-1}^{+1} r^2 \, dr$$

$$= 2\pi \int_0^\infty V(r) \left[\frac{\exp(-iqr) - \exp(iqr)}{-iqr} \right] r^2 dr$$

$$= \frac{4\pi}{q} \int_0^\infty r\, V(r) \, \sin qr\, dr. \tag{19.38}$$

The Born approximation for the differential cross-section then becomes

$$\frac{d\sigma}{d\Omega} = \frac{4m^2}{\hbar^4 q^2} \left| \int_0^\infty r\, V(r) \, \sin(qr) dr \right|^2, \tag{19.39}$$

which is independent of ϕ but depends on the scattering angle, θ, through q. A little simple trigonometry, based on Fig. 19.3, shows that

$$q = 2k \, \sin\frac{\theta}{2} \qquad \text{since } k = k'. \tag{19.40}$$

Note that here, θ is the usual polar angle with respect to the direction of the incident particle, $\underline{k}$, not with respect to the wave vector transfer $\underline{q}$.

Example 19.1 The screened Coulomb potential: the scattering of an electron by an atom is well represented by the potential energy

$$V(r) = -\frac{Ze^2}{r} \, \exp(-\beta r), \tag{19.41}$$

where the so-called **screening factor**, $\exp(-\beta r)$, models the screening effect of the atomic electrons; when the projectile electron is far from the atom it 'sees' a reduced nuclear charge. Only when it penetrates close to the nucleus does it feel the full effect of the charge Ze on the nucleus.

Evaluating the integral which appears in the expression for the cross-section (19.39):

$$
\int_0^\infty V(r)\, r \sin qr\, dr = -Ze^2 \int_0^\infty \exp(-\beta r)\, \sin qr\, dr
$$

$$
= -Ze^2 \int_0^\infty \frac{1}{2i}\left[\exp\left[(iq-\beta)r\right] - \exp\left[-(iq+\beta)r\right]\right] dr
$$

$$
= -Ze^2 \frac{1}{2i}\left[\frac{\exp(iq-\beta)r}{(iq-\beta)} + \frac{\exp -(iq+\beta)r}{(iq+\beta)}\right]_0^\infty
$$

$$
= -Ze^2 \frac{q}{\beta^2 + q^2}. \tag{19.42}
$$

Substituting into the expression for the Born cross-section yields

$$
\frac{d\sigma}{d\Omega} = (Ze^2)^2 \frac{4m^2}{\hbar^4 q^2} \frac{q^2}{(\beta^2+q^2)^2} = (Ze^2)^2 \frac{4m^2}{\hbar^4} \frac{1}{(\beta^2+q^2)^2}, \tag{19.43}
$$

where

$$
q = 2k\sin(\theta/2) = \frac{2mv}{\hbar}\sin(\theta/2). \tag{19.44}
$$

We can now obtain the cross-section for scattering by the Coulomb potential with no screening by taking the limit $\beta \to 0$. Note that the screening factor in the previous calculation is essential to make the integral convergent. We obtain

$$
\frac{d\sigma}{d\Omega} = (Ze^2)^2 \frac{4m^2}{\hbar^4 q^4} = \left(\frac{Ze^2}{2mv^2}\right)^2 \frac{1}{\sin^4\theta/2}. \tag{19.45}
$$

You may recognise this formula as the **Rutherford scattering cross-section**. It just so happens that the result of the classical calculation of the Coulomb scattering cross-section agrees precisely with the quantum-mechanical result in the Born approximation. So Rutherford's back-of-the-envelope calculation, which was the foundation stone for the nuclear picture of the atom, was in some sense a lucky fluke.

19.3.2 Two-Body Scattering

In our analysis so far, we have assumed that a beam of particles is scattered by a fixed scattering centre, with the interaction represented by a potential energy term, $V(\underline{r})$. This is a good approximation in many problems of physical interest, such as electron–atom scattering, where we can regard the target particle as infinitely heavy, but is not always the case. Let us consider a two-particle system whose mutual potential energy depends only on their relative separation, so that the Hamiltonian is

$$
\hat{H} = -\frac{\hbar^2}{2m_1}\nabla_1^2 - \frac{\hbar^2}{2m_2}\nabla_2^2 + V(\underline{r}_1 - \underline{r}_2). \tag{19.46}
$$

We can define the usual centre of mass and relative position vectors

$$
\underline{R} = \frac{m_1\underline{r}_1 + m_2\underline{r}_2}{(m_1+m_2)} \qquad \text{and} \qquad \underline{r} = \underline{r}_1 - \underline{r}_2, \tag{19.47}
$$

respectively. Equivalently, we may write

$$\underline{r}_1 = \underline{R} + \frac{m_2}{(m_1 + m_2)}\,\underline{r} \quad \text{and} \quad \underline{r}_2 = \underline{R} - \frac{m_1}{(m_1 + m_2)}\,\underline{r}. \tag{19.48}$$

By expressing the gradient operators

$$\underline{\nabla}_1 = \left(\frac{\partial}{\partial x_1}, \frac{\partial}{\partial y_1}, \frac{\partial}{\partial z_1}\right) \quad \text{and} \quad \underline{\nabla}_2 = \left(\frac{\partial}{\partial x_2}, \frac{\partial}{\partial y_2}, \frac{\partial}{\partial z_2}\right) \tag{19.49}$$

in terms of the operators

$$\underline{\nabla}_R = \left(\frac{\partial}{\partial X}, \frac{\partial}{\partial Y}, \frac{\partial}{\partial Z}\right) \quad \text{and} \quad \underline{\nabla} = \left(\frac{\partial}{\partial x}, \frac{\partial}{\partial y}, \frac{\partial}{\partial z}\right) \tag{19.50}$$

by, for example, writing

$$\frac{\partial}{\partial x_1} = \frac{\partial X}{\partial x_1}\frac{\partial}{\partial X} + \frac{\partial x}{\partial x_1}\frac{\partial}{\partial x} = \frac{m_1}{(m_1 + m_2)}\frac{\partial}{\partial X} + \frac{\partial}{\partial x} \tag{19.51}$$

and so on, it is possible to cast the Hamiltonian in the form

$$\hat{H} = -\frac{\hbar^2}{2M}\nabla_R^2 - \frac{\hbar^2}{2\mu}\nabla^2 + V(\underline{r}), \tag{19.52}$$

where $M = (m_1 + m_2)$ and $\mu = m_1 m_2/(m_1 + m_2)$ is the reduced mass. We see that $\hat{H}$ is the sum of two commuting terms:

$$\hat{H} = \hat{H}_{\text{CM}} + \hat{H}_{\text{rel}} \quad \text{where} \quad \hat{H}_{\text{CM}} = -\frac{\hbar^2}{2M}\nabla_R^2. \tag{19.53}$$

$\hat{H}_{\text{CM}}$ just describes the free motion (kinetic energy) of the centre of mass. In the CM frame, the centre of mass is at rest, and the Hamiltonian simply reduces to

$$\hat{H}_{\text{rel}} = -\frac{\hbar^2}{2\mu}\nabla^2 + V(\underline{r}), \tag{19.54}$$

which is *identical in form to the Hamiltonian of a single particle moving in the fixed potential, $V(\underline{r})$*. Thus the CM cross-section for two-body scattering can be obtained immediately from the solution to the problem of a single particle of mass μ scattering from a fixed potential.

19.4 Time-Independent Approach to Scattering

We will now develop a more thorough formalism for treating scattering via the time-independent approach.

19.4.1 Relation of Wave Packet to Time-Independent Approach

The scattering problem looks a little bit different from typical problems in quantum mechanics where one solves for eigenstates and eigenvalues of the given system. Physically, scattering is about wave packets that approach each other, collide and then states going off in new directions. In the previous section this physical picture was implemented with some handwaving assumptions to arrive at an expression for the probability amplitude of the scattered particle. We now want to understand how this problem should be properly posed starting from eigenstates of the Hamiltonian $\hat{H}$ and show how they relate to the scattering amplitude in which one is ultimately interested. The discussion will again be restricted to the case of a particle scattering off a potential. The first claim, which will be shown to be correct in Section 19.4.3, is that the eigenstates of the Hamiltonian at large distances $\underline{r}$ have the asymptotic form

$$\psi_{\underline{k}}^{+}(\underline{r}) \sim \exp(i\underline{k} \cdot \underline{r}) + f_{\underline{k}}(\hat{r})\frac{\exp(ikr)}{r}, \tag{19.55}$$

so that

$$\hat{H}\psi_{\underline{k}}^{+} = E\psi_{\underline{k}}^{+}. \tag{19.56}$$

These eigenstates have two terms, the first is a plane wave and the second is called the outgoing spherical wave.

The second claim, also shown in Section 19.4.3, is that the particle undergoing scattering can be expanded as a linear combination of these eigenstates (19.55), with the expansion coefficients exactly the same as for just the first plane wave terms without the outgoing components. Thus, rather than the expansion in (19.20) in terms of plane wave states, one can now expand the state in terms of eigenstates of the full Hamiltonian with exactly the same expansion coefficients as in (19.20), to give

$$\psi_{\underline{k}_0}(\underline{r},t) = \frac{1}{(2\pi)^3} \int g_{\underline{k}_c}(\underline{k}) \exp[-i(\underline{k} \cdot \underline{r}_i + \omega(k)(t - t_i))]\psi_{\underline{k}}^{+}(\underline{r})d^3k. \tag{19.57}$$

In other words the outgoing components of (19.55) in (19.57) do not alter the behaviour of the incoming wave packet. These two claims together establish the connection between eigenstates of the Hamiltonian and scattering, with the function $f_{\underline{k}}(\hat{r})$ in (19.55) precisely the scattering amplitude. The details of these claims will now be established in what follows.

19.4.2 The Scattering Amplitude

Suppose the region where there is an interaction between the particles and the scattering potential is *finite*. Then the wave function at large distances from the scattering centre can be written as the sum of an *incident plane wave* and a *scattered* wave with a *spherical envelope*:

$$\psi(r, \theta, \phi) \overset{r \to \infty}{\to} \exp(ikz) + f(\theta, \phi) \frac{\exp(ikr)}{r} \equiv \psi_{\text{in}} + \psi_{\text{scatt}}. \tag{19.58}$$

This satisfies the time-independent Schrödinger equation for large r. The quantity $f(\theta, \phi)$ is called the **scattering amplitude**.

Probability Current Density: If we denote Schrödinger's equation

$$-\frac{\hbar^2}{2m}\nabla^2\Psi(\underline{r},t) + V(\underline{r})\,\Psi(\underline{r},t) = i\hbar\frac{\partial\Psi(\underline{r},t)}{\partial t} \tag{19.59}$$

by (SE) and its complex conjugate by (SE)*, then by taking the combination

$$\Psi^*\,(\text{SE}) \,-\, \Psi\,(\text{SE})^* \tag{19.60}$$

we get a continuity equation

$$\frac{\partial}{\partial t}\rho \,+\, \underline{\nabla}\cdot\underline{j} \,=\, 0, \tag{19.61}$$

where

$$\rho \,=\, \Psi^*\,\Psi \qquad \text{and} \qquad \underline{j} \,=\, -\frac{i\hbar}{2m}\left(\Psi^*\underline{\nabla}\Psi \,-\, \Psi\underline{\nabla}\Psi^*\right). \tag{19.62}$$

Integrating over a volume V, bounded by a closed surface S, and using the divergence theorem, we obtain

$$\frac{\partial}{\partial t}\int_V \rho\,d^3r \,=\, -\int_V \underline{\nabla}\cdot\underline{j}\,d^3r \,=\, -\int_S \underline{j}\cdot d\underline{S}. \tag{19.63}$$

Thus we may identify ρ as the **probability density** and $\underline{j}$ as the **probability current density**.

Incident Flux: This is obtained from the probability current density – the number of incident particles crossing a unit area in unit time. As $z \to -\infty$, $\Psi_{in}(\underline{r},t) \to \psi_{in}$ $\exp(-iEt/\hbar)$, so

$$\underline{j}_{in}(\underline{r},t) = -\frac{i\hbar}{2m}\left(\Psi_{in}^*\underline{\nabla}\Psi_{in} - \Psi_{in}\underline{\nabla}\Psi_{in}^*\right) \,=\, \frac{\hbar k}{m}\,\underline{e}_3,$$

i.e. $\qquad \left|\underline{j}_{in}(\underline{r},t)\right| = \frac{\hbar k}{m} \,=\, \left|\underline{v}\right|. \tag{19.64}$

Outgoing or Scattered Flux: At large r (i.e. $r >$ size of incoming wave packets), we have, in spherical polars:

$$\underline{\nabla} = \underline{e}_r\frac{\partial}{\partial r} \,+\, \underline{e}_\theta\frac{1}{r}\frac{\partial}{\partial\theta} \,+\, \underline{e}_\phi\frac{1}{r\sin\theta}\frac{\partial}{\partial\phi},$$

so $\quad \underline{j}_{scatt}(\underline{r},t) = -\frac{i\hbar}{2m}\left(\Psi_{scatt}^*\underline{\nabla}\Psi_{scatt} - \Psi_{scatt}\underline{\nabla}\Psi_{scatt}^*\right)$

$$= -\frac{i\hbar}{2m}\left(\Psi_{scatt}^*\frac{\partial\Psi_{scatt}}{\partial r} - \Psi_{scatt}\frac{\partial\Psi_{scatt}^*}{\partial r}\right)\underline{e}_r \,+\, O(1/r^3)$$

$$= -\frac{i\hbar}{2m}\left|f(\theta,\phi)\right|^2 \times \frac{2ik}{r^2}\underline{e}_r \,+\, O(1/r^3),$$

i.e. $\qquad \left|\underline{j}_{scatt}(\underline{r},t)\right| \overset{r\to\infty}{\to} \frac{\hbar k}{m}\frac{\left|f(\theta,\phi)\right|^2}{r^2}. \tag{19.65}$

This is the number of scattered particles crossing a unit area per unit time.

For an element of area dA at a distance r from the scattering centre, the element of solid angle $d\Omega$ subtended by dA is given by $dA = r^2 d\Omega$.

- Therefore the number of particles scattered into $d\Omega$ per unit time is

$$\left| \underline{j}_{\text{scatt}}(\underline{r},t) \right| r^2 d\Omega \;=\; \frac{\hbar k}{m} \left| f(\theta,\,\phi) \right|^2 d\Omega. \tag{19.66}$$

- Hence the scattered flux we desire is simply $\dfrac{\hbar k}{m} \left| f(\theta,\,\phi) \right|^2$.

Thus we obtain the important result that the differential cross-section is given by

$$\frac{d\sigma}{d\Omega} \;=\; \left| f(\theta,\,\phi) \right|^2. \tag{19.67}$$

19.4.3 Green Function Method for Calculating $f(\theta,\,\phi)$

We first rewrite the (time-independent) Schrödinger equation as

$$\left(\nabla_r^2 + k^2 \right) \psi(\underline{r}) \;=\; U(\underline{r})\,\psi(\underline{r}), \tag{19.68}$$

where $k^2 \equiv (2mE)/\hbar^2$ and $U(\underline{r}) \equiv (2m\,V(\underline{r}))/\hbar^2$. Our task is to solve this equation subject to the **boundary condition**

$$\psi(\underline{r}) \stackrel{r\to\infty}{\to} \exp(ikz) \;+\; f(\theta,\,\phi)\,\frac{\exp(ikr)}{r}. \tag{19.69}$$

To do this, we first define a **Green function** $G_k(\underline{r},\underline{r}')$ for the differential operator $\left(\nabla^2 + k^2 \right)$:

$$\left(\nabla_r^2 + k^2 \right) G_k(\underline{r},\underline{r}') \equiv \delta^{(3)}(\underline{r} - \underline{r}'). \tag{19.70}$$

Multiply both sides of this equation by $U(\underline{r}')\,\psi(\underline{r}')$ and integrate with respect to $\underline{r}'$:

$$\left(\nabla_r^2 + k^2 \right) \int G_k(\underline{r},\underline{r}')\,U(\underline{r}')\,\psi(\underline{r}')\,d^3r' = \int \delta^{(3)}(\underline{r} - \underline{r}')\,U(\underline{r}')\,\psi(\underline{r}')\,d^3r'$$
$$= U(\underline{r})\,\psi(\underline{r}). \tag{19.71}$$

Comparing with Eq. (19.68), we see that a *formal* solution is

$$\psi(\underline{r}) \;=\; \int G_k(\underline{r},\underline{r}')\,U(\underline{r}')\,\psi(\underline{r}')\,d^3r'. \tag{19.72}$$

To this, we may add any solution of the *homogeneous* equation

$$\left(\nabla_r^2 + k^2 \right) h(\underline{r}) \;=\; 0. \tag{19.73}$$

In particular

$$\left(\nabla_r^2 + k^2 \right) \exp(ikz) \;=\; 0. \tag{19.74}$$

So the appropriate general solution that fits the boundary conditions is

$$\psi(\underline{r}) \;=\; \exp(ikz) \;+\; \int G_k(\underline{r},\underline{r}')\,U(\underline{r}')\,\psi(\underline{r}')\,d^3r', \tag{19.75}$$

which is an **integral equation** for $\psi(\underline{r})$.

What is $G_k(\underline{r}, \underline{r}')$?

Consider the slightly simpler problem (i.e. set $\underline{r}' = 0$)

$$\left(\nabla_r^2 + k^2\right) G_k(\underline{r}) = \delta^{(3)}(\underline{r}). \tag{19.76}$$

Define a Fourier transform as

$$G_k(\underline{r}) = \frac{1}{(2\pi)^3} \int d_k(\underline{\tilde{k}}) \exp(i\underline{\tilde{k}} \cdot \underline{r}) \, d^3\tilde{k},$$

and recall that $\quad \delta^{(3)}(\underline{r}) = \frac{1}{(2\pi)^3} \int \exp(i\underline{\tilde{k}} \cdot \underline{r}) \, d^3\tilde{k}.$

Then

$$\left(\nabla_r^2 + k^2\right) G_k(\underline{r}) - \delta^{(3)}(\underline{r}) = \frac{1}{(2\pi)^3} \int \left[\left(-\tilde{k}^2 + k^2\right) d_k(\underline{\tilde{k}}) - 1\right] \exp(i\underline{\tilde{k}} \cdot \underline{r}) \, d^3\tilde{k} \equiv 0, \tag{19.77}$$

which holds $\forall \underline{r}$, therefore

$$d_k(\underline{\tilde{k}}) = \frac{1}{k^2 - \tilde{k}^2}$$

and $\quad G_k(\underline{r}) = \frac{1}{(2\pi)^3} \int \exp(i\underline{\tilde{k}} \cdot \underline{r}) \frac{d^3\tilde{k}}{k^2 - \tilde{k}^2}.$

The evaluation of this integral is rather technical. However, the angular part is quite straightforward. In spherical polars (taking $\underline{r}$ parallel to $\underline{e}_3'$, i.e. *not* the same $\underline{e}_3$ used above in the context of $\underline{j}_{in}(\underline{r}, t)$):

$$\underline{\tilde{k}} \cdot \underline{r} = \tilde{k} r \cos\theta \qquad d^3\tilde{k} = \tilde{k}^2 \, d\tilde{k} \, \sin\theta \, d\theta \, d\phi. \tag{19.78}$$

Therefore

$$\begin{aligned}
G_k(\underline{r}) &= \frac{1}{(2\pi)^2} \int_{-1}^{+1} d(\cos\theta) \int_0^\infty \frac{\tilde{k}^2 \, d\tilde{k}}{k^2 - \tilde{k}^2} \exp(i\tilde{k}r\cos\theta) \\
&= \frac{1}{(2\pi)^2} \int_0^\infty \frac{\tilde{k}^2 \, d\tilde{k}}{k^2 - \tilde{k}^2} \left[\frac{\exp(i\tilde{k}r\cos\theta)}{i\tilde{k}r}\right]_{-1}^{+1} \\
&= -\frac{i}{4\pi^2 r} \int_0^\infty \frac{\tilde{k} \, d\tilde{k}}{k^2 - \tilde{k}^2} \left[\exp(i\tilde{k}r) - \exp(-i\tilde{k}r)\right] \\
&= -\frac{i}{4\pi^2 r} \int_{-\infty}^\infty \frac{\tilde{k} \, d\tilde{k}}{k^2 - \tilde{k}^2} \exp(i\tilde{k}r),
\end{aligned} \tag{19.79}$$

which is not well-defined because of the poles at $\tilde{k} = \pm k$.

Consider the following well-defined function, where $\tilde{k}$ is in general complex and ϵ is real and positive:

$$G_k^+(\underline{r}) \equiv \lim_{\epsilon \to 0} \frac{-i}{4\pi^2 r} \oint_C \frac{\tilde{k} \, d\tilde{k}}{(k^2 + i\epsilon) - \tilde{k}^2} \exp(i\tilde{k}r). \tag{19.80}$$

The integrand has poles as shown in Fig. 19.4 at $\tilde{k} = \pm\sqrt{k^2 + i\epsilon} \approx \pm(k + i\epsilon/2k)$.

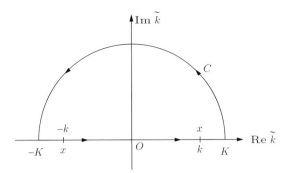

The complex plane displaying the poles of the function (19.80).

If we choose C to be the closed semi-circular contour shown, then

$$\oint_C = \int_{-K}^{+K} + \int_{\text{semi-circle}} . \qquad (19.81)$$

On the semi-circle we have $\text{Im } \tilde{k} > 0$, therefore $\exp(i\tilde{k}r) \to 0$ as $K \to \infty$, and hence $\oint_C = \int_{-K}^{+K}$. Cauchy's residue theorem tells us that:

$$\oint_C = 2\pi i \times \sum (\text{residues of poles enclosed by } C) \qquad (19.82)$$

$$= 2\pi i \left[\frac{-\tilde{k} \, \exp(i\tilde{k}r)}{\sqrt{k^2 + i\epsilon} + \tilde{k}} \right]_{\tilde{k} \,=\, \sqrt{k^2 + i\epsilon}}$$

$$= -\pi i \exp \left(i\sqrt{(k^2 + i\epsilon)}r \right). \qquad (19.83)$$

Thus

$$G_k^+(\underline{r}) = -\frac{1}{4\pi} \frac{\exp(ikr)}{r}. \qquad (19.84)$$

By choosing the opposite sign for the $i\epsilon$ term, we could similarly define

$$G_k^-(\underline{r}) \equiv \lim_{\epsilon \to 0} \frac{-i}{4\pi^2 r} \oint_C \frac{\tilde{k} \, d\tilde{k}}{(k^2 - i\epsilon) - \tilde{k}^2} \exp(i\tilde{k}r) = -\frac{1}{4\pi} \frac{\exp(-ikr)}{r}. \qquad (19.85)$$

Then replacing r by $|\underline{r} - \underline{r}'|$ we have

$$G_k^\pm(\underline{r}, \underline{r}') = G_k^\pm(|\underline{r} - \underline{r}'|) = -\frac{1}{4\pi} \frac{\exp(\pm ik|\underline{r} - \underline{r}'|)}{|\underline{r} - \underline{r}'|}, \qquad (19.86)$$

and the corresponding 'solutions' to the Schrödinger equation are

$$\psi^\pm(\underline{r}) = \exp(ikz) - \frac{1}{4\pi} \int \frac{\exp(\pm ik|\underline{r} - \underline{r}'|)}{|\underline{r} - \underline{r}'|} U(\underline{r}') \, \psi^\pm(\underline{r}') \, d^3 r'. \qquad (19.87)$$

This integral equation for ψ may be used as the basis for an iterative solution to the scattering problem.

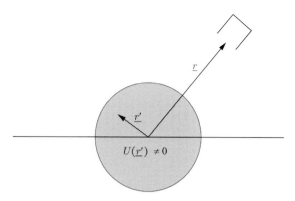

Fig. 19.5   The scattering centre and the region around it.

Asymptotic Behaviour – Fitting the Boundary Conditions

When $r \gg r'$ in Fig. 19.5, we can write

$$\left|\underline{r} - \underline{r}'\right| = \left\{r^2 - 2\underline{r} \cdot \underline{r}' + r'^2\right\}^{1/2} = r\left\{1 - \frac{2\underline{r} \cdot \underline{r}'}{r^2} + \frac{r'^2}{r^2}\right\}^{1/2}$$

$$= r - \frac{\underline{r} \cdot \underline{r}'}{r} + O(r^{-1}).$$

Now define $\hat{r} = \dfrac{\underline{r}}{r}$, so

$$\frac{\exp(\pm ik|\underline{r} - \underline{r}'|)}{|\underline{r} - \underline{r}'|} \overset{r \to \infty}{=} \frac{\exp(\pm ikr)}{r}\exp(\mp ik\underline{r}' \cdot \hat{r}) + \text{higher-order terms}$$

and

$$\psi^{\pm}(\underline{r}) \overset{r \to \infty}{=} \exp(ikz) - \frac{\exp(\pm ikr)}{4\pi r}\int \exp(\mp ik\underline{r}' \cdot \hat{r}) U(\underline{r}')\psi^{\pm}(\underline{r}')\, d^3r',$$

from which we deduce that the required Green function is clearly $G_k^+(\underline{r},\underline{r}')$. Hence

$$\psi^{+}(\underline{r}) \overset{r \to \infty}{=} \exp(ikz) - \frac{\exp(ikr)}{4\pi r}\int \exp(-ik\underline{r}' \cdot \hat{r}) U(\underline{r}')\,\psi^{+}(\underline{r}')\, d^3r', \qquad (19.88)$$

from which we may immediately identify the scattering amplitude

$$f(\theta,\phi) = -\frac{1}{4\pi}\int \exp(-ik\underline{r}' \cdot \hat{r}) U(\underline{r}')\,\psi^{+}(\underline{r}')\, d^3r'. \qquad (19.89)$$

Note: Physically, the Green function G^- describes propagation *towards* the scattering centre and therefore does not satisfy the required boundary conditions.

Wave-Packet Scattering

Returning to the discussion at the start of this section on the relation between eigenstates of the Hamiltonian and wave packet states in scattering, (19.88) is the asymptotic limit of the scattering eigenstates of the Hamiltonian, and is of the form of the first claim Eq. (19.55).

We will now prove the second claim that the expansion of the wave packet state in terms of plane waves in (19.20) has the same expansion when using the energy eigenstates associated with (19.88). In other words, the outgoing wave components of (19.88) have no contribution to the initial incoming wave packet state that approaches the scattering centre. For this it is useful to use the exact expression for the outgoing component, the second term in (19.87), for the + or outgoing wave, and not just the asymptotic behaviour. We write the integral for the wave packet convolution

$$\psi_{scatt}(\underline{r},t) \sim \frac{1}{(2\pi)^3} \int d^3k \; g_{\underline{k}_c}(\underline{k}) \exp(-i(\underline{k}\cdot\underline{r}_i + i\omega(k)(t-t_i)))[\exp(i\underline{k}\cdot\underline{r}) + \psi_{\underline{k}}^+(\underline{r})],$$

(19.90)

where

$$\psi_{\underline{k}}^+(\underline{r}) = -\frac{1}{4\pi} \int \frac{\exp(\pm ik|\underline{r}-\underline{r}'|)}{|\underline{r}-\underline{r}'|} U(\underline{r}') \, \psi_{\underline{k}}^+(\underline{r}') \, d^3r'$$

(19.91)

and $g_{\underline{k}_c}(\underline{k})$ is given in (19.4). Thus the outgoing part of the wave function will be

$$\psi^+(\underline{r},t) \sim \int \frac{dk_x dk_y dk_z}{(2\pi)^3} \exp\left(\frac{k_x^2 + k_y^2 + (k_z - k_c)^2}{\Delta k^2}\right) \exp(-i\underline{k}\cdot\underline{r}_i)$$

$$\times \exp\left(-i\hbar\frac{k_x^2 + k_y^2 + k_z^2}{2m}(t - t_i)\right) \int d^3r' \frac{\exp(\pm ik|\underline{r}-\underline{r}'|)}{|\underline{r}-\underline{r}'|} U(\underline{r}') \, \psi_{\underline{k}}^+(\underline{r}').$$

If the wave packet spread is well below the extent of the region where the interaction potential is significant, and provided there are no narrow resonances for which the scattering amplitude varies considerably with $\underline{k}$, then it can be assumed that $\psi_{\underline{k}}^+(\underline{r}')$ varies slowly over the dominant region of the above integral, so it can be factored out of this integral. Accounting for that and now extracting from the above just the terms involved in the k-integration gives

$$\sim \int \frac{dk_x dk_y dk_z}{(2\pi)^3} \exp\left(\frac{k_x^2 + k_y^2 + (k_z - k_c)^2}{\Delta k^2}\right) \exp(-i\underline{k}\cdot\underline{r}_i)$$

$$\times \exp\left(-i\hbar\frac{k_x^2 + k_y^2 + k_z^2}{2m}(t - t_i)\right) \frac{\exp(\pm ik|\underline{r}-\underline{r}'|)}{|\underline{r}-\underline{r}'|},$$

(19.92)

which we will show is negligible for time $t = t_i$ when the wave packet is peaked at $\underline{r}_i$ up to the time when the wave packet begins to overlap with the potential around the origin. Doing first the k_x and k_y integrals, they are peaked around 0 to give just the z-direction integration

$$\sim \int dk_z \exp\left[-\frac{(k_z - k_c)^2}{\Delta k^2} - i(k_z r_{iz} - k_z|\underline{r}-\underline{r}'| + \frac{\hbar k_z^2(t - t_i)}{2m})\right]$$

$$\sim \exp\left[-\frac{(r_{iz} + \hbar k_c(t - t_i)/m - |\underline{r}-\underline{r}'|)^2}{\left(\frac{1}{\Delta k^2} + \frac{i\hbar(t-t_i)}{2m}\right)}\right],$$

(19.93)

where the last line follows from doing the Gaussian integration. We see at $t = t_i$ (since the wave packet is configured to start to the left of the origin initially), that we have

$\sim \exp(-\Delta k^2(|r_{iz}| + |\underline{r} - \underline{r}'|)^2) \ll 1$, since r_{iz} is a large distance along the negative z-axis away from the origin where the potential is. The time when the wave packet approaches the potential around the origin is when $-\hbar k_c t_i/m \approx |r_{iz}|$ (recall $t_i < 0$ and the wave packet approaches the origin at $t = 0$) so that $r_{0z} - \hbar k_c t_i/m \lesssim a$, where a is the characteristic extent of the scattering potential. Moreover, in order for the wave packet to remain localised, it means its spread $1/\Delta k \lesssim a$ and that it remains so up to the time it reaches the scattering potential, so $1/\Delta k^2 - i\hbar t_i/2m \approx 1/\Delta k^2$. For time t beyond this point, (19.93) is not suppressed, which is expected since once the interaction with the potential becomes significant, the outgoing component is expected to develop. However, for time t before then $\Delta k^2(r_{0z} + \hbar k_c(t - t_i)/m - |\underline{r} - \underline{r}'|)^2 \gg 1$ and so the Gaussian in (19.93) remains suppressed. This shows that the outgoing component of the scattering eigenstate plays a negligible role in the evolution of this state and only in the other incoming component is it important up to the point the incoming wave packet begins to overlap with the scattering potential.

Born Again

As a first approximation, assume that the scattered wave is negligible compared to the incident plane wave and approximate $\psi^+(\underline{r}')$ by $\exp(ikz')$ in the integral on the right-hand side of Eq. (19.89). Then

$$f_B(\theta, \phi) = -\frac{1}{4\pi} \int \exp(ikz' - ik\underline{r}' \cdot \hat{r}) \, U(\underline{r}') \, d^3r'. \tag{19.94}$$

For a real potential we have energy conservation, so as shown in Fig. 19.6 $|\underline{k}| = |\underline{k}'|$, thus $\underline{k} = k \, \underline{e}_3$ and $\underline{k}' = k' \hat{r} = k \hat{r}$ and we have **elastic scattering**. Therefore

$$kz' = \underline{k} \cdot \underline{r}' \quad \text{and} \quad k\underline{r}' \cdot \hat{r} = \underline{k}' \cdot \underline{r}'. \tag{19.95}$$

Thus

$$f_B(\theta, \phi) = -\frac{1}{4\pi} \int \exp(i(\underline{k} - \underline{k}') \cdot \underline{r}') \, U(\underline{r}') \, d^3r' = -\frac{1}{4\pi} \int \exp(i\underline{q} \cdot \underline{r}') \, U(\underline{r}') \, d^3r', \tag{19.96}$$

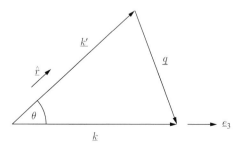

Fig. 19.6   The wave vectors for the initial and final state in elastic scattering.

where $q \equiv \underline{k} - \underline{k}'$, which is the **first Born approximation** to the scattering amplitude. For a *central* potential, we can write $U(\underline{r}') = U(r')$ and perform the angular integral as before to obtain

$$f_B(\theta) = -\frac{1}{q} \int_0^\infty r \sin(qr) U(r) \, dr \quad \text{with} \quad U(r) = \frac{2mV(r)}{\hbar^2}, \tag{19.97}$$

giving for the differential cross-section

$$\left(\frac{d\sigma}{d\Omega} \right)^{Born} \equiv |f_B(\theta)|^2 = \frac{4m^2}{\hbar^4 q^2} \left| \int_0^\infty r \, \sin(qr) V(r) \, dr \right|^2, \tag{19.98}$$

in agreement with the result obtained in (19.39). Exploration of the range of validity of the Born approximation is left as a problem at the end of this chapter.

Centre of Mass Cross-Section: The above result holds for a single particle scattering off a *potential* $V(r)$. For two-particle scattering one must replace the particle mass by the reduced mass μ, i.e.

$$m \to \mu = \frac{m_1 m_2}{(m_1 + m_2)}. \tag{19.99}$$

The Born Series – a Systematic Expansion in the Scattering Potential $U(r)$

The power of the Green function formulation is illustrated by the ease with which it generates higher-order approximations to the scattering amplitude. We have previously shown that

$$\psi^+(\underline{r}) = \exp(i\underline{k} \cdot \underline{r}) + \int G_k(\underline{r}, \underline{r}') U(\underline{r}') \psi^+(\underline{r}') \, d^3 r', \tag{19.100}$$

which is an *integral equation* for $\psi^+(\underline{r})$. The *first* term of an iterative solution for $f(\theta, \phi)$ was obtained by substituting $\exp(i\underline{k} \cdot \underline{r}')$ for $\psi^+(\underline{r}')$ on the right-hand side of Eq. (19.89). We can systematically obtain higher-order terms by first trivially rewriting Eq. (19.100) as

$$= \exp(i\underline{k} \cdot \underline{r}') + \int G_k(\underline{r}', \underline{r}'') U(\underline{r}'') \psi^+(\underline{r}'') \, d^3 r''. \tag{19.101}$$

Substituting this into the right-hand side of Eq. (19.89) (and using Eq. (19.95)) we get

$$f(\theta, \phi) = -\frac{1}{4\pi} \int \exp(-i\underline{k}' \cdot \underline{r}') U(\underline{r}') \exp(i\underline{k} \cdot \underline{r}') d^3 r'$$

$$- \frac{1}{4\pi} \int \int \exp(-i\underline{k}' \cdot \underline{r}') U(\underline{r}') G_k(\underline{r}', \underline{r}'') U(\underline{r}'') \psi^+(\underline{r}'') d^3 r'' d^3 r'. \tag{19.102}$$

The **second Born approximation** is obtained by the approximation $\psi^+(\underline{r}'') \approx \exp(i\underline{k} \cdot \underline{r}'')$ in the last line. We may obtain the third Born approximation by substituting the exact result for $\psi^+(\underline{r}'')$ in the right-hand side of Eq. (19.102). And so on …

Clearly, higher-order terms involve higher powers of the scattering potential $U(r)$. In this way we generate a perturbation series, the **Born series**, which (is not just a bunch of movies) involves a sequence of virtual scattering events – in direct analogy to the results

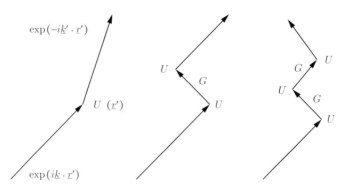

Fig. 19.7 Representative diagrams from the Born series, showing a sequence of virtual scattering events.

obtained in time-dependent perturbation theory in Chapter 18. Each term in the series can be represented by a diagram similar to those sketched in Fig. 19.7.

As before, one can read off a set of rules for obtaining the scattering amplitude from diagrams: each *external line* is associated with a free particle wave function, e.g. $\exp(\pm ik \cdot r'')$, each *internal line* with a 'propagator', e.g. $G_k(r', r'')$, and each *vertex* with a virtual scattering event or interaction with $U(r')$. Finally, we must integrate over all possible positions where each interaction can 'occur'.

Note: We have employed a variant of the delta-function normalisation for the incident plane wave: namely, $C \exp(ikz)$ with $C = 1$. The same result for $f(\theta, \phi)$ is obtained whatever choice is made for C.

19.4.4 Partial Wave Analysis

The Born series is most useful at *high energies*, i.e. when the energy, E, of the particles in the incoming beam is much greater than the scattering potential $V(r)$, so a perturbation series in $V(r)$ converges rapidly. Partial wave analysis is in principle exact at all energies, but is most useful at *low energies*, so it is complementary to the Born series.

General Solution of the Time-Independent Schrödinger Equation

We will restrict our analysis to **spherically symmetric potentials**, $V(r)$, for which we may write the general solution of the TISE as

$$\psi(r) = \sum_{l=0}^{\infty} \sum_{m=-l}^{l} R_l(r)\, Y_{lm}(\theta, \phi), \tag{19.103}$$

where $Y_{lm}(\theta, \phi)$ are the usual spherical harmonics, and the functions $R_l(r)$ satisfy the radial equation

$$\left[\frac{d^2}{dr^2} + \frac{2}{r}\frac{d}{dr} - \frac{l(l+1)}{r^2} + k^2 \right] R_l(r) = U(r)\, R_l(r), \tag{19.104}$$

with $k^2 \equiv (2mE)/\hbar^2$ and $U(r) \equiv (2m\,V(r))/\hbar^2$ as usual.

Short-Range Potential

In this case, there exists a region $r > r_0$ where $U(r) = 0$, so $R_l(r)$ satisfies

$$\left[\frac{d^2}{dr^2} + \frac{2}{r} \frac{d}{dr} - \frac{l(l+1)}{r^2} + k^2 \right] R_l(r) = 0. \qquad (19.105)$$

The general solution of this equation is

$$R_l(r) = A_l\, j_l(kr) - B_l\, n_l(kr) \qquad (r > r_0), \qquad (19.106)$$

where A_l and B_l are constants, and $j_l(kr)$ and $n_l(kr)$ are the spherical Bessel functions and Neumann functions of order l, respectively. Fortunately, we don't need many detailed properties of these functions; we require only their behaviour for large kr and small kr.

Free-Particle Solution, $U(r) = 0$

For the case where the potential is zero everywhere, $U(r) = 0\ \forall r$, we must have $B_l = 0$ because the Neumann functions are singular at $r = 0$. Note that we don't need to impose $B_l = 0$ for the solution for the short-range potential in the region $r > r_0$, because this region does not include the origin. Therefore the quantities B_l/A_l will provide us with information about the scattering amplitude.

A free particle propagating with momentum $\hbar \underline{k}$ is represented by

$$\psi(\underline{r}) = \exp(ikz) = \exp(ikr\cos\theta) \qquad (19.107)$$

in spherical polars, with the z-axis chosen parallel to $\underline{k}$. We may expand $\psi(r)$ in terms of spherical harmonics as in Eq. (19.103). Since there is no dependence on ϕ, and $Y_{lm}(\theta, \phi) \propto \exp(im\phi)$, only terms having $m = 0$ will contribute to the sum:

$$\exp(ikz) = \sum_{l=0}^{\infty} A_l\, Y_{l0}(\theta, \phi)\, j_l(kr)$$

$$= \sum_{l=0}^{\infty} a_l\, P_l(\cos\theta)\, j_l(kr), \qquad (19.108)$$

where $P_l(\cos\theta)$ are the usual Legendre polynomials.

We may determine the constants a_l by projection in the usual way using the orthogonality properties of the Legendre polynomials:

$$\int_{-1}^{1} P_l(x)\, P_{l'}(x)\, dx = \frac{2}{2l+1}\, \delta_{ll'}, \qquad (19.109)$$

which gives

$$\int_{-1}^{1} P_l(\cos\theta)\, \exp(ikr\cos\theta)\, d(\cos\theta) = a_l\, \frac{2}{2l+1}\, j_l(kr). \qquad (19.110)$$

The left-hand side of Eq. (19.110) can be integrated by parts to give, at large r:

$$\left[\frac{1}{ikr}\exp(ikr\cos\theta)\,P_l(\cos\theta)\right]_{-1}^{+1} - \int_{-1}^{+1}\frac{1}{ikr}\exp(ikr\cos\theta)\,P_l'(\cos\theta)\,d(\cos\theta)$$

$$= \frac{1}{ikr}\left[\exp(ikr) - (-1)^l\exp(-ikr)\right] + O(1/r^2), \qquad (19.111)$$

where we used $P_l(1) = 1$ and $P_l(-1) = (-1)^l$. On the right-hand side we use the asymptotic behaviour of the Bessel functions

$$j_l(kr) \overset{r\to\infty}{\to} \frac{1}{kr}\sin(kr - l\pi/2). \qquad (19.112)$$

Comparing sides as $r \to \infty$, we get

$$\text{left-hand side} \overset{r\to\infty}{\to} \frac{1}{ikr}\left[\exp(ikr) - (-1)^l\exp(-ikr)\right]$$

$$\text{right-hand side} \overset{r\to\infty}{\to} \frac{1}{kr}\left[\sin(kr - l\pi/2)\right]\frac{2a_l}{2l+1}$$

$$= \frac{1}{ikr}\left[\exp(ikr)(-i)^l - (i)^l\exp(-ikr)\right]\frac{a_l}{2l+1},$$

where we used $\exp(\pm i\pi l/2) = (\exp(\pm i\pi/2))^l = (\pm i)^l$. Therefore

$$a_l = (2l+1)\,i^l, \qquad (19.113)$$

and so

$$\exp(ikz) = \exp(ikr\cos\theta) = \sum_{l=0}^{\infty}(2l+1)\,i^l\,P_l(\cos\theta)\,j_l(kr). \qquad (19.114)$$

This is called the **partial wave expansion** of a plane wave. It's the decomposition of a free-particle state of momentum $\hbar k$ into a sum of eigenstates of the angular-momentum-squared operator $\hat{L}^2$ having $L_z = m\hbar = 0$.

Solution with $U(r) \neq 0$ Everywhere

The solutions must have the same *angular* dependence as the free-particle solution because the potential $V(r)$ is independent of θ and ϕ by assumption. Thus we may write

$$\psi(\underline{r}) = \sum_{l=0}^{\infty}(2l+1)\,i^l\,P_l(\cos\theta)\,R_l(r). \qquad (19.115)$$

For $r > r_0$, where $U(r) = 0$, performing a partial wave expansion, we have from Eq. (19.106):

$$\psi(\underline{r}) = \sum_{l=0}^{\infty} (2l + 1)\, i^l\, P_l(\cos\theta)\, [\, A_l\, j_l(kr) - B_l\, n_l(kr)\,]$$

$$\overset{r\to\infty}{\to} \sum_{l=0}^{\infty} (2l + 1)\, i^l\, P_l(\cos\theta)\, \frac{1}{kr}\, [\, A_l\, \sin(kr - l\pi/2) + B_l\, \cos(kr - l\pi/2)\,]$$

$$\overset{r\to\infty}{\to} \sum_{l=0}^{\infty} (2l + 1)\, i^l\, P_l(\cos\theta)\, \frac{1}{kr}\, C_l\, \sin(kr - l\pi/2 + \delta_l), \tag{19.116}$$

where $C_l \cos\delta_l = A_l$ and $C_l \sin\delta_l = B_l$, hence $\tan\delta_l = B_l/A_l$. The quantity δ_l is called a **phase shift**.

Fitting the Solution to the Asymptotic Boundary Conditions

Recall that the asymptotic solution of the TISE is

$$\psi(\underline{r}) \overset{r\to\infty}{\to} \exp(ikz) + f(\theta, \phi)\, \frac{\exp(ikr)}{r} \equiv \psi_{\text{in}} + \psi_{\text{scatt}}. \tag{19.117}$$

By substituting Eq. (19.114) and Eq. (19.112) for $\exp(ikz)$, and Eq. (19.116) for $\psi(\underline{r})$, into Eq. (19.117), and recalling that $f(\theta, \phi)$ is independent of ϕ for a central potential, we may identify

$$f(\theta)\, \frac{\exp(ikr)}{r} = \sum_{l=0}^{\infty} (2l + 1)\, i^l\, P_l(\cos\theta)\, \frac{1}{kr}\, C_l\, \sin(kr - l\pi/2 + \delta_l)$$

$$- \sum_{l=0}^{\infty} (2l + 1)\, i^l\, P_l(\cos\theta)\, \frac{1}{kr}\, \sin(kr - l\pi/2). \tag{19.118}$$

We can now identify the unknown constants C_l and δ_l by equating coefficients of $\exp(\pm ikr)/r$:

$$\frac{e^{-ikr}}{r} : \sum_{l=0}^{\infty} (2l + 1)i^l P_l(\cos\theta)\frac{1}{2ik}\left[C_l \exp(il\pi/2)\exp(-i\delta_l) - \exp(il\pi/2)\right] \quad = 0,$$

$$\frac{e^{ikr}}{r} : \sum_{l=0}^{\infty} (2l + 1)i^l P_l(\cos\theta)\frac{1}{2ik}\left[C_l \exp(-il\pi/2)\exp(i\delta_l) - \exp(-il\pi/2)\right] \quad = f(\theta).$$

The first of these results gives

$$C_l = \exp(i\delta_l), \tag{19.119}$$

whilst the second gives, on noting that $i^l \exp(-il\pi/2) = i^l\,(-i)^l = 1$:

$$f(\theta) = \frac{1}{2ik} \sum_{l=0}^{\infty} (2l + 1)\, P_l(\cos\theta)\, \left(e^{2i\delta_l} - 1\right)$$

or $\quad f(\theta) = \frac{1}{k} \sum_{l=0}^{\infty} (2l + 1)\, \exp(i\delta_l)\, \sin\delta_l\, P_l(\cos\theta). \tag{19.120}$

The δ_l are called **phase shifts**, and Eq. (19.120) is known as the partial wave expansion of the scattering amplitude.

Total Cross-Section

Before trying to interpret the physical meaning of the phase shifts, let us obtain the (much simpler!) result for the total cross-section σ_T. Recall that

$$\frac{d\sigma}{d\Omega} = |f(\theta)|^2 \qquad \text{and} \qquad \sigma_T = \int \frac{d\sigma}{d\Omega}\, d\Omega. \qquad (19.121)$$

Therefore

$$\frac{d\sigma}{d\Omega} = \frac{1}{k^2} \left\{ \sum_{l=0}^{\infty} (2l+1)\, \exp(i\delta_l)\, \sin\delta_l\, P_l(\cos\theta) \right\}$$

$$\times \left\{ \sum_{l'=0}^{\infty} (2l'+1)\, \exp(-i\delta_{l'})\, \sin\delta_{l'}\, P_{l'}(\cos\theta) \right\}$$

$$\text{and} \quad \sigma_T = 2\pi \int_{-1}^{+1} \frac{d\sigma}{d\Omega}\, d(\cos\theta).$$

But the orthogonality relations

$$\int_{-1}^{1} P_l(\cos\theta)\, P_{l'}(\cos\theta)\, d(\cos\theta) = \frac{2}{2l+1}\, \delta_{ll'} \qquad (19.122)$$

tell us that all the cross terms in the sums vanish, and thus

$$\sigma_T = \frac{4\pi}{k^2} \sum_{l=0}^{\infty} (2l+1)\, \sin^2\delta_l = \sum_{l=0}^{\infty} \sigma_l, \qquad (19.123)$$

where the quantities σ_l are known as **partial wave cross-sections**.

The Optical Theorem

Note that since

$$f(\theta) = \frac{1}{k} \sum_{l=0}^{\infty} (2l+1)\, \exp(i\delta_l)\, \sin\delta_l\, P_l(\cos\theta) \qquad (19.124)$$

and $P_l(1) = 1$, we have

$$f(\theta = 0) = \frac{1}{k} \sum_{l=0}^{\infty} (2l+1)\, \exp(i\delta_l)\, \sin\delta_l. \qquad (19.125)$$

Taking the imaginary part of this, we obtain

$$\text{Im}\, f(0) = \frac{1}{k} \sum_{l=0}^{\infty} (2l+1)\, \sin^2\delta_l = \frac{k}{4\pi}\, \sigma_T. \qquad (19.126)$$

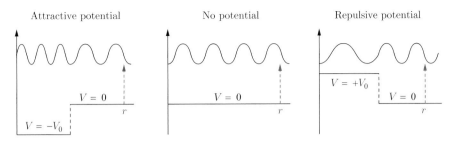

Fig. 19.8   The diagrams show radial wave functions, $u_l(r) = r R_l(r)$. The attractive potential pulls in the wave giving positive δ_l, while the repulsive potential pushes out the wave for negative δ_l.

This very important result, which relates the **total cross-section** σ_T to the *forward part of the scattering amplitude* $f(0)$, is called the **optical theorem**.

What Do the Phase-Shifts δ_l Represent?

Making a change of variable $u_l(r) = r R_l(r)$ in the radial equation gives

$$\left[\frac{d^2}{dr^2} - \frac{l(l+1)}{r^2} + k^2 - U(r) \right] u_l(r) = 0. \tag{19.127}$$

If we require that $R_l(r)$ is regular (well-behaved) at the origin, we obtain the boundary condition $u_l(0) = 0$. From the previous section:

when $U(r) \equiv 0$: $u_l(r) \quad \propto \quad r\, j_l(kr) \stackrel{r \to \infty}{\to} \sin(kr - l\pi/2)$

$$\stackrel{r \to 0}{\to} r^{l+1} \qquad\qquad (\text{since } j_l(r) \stackrel{r \to 0}{\to} r^l) \, ;$$

when $U(r) \neq 0$: $u_l(r) \stackrel{r \to \infty}{\to} \sin(kr - l\pi/2 + \delta_l)$

$$\stackrel{r \to 0}{\to} r^{l+1}.$$

For an **attractive potential**, $\left| k^2 - U(r) \right| > k^2$. Therefore the magnitude of $d^2 u_l/dr^2$ is *increased* for a given l, so $u_l(r)$ oscillates more rapidly, thus $u_l(r)$ is *pulled in* towards the origin, as shown in the first plot in Fig. 19.8. Hence, the phase shift δ_l is *positive*.

For a **repulsive potential**, $\left| k^2 - U(r) \right| < k^2$. Therefore the magnitude of $d^2 u_l/dr^2$ is *decreased* for a given l, hence $u_l(r)$ is *pushed out* away from the origin, as shown in the last plot of Fig. 19.8. Hence, the phase shift δ_l is *negative*.

Applications

The partial wave expansion is useful because, at low energies, the angular momentum quantum number l is bounded, so that only a few partial waves contribute. A semiclassical argument for this is as follows. Consider an incoming particle with speed v in the direction

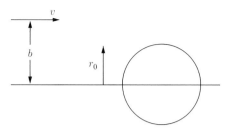

Fig. 19.9 An incoming particle at impact parameter b from the centre of the potential.

of the beam, a distance b (the impact parameter) from the parallel axis as shown in Fig. 19.9, which passes through the centre of the short distance potential with range r_0.

The angular momentum of the particle about the beam axis is

$$L = mvb = m\frac{\hbar k}{m}b = \hbar k b = l\hbar \quad \Rightarrow \quad l = kb. \tag{19.128}$$

For the case when $b > r_0$, we expect very little scattering, so the maximum value of l we need to consider is

$$l_{max} \approx k r_0. \tag{19.129}$$

Example 19.2 The square-well potential: the potential, shown in Fig. 19.10, is

$$V(r) = \begin{cases} -V_0 & r < r_0, \\ 0 & r > r_0. \end{cases} \tag{19.130}$$

- **For $r < r_0$:**

$$\left[\frac{d^2}{dr^2} + \frac{2}{r}\frac{d}{dr} - \frac{l(l+1)}{r^2} + k_1^2\right] R_l(r) = 0,$$

where $\qquad k_1^2 = k^2 + \frac{2mV_0}{\hbar^2} \equiv \frac{2m}{\hbar^2}(E + V_0).$

The regular solution at $r = 0$ is

$$R_l(r) = a_l\, j_l(k_1 r). \tag{19.131}$$

- **For $r > r_0$:**

$$\left[\frac{d^2}{dr^2} + \frac{2}{r}\frac{d}{dr} - \frac{l(l+1)}{r^2} + k^2\right] R_l(r) = 0. \tag{19.132}$$

The solution (which needn't be regular at $r = 0$) is

$$\begin{aligned} R_l(r) &= A_l\, j_l(kr) - B_l\, n_l(kr) \\ &= C_l \cos\delta_l\, j_l(kr) - C_l \sin\delta_l\, n_l(kr). \end{aligned} \tag{19.133}$$

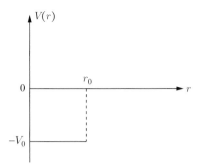

Fig. 19.10 Square-well potential.

Determining the Phase Shifts, δ_l

We can determine δ_l by matching the two solutions at $r = r_0$. Consider the logarithmic derivative

$$\frac{d}{dr}\{\ln R_l(r)\} = \frac{1}{R_l(r)}\frac{dR_l(r)}{dr}, \tag{19.134}$$

then

$$\gamma_l(k) \equiv \frac{k_1\, j_l'(k_1 r_0)}{j_l(k_1 r_0)} = k\, \frac{\cos\delta_l\, j_l'(kr_0) - \sin\delta_l\, n_l'(kr_0)}{\cos\delta_l\, j_l(kr_0) - \sin\delta_l\, n_l(kr_0)}, \tag{19.135}$$

where we use the standard notation

$$j_l'(\rho) \equiv \frac{d}{d\rho}\, j_l(\rho) \quad \text{etc.} \tag{19.136}$$

We can now solve for $\tan\delta_l$ to obtain

$$\tan\delta_l = \frac{\gamma_l(k)\, j_l(kr_0) - k\, j_l'(kr_0)}{\gamma_l(k)\, n_l(kr_0) - k\, n_l'(kr_0)}. \tag{19.137}$$

1. In the limit $\rho \to 0$, the spherical Bessel and Neumann functions have simple behaviour:

$$j_l(\rho) \overset{\rho\to 0}{\to} \frac{\rho^l}{1.3.5\ldots(2l+1)} \quad \text{and} \quad n_l(\rho) \overset{\rho\to 0}{\to} -\frac{1.3.5\ldots(2l-1)}{\rho^{l+1}}. \tag{19.138}$$

   Thus when $kr_0 \ll 1$ we obtain

$$\tan\delta_l \propto (kr_0)^{2l+1}. \tag{19.139}$$

   This result is not restricted to the square-well potential, but in fact holds for all smooth potentials.

2. For certain values of the energy, the denominator in Eq. (19.137) will *vanish* so that $\tan\delta_l \to \infty$. Then the phase shift, δ_l, passes through $\pi/2$, or more generally, through $(n+\frac{1}{2})\pi$. When $\delta_l = \pi/2$, the partial wave cross-section, $\sigma_l(k)$, has a maximum

$$\sigma_l(k) = \frac{4\pi}{k^2}(2l+1)\sin^2\delta_l(k) \to \frac{4\pi}{k^2}(2l+1), \tag{19.140}$$

   i.e. the largest value it could have at that energy.

S-wave Scattering

Consider the case where $kr_0 \ll 1$, so we can safely ignore all $l \geq 1$.

When $r < r_0$:

$$R_0(r) = a_0 \, j_0(k_1 r) \qquad \text{where} \quad j_0(\rho) = \frac{\sin \rho}{\rho},$$

therefore $\qquad u_0(r) \equiv r \, R_0(r) = \frac{a_0}{k_1} \sin(k_1 r).$

When $r > r_0$:

$$R_0(r) = C_0 \, \{\cos \delta_0 \, j_0(kr) - \sin \delta_0 \, n_0(kr)\} \text{ where } n_0(\rho) = -\frac{\cos \rho}{\rho},$$

therefore $\qquad u_0(r) = \frac{C_0}{k} \{\cos \delta_0 \, \sin(kr) + \sin \delta_0 \, \cos(kr)\} = \frac{C_0}{k} \sin(kr + \delta_0).$

Matching at $r = r_0$ for $u_0(r)$ and $\dfrac{du_0(r)}{dr}$ yields

$$\frac{k \, \cos(kr_0 + \delta_0)}{\sin(kr_0 + \delta_0)} = \frac{k_1 \, \cos(k_1 r_0)}{\sin(k_1 r_0)},$$

$$\frac{k}{k_1} \, \tan(k_1 r_0) = \tan(kr_0 + \delta_0)$$

$$\Rightarrow \qquad \delta_0 = \tan^{-1}\left[\frac{k}{k_1} \, \tan(k_1 r_0)\right] - kr_0, \tag{19.141}$$

which is an *exact* result for the S-wave phase shift, δ_0. Now

$$k_1^2 = k^2 + \frac{2mV_0}{\hbar^2} \equiv \frac{2m}{\hbar^2} (E + V_0). \tag{19.142}$$

At very low energies $k/k_1 \ll 1$ and provided kr_0 is such that $\tan(k_1 r_0) \leq 1$, then

$$\delta_0 \approx \frac{k}{k_1} \tan(k_1 r_0) - kr_0 \approx \sin \delta_0,$$

or $\qquad \sin \delta_0 \approx kr_0 \left(\frac{\tan(k_1 r_0)}{k_1 r_0} - 1\right),$

which yields for the total cross-section

$$\sigma_T \approx \frac{4\pi}{k^2} \sin^2 \delta_0 = 4\pi r_0^2 \left(\frac{\tan(k_1 r_0)}{k_1 r_0} - 1\right)^2,$$

whilst $\qquad \dfrac{d\sigma}{d\Omega} \approx \dfrac{1}{k^2} \delta_0^2 = r_0^2 \left(\dfrac{\tan(k_1 r_0)}{k_1 r_0} - 1\right)^2.$

Notes:

1. $\dfrac{d\sigma}{d\Omega}$ is isotropic.
2. $\sigma_T \to 0$ when $\tan k_1 r_0 = k_1 r_0$, i.e. $\delta_0 = \pi, 2\pi, \dots$. This is known as the Ramsauer–Townsend effect, and is seen in electron scattering from rare gas atoms at for example 0.7 eV.

3. Resonance effects occur when $k_1 r_0 = \pi/2, 3\pi/2, \ldots$, although there the derivation of the result for σ_T can't be trusted because we assumed that $\tan(k_1 r_0) \leq 1$.

General Procedure for Determining δ_l

For an arbitrary central potential $V(r)$, we may determine the phase shifts δ_l as follows:

1. Solve

$$\left[\frac{d^2}{dr^2} + \frac{2}{r}\frac{d}{dr} - \frac{l(l+1)}{r^2} + k^2 - U(r) \right] R_l(r) = 0 \qquad (19.143)$$

either analytically or, if necessary, numerically.

2. Having determined $R_l(r)$, compare $R_l(r)$ with $j_l(kr)$ in the large r ($r \to \infty$) region to identify δ_l.

   OR, where feasible, compute δ_l from the *general formula*

$$\sin \delta_l = -k \int_0^\infty r^2 R_l(r) U(r) j_l(kr) \, dr. \qquad (19.144)$$

Derivation of the General Formula

To derive the above expression for $\sin \delta_l$, first define the **linear operator** $\mathcal{L}$:

$$\mathcal{L} = \left(\frac{d^2}{dr^2} + \frac{2}{r}\frac{d}{dr} - \frac{l(l+1)}{r^2} + k^2 \right) = \left(\frac{1}{r}\frac{d^2}{dr^2} r - \frac{l(l+1)}{r^2} + k^2 \right),$$

then $\mathcal{L} R_l(r) = U(r) R_l(r)$ and $\mathcal{L} j_l(kr) = 0,$

where we have normalised $R_l(r)$ so that

$$R_l(r) \overset{r \to \infty}{\to} \frac{\sin(kr - l\pi/2 + \delta_l)}{kr}. \qquad (19.145)$$

Now consider the integral

$$\int_0^\infty r^2 [j_l \mathcal{L} R_l - R_l \mathcal{L} j_l] \, dr = \int_0^\infty r^2 \left[j_l \frac{1}{r}\frac{d^2}{dr^2}(rR_l) - R_l \frac{1}{r}\frac{d^2}{dr^2}(rj_l) \right] dr, \qquad (19.146)$$

$$= \int_0^\infty r^2 j_l U(r) R_l \, dr, \qquad (19.147)$$

where $j_l \equiv j_l(kr)$ and $R_l \equiv R_l(r)$. The right-hand side of Eq. (19.146) may be integrated by parts:

$$\int_0^\infty \left[(rj_l) \frac{d^2}{dr^2}(rR_l) - (rR_l) \frac{d^2}{dr^2}(rj_l) \right] dr$$

$$= \left[(rj_l) \frac{d}{dr}(rR_l) - (rR_l) \frac{d}{dr}(rj_l) \right]_0^\infty - \int_0^\infty \left[\frac{d}{dr}(rj_l) \frac{d}{dr}(rR_l) - \frac{d}{dr}(rR_l) \frac{d}{dr}(rj_l) \right] dr.$$

$$(19.148)$$

The second term on the right-hand side vanishes identically, while the contribution of the lower limit ($r = 0$) of the first term gives zero, giving the following for the right-hand side of Eq. (19.146):

$$\lim_{r \to \infty} \left[(r j_l) \frac{d}{dr}(r R_l) - (r R_l) \frac{d}{dr}(r j_l) \right].$$

$$(19.149)$$

Now substitute the asymptotic behaviour

$$r j_l(kr) \overset{r \to \infty}{\to} \frac{\sin(kr - l\pi/2)}{k} \qquad \text{and} \qquad r R(r) \overset{r \to \infty}{\to} \frac{\sin(kr - l\pi/2 + \delta_l)}{k}$$

$$(19.150)$$

into Eq. (19.149) to obtain the following expression for the right-hand side of Eq. (19.146):

$$\frac{1}{k} \left[\sin(kr - l\pi/2)\, \cos(kr - l\pi/2 + \delta_l) - \sin(kr - l\pi/2 + \delta_l)\, \cos(kr - l\pi/2) \right] = -\frac{1}{k}\sin\delta_l.$$

Finally, equating the above result to expression Eq. (19.147), we obtain the result stated in Eq. (19.144):

$$\int_0^\infty r^2\, R_l(r)\, U(r)\, j_l(kr)\, dr \;=\; -\frac{1}{k}\, \sin\delta_l.$$

$$(19.151)$$

Partial Wave Expansion and the Scattering Matrix

From Eq. (19.119) and the expression for $f(\theta)$ immediately preceding it, we see that the effect of the scattering potential $U(r)$ is simply to attach a momentum-dependent **phase factor** $\exp(2i\delta_l(k))$ to each outgoing partial wave with angular momentum l. The amplitude of the wave is unchanged. This doesn't mean there is no scattering because the angular distribution is altered by this phase shift. The quantity

$$S_l(k) \;=\; \exp(2i\delta_l(k))$$

$$(19.152)$$

is called the **partial wave S-matrix element** for angular momentum l.

The Scattering Operator $\hat{S}$

$\hat{S}$ is defined as the $t \to \infty$ limit of the evolution operator $\hat{U}(t, -t)$, and is therefore a function of the Hamiltonian operator. The matrix elements of $\hat{S}$ between initial and final momentum eigenstates $|\underline{p}_i\rangle$ and $|\underline{p}_f\rangle$ give the scattering amplitude

$$S_{fi} \;\equiv\; \langle \underline{p}_f | \hat{U}(t, -t) | \underline{p}_i \rangle.$$

$$(19.153)$$

This is the probability amplitude for a particle with initial momentum $\underline{p}_i$ at large negative times, to be scattered to have momentum $\underline{p}_f$ at large positive times. It can be shown that this S_{fi} is equal to the scattering amplitude $f(\theta, \phi)$ we have calculated in the time-independent formalism.

In partial wave analysis we work instead in the common eigenbasis of energy $E = \hbar^2 k^2 / 2m$, angular momentum l and z-component of angular momentum, $m = 0$. All of these quantities are **conserved** for a spherically symmetric potential, therefore the

S-matrix must be *diagonal in this basis*. Since $\hat{S}$ is unitary its eigenvalues $S_l(k)$ must be complex numbers of magnitude unity, i.e.

$$S_l(k) \; = \; \exp(i\theta), \tag{19.154}$$

therefore each scattered partial wave is simply modified by a phase. Here, $\theta = 2\delta_l$.

If we work in some other basis, e.g. the momentum basis above, the matrix elements of $\hat{S}$, i.e. $\langle \underline{p}_f | \hat{U}(t, -t) | \underline{p}_i \rangle$, are still elements of a unitary matrix, but it is no longer a diagonal matrix because (linear) momentum is not conserved in the scattering process.

19.A Appendix: Solutions of the Radial Equation for a Free Particle

We now detail some properties of the solutions of the radial equation for a free particle. We restrict ourselves to those properties which prove useful in partial wave analysis.

The radial equation for a free particle is

$$\left[\frac{d^2}{dr^2} + \frac{2}{r}\frac{d}{dr} - \frac{l(l+1)}{r^2} + k^2 \right] R_l(r) \; = \; 0. \tag{19.155}$$

If we introduce the variable $\rho = kr$, this becomes

$$\left[\frac{d^2}{d\rho^2} + \frac{2}{\rho}\frac{d}{d\rho} - \frac{l(l+1)}{\rho^2} + 1 \right] R_l(\rho) \; = \; 0. \tag{19.156}$$

Unlike many similar-looking equations in quantum mechanics, this can be solved in terms of simple functions. The solutions are known as **spherical Bessel functions** and **spherical Neumann functions**. The regular solution, for a given value of the positive integer l, is the spherical Bessel function of order l:

$$j_l(\rho) \; = \; (-\rho)^l \left(\frac{1}{\rho}\frac{d}{d\rho} \right)^l \left(\frac{\sin\rho}{\rho} \right). \tag{19.157}$$

It's *regular* because it's well behaved at the origin, $\rho = 0$. In fact, $j_l(0) = 1$ for $l = 0$, and $j_l(0) = 0$ for $l > 0$. The irregular solution, which is divergent at the origin, is

$$n_l(\rho) \; = \; -(-\rho)^l \left(\frac{1}{\rho}\frac{d}{d\rho} \right)^l \left(\frac{\cos\rho}{\rho} \right), \tag{19.158}$$

and is known as the spherical Neumann function of order l. (See the book by Shankar for an elegant detailed derivation of these results.)

The first few functions, i.e. those useful in partial wave analysis, are listed below:

$$j_0(\rho) = \frac{\sin \rho}{\rho},$$

$$j_1(\rho) = \frac{\sin \rho}{\rho^2} - \frac{\cos \rho}{\rho},$$

$$j_2(\rho) = \left(\frac{3}{\rho^3} - \frac{1}{\rho}\right) \sin \rho - \frac{3}{\rho^2} \cos \rho,$$

$$n_0(\rho) = -\frac{\cos \rho}{\rho},$$

$$n_1(\rho) = -\frac{\cos \rho}{\rho^2} - \frac{\sin \rho}{\rho},$$

$$n_2(\rho) = -\left(\frac{3}{\rho^3} - \frac{1}{\rho}\right) \cos \rho - \frac{3}{\rho^2} \sin \rho. \qquad (19.159)$$

The $\rho \to 0$ and $\rho \to \infty$ limits of $j_l(\rho)$ and $n_l(\rho)$ are particularly useful.

1. Behaviour near the origin. For $\rho \ll l$:

$$j_l(\rho) \overset{\rho \to 0}{\to} \frac{\rho^l}{1.3.5 \ldots (2l + 1)} \equiv \frac{\rho^l}{(2l + 1)!!} \qquad (19.160)$$

and

$$n_l(\rho) \overset{\rho \to 0}{\to} -\frac{1.3.5 \ldots (2l - 1)}{\rho^{l+1}} \equiv -\frac{(2l - 1)!!}{\rho^{l+1}}. \qquad (19.161)$$

The spherical Bessel functions $j_l(\rho)$ vanish at the origin for $l > 0$ (despite the apparent singularities as $\rho \to 0$ in the expressions on the previous page), whilst the spherical Neumann functions $n_l(\rho)$ diverge as $\rho \to 0$.

2. For $\rho \gg l$ we have the asymptotic expressions

$$j_l(\rho) \overset{\rho \to \infty}{\to} \frac{1}{\rho} \sin\left(\rho - \frac{l\pi}{2}\right) \qquad (19.162)$$

and

$$n_l(\rho) \overset{\rho \to \infty}{\to} -\frac{1}{\rho} \cos\left(\rho - \frac{l\pi}{2}\right). \qquad (19.163)$$

Physically, these limits are so simple because the so-called **centrifugal barrier** term $-l(l + 1)/\rho^2$ vanishes as $\rho \to \infty$.

Summary

- The scattering process is characterised by the scattered and incident flux with the ratio of these quantities the differential cross-section. (19.1)
- The incoming particle which is to scatter is a wave packet state localised in space but moving in time and having some central value for its momentum. To a good approximation, such a state can be replaced by the associated plane wave state at the central value of the momentum. (19.2, 19.3)

- The Born approximation is the simplest expression for the elastic scattering cross-section computed to first order in the scattering potential. (19.3.1)
- For a central potential the calculation of the differential cross-section simplifies. The Born approximation for the Coulomb potential leads to the Rutherford scattering cross-section. (19.3.1)
- The time-independent approach to scattering computes eigenstates of the Hamiltonian with appropriate boundary conditions from which the relevant scattering amplitudes can be extracted. (19.4)
- Green's function methods can be used for computing the scattering amplitude. (19.4.3)
- Relation between scattering eigenstates of the Hamiltonian and wave packets used for scattering are obtained. (19.4)
- An alternative method for computing the scattering amplitude is partial wave analysis, which is most useful for low-energy scattering. This method is used to derive the optical theorem which relates the total cross-section to the forward part of the scattering amplitude. (19.4.4)

Further Reading

Further treatment of scattering can be found in the following textbooks:

Merzbacher, E. (1968). *Quantum Mechanics*, 3rd ed. Wiley, Chichester.

Schiff, L. L. (1981). *Quantum Mechanics*, 3rd ed. McGraw-Hill, New York.

For an elegant derivation of spherical Bessel and Neumann functions, please see

Shankar, R. (1994). *Principles of Quantum Mechanics*, 2nd ed. Plenum Press, New York. (Also discussed in Chapter 11 of the current volume.)

Problems

19.1 Starting with the momentum space wave function

$$\psi(k_x, k_y, k_y, t) = \left(\frac{2^{3/2}\pi^{1/2}}{\Delta k}\right)^{3/2} \exp\left(-\frac{k_x^2 + k_y^2 + (k_z - k_c)^2}{\Delta k^2}\right)$$
$$\times \exp\left[-i\hbar\left(\frac{k_x^2 + k_y^2 + k_z^2}{2m}\right)t\right],$$

show that it is normalised.

Fourier transform and show that the normalised position space wave function is

$$\psi(x, y, z - z_0, t) = \frac{1}{\pi^{3/4}2^{3/4}\Delta k^{3/2}}\left(\frac{1}{\frac{1}{\Delta k^2} + \frac{i\hbar t}{2m}}\right)^{3/2} \exp\left[ik_c(z - z_0) - \frac{i\hbar k_c^2}{2m}t\right]$$
$$\times \exp\left[-\frac{x^2 + y^2 + (z - z_0 - \frac{\hbar k_c t}{m})^2}{4\left(\frac{1}{\Delta k^2} + \frac{i\hbar t}{2m}\right)}\right].$$

Calculate the expectation values of position and square of position for the three coordinate directions using the position space wave function.

Calculate the expectation values of momentum and square of momentum for the three coordinate directions, using the position space wave function. Do the same calculation with the momentum space wave function and show that the results are the same.

Calculate the uncertainties Δx, Δy, Δz, Δp_x, Δp_y and Δp_z.

19.2 Calculate scattering using the Born approximation.

- A particle of mass m and momentum $\underline{p} \equiv \hbar \underline{k}$ is scattered by the potential

$$V(r) = V_0 \, \exp(-ar).$$

Show that, in the (first) Born approximation, the differential and total cross-section is given by

$$\left(\frac{d\sigma}{d\Omega} \right) = \left(\frac{4V_0 ma}{\hbar^2} \right)^2 \frac{1}{(a^2 + K^2)^4},$$

where $K = 2k \sin \theta/2$ is the magnitude of the wave vector transfer and

$$\sigma_T = \frac{64\pi m^2 V_0^2}{3a^4 \hbar^4} \left\{ \frac{16k^4 + 12a^2 k^2 + 3a^4}{(a^2 + 4k^2)^3} \right\}.$$

Hint: The required integral may be obtained by parametric differentiation of the integral

$$\int_0^\infty \sin(Kr) \exp(-ar) dr.$$

- Evaluate the differential cross-section in the Born approximation for the potential

$$V(r) = V_0/r^2.$$

What happens to the total cross-section for this potential?
You may assume that

$$\int_0^\infty \frac{\sin x}{x} dx = \pi/2.$$

19.3 Show that in a classical elastic two-body collision between particles of mass m_1 and m_2, the LAB frame scattering angle, θ, and the CM frame scattering angle, θ^*, are related by

$$\tan \theta = \frac{\sin \theta^*}{\rho + \cos \theta^*} \quad \text{where} \quad \rho = m_1/m_2,$$

and hence that the LAB and CM frame differential cross-sections are related by

$$\left(\frac{d\sigma}{d\Omega} \right)_{LAB} = \frac{\left(1 + \rho^2 + 2\rho \cos \theta^* \right)^{3/2}}{|1 + \rho \cos \theta^*|} \cdot \left(\frac{d\sigma}{d\Omega} \right)_{CM}.$$

19.4 The wave function $\psi(\underline{r})$ that describes the scattering of a particle by a short-range potential has the form

$$\psi(\underline{r}) \;\overset{r\to\infty}{=}\; \exp(ikz) \;+\; \frac{\exp(ikr)}{r}\, f(\theta,\,\phi)$$

in the asymptotic region. Evaluate for this wave function the radial component of the probability current density $j_r(r,\,\theta\,\phi)$.

Conservation of probability requires that the total probability current leaving any closed region of space be zero asymptotically. By integrating $j_r(r,\,\theta\,\phi)$ over the surface of a large sphere centred on the origin and using this condition, show that

$$\operatorname{Im} f(\theta = 0) \;=\; \frac{k}{4\pi}\,\sigma_T,$$

where σ_T is the *total cross-section*. This important result, known as the *optical theorem*, is true quite generally.

19.5 Derive the Green function $G_k^+(\underline{r})$ which satisfies $\left(\nabla^2 + k^2\right) G_k^+(\underline{r}) = \delta^{(3)}(\underline{r})$ without using Fourier transformation and complex analysis.

First solve for $r \neq 0$. We need a rotationally invariant solution $G_k^+(\underline{r}) = \psi(r)$ which satisfies the Schrödinger equation for a free particle. In spherical polar coordinates:

$$\frac{1}{r^2}\frac{\partial}{\partial r}\left(r^2 \frac{\partial}{\partial r}\psi(r)\right) + k^2\psi(r) \;=\; 0 \qquad \text{when } r \neq 0 \text{ and } k^2 = 2mE/\hbar^2.$$

Using the substitution $\psi(r) = U(r)/r$, show that the general solution of this ordinary differential equation is

$$\psi(r) \;=\; A\,\frac{\exp(ikr)}{r} \;+\; B\,\frac{\exp(-ikr)}{r}.$$

Show that the first and second terms in $\psi(r)$ tend towards eigenstates of the momentum operator, with eigenvalues $\pm\hbar k\underline{e}_r$, respectively, as $r \to \infty$. (Explain this geometrically.) Thus, for an *outgoing* spherical wave, we must have $B = 0$.

To determine A, evaluate $\left(\nabla^2 + k^2\right)\psi(r)$, being careful about what happens at $r = 0$.

Hint: Note that $\nabla^2(\phi\chi) = (\nabla^2\phi)\chi + 2\nabla\phi\cdot\nabla\chi + \phi(\nabla^2\chi)$ and $\nabla^2(1/r) = -4\pi\,\delta^{(3)}(\underline{r})$. (The committed should prove the latter identity by applying the divergence theorem to the vector field $\underline{\nabla}(1/r)$. Integrate over an infinitesimal sphere centred on the origin.)

19.6 A rough estimate as to the region of validity of the Born approximation for the scattering amplitude may be obtained by requiring that the scattered wave is small compared with the incident plane wave, both being evaluated at the origin. Show that this leads to the criterion

$$\frac{2\mu}{\hbar^2 k}\left| \int_0^\infty \exp(ikr)\,\sin(kr)\,V(r)\,dr \right| \;\ll\; 1.$$

Show also that for the Yukawa potential

$$V(r) = V_0 \frac{\exp(-r/a)}{r/a}$$

this leads to the conditions

$$\frac{2\mu |V_0| a^2}{\hbar^2} \ll 1 \quad \text{for} \quad ka \ll 1,$$

$$\frac{|V_0| a \log(2ka)}{\hbar v} \ll 1 \quad \text{for} \quad ka \gg 1,$$

where v is the velocity of the projectile.

19.7 By calculating the amplitude for scattering from a central potential, $V(r)$, in the second Born approximation, show that when $V(r) > 0$, $\forall r$, the first Born approximation gives too large an estimate for the differential cross-section at low energies (i.e. as $k \to 0$).

19.8 Evaluate the phase shifts for the *hard sphere* potential

$$V(r) = \begin{cases} +\infty & r < a, \\ 0 & r \geq a, \end{cases}$$

and show that the total cross-section tends to $4\pi a^2$ in the zero-energy limit. Is this what you might expect classically?

What happens for the attractive potential? In other words for

$$V(r) = \begin{cases} -\infty & r < a, \\ 0 & r \geq a. \end{cases}$$

19.9 Find the phase shifts for scattering by the potential

$$U(r) = \frac{A}{r^2},$$

where the constant $A > 0$.

You may assume that the asymptotic behaviour

$$j_l(\rho) \overset{\rho \to \infty}{\to} \frac{1}{\rho} \sin\left(\rho - \frac{l\pi}{2}\right)$$

holds when l is non-integer, and also that

$$j_l(\rho) \overset{\rho \to 0}{\to} \frac{\rho^l}{1.3.5 \dots (2l+1)}.$$

Hint: Write down the radial equation for this problem, and note that the potential $U(r)$ and the term containing $l(l+1)$ have the same r dependence.

19.10 Show that the differential cross-section for scattering of particles at energies sufficiently low to neglect d and higher waves has the form

$$\frac{d\sigma}{d\Omega} = \frac{A + B\cos\theta + C\cos^2\theta}{k^2},$$

and determine A, B and C in terms of the s- and p-wave phase shifts.

Write down the formula for the total cross-section.

19.11 If $R_l(r)$, $\overline{R}_l(r)$, δ_l and $\overline{\delta}_l$ are the respective radial functions and phase shifts for the potentials $U(r)$ and $\overline{U}(r)$, show that

$$\sin(\delta_l - \overline{\delta}_l) = -k \int_0^\infty r^2 \overline{R}_l(r) \left\{ U(r) - \overline{U}(r) \right\} R_l(r) \, dr,$$

and hence in the approximation that $\left| U(r) - \overline{U}(r) \right| \ll |U(r)|$:

$$\delta_l - \overline{\delta}_l = -k \int_0^\infty r^2 R_l^2(r) \left\{ U(r) - \overline{U}(r) \right\} dr.$$

By making the substitutions

$$U(r) \rightarrow \lambda\, U(r),$$
$$\overline{U}(r) \rightarrow (\lambda + d\lambda)\, \overline{U}(r),$$

and considering $\dfrac{d}{d\lambda}\, \delta_l$, prove the results

$$\begin{cases} \delta_l > 0 & \text{if} \quad U(r) < 0 \quad \text{(attractive)}, \\ \delta_l < 0 & \text{if} \quad U(r) > 0 \quad \text{(repulsive)}. \end{cases}$$

Appendix: Gaussian/SI Electromagnetic Units

Here the relations between Gaussian cgs and SI units are given for electromagnetic quantities.

Table A.1 Units of electromagnetic quantities in Gaussian cgs and SI units

Quantity	Gaussian cgs	SI
charge	$cm^{3/2}g^{1/2}s^{-1}$ (statC)	Coulomb
electric field	$cm^{-1/2}g^{1/2}s^{-1}$ (statV/cm)	Volt/m
magnetic field	$cm^{-1/2}g^{1/2}s^{-1}$ (Gauss)	Tesla
vector potential	$cm^{1/2}g^{1/2}s^{-1}$	Volts $\cdot$ s/m
vacuum permittivity – ϵ_0	–	8.854×10^{-12} $Coulomb^2 kg^{-1} m^{-3} s^2$
vacuum permeability – μ_0	–	1.257×10^{-6} $Coulomb^{-2} kg \cdot m$

Note that the speed of light $c = 1/\sqrt{\epsilon_0 \mu_0}$.

Table A.2 Electromagnetic expressions in Gaussian cgs and SI units

Expression	Gaussian cgs	SI
Maxwell's equations	$\underline{\nabla} \cdot \underline{E}^{G} = 4\pi\rho^{G}$	$\underline{\nabla} \cdot \underline{E}^{SI} = \rho^{SI}/\epsilon_0$
	$\underline{\nabla} \cdot \underline{B}^{G} = 0$	$\underline{\nabla} \cdot \underline{B}^{SI} = 0$
	$\underline{\nabla} \times \underline{E}^{G} = -\frac{1}{c}\frac{\partial \underline{B}^{G}}{\partial t}$	$\underline{\nabla} \times \underline{E}^{SI} = -\frac{\partial \underline{B}^{SI}}{\partial t}$
	$\underline{\nabla} \times \underline{B}^{G} = \frac{4\pi}{c}\underline{J}^{G} + \frac{1}{c}\frac{\partial \underline{E}^{G}}{\partial t}$	$\underline{\nabla} \times \underline{B}^{SI} = \mu_0\underline{J}^{SI} + \frac{1}{c^2}\frac{\partial \underline{E}^{SI}}{\partial t}$
Vector potential relations	$\underline{E}^{G} = -\underline{\nabla}\phi^{G} - \frac{1}{c}\frac{\partial \underline{A}^{G}}{\partial t}$	$\underline{E}^{SI} = -\underline{\nabla}\phi^{SI} - \frac{\partial \underline{A}^{SI}}{\partial t}$
	$\underline{B}^{G} = \underline{\nabla} \times \underline{A}^{G}$	$\underline{B}^{SI} = \underline{\nabla} \times \underline{A}^{SI}$
Energy density	$\rho^{G} = \frac{1}{8\pi}(\underline{E}^{G2} + \underline{B}^{G2})$	$\rho^{SI} = \frac{1}{2}(\epsilon_0\underline{E}^{SI2} + \underline{B}^{SI2}/\mu_0)$
Lorentz force	$\underline{F}^{G} = q^{G}(\underline{E}^{G} + \frac{1}{c}\underline{v} \times \underline{B}^{G})$	$\underline{F}^{SI} = q^{SI}(\underline{E}^{SI} + \underline{v} \times \underline{B}^{SI})$
Coulomb potential	$V^{G} = \frac{q_1^{G}q_2^{G}}{r}$	$V^{SI} = \frac{q_1^{SI}q_2^{SI}}{4\pi\epsilon_0 r}$

Conversion of quantities:

$$\underline{E}^{G} = \sqrt{4\pi\epsilon_0}\,\underline{E}^{SI}, \qquad \underline{B}^{G} = \sqrt{\frac{4\pi}{\mu_0}}\,\underline{B}^{SI}, \qquad \underline{A}^{G} = \sqrt{\frac{4\pi}{\mu_0}}\,\underline{A}^{SI} \tag{A.1}$$

and

$$q^{G} = q^{SI}/\sqrt{4\pi\epsilon_0}. \tag{A.2}$$

Index